Published by	:	**Daya Publishing House®**
		A Division of
		Astral International Pvt. Ltd.
		House No. 96, Gali No. 6,
		Block-C, 30ft Road, Tomar Colony, Burari
		New Delhi-110 084
		E-mail: info@astralint.com
		Website: www.astralint.com
Sales Office	:	4760-61/23, Ansari Road, Darya Ganj
		New Delhi-110 002 Ph. 011-23245578, 23244987
Laser Typesetting	:	**Rajender Vashist**
		Delhi - 110 059
Printed at	:	**Thomson Press Pvt. Ltd.**

PRINTED IN INDIA

ANIMAL GENETICS AND BREEDING

Dr. Arun Kumar Tomar
M.Sc. (Ag.), Ph.D., A.R.S.
Head, Division of Animal Genetics & Breeding
Central Sheep & Wool Research Institute,
Avikanagar, Rajasthan - 304501

Prof. Sukhvir Singh Tomar
M.Sc. (Ag.), Ph.D.
Former Principal Scientist (AGB),
NDRI, Karnal (Haryana)

Dr. Rajbir Singh
M.Sc., Ph.D.
Professor (AGB) cum Dean,
College of Vet. Sci. & A.H.
S.V.B.P. University of Agr. & Technology,
Modipuram, Meerut (U.P.)

2015

Daya Publishing House®
A Division of
Astral International (P) Ltd
New Delhi 110 002

ANIMAL GENETICS
AND
BREEDING

ABOUT THE AUTHORS

Dr. Arun Kumar Tomar is B.Sc. (Ag.) and M.Sc. (Ag.) in A.H. & Dairying from J.V.College, Baraut, Bagpat (U.P.) and Ph.D. (Animal Breeding) from CCS, HAU, Hisar. He has worked in the capacity of Scientist (AGB) at C.S.W.R.I. Avikanagar (Rajasthan), Senior Scientist and Principal Scientist (AGB) at Project Directorate on Cattle, Meerut (U.P.). He is presently working as Head, Division of Animal Genetics & Breeding, Central Sheep & Wool Research Institute, Avikanagar Rajasthan. He has published some books & booklets and many research papers and has experience of editorial work in the capacity of Assistant Editor of Research Journal.

Professor Sukhvir Singh Tomar is B.Sc. (Ag.) and M.Sc. (Ag.) in A.H. and Dairying from J.V. College Baraut, Baghpat (UP) and Ph.D. (A.G.B.) from N.D.R.I. Karnal. He is retired Principal Scientist (AGB) from N.D.R.I.,Karnal (I.C.A.R.) having 40 years of research and teaching experience, has published about 300 research articles and 3 text books on population genetics and animal breeding. He has long experience for about 28 years as Editor of Research Journal and guided a number of M.Sc. and Ph.D. students. He is still actively engaged in research and publication work.

Dr. Rajbir Singh is B.Sc. (Ag.) from J.V. College Baraut, Baghpat (UP), M.Sc. (Ag.) in A.H. and Dairying and Ph.D. (A.G.B.) from N.D.R.I. Karnal. He is presently working as Professor of Animal Science and Dean, College of Veterinary Science and Animal Husbandry, S.V.B.P. University of Agriculture and Technology, Modipuram, Meerut. He has long experience of teaching to graduate and post-graduate classes. He has guided M.Sc. and Ph.D. students and published number of research papers.

Preface

Man has domesticated though a mere handful of about 40 useful animal species out of innumerable ones having desirable characteristics for mankind and has also been making continuous efforts to develop various breeds and types of animals through selection and breeding practices as well as to know new methods of improvement for various animal and plant species to serve various purposes like milk, meat, eggs, fibre, hides/skin, fuel, manure, draft power etc. As a result a new branch of science, called Genetics had emerged during 20th century with tremendous advancement to exploit the genetic variability for permanent improvement in production and reproduction efficiency of animal and plant populations. The genetics has gained so much popularity that it has found its place in the education curriculum of Agriculture and Veterinary streams throughout the world.

The various courses and their syllabus for teaching to B.V.Sc. & A.H. and B.Sc. (Ag.) students have been nationalized recently by the *Veterinary Council of India* and the *Indian Council of Agricultural Research,* respectively. These nationalized courses are now being taught in all the Veterinary Colleges and Agricultural Colleges under State Agricultural Universities and other colleges affiliated to other traditional Universities all over India. One of these prescribed course is animal genetics and animal breeding with the aim to learn and develop new methods of livestock improvement.

The preent book has been written with the objective to cover the syllabus of courses prescribed at country level by V.C.I. and I.C.A.R. for B.V.Sc. & A.H students and for B.Sc. (Ag.) students of Indian Universities on Animal Genetics, Population Genetics and Animal Breeding, particularly in Indian context. Hope this book will be of great help and great use in general to all interested in the subject and particularly to the under-graduate and post-graduate students, to the teachers and for those who appear in All India Competitive Examination of JRF,

SRF, NET, SET, and others.

This book has covered all the topics of the subject of animal genetics and breeding prescribed in the syllabus. The entire subject matter has been spread over 27 chapters. The first 10 chapters of the book have been devoted to *Principles of Animal Genetics*, next 9 chapters to *Population Genetics* concerning with the genetic structure of population for qualitative and quantitative characters and last 8 chapters to *Animal Breeding* covering the methods of exploitation of genetic variation for the genetic improvement of farm animals.

The most *attractive feature* of this book is that the subject matter has been compiled, arranged, presented and expressed in very simple language. Moreover, the numerical solved problems and the numerical exercises to be done by the students have also been given at the end of different parts of the book. Thus, this book has covered both the theoretical as well as practical aspects as per requirement of the courses.

The authors do not claim any originality, as the subject matter has been collected and compiled from various sources, arranged and presented as per requirement of the students of Indian Universities and for the candidates appearing in various competitive examinations. The subject matter has been expressed in simple language so as the concept of the subject become clear.

The authors are highly thankful to those who extended their help, cooperation and suggestions during the course of writing and finalizing the manuscript of this book.

We welcome the reader to bring in our knowledge the mistakes that are likely to occur and send their suggestions for further improvement of the book.

Dr. Arun Kumar Tomar

Prof. Sukhvir Singh Tomar

Dr. Rajbeer Singh

Contents

Introduction

The life is so greatly diversified into its various forms that the different forms are broadly first classified into two main groups, namely a*nimal kingdom and plant kingdom.* Further, a large number of distinctly separate groups exist in each of the two kingdoms. All the individuals of each distinct group within a kingdom have some common characteristics. Based on the common characteristics, the individuals belonging to different distinct groups within each kingdom of life have been classified in a hierarchical manner into *phylum, order, class, families, genera and species.*

Species is the basic unit of biological classification. Each species has its different genetic material from other species. The genetic difference between species is due to the restriction of exchange of genetic material (genes flow) between any two species except rare cases. This restriction of gene flow among the individuals of two species is called as *genetic isolation.* The genetic isolation is due to different genetic make up (chromosomes) of different species and the different karyotype of their chromosomes (grouping of chromosomes according to their number, size and shape). Thus, the restriction of gene flow (gene exchange) results into the common gene pool of a species which is a basic and essential factor to the recognition of the species. The genetic isolation leads to reproductive *isolation* which is caused by a number of mechanisms like preventing the mating and fertilization of parents, survival and fertility of progeny, if could be produced. Therefore, the important factor in speciation is the genetic isolation of gene pools. This is the reality and biological essence of considering the species as a unit of biological classification.

The individuals of a species show two main tendencies. The first is the tendency of *"like begets like"* in each species and second tendency is the existence of *discrete groups within a species*. "Like begets like" implies that organisms of a species give rise to its own type viz. women give birth to human babies, cat to kittens, pea seeds will grow pea plants, wheat seeds to wheat plants, mango to mango, etc. This tendency of like begets like, is the unique ability which is the result of transmission of *genetic material* or *hereditary material* by the descent (parent) to their descendents (progeny). The transmission of genetic material is accomplished through the process of sexual reproduction which involves gamete formation by meiotic cell division, mating of animals of opposite sex and union of gametes of opposite sex (fertilization). The genetic material which is transmitted is contained in the gametes, embodies the necessary information required to guide the formation of a newly produced living individual of the species own type, development of its various characteristics and maintenance of its life.

Discrete groups in a species: Inspite of the tendency of "like begets like" in a species, the different individuals of a species having some of the similar characteristics (mostly the physical traits) to some extent, also show dissimilarities to various extent. The similarties and dissimilarities form the distinct groups within a species resulting to divide a species into sub-species, breeds/varieties/races throughout the world.

There is thus a second tendency among the individuals belonging to a particular species (breed, herd/flock/strains/lines) that they differ from each other to the extent that each and every individual is unique and *no two individuals are identical*, how closely related they are, except identical twins and clones. Therefore, any two individuals related either through the descent (direct relatives, parent offspring) or as the collateral relatives (full sibs), they resemble each other only to some extent, and hence the resemblance is not exact. This implies that there is also the *tendency of differences* among individuals of similar heredity besides similarities. The related individuals thus have the similarities and also the differences among themselves in their characteristics.

Therefore, there is further grouping of a species, based on differences, into breeds / races, and also into ecotypes or demes (local population). These local populations may be the different strains, lines within a strain as in case of poultry, and herds of cattle or flocks of sheep and goat within which mating occur as a requirement of sexual reproduction. Thus, the individuals of a species living in a locality mate together and reproduce. The sexual reproduction exchange the genes more frequently among the individuals residing in the same locality compared to the individuals of the same species located distantly. The individuals belonging to a species thus residing in the same area share more genes in common due to the gene exchange among themselves but may differ in gene frequencies of the gene pool of other sub-populations of the same species residing distantly. There are number of reasons *viz.* geographical isolation by distance, slow migration, territoriality or preferential mating/marriages among humans, etc.

The variation within a species can be categorized as: *(i) Group variation*: The individuals of different breeds within a species, lines and herds/flocks of a breed form different groups and differ from each other in most of their characteristics. These variations are called as group variation (*breed variation, herd variation and line variation,* sire line etc). *(ii) Individual variation*: The closely related individuals, even of the same sire-dam families, show variations among themselves. This variation among individuals of the same group is called as the individual variation. *(iii) Within individual variation*: The phenotypic values for any quantitative trait recorded on an individual at different times are different. The variation between records of the same individual taken at different times is called within individual variation or intra individual variation.

The genetic material passed on (transmitted) by the parent to all of the progenies is similar to produce "like begets like" within a species as well as to produce similarities (resemblance) among relatives. However, at the same time, the genetic material passed on to the progeny is not exactly identical but differ in some way to produce dissimilarities (differences) observed even among close relatives.

The differences in genetic material transmitted even among close relatives are caused by three genetic processes. One process is the *change in genetic material* (*mutation*) and its transmission into progeny. The second is the *crossing over* (exchange of the parts of homologous chromosomes) and *segregation of genes* during meiosis which is a requirement of gamete formation in sexually mature animal of both the sexes. This process of crossing over produces two types of the gametes (parental and recombinants) and the genes present on homologous chromosomes are segregated in two gametes having half of the total DNA in each gamete. On mating each parent transmits only one of the two gametes (carrying only half of the total DNA) to each of its progeny. These form the new combinations of existing genes The third process is the *recombination of genes* involving mating of animals of opposite sexes and uniting two gametes, one each from both the sexes on fertilization to form zygote. This zygote gives rise to a new individual, of next generation, with new and different gene combinations than its parents had and its collateral relatives will have.

The phenomenon of crossing over by exchanging homologous parts of chromosome to cause genes segregation is a chance event because it is not certain that which portion of chromosome will have crossing over, with what rate the crossing over will occur and further which half of the total DNA content will go into which of the two gametes formed by meiotic cell division. Further, the union of two sex gametes from mating of two parents to form zygote (fertilization) is also not certain but also a random process and hence a chance event. Thus, all the genetic phenomenon (known as the genetic segregation and recombination), involving mutual crossing over of some homologous parts of chromosomes causing genes segregation in gametes and union of gametes of opposite sex on fertilization forming zygote are all chance events. All these genetic mechanisms, of crossing over (gene segregation), halving nature of inheritance and sampling nature of inheritance (genetic recombination) on fertilization to form zygote, produce the gene combinations which are different (causing genetic differences) among progeny and parents, among the different progenies of same parents (full sibs and half sibs, the collateral relatives) and among other collateral relatives. Therefore, all these above genetic phenomenon cause the genetic differences among relatives. This explains the phenomenon of *genetic dissimilarities among relatives* besides the effects of mutations and of the environmental factors.

The genetic material, its distribution in chromosomes, its particulate nature, chromosome theory of inheritance, cell division (meiosis), fertilization and its transmission from parents to offspring are collectively called as the *physical basis of inheritance*. The chemical structure of the genetic material (nucleosome concept or DNA structure) and the DNA function (replication and protein synthesis) form the *chemical basis of inheritance*. The physical and chemical basis of inheritance together with transmission of genetic material from parents to offspring (sexual reproduction) is responsible for maintenance of the genetic continuity from generation to generation and called as the *bridge of inheritance.*

The Geneticists and Animal Breeders have always been trying since long to exploit the genetic variation within a species (breed variation, herd variation, individual variation and within individual variation), by the application of the principles of genetics and population genetics using biometrical techniques (animal breeding) to make the domestic animals more useful and economic for the welfare of mankind. The *mechanism involved for resemblance and differences among relatives* and methods to exploit these mechanisms are the important aspects to revolve around the complete subject matter contained in this book. The complete subject matter in this book has therefore been accordingly divided into three parts viz. animal genetics, population genetics and animal breeding

The mystery of similarities and differences had been solved in the beginning of 20th century and new branch of biological science, called as *Genetics* has emerged which deals with all these mysterious phenomenon and it has gained tremendous advancements.

 Genetics is obviously based on the *sexual reproduction of organisms* which is important for a geneticist for several reasons. *First* is that the man, higher animals and plants reproduce by this process. *Secondly,* the sexual reproduction generates innumerable genetic combinations through crossing over depending upon the number of heterozygous loci and independent assortment of gametes. The chromosomal differences between opposite sexes have fostered genetic differences during evolution. *Thirdly,* it provides better opportunities for understanding the mechanism of inheritance than asexual reproduction, for the reason that the different individuals can be mated/crossed and their consequences can be studied though successive generations. *Fourth,* the small size, précised organization and activity of reproductive cells (sperms and ova) and the product of their fusion (zygote) are considered a remarkable method of organic multiplication. The eggs are little larger than sperms as they store materials for early development. These minute and organized materials (gametes) form the physical basis for the transmission of inherited qualities and are called the germ plasm.

The study of the principles of inheritance (heredity) in animal species including physical and chemical basis of inheritance and that how the characters are transmitted (inherited) from parents to offspring, showing similarities and differences based on the outcome of the individual mating is the subject matter of *animal genetics.* The study of variation of both types (genetic and environmental) along with their partitioning into different components based on the outcome of the mating at population level using the biometrical techniques is the subject matter of *population genetics.* The application of the principles of animal genetics and population genetics using biometrical techniques to exploit genetic variation for the improvement of animals is known as the *animal breeding.*

PART I
GENETICS

Chapter 1
History of Genetics and Animal Breeding

A newly produced living individual (progeny) arises from a pre-existing two similar individuals (parents) which also had arisen from the earlier pre-existing two individuals (called as grand parents which are four in number) and so on. Thus, each living individual has the ability to produce its similar (own) type of new individuals and so there is continuity of being produced a new living individual of next generation of its parental type. This continuity of producing of new living individuals is called the *life*. The study of various aspects of life is called the *biology* or life science. An object having the properties of life is called *organism*.

The study of various aspects of life has now grown into a separate subject *'genetics'* which was earlier considered as a branch of biology. Pai (1974) in his book "Foundations of Genetics" had rightly considered the genetics as the study of life. He argued that besides heredity, all the characteristics like reproduction, response to environmental stimuli, evolution and the activities to carry out biochemical reactions and physi-ological functions that characterize the living being from the non living ones are genetically controlled. All the aspects of life (covered under biology) are genetically controlled. The subject matter of *genetics* is the genetic control of different aspects of life. Therefore, the term genetics should be considered as the synonym or parallel to the biology rather than its branch. In this context, it would be better to call the *genetics* as the upper level of biology. *There is end of life when the genes stop functioning or start improper functioning.*

1.1 GENETICS

The term genetics was coined by William Bateson (1906) for the study of heredity and variation, from a Greek word *gen* which means 'to become' or 'to grow into'. The heredity involves the transmission of genetic material (genes) from one generation to the next. The variation includes the differences observed within each species. The differences among individuals of a species are hereditary as well as environmental in origin. The hereditary differences are called genetic differences which are brought by two ways, one by transmission of changed genetic material (called as mutation) and second by new combinations of existing genes which arises due to crossing over during meiosis (gamete formation) and recombination of gametes on fertilization. The heredity, the genetic material (genes) determines the basic pattern of a character and the environment modifies the development of a character to different degrees. The heredity and variation have the significant role in the formation of new species that are reproductively isolated. The production of new species is called as the *speciation* and a part of the evolution.

1.2 HISTORICAL DEVELOPMENTS OF GENETICS

In pre-historic period, the selection and hybridization were used by both plant and animal owners for improvement of their stock but they did not have the knowledge of heredity. In antiquity, at least up to the middle of eighteenth century, it was a common thinking among biologists that it was not necessary for the biological material to be transmitted between generations and that the concept of spontaneous generation prevailed according to which many organisms particularly the small primitive could arise spontaneously from various combinations of decaying matter.

Most of the work done by various workers prior to the work of Mendel, remained limited to understand the process of reproduction, cell theory and development of various organs of body. Therefore, the historical developments related to genetics have been divided into 3 main parts *viz.* pre-Mendelian work, Mendelian era and the post-Mendelian era of genetics.

(1) Pre-Mendelian Work: Harvey, W. (1578-1657) made experiment on deer and in 1651 proposed that a new organism is formed from egg with a mystical influence of the male semen, called *aura seminalis*.

Regnier de Graaf (1641-73) suggested that both the parents should contribute to heredity.

Robert Hooke (1665) gave the idea of *cell* by working on the texture of cork after finding hollow space which means cella in Greek language.

Leeuwenhoek, A.V. (1632-1723) in the year, 1677, discovered the innumerable "animalcula" in mammalian semen which are now called as spermatozoa.

Grew (1682) reported the reproductive organs of the plants. Rudolf Camerarius in 1694 described the *sexual reproduction* in plants and produced plant hybrids.

Swammerdam (1637-1680), Dutch scientist in 1679 advanced the *preformation* theory that the sex cells (sperm or egg) contained within itself the entire organism in perfect miniature form and an individual is developed as a simple enlargement of a minute but preformed individual. This theory was supported by Leewenhoek, Malpighi (1673), Hartsoeker (1695), Reaumur (1720-1793), Spallanzani (1729-1799) and others.

Linnaeus (1707-1778), the founder of systematics, during eighteenth century, proposed the theory of *"fixing of species"* which stated that individuals of one species give rise to individuals of the same type.

Wolff, C.F. (1738-1794), a German scientist replaced preformation theory and proposed the *epigenesis* which stated that germ cells contain undifferentiated living substances capable of forming organized body after fertilization. He believed that the tissues and organs formed a fresh (*de novo*) through mysterious vital forces.

Lamarck (1744-1829), French biologist explained that small hereditary agents have the extraordinary ability to respond directly to the environment. The excessive use or disuse of any organ lead to an altered inheritance in the descendents and this was termed as theory of *inheritance of acquired characters*.

Von Baer (1792-1876) gave the idea of development of tissues and organs through a gradual transformation of specialized tissue and this was known as *transformation theory*. He also discovered the mammalian egg in 1828.

Charles Darwin (1809-1882) in 19th century believed in theory of inheritance of acquired characters and proposed the *theory of evolution* based on *natural selection*. In 1859 he published a book *"Origin of Species."* Darwin (1868) proposed the theory of *pangenesis* according to which every part of body produces very small invisible bodies called *"pangenes"* and contributes their copies through blood stream to a sex cell and conveys their own nature to ovum and spermatozoa. Thus, the character acquired by the parent might be passed on to the offspring.

It had long been thought by a long line of biologists based on the hybridization of two differently appearing parental stocks that heredity is a *blending process*. Kolreuter (1733-1806) in 1760, Weismann (1834-1914) and others failed to obtain clear cut results with inter specific hybrids. They noted that hybrids between species show a uniform appearance but their offspring showed considerable diversity. Similar observations were noted by many botanists (Gartner, 1772-1850; Naudin, 1815-1899; and C. Darwin (1809-1882).

The details of *cell structure and cell division* were made known by the notable work of many biologists: Schleiden (1804-81) in 1838; Schwann (1810-82) in 1839; Negeli (1817-91); Virchow (1821-1902). The idea of *cell theory* was first given by Dutrochet, H.J., Frenchman, and others in the beginning of 19th century (1802 – 26) but Schleiden, M.L., German botanist and Schwann, T., German zoologist in 1839 outlined the basic features of cell theory to the discovery of earlier workers. The *cell nucleus* was first named and described by Robert Brown (1773-1858) in1833 who established that the nucleus is a fundamental and constant component of the cell.

(2) Mendelian era: Gregor Johann Mendel (1822-1891) - The Father of Genetics: Mendel's work started a new and real era of genetics. Based of the experiments, Mendel (1865) first noted that there was no blending or dilution of inheritance and the inherited characters remained unchanged. He proposed the *particulate theory of inheritance* and formulated the fundamental laws of heredity (the law of segregation and recombination). Mendel published his results in 1865. However, there had been no response to Mendel's discoveries and his paper was overlooked until 1900 because of some reasons. The publication of Mendel's paper of 1865 is the outstanding event in the history of genetics because he worked out the one pair (monohybrid) and two pair (dihybrid) ratios which proved to be basic to the study of genetics.

Schneiden (1873) gave the first account of mitosis and named *chromatin* to the dark staining nuclear threads.

Flemming (1879) described the longitudinal splitting of chromosome during nuclear division and a migration of the daughter halves to the daughter nuclei. In the year 1882, he discovered the *lamp brush chromosomes* and coined the term *mitosis*.

Van Beneden (1845-1910) in the year 1883 showed that gametes are haploid and body cells are diploid.

Raux (1883) pointed out *linear structure of chromosomes* and their point-by-point division into equal longitudinal halves. He saw this as a strong argument in favour of identifying the chromosomes as the bearer of the units of heredity. These ideas were at once adopted by Weismann and put into a theory of heredity and development.

Waldeyer (1888) coined the term *chromosome* to the chromatin bodies observed in the nucleus of plant cells. The chromosomes present in the nucleus are rod shaped bodies or thread-like structures containing many genes which are responsible for passing of hereditary units from parents to offspring.

Hertwig, Oscar (1875) discovered the *process of fertilization* by fusion of nuclei from two gametes (sperm and egg) in Sea Urchin. In the year 1892, he published a monograph on "The cell and the tissue" to make an attempt of general synthesis of biological phenomena based on characteristics of cell, its structure and function. He thus created a new branch of biology, the *cytology*. The discovery of fertilization leads to the doctrine that the *cell nucleus is the bearer of the physical basis of heredity*. The somatic cells have a double or diploid chromosome or hereditary constitution whereas the reproductive cells or sex cells or gametes have the single (haploid) number of chromosomes or hereditary constitution.

Strasburger (1844-1912) showed union of nuclei from two gametes in plants similar to that shown by Hertwig and termed *nucleoplasm and cytoplasm* to the protoplasmic material in the nucleus and its surrounding cell body, respectively. The reduction cell division (meiosis) was discovered in 1888.

Wiesmann (1834-1914) counteracted the theory of pangenesis and inheritance of acquired characters by proposing the *"theory of Germ- plasm"* in (1884). He developed the concept of the continuity of substance (germ plasm) from parent to offspring, based on his experiment on mice for the presence of complete tail in new born mice even after denuding their tails for 22 generations. He proposed that multicellular organism give rise to two types of tissue, namely somatoplasm and germplasm.

Galton, Sir Francies (1822-1911), cousin of Charles Darwin tried to understand the heredity and published a book "Hereditary Geniucs" in 1869. Galton (1871) showed that acquired characters are not inherited through blood and rejected the Darwin's theory of pangenesis. He further (1876) showed that there are two types of twin *viz.* identical twin and fraternal twin, in1893 he proved that a single cell give rise to an individual which maintains its individuality and in 1899 gave the theory of natural inheritance describing that the inheritance is particulate mainly, if not solely.

Boveri, T. (1892) described synapsis and meiosis in *Ascaris.*

(3) Post Mendelian era of genetics: The early 20th century was the period during which a whole series of investigations laid the foundation of our knowledge of the behaviour (accurate partitioning and separation) of chromosomes in mitosis and meiosis (Bovery 1862-1915; Henking 1891; Mongomery 1873-1912 and others).

In the year 1900, three scientists, Erich Von Tschermak of Austria, Hugo De Vries of Netherland (Holland) and Carl Correns of Germany, worked independently of each other and rediscovered Mendel's laws and after reviewing the Mendel's law of heredity, recognized their significance.

Montgomery (1901) had concluded that in the zygote the members of each pair of chromosome are respectively of maternal and paternal origin.

De Vries (1901) coined the term *mutation* for sudden spontaneous change in the hereditary material of Oenothera plant. DeVries (1902) gave mutation theory of evolution which stated that variability is not always continuous but sometimes it is sudden and inherited as mutation and give rise to development of new species.

Mc Clung, C.E.(1902) described the sex chromosomes in Grasshopper.

William Bateson (1902-09) translated the Mendel's paper. He defended Mendelism and gave the terms *genetics, allelomorph, homozygote, heterozygote, F_1, F_2 and epistatic genes.*

Sutton and Boveri (1902) elucidated the *chromosomal theory of inheritance* and showed parallelism between chromosomes and Mendelian factors indicating that chromosomes are the vehicles of heredity. The fact that genes are present in chromosomes was fully accepted in 1902.

Sutton (1902) produced new evidence that in the early stages of meiosis there is synapsis (temporary union of corresponding maternal and paternal chromosomes) and after synapsis, the members of a pair of gene pair pass to different cells.

Sutton (1903) studied the chromosomes in heredity and emphasized the importance of the fact that the diploid chromosome group (diploid cell) possesses two morphologically similar sets and that during meiosis every gamete receives only one chromosome of each homologous pair (haploid set).

The *gene hypothesis* given by Sutton was supported by T.H. Morgan and his students (A.H. Strutevent, C.B. Bridges and H.J. Muller). They all agreed that chromosomes are composed of smaller units called genes which are carrier of hereditary characters.

Bateson and Punnet (1906) working on pea observed that all the characters in plants do not assort independently but some of them are transmitted together and described *linkage.*

Johenson (1909) called purelines to the progeny produced by self fertilization. In 1911, he coined the terms *gene, genotype and phenotype.*

Nelson-Ehle (1909) described the inheritance of quantitative characters working on seed colour in wheat which was based on the *multiple factor hypothesis* and these multiple factors were later called as *polygenes* by K.Mather (1949)

Morgan, T.H. (1910) proposed the mechanism of *sex linkage* for white eye in Drosophila and gave the concept of linear arrangement of genes on chromosomes, constructed the gene maps of Drosophila and supported the chromosome theory of heredity.

Morgan and his students (Bridges, Muller, and Strutevent, 1915) published a book "Mechanism of Mendelian Heredity." Morgan (1933) established gene concept.

Bridges, C.B. (1917-23) discovered different types of *chromosomal abnormalities*, described genic balance theory of sex determination, and non-disjunction as a proof of *chromosome theory of sex determination*.

Bernstein, F. (1925) described the determination of *ABO blood group in men* by multiple alleles whereas the Rh blood group was discovered by Landsteiner, K and Weiner, A.S. (1940).

Muller, H.J. (1927) discussed the *artificial induction of mutation by X- rays* in animals. He contributed on radiation genetics and got Nobel Prize in 1946.

Stern, S. (1931) gave a *cytological proof of crossing over* in Drosophila.

Schlesinger, M. (1934) had shown that bacteriophages are composed of DNA and protein.

Landsteiner and Wiener (1940) discovered *Rh blood group in man*.

Beadle, G.W. and Tatum, E.I. (1941) published on biochemical genetics of Neurospora.

Chargaff, E. (1950) showed that *A = T and C = G* in DNA.

Harshey and Chase (1952) worked on bacteriophage and showed that DNA enters the host cells and acts as hereditary material.

Watson, J.D. and Crick, F.H.C. (1953) gave the *double helix model of DNA molecule* and got noble prize in 1962.

Benzer (1955) worked on *fine structure of gene* and that gene has many mutable sites.

Beadle, Tatum and Laderberg (1958) proposed *one gene - one enzyme theory*.

Crick (1958) enunciated the central dogma in molecular genetics and in 1961 suggested the *three letter code*.

Harries, H. (1970) gave DNA-RNA hybridization technique, Khorana, H.G. (1971) synthesized the gene artificially, and McClintock, M. (1983) gave the jumping gene concept.

1.3 HISTORY OF ANIMAL BREEDING

The social and scientific progress rested on civilization and modern agriculture which in turn is a consequence of domestication of animals and plants. There are about 1, 00, 000 species of animals in literature but could be domesticated only a handful species (about 40) including poultry by man to serve various purposes. The exact time of domestication is not known but it must have started at the end of the Old Stone Age, *Paleolic period (1000 - 8000 BC)* which was the period of using of stone and bone implements without agriculture or in the beginning of New Stone Age, *Neolithic period (8000 - 6000 BC)* which extended up to about BC 3000 when man used to live in huts or wooden houses. However, domestication had been certain long before the time of recording historical data.

The use of cow ghee in religious ceremonies had been mentioned in Vedas (Rig Veda) and there are mentions of the cow *"Kam Dhenu"* in ancient writings. In Shiva Temples are found the references of sculptures of *"Nandi Bull"*. The cow has been said to be the mother of *Rudras,* the daughter of Vasus and the sister of *Aditayas* in vedas. In Hindu mythology, the male buffalo has been referred to as the vehicle of the God of death, *Yama"* and also regarded as the physical form of the demon *Mahisasur,* a symbol of inertia- *"Tamas".*

The starting of domestication had been certainly need based. Man thought the need of certain animals and attempted to domesticate them. The needs of man could have been for religious rites (as sacrifice to the God) and for meeting his requirements like meat for food, skin for clothing, etc. This could have started the domestication of animals of economic value like, sheep, goat, cattle, pig and fowl. Attraction of some animals (dogs) towards waste food available around man's dwellings could also have been the reason of domestication, as these animals acted as scavengers of man's surroundings and hence they slowly became pet animals.

In ancient time, there was human culture of food collection by hunting the animals which followed the culture of plant cultivation. This favoured rapid domestication of a number of animal species. There had been then the selection between species to serve human culture and needs *viz.* milk and draft animals, scavengers, meat and cloth animals, etc.

1.3.1 DOMESTICATION AND ANIMAL BREEDING

Domestication vs Evolutionary forces: The domestication had intensified the *forces of evolution* by shifting the emphasis to small population size, mating systems, artificial selection and migration of animals in small numbers. Different forces of evolution played their significant role. Before domestication, the wild animals used to move widely in groups and bred randomly. After domestication, the animals were started confined in captivity within human environment. Under domestication, the animals were more narrowly restricted by herding or fencing them to live, grow and reproduce in the restricted area. The concentration on a few good animals for breeding inevitably led to some *inbreeding.* This together with *small population size* (random genetic drift) under domestication resulted in an increase of inbreeding and relatedness among the domestic animals. The inbreeding together with *natural selection* had been the important force to fix a particular type, colour and conformation and to produce diverse races of domestic animals. This had thus resulted into different breeds of today.

Based on the idea of resemblance among relatives, man started breeding the most desirable animals and castrating others. The castration is an ancient practice. Man chose the breeding pattern as to which animal should be kept for breeding *(selective breeding).* No doubt that man had practiced the *artificial selection* for thousands of years. This had resulted into the genes of favoured animals to become more abundant in the population. The selection thus became known to be the main force of changing the genetic structure of population.

Improvement gradually started for a specialized function *viz.* increased milk/ egg/wool production, draught, speed, etc for meeting human requirements. This developed the breeds and varieties by morphological diversification and imposed reproductive isolation through geographical isolation by distance (determined by dense forests, mountains, hilly rivers, deserts, etc.), disease prevalence and natural selection. The artificial selection had played its role in changing the genetic structure to meet the needs of different geographical areas. The *divergent selection* divides a well established breed into different types. Thus, the domestic species of animals diverged for different functions. Short horn cattle had diverged into beef and milk to the extent that they converged with other breed selected for the same function. The dairy Short horn breed now resembles the Friesian breed more that it resembles the beef Short horn (which now resembles the Angus breed). The dogs have diverged into types selected for herding, bird hunting, trail hunting, guarding and draft purposes; horses selected for speed and riding; chickens selected for fighting, meat and egg production.

Migration of animals is known since beginning of domestication. Different species have migrated in small numbers from their centers of domestication to other countries having different cultures and environmental conditions. Thus, *migration* and *genetic drift* had played their role. Man had transported the animals far away beyond the area where they were born (native land) and could have wondered before domestication. A number of examples may be cited. The invading armies and the migrating people brought with them the animals from their native lands. Alexander, The Great, brought large number of livestock with his armies from Greece to India. Ganghis Khan during 13th century with his armies travelled with enormous reserves of cavalry horses all the way from China to Central Europe. Likewise, cattle were brought from Holland to England across water, Merino sheep from Spain to many places, Angora goats from Turkey to USA and South Africa, Zebu cattle from India to Brazil and USA, Short horn cattle from England to Argentina and Australia, etc.

The migration of animals to far off places had resulted in an increased amount of *out breeding* (grading up and crossbreeding). This had resulted into a change in genetic structure of population of domestic animals. In view of the impact of domestication, there was a change in population size, mating system and artificial selection and that too divergent selection.

1.3.2 ANIMAL BREEDING OF PRE-MENDELIAN ERA

More than thousand years ago, the Arabs in their horse breeding used the genealogies to breed them but the manner of using is not known. Lush (1945) presumed that like modern Arabs, the pedigrees could have been kept only in female line as an aid to selection and taking care to avoid close inbreeding.

The Romans of the time of Varo and Cato (2000 years ago) made many comments about the kinds and types of animals to be selected for breeding purposes, although they made no attempt to record long pedigrees for their livestock. The Roman, Varo advocated the idea of *progeny testing*.

Pure Breeding: The use of pedigrees was started in rural England late in the 18 th century. Robert Bakewell (1725-1795) of England is remembered for setting the pattern of modern animal breeding and called as the founder of animal breeding. The *pedigree breeding* was established in his time and he made it popular. He made efforts to produce farm animals (cattle, sheep and horses) with increased efficiency. He developed theories, tested them with experiments and laid down the following principles-

- Like begets (produce) like
- Breed the best to the best and introduced *inbreeding* as a tool in livestock improvement
- Inbreeding produces prepotency and refinement, and development of relatively true breeding strains
- He initiated sire testing by *leasing the sire* to other livestock breeders. He started *ram-letting* instead of selling his best males, for using the males for one year period at a time and taking the best males back. Thus, he used the idea of *progeny testing* of Roman, Varo.

Robert Bakewell laid the foundation for the Shire horses, Longhorn cattle, and Leicester sheep. His success attracted many *imitators* (who follow examples). Many ambitious stockmen from many parts of England went to work with him so as to know his methods. The greatest contribution of Bakewell to breeding methods was that he appreciated the inbreeding as most effective tool to produce refinement and fixing type.

Collings (1775) learned from Bakewell and laid down the foundation of Short horn breed of cattle.

Thomas Bates (1775-1849) studied the methods of Collings and started keeping records on feed consumption for beef and milk production. He developed a large herd of cattle by *selection and inbreeding*.

In the 17th century, the Arab horses came to Britain and crossed with local available horses and a belief in *purity of pedigree* was adapted.

Buffon (1708-88), French naturalist advocated systematic *crossbreeding*. With the efforts of Bakewell followers, there soon became groups of animals closely related to each other and similar in type all over England. The modern breeds were developed from these groups of animals, though these were not formally organized as such until later.

The methods of improvement in breeding stock of Bakewell and his followers came to be known to the people of other lands and importation of breeding stock to those lands was started. This became a source of income to British stockmen for which they were encouraged for further improvement of breeding stock to attract foreign customers.

Breed association and pedigree recording: With the passage of time, the number of animal breeders and generations of pedigreed animals increased. However, no one could remember all the animals far back in the pedigree and hence the statements and records of others were unbelievable. Therefore, the herd books were formed to supply correct information.

The first **stud herd book** was "An introduction to the General Stud Book" for the thorough bred horse and it was started in 1791. This contained the records of the pedigrees of the horses winning important races.

Stud herd books for horses were founded in France in 1826, in Germany in 1827 and in Australia in 1847.

The first *cattle herd book* "Short horn herd book"" was published by Coates in 1822. An English Hereford herd book was published in 1846 and a Polled herd book for Aberdeen Angus cattle in 1862.

The first cattle herd book in France was established in 1855, the first German herd book in 1864, the first Dutch one in 1874 and the first Danish one in 1881.

The first *swine herd book* in the world was that of the American Berkshire Association in 1876.

In continental countries of Europe, pure breeding and registration work were generally organized at a later date than in Britain. In Germany and adjoining lands, extensive efforts for improvement were made first in sheep breeding, then in horse breeding and in cattle breeding.

Measurement of wool fineness with a micrometer was started as long ago as 1779 by Dauben Ton. Abilgard in 1802 started marking (identification) sheep individually so that their production could be recorded and used as a basis of selection.

After the turn of 18th century, there was no intellectual background to the efforts of animal breeders in Europe and there was resurgence (revival) of the theory of pure breeding (pedigree breeding). It was common practice during 1800 to cross extensively in accordance with the idea expressed by Buffon (1780) that perfection could be attained only through widespread crossing and mixing all individuals with desirable points. Then for a half century, the trend changed to follow *pure breeding* for improving a breed from within itself.

Krunitz (1845) wrote that the English improve a race from within the same race and mating them together and to keep the stock unmixed to produce a race of desired qualities. This procedure developed to the theory of *"racial constancy"* according to which each animal transmits its racial characteristics rather than of others.

In 1860, attention was devoted to the individual than to the "racial constancy'. During 19th century, a *"doctrine of indigenous breeds"* was popular in Norway, which held that locally adapted types of animals were the best to use.

In 1859, Darwin published "The origin of species" stressing the effectiveness of natural and artificial selection.

In 1865, Mendel published his results on the inheritance of garden pea but it had no immediate impact on animal breeding.

The *production of all cows* had been recorded continuously since 1871 in Strickof Agr. School, Zurich, Switzerland. The first *cow testing association* in the world was established in 1872 in Denmark.

Allen, R.L. (1847) mentioned in his book "*Domestic Animals*" that every country and almost every district has its peculiar breed of cattle.

Galton, Sir Francies (1822-1911), cousin of Charles Darwin, introduced the *quantitative approach* to variation among relatives. Galton used statistical methods for scientific discovery of continuous variation, concentrating on pedigree analysis and twin analysis on human population and natural population. Galton made the following contributions- .

- 1869- He published a book "*Hereditary Geneiucs*". He discussed that human intelligence has hereditary differences between individuals and families.
- 1871- He showed that acquired characters are not inherited,
- 1876- He analyzed twin data and showed two type of twins *viz.* identical and fraternal, which helped to study the importance of heredity and environment to affect human characteristics,
- 1881- He worked on parent-offspring correlation for human stature (height) and indicated the incomplete determination of heredity explaining the differences in terms of environmental variation. He pointed out that the parents with extreme phenotypic values tend to have their children close to the average value of the population and hence the children of extreme parents *regress* towards the population mean and he coined the term "*regression*" and gave the law of filial regression after graphical representation of the linear relation between parent and children. He also showed that human characteristics can be described in the form of normal curve or frequency curve which can be described in terms of mean, variance and coefficient of variation and that the spread of curve depends on the number of factors causing variation.
- 1899- He gave the theory of natural inheritance describing that the inheritance is particulate mainly if not solely.
- 1893- He proved that a single cell give rise to an individual which maintains its individuality.

Karl Pearson, student of Galton, worked (1894-99) on frequency distribution and published statistical methods for standard deviation, variance, chi-square. Pearson and Galton assured that continuity of variation was an expression of blending heredity.

1.3.3 ANIMAL BREEDING OF POST-MENDELIAN ERA

Hugo devries, Correns and Tschermark (1900) independently discovered Mendel's law of heredity. William Bateson (1902) translated Mendel's paper, defended Mendelism and coined the terms *allelomorph, homozygote and heterozygote*.

Johansen (1903) proposed the *pureline concept* indicating that selection was effective in changing the mean of a population for quantitative characters only for genetically heterozygous population but ineffective for purebreds because the variation in pure lines must have been due to environmental factors.

Castle, W.E. (1903) of United States worked on the inheritance pattern of coat colour in mice and supported Mendelism (gametic purity) rather than blending basis of heredity. He published the effect of selection against recessive genes and concluded that the race remains constant after selection is stopped which is in fact similar to Hardy-Weinberg law.

Yule (1902) and Pearson (1904) proposed the genetic equilibrium principle for special case of equal gene frequency (p = q = 0.5) and Castle (1903) for other values. Pearson showed that F $_2$ ratio of 1:2:1 is maintained indefinitely in a random mating large population and thus inadvertently supported the law of segregation inspite of the fact that he rejected Mendelism.

George Harrison Shull (N.Y.) and Edward Murrey East with H.K. Hayer (Jamaica), the *inbreeding experiment* was started on maize in 1905, based on Johansen's pure line concept and the biometric concept of Galton and Pearson. This opened the field of quantitative theory and applied plant breeding.

East (1907) developed a series of pure lines from heterozygous population of maize as a result of artificial selection and inbreeding based on Johansen's pureline concept.

Shull (1908) reported the decrease in size and vigour in maize when F $_1$ were inbred.

Hardy, G.H. (England) and Weinberg, W. (Germany) in 1908 independently showed that *genetic equilibrium principle* is applicable for any value of gene frequency. Weinberg showed the *utility of equilibrium principle* to prove Mendelian heredity in human families, independently assorting non allelic loci, correlation between close relatives and method of partitioning variance between genetic and environmental sources.

Nelson-Ehle (1909) worked on seed colour in wheat and explained the character to be governed by three pairs of alleles with additive and non dominant effects *(multiple factors hypothesis)* and these multiple factors were later called as *polygenes* by K. Mather (1949). This reconciled the blending hypothesis and lead to a basis of *quantitative genetic theory* and *selection theory.*

Morgan (1910) proposed the *mechanism of sex linkage* for white eye in Drosophila and this led to the concept of *linear arrangements of genes* on chromo-somes.

Raymond Pearl (1912-14) who was poultry breeder erroneously concluded that there is no change in the frequency of heterozygote on brother sister mating and devised a coefficient of inbreeding. He also demonstrated the *effectiveness of pedigree selection* compared to mass selection in poultry.

H.S. Jennings and H.D. Fish independently contradicted with the findings of R. Pearl and said that sib mating leads to a decline in heterozygosity like selfing but with low rate. Jennings (1916) proposed the mathematical *theory of inbreeding.* Shull (1916) coined the term *heterosis* to describe the vigour of F $_1$ hybrids due to heterozygosity.

Morgan, Bridges, Muller and Sturtevant (1915) published a book *"Mechanism of Mendelian Heredity"*

The first commercial "Crossed Corn" was produced by East and D.F. Jones (1917). The heterosis was explained by Jones due to linked dominant genes controlling the increased vigour and proposed the "double cross" system to produce commercial maize *viz.* A x B and C x D.

Bernstein (1925) applied the HW principles to the proportions of human ABO blood group phenotypes and postulated the correct mode of single locus heredity rather than two loci proposed earlier.

The Mendelians expressed their results with genotype and phenotype frequencies while the Biometricians (Francis Galton and Karl Pearson) used the correlation and regression. The gap between Mendelian and Biometrician was bridged by R.A. Fisher (England) and Sewell Wright (United States) by demonstrating that Mendelian frequencies are the basis of biometrical correlations.

Sewell Wright was the student of Castle. He made the following contributions-

- 1917, Wright published a number of papers on *colour inheritance* in mammals using Hardy-Weinberg Law to calculate the genotype frequencies in populations. He compared the observed and expected frequency data based on random mating using Hardy-Weinberg principle for *colour inheritance* in Short horn cattle and also rejected the single locus hypothesis for the *human eye colour inheritance.*

- 1921, Wright designed a method to estimate the extent of influence for each of a number of causing factors on a trait and invented the *method of path coefficient.* The path coefficient is useful to analyze (by subdividing correlation in a casual scheme) the relative importance of interacting factors (hereditary and environmental factors) to determine the measured effects and correlation between relatives under various mating systems connecting zygotes and gametes. He applied path coefficient analysis to estimate the measure of relatedness, called as the *coefficient of relationship* and proposed the formula to estimate the *inbreeding coefficient*

- 1921, Wright published a series of 5 papers under *"Systems of Mating"* covering the biometrical relations between parent and offspring, and the *effects of inbreeding, assortative mating,* and *selection* on the genetic composition of a population

- 1922, Wright stated that the *fundamental effect of inbreeding* is to increase the homozygosity and leads to differentiation among families due to chance fixation of different combinations while crossing leads to improvement as each family provides some dominant factors which are lacking in other families.

- 1929, he introduced a model criticizing Fishers theory suggesting that frequency of dominant modifiers decreases as it approached fixation.

- 1931, Wright published a paper on *"Evolution of Mendelian populations"* mentioning the *effects on gene frequencies in populations of known forces* which have directed effects (migration, selection and recurrent mutation) and non directed effect (random fluctuations due to sample size in effective population size) and concluded that a balance is maintained among these forces.

- 1931, Wright coined the term *genetic drift* based on the outcome of random sampling of gametes in small populations and considered the random drift important in evolution through natural selection which acts on gene combinations created by random drift indicating that random drift is a raw material for selection rather than an alternative to natural selection.

- 1936, he gave the theory of *discriminant function*.

Fisher, R.A. (1890-1962) was a Mathematician in Cambridge (England) and his main contributions are the technique of the *analysis of variance* and other theories of population genetics on the supposition of Mendelism which bridged the gap between Mendelians and biometricians (Galton and Karl Pearson).

- 1913, Fisher worked on criterion for fitting of frequency curve and *test of significance of the mean* of a sample for normally distributed data.
- 1914, derived the distribution of Galton's correlation coefficient using Pearson's data on human stature and formed table of correlation in association with Pearson.
- 1918, published a paper "The *correlation between relatives* on the supposition of Mendelian inheritance" and demonstrated that correlation between relatives could be accounted for by Mendelian concept of heredity rather than blending inheritance. He examined the correlation between relatives in terms of gene interaction, multiple alleles, linkage and assortative mating. He conceived that correlation is the result of chain of events.
- 1918, Fisher worked on a measure of the variability in terms of standard deviation and partitioning of the variance due to the independent causing factors by noting that $\sigma^2_Y = \sigma^2_{X1} + \sigma^2_{X2}$. This helped to partition the total phenotypic variance into its genetic and environmental components. Fisher (1918) was first to divide the genetic variance into additive, dominance and non additive parts.
- 1922, discussed the consequences of Mendelism for quantitative traits in terms of the interaction of selection, mutation, dominance, random drift and assortative mating and demonstrated the elimination of an allele when selection favours one homozygote while the selection in favour of heterozygote leads to a condition of stable equilibrium *(balanced polymorphism)*.
- 1922-29, he published a number of papers in support of the views that
  - (i) the survival of a rare gene depends upon chance rather than selection,
  - (ii) low mutation rate balance the effects of adverse selection in a large population more easily because a mutant allele would be fixed at low frequencies in a large than in a small population,
  - (iii) selection eliminates one allele when heterozygote are intermediate between the homozygote at a locus but selection is ineffective to eliminate deleterious recessive allele at low frequencies when there is complete dominance,
  - (iv) the evolution is more concerned in large population in which the variability is more due to the presence of deleterious alleles, and that
  - (v) the natural selection leads to evolution at a slow rate.
- 1929, Fisher proposed the theory of *"Evolution of Dominance"* which states

that if an unfavourable gene is produced continuously by mutation, the other genes which modify the action of the mutant gene will eventually be selected and the resulting phenotype will resemble to that of favoured wild type. However, the assumption of selective advantage of genes modifying dominance was questioned by Wright (1930) and Haldane (1930). Fisher (1929) showed that intensity of selection increases when modifying alleles approached fixation and further Fisher (1949) suggested that selection intensity is proportional to mutation rate.

- 1930, Fisher gave the *fundamental theorem of natural selection* with the aim to determine the rate of increase (Malthusian parameter) or mean selective value (as termed by Wright) or Darwinian fitness of genotype. Fisher (1930) further established that variation irrespective of its origin is more important and suggested that the basis of the process of evolution is not the mutation theory but it is the theory of natural selection.

Haldane J.B.S. Cambridge University, England made the following contributions-

- 1912, worked on *linkage in mammals*
- He published 9 papers upto 1932, on "Mathematical theory of *Natural and Artificial Selection*". Haldane believed that selection is the most important factor of evolution and defended the principle of natural selection, along with the importance of chromosomal changes and polyploidy in evolution. Haldane agreed with Fisher about the theory of natural selection and with Wright about the evolution theory that evolution is slow in small population.
- 1932, Haldane disagreed on the theory of evolution of dominance put forth by Fisher indicating that enough modifiers are not available in natural populations.
- Haldane's contributions had also been on the *causes of evolution* (1932),
    - *(i)* Description of G-E interaction (1946),
    - *(ii)* Measurement *of natural selection* (1954) and
    - *(iii)* Cost of natural selection (1957).
- Haldane (1962) worked the *frequencies of genes with different genetic systems* (autosomal dominant, recessive, sex linked and partially sex linked genes) including equilibrium states in a population.

Sprague and Tatum (1942) gave the concept of general and specific combining ability (GCA and SCA).

The diallel theory for estimation of the components of variation was developed by Jinks and Hayman in 1960's. The procedure for diallel analysis was proposed by Hayman (1954) and Griffing (1956).

Lush, J.L. of U.S.A. made significant contribution in the field of animal breeding.

- Lush (1937) coined the terms *heritability and repeatability* of characters and most probable producing ability of dairy cows in 1945, with methods of their estimation.
- Lush (1945) suggested that the breeding stock should be selected under such

similar environmental conditions where the selected animals are to be used and live to perform, to avoid the effect of *G-E interaction.*

- Lush *et al.* (1948) proposed a method of estimating sib correlation for *heritability of viability in fowl.*

- Lush (1950) used intra-herd regression of daughter on dam to study the *inheritance of susceptibility of mastitis* in dairy cattle and also the repeatability based on regression of future performance on previous year performance.

Johansson and Hansson (1940), Ludwick and Peterson (1943), and Mahadevan (1951) proposed the formulae to estimate the *persistency of lactation.*

Hazel (1943) first had shown the method of estimating *genetic correlation* from cross covariance of one trait in progeny and other trait in parent. Hazel (1943) extended the application of Fisher's discriminant function (1936) in animal improvement by giving the *selection index theory* (genetic basis for constructing the selection index).

Muller (1950) used the term *genetic load* to describe the effect of mutant gene on human population, as the reduction in average fitness of the population.

Robertson and Rendel (1950) proposed the formula for *accuracy of selection* as the square root of heritability based on single and multiple records. Robertson (1959) proposed the formula to estimate the *standard error of genetic correlation.* Rendel and Robertson (1959) gave the method to measure *genetic change* from 4 paths *viz.* sire to produce cows, sire to produce bulls, dam to produce cow and dam to produce bulls.

Smith (1962) developed the method of estimating *genetic change* based on the comparison of mean performance of paternal half sisters in different years.

Different *sire indices* based on different criteria were constructed by different workers *viz.* simple daughter average index (Edward, 1932), equi-parent index (Hansson, 1913) which was later on called as Yapp's index, corrected daughter average index (Krishnan, 1956), stable mate or contemporary comparison index (Searle, 1964), contemporary daughter average index or dairy search index (Sunderasan *et al.* (1965) and BLUP method was developed by Henderson (1949, 1973).

Kampthorne and Tandon (1953) developed the method of estimating *heritability* from data with different number of daughters per dam.

Kempthorne and Nordskog (1959) gave the procedure to construct the *restricted selection index,* by putting restriction, so that some linear functions of genotypic values would remain constant, equal to the number of traits that are not allowed to change.

Tomar, N.S. (1965) proposed the formula to estimate the *breeding efficiency* for the lifetime of a cow and a buffalo based on the numbers of calving in her lifetime.

Tomar, S.S. during 1971-98 had made the following significant contributions in the field of Animal Breeding Research-

- 1971, worked out the procedure to predict body weight of cows from body measurements under field condition,

- 1981, worked out the formula to estimate the *standard error of mean* for threshold characters,
- 1981, suggested the procedure to estimate the *breeding value of young sires* at an early age based on mate's performance,
- 1982, suggested the possibility of mutant allele (lethal) responsible for athelia condition in Sahiwal heifer and its pleiotrophic effects on reducing birth weight, growth rate and survivability causing death in early age before sexual maturity.
- 1985, started *new area of research* on sire evaluation and selection programme. The procedures and formulae were developed for estimating the *replacement rate, replacement index and selective value* of dairy animals. This had helped the animal breeders to avoid the wastage of animal breeding data and formulating the proper mating plan for sire evaluation, to remove non-orthogonality of data and removal of biasness in data analysis, to decide the optimum number of services of each bull under P.T. programme, to give chance to a more number of sires to prove their genetic worth in progeny testing programme, and to retain a dairy animal in the herd for her replacement. Many imitators from different research organizations followed these methods.
- 1991 and 1998, proposed the *method of estimating the heritability of threshold data* without transformation,
- During his carrier as a Research Scientist, made a number of recommendations on adjustment of data recorded on sick animals before genetic analysis, selection of cows based on udder measurements, optimum length of service period and dry period in cows, feasibility of weaning in dairy cows, and maternal effect on birth weight in cattle and sheep.

1.4 APPLICATIONS OF GENETICS

The knowledge of genetics has been proved to be of great and vital importance to the mankind having deep impact on the socio-cultural aspects besides helping in improvement of domestic animals and agricultural crops.

1. **Improvement of animals:** The knowledge of genetics has made a significant contribution in the improvement of different livestock and avian species like cattle, buffaloes, sheep, goat, pigs, poultry etc. The animal breeders have exploited the knowledge of genetics by conducting various experiments related to breeding systems and selection to make use of the existing genetic variability. This has increased in development of new breeds for higher production of milk, meat and egg and it has come out with white revolution.

2. **Improvement of agricultural crops:** Like improvement of livestock and poultry production, genetics has made also the notable contributions in making improvement and development of new high yielding varieties of food crops and horticultural crops etc. The disease resistant varieties of crop plants have been developed with the knowledge of genetics. The green revolution for improvement of yield of crops and fruits is the outcome of the

applications of the genetic principles. It is not only the improvement in the quantity but the quality of food has also been improved *viz.* seedless varieties with higher juice percentage in orange etc. have been developed.

3. **Application in medical science:** Every character has its genetic base and likewise most of the diseases in men are inherited like, diabetes, hemoglobin abnormalities, hemophilia, tuberculosis, colour blindness, blood pressure and other heart troubles, some types of blindness and deafness, idiocy and some genetic defects. The present day knowledge of genetics has helped to develop the techniques of genetic surgery and genetic engineering for correcting various disorders. The frequency of genetic defects can also be reduced in a population by use of the principles of genetics.

4. **Application in solving legal disputes:** The blood group in men is genetically controlled. Likewise, the DNA finger printing has genetic base as the DNA of two individuals in not identical. This knowledge is helpful in solving many legal issues like dispute parentage, divorce, legal estate, etc.

5. **Removal of misconceptions:** The knowledge of genetics has helped to understand the reality of some false beliefs which prevailed world wide in human society, like blood relation and blood relatives, blood will tell, bloodline and a very bad phrase "bad blood' being used in the society. The knowledge of genetics has made it clear that it is not the blood but it is DNA of reproductive cells (gametes) which is carrier of genetic information and transmitted from parents to offspring. Thus blood has nothing to do with all these phrases.

6. Genetics is related with other sciences.

Chapter 2

Mendel's Law of Heredity

The science of genetics began with the work of Mendel which was published in 1865, based on his experiment started in 1857 on garden pea, in the proceedings of the Brunn Society of Natural Sciences. He was the first and pioneer worker to discover the law of heredity and hence called the *"father of genetics"*.

Gregor Johann Mendel was born in 1822 in a poor farmer family in a place now in Czechoslovakia. He could not fulfill his desire of proper education due to his poverty and hence went to Augustinian monastery at Brunn. He was appointed a priest in 1847 and subsequently sent in 1851 for training in physics, mathematics and natural sciences to the University of Vienna. He became a teacher after his return to Brunn in 1854 and started his experiment in 1857 on garden peas (*Pisum sativum*) in the monastery garden.

2.1 MENDEL'S EXPERIMENT

Mendel's choice to start his experiment on garden peas had been for the following reasons

- The peas were easily available in a wide array of different forms and colours,
- The plants were easy to grow and took only a single growing season to have one generation,
- Pea was a sexually reproducing and self fertilized plant for which it could not pose any problem as he used selfing as a technique in his experiment and for cross fertilization experiment it was easy to emasculate and to cross fertilize,
- Pea had many sharply defined inherited characters each with two alternate appearances (phenotypes or characters having clear cut differences).

Mendel had chosen 7 characters individually and crossed two varieties each having different appearances of the character. These seven characters in garden peas were:

Characters	Phenotypes (Dominant or recessive)
Seed characters:	
(1) Form of ripe seeds –	Round or wrinkled surface
(2) Colour of seed cotyledons-	Yellow or green
(3) Colour of seed coat-	Gray or white
Pod characters:	
(4) Form of ripe pods-	Full or wrinkled
(5) Colour of ripe pods-	Green or yellow
Stem characters:	
(6) Length of stem-	Tall (6-7 ft) or dwarf (¾ - 1 ft)
(7) Flower position on stem-	Axial or terminal

Mendel made the monohybrid crosses *i.e.* a cross having two different appearances of the same character. The cross fertilization was done and selfing was avoided so as both the distinguished appearances of the same character were united. Thus hybrids were produced. The experiment was repeated for several generations for all the seven characters. The reciprocal crossings were also made in all the experiments.

The original pure breed plants used for hybridization were called the parental generation denoted by P_1 generation whereas their offspring were referred as first filial or F_1 generation (hybrids). The succeeding generations descended from this cross (F_1) were called F_2 generation and so on.

The experimental findings afforded evidence that the hybrids (F_1's) are not exactly intermediate in appearance between the two parental varieties but resemble to one of the two parental characters for all the seven crosses. The character which expressed itself in the hybrid unchanged is called as *dominant character* whereas that which was not expressed (remained hidden or latent) in the hybrid (F_1) is called as *recessive character*. It was interesting to note that reciprocal crossings did not change the results and gave identical results.

On analyzing the results of F_2 generation produced by selfing of F_1,s, it was observed that the characters of both original varieties (P_1 and P_2) appeared in F_2 generation *i.e.* both appearances (phenotypes) of the character (dominant and recessive) were observed in F_2. Further, counting of the seeds, pods, and plants for two contrasting forms of the same character in F_2, it was observed that the individuals with recessive appearance were less frequent (about ¼ of the total individuals in F_2) and the individuals with dominant appearance were most frequent (about ¾ of the total individuals in F_2 population).

Mendel made the following conclusions

1. There are entities "hereditary determinants or factors" and each adult pea plant had two factors, one from each parent, for each characteristic. These Mendelian factors were later called as genes (Johanssen, 1903) and the two alternative form of gene as alleles.

2. Out of two alternate forms of character of parental variety, only one appeared in F_1 and the reciprocal crosses gave identical results.

3. The character which disappeared or remained hidden (recessive) in F_1, reappeared in F_2, but its frequency was as low as about ¼ of the total progeny in F_2.

2.2 MENDEL'S LAW OF INHERITANCE

On the basis of Mendel's work on garden pea and his findings, the following laws of inheritance were formulated:

1. LAW OF DOMINANCE

From the evidence of F_1 and F_2 results it was clear that there were two determining agents (factors or genes) responsible for each character (trait). This showed that genes or factors were in pairs and it is sometimes referred to as *law of paired factors*.

Secondly, the appearance of one form of a character could be hidden in F_1 but not destroyed/modified because it reappeared again in F_2. This phenomenon by which one form of character appears and the other alternate form remain hidden, though the factors for both the forms of character are present, is called the *dominance* and sometimes referred as *law of dominance*. The reappearance of a recessive/ hidden character in F_2 disproved the blending inheritance theory and hence it was a notable observation to the hereditary theory. The allele which can phenotypically express itself in the heterozygote as well as in homozygote is called a dominant gene or dominant character whereas the gene (allele) whose expression remains hidden in F_1 and can only express itself in homozygote is called a recessive gene (factor or character). The phenomenon of dominance is very common but it is not universal.

Importance of dominance: The phenomenon of dominance is of practical importance. The mutant alleles are recessive and detrimental to the organism while the dominant alleles produce normal character. For example, idiocy in man is caused by a recessive mutant gene besides being caused by other factors. Thus, a man with heterozygous genotype is normal without showing any sign of idiocy but carrier for idiocy allele. Therefore, the marriage of two heterozygous parents will produce some idiot children, in spite of the fact that none of the parent showed any sign of idiocy.

2. LAW OF SEGREGATION

Monohybrid cross: Mendel noted that self fertilization of F_2 plants with recessive character (*viz.* wrinkled seed, *rr*) always bred true. On the other hand, the self fertilization of F_2 plants with dominant character behaved differently in the way that some of them breed true and others did not. The plants which did not breed true produced two types of plants- some with dominant characters (smooth or round seed, *RR*) and others with recessive character (wrinkled seeds, *rr*) in a ratio of 3:1. Thus, F_2 individuals were of two types- one of pure producers, made up of identical genes (homozygote, *RR* and *rr*; *homo* means the same) and others of hybrid or mixed producers (heterozygote, *Rr*; *hetero* means different). Diagrammetically, the cross can be shown as under for seed form (shape):

Parents		(P$_1$) Smooth x (P$_2$) Wrinkled
		↓
		RR rr
F$_1$	(cross of P$_1$ x P$_2$)	Smooth (hybrid), Rr
F$_2$	(selfing of F$_1$)	3 Smooth : 1 Wrinkled
		(RR, Rr) (rr)
F$_3$	(selfing of F$_2$)	($_{1/3}$ breed true : All breed true
		+ $_{2/3}$ hybrid)

Therefore, the hybrid plants (heterozygote), which did not breed true and produced two types of plants with dominant and recessive characters, must contain factors (genes) for producing both the phenotypes of the character (*viz.* seed shape), one each for dominant (round) and recessive appearance (wrinkle). Likewise, the true breeding plants must also have two genes because pure lines are expected from each type (*RR* and *rr*) upon their self fertilization. Mendel realized that these results could be possible only if two genes of a pair remained distinct in the hybrid without changing their identity, in the hybrids, and the hybrids possess both types of parental factors (dominant and recessive) which produce two types of plants.

The situation discussed above is possible only if the two factors (*R* and *r*) responsible for the two parental types (round and wrinkle) segregate (separate) from each other during gamete formation and one gamete carries gene (*R*) for dominant character (round) while other gamete carries gene (*r*) for recessive character (wrinkle). Thus, there are two kinds of equally frequent pollen (*R* and *r*) and two kinds of equally frequent eggs (*R* and *r*). Therefore, members of a given pair of gene (*R* and *r*) segregate (separate) from each other during gamete formation, into equal numbers of gametes. The recessive gene (*r*) expresses itself in F$_2$ generation. Mendel called to this reappearance of the parental recessive character (wrinkle) as the segregation (separation) of the members of a given pair of gene (*R* and *r*). It is known as the principle of segregation.

The *law of segregation* states that genes exist in pairs in the cell of the individual and the two members of a gene pair segregate (separate) from each other during gamete formation so that each gamete receives one gene of each pair. *The law of segregation is applied universally.* The genes are consequently not blended or mixed to each other in the hybrid, but they preserve their identities and transmitted unchanged in hybrids, segregate at the time of gamete formation and reappears in the offspring. This stability of genes and their transmission in unchanged condition between cells and generation is termed as *purity of genes* or gametes. The two alleles (dominant and recessive) of a gene pair remain together through generations without permanent influence on each other.

Test of homozygous and heterozygous dominant

Mendel used two types of tests to distinguish between homozygous and heterozygous dominants. These were self-fertilization and back cross (test cross).

An individual heterozygous for one pair of gene (R and r) is called *monohybrid* and a cross of two individuals that are heterozygous for one pair of gene is called a *monohybrid cross*. This is the cross between two pure breeding individuals which differ from each other only in one character and the resulting hybrid is called monohybrid. The *dihybrid* or *polyhybrid* crosses are the cross between two pure breeding individuals which differ in two or more than two characters, respectively.

(i) Self-fertilization: The self-fertilization of dominant individuals produces individuals only with dominant character. On the contrary, the self fertilization of heterozygous individual produces two types of individuals- one type are those having dominant character and other type are those having recessive character, in a ratio of 3 dominants: 1 recessive. Therefore, the appearance of recessive character from self fertilization of hybrid individuals (heterozygous) or from self fertilization of homozygous recessive individuals indicated that the plants which produced two types of plants were heterozygous. These crosses can be illustrated taking dominant gene as R and the recessive gene as r:

(i) Parents	RR	x	rr
		$\downarrow$	
F$_1$	Rr (Dominant character)		
(ii) Parents	Rr	x	Rr
		$\downarrow$	
F$_1$	¼ RR: ½ Rr: 1/4 rr		
	(3/4 Dominant trait) (¼ Recessive trait)		
(iii) Parents	rr	x	rr
		$\downarrow$	
F$_1$	rr (Recessive. trait)		

(ii) Back cross or Test cross: The crossing of heterozygous progeny (Rr) with any one of its parent (dominant, RR or recessive, rr) is called as *back cross* whereas the *test cross* is a form of back cross when the heterozygous progeny (Rr) is mated to a recessive homozygous parent (rr). A breeding test performed to determine the genotype of an individual with dominant phenotype is called the *testcross*. The test cross is performed by mating of the individual under test (dominant phenotype) with a recessive homozygote.The crossing of homozygous dominant individual with homozygous recessive individual produces all the progeny with dominant character as:

$$RR \quad x \quad rr$$

$$\downarrow$$

Rr (Dominant Phenotype)

The *test cross* (Heterozygous x Recessive homozygous) results in a 1:1 ratio of dominant and recessive phenotype as:

Parents   Rr   x   rr

$$\downarrow$$

Eggs (female gametes)

	r	r	
Pollen R	Rr	Rr	(All with dominant phenotype)
(male gametes)			
r	rr	rr	(All with recessive phenotype)

The results of test cross provided very useful information in providing a tool for *determining the type of gametes produced by the parents* and therefore, the recessive back cross is often referred to as a *test cross*.

The backcross of hybrid F_1 with homozygous dominant parent produces all progeny with dominant phenotype and hence can not be taken as a test cross. This has been shown as under:

Parents   Rr   x   RR

$$\downarrow$$

Progeny   RR,   RR: Rr, Rr (All with dominant phenotype)

Therefore, recessive back cross (test cross) has great importance because it produces identical phenotypic and genotypic ratios.

Dihybrid cross: After completing seven monohybrid crosses, Mendel crossed varieties of peas which differ in two characters (dihybrid cross). Such a cross can be illustrated by a cross between pea plants pure breeding for round seed and yellow cotyledons (both characters dominant) with those pure breeding for wrinkled seed and green cotyledons (both characters recessive). The resulting hybrids (F_1) were having both dominant characters. Self fertilization of F_1 plants (heterozygous for both the characters) gave somewhat complex results. The total numbers of F_2 plants were obtained as:

Round seeds = 423   315 round - yellow
                           108 round – green   416 = yellow cotyledons
Wrinkled seeds = 133   101 wrinkled-yellow
                           32 wrinkled-green   140 = green cotyledons

Mendel first checked to test the ratio as per monohybrid cross taking F_2 as monohybrid for the two characters separately, totaling the progeny for one character at a time ignoring the second character. He obtained 3:1 ratio for both the characters separately. The analysis of F_2 data of the above dihybrid cross indicated 3:1 ratio with respect to each single gene pair in which dominance occurs. Thus, there was segregation of genes and hence purity of gametes. The law of segregation of genes was thus proved among F_2 offspring of dihybrid. Such a cross can be shown as under:

Parents	Round x wrinkled	Yellow x green
	↓	↓
F_1 Hybrid	Round	Yellow
F_2 Phenotype	round, wrinkled	yellow, green
No. observed	423 133	416 140
Ratio	3:1	3:1

The segregation of two gene pairs can be treated as independent events. The law of segregation holds true for single gene pair as well as for two gene pairs.

3. LAW OF INDEPENDENT ASSORTMENT

Combining two sets of results in a single dihybrid cross for two different pairs of genes affecting two different characters (round vs wrinkled seeds and yellow vs green cotyledons), Mendel found that the two pairs of genes behaved independently of one another *i.e.* each pair of genes segregated / assorted independently in a manner that other pair of genes was not there. This means that the chance for a plant to have smooth or wrinkled seeds do not interfere with the appearance of another character *viz.* cotyledon colour (yellow or green). Thus, the appearance of one character (seed shape) was independent of the appearance of other character (seed cotyledon colour). Mendel called it as the law of independent assortment, due to independence of these two pairs of genes. This law of I.A. states that *members of different pairs of genes segregate (assort) independently of each other at the time of gamete formation.*

The independent assortment of gametes for two pairs of genes is based on random union of gametes. From Mendel's law of segregation, it is known that

R gametes = r gametes = ½; and

Y gametes = y gametes = ½

The results of independent assortment of two pairs of genes among F_2 progeny can be obtained by any of the following three methods-

Punnet's *Gametic Checkerboard method*: Half the gametes contain R gene and other half contain r, because R segregates from r. Similarly, half the gametes contain Y and another half contain y for the same reason that Y segregates from y. Putting them together half of the half R gametes will contain Y gametes (*i.e.* ½ Y of ½ R = ¼ RY, another half of R gametes will contain y (*i.e.* ½ y of R = ¼ Ry), and likewise half of r gametes will contain Y (½ Y of ½ r = ¼ rY) whereas other half of r gametes will contain y gametes (½ y of ½ r = ¼ ry). Therefore, dihybrid F_1 individuals produce 4 types of gametes in each of the two sexes in equal proportions *viz.* RY, Ry, rY, ry. Consequently, the probability of forming 4 gametes being (by product rule):

R and Y $= \frac{1}{2} \times \frac{1}{2} = \frac{1}{4}$ RY; R and y $= \frac{1}{2} \times \frac{1}{2} = \frac{1}{4}$ Ry

r and Y $= \frac{1}{2} \times \frac{1}{2} = \frac{1}{4}$ rY ; r and y $= \frac{1}{2} \times \frac{1}{2} = \frac{1}{4}$ ry

Again the product rule of probability will be applied during zygote formation, because any gamete is free or independent to unite with any gamete of opposite sex. Thus the probability of the zygote with RRYY will be $\frac{1}{4}$ RY x $\frac{1}{4}$ RY = 1/16 RRYY and likewise will be the probability for all the 4 x 4 =16 combinations of zygotes.

The F_2 ratio closely fit the 3:1 ratio of crosses involving single pair in which dominance occurs. Both set of results are combined in a dihybrid cross, when each gene pair acts independently. Therefore, if a seed has a $\frac{3}{4}$ chance to be round (smooth) with a $\frac{3}{4}$ chance of being yellow cotyledons, its chance of being both round and yellow at the same time in a dihybrid, will be $\frac{3}{4} \times \frac{3}{4} = 9/16$ and likewise the ratio of each phenotypic combination may be obtained by multiplying the probabilities of the individuals phenotype. A phenotypic ratio of 9:3:3:1 among the F_2 progeny is obtained as:

$\frac{3}{4}$ round x $\frac{3}{4}$ yellow = 9/16 round-yellow

$\frac{3}{4}$ round x $\frac{1}{4}$ green = 3/16 round-green

$\frac{1}{4}$ wrinkled x $\frac{3}{4}$ yellow = 3/16 wrinkled-yellow

$\frac{1}{4}$ wrinkled x $\frac{1}{4}$ green = 1/16 wrinkled-green

Digrametically, the F_2 dihybrid cross ratio (9:3:3:1) can be illustrated as:

Parents Round –Yellow x Wrinkled-green

(RR YY) (rr yy)

F_1 Round-Yellow

(Rr Yy)

F_2 Female gametes (ova or eggs)

		$\frac{1}{4}$ RY	$\frac{1}{4}$ Ry	$\frac{1}{4}$ rY	$\frac{1}{4}$ ry
Male gametes (sperms)	$\frac{1}{4}$ RY	Round Yellow RRYY	Round Yellow RR Yy	Round Yellow Rr YY	Round Yellow Rr Yy
	$\frac{1}{4}$ Ry	Round Yellow RR Yy	Round green RR yy	Round Yellow Rr Yy	Round green Rr yy
	$\frac{1}{4}$ rY	Round Yellow Rr YY	Round Yellow Rr Yy	wrinkled Yellow rr YY	wrinkled Yellow rr Yy
	$\frac{1}{4}$ ry	Round Yellow Rr Yy	Round green Rr yy	wrinkled Yellow rr Yy	wrinkled green rr yy

Branching method: There is also alternate method called *branched* or forked-line method to represent different combination used to obtain F_2 ratio by drawing branched diagram and then using the product rule considering the two monohybrid cross *i.e.* Rr x Rr and Yy x Yy operating together. If one allele of a gene pair is dominant, it results in a 3:1 ratio from both the monohybrid cross. Both the pairs of genes being independent, each monohybrid allele will segregate and will combine with any of the other. It can be shown as:

F_1 Round-Yellow (Rr Yy)

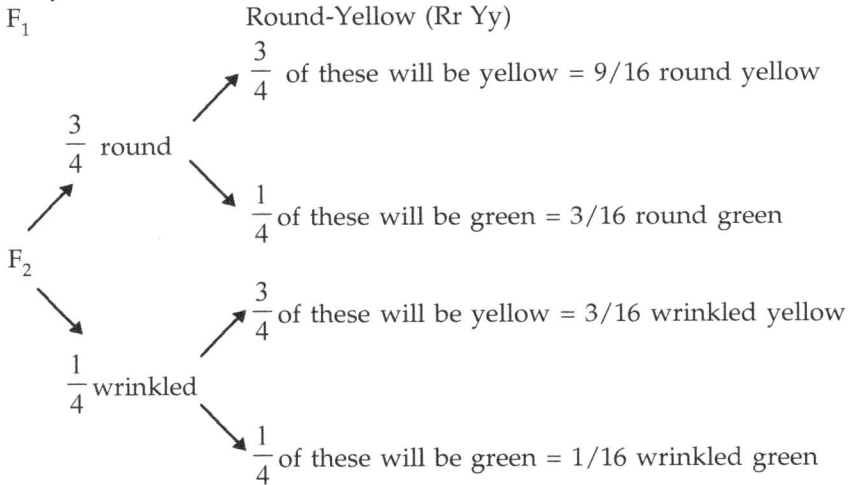

$\frac{3}{4}$ round

$\frac{3}{4}$ of these will be yellow = 9/16 round yellow

$\frac{1}{4}$ of these will be green = 3/16 round green

F_2

$\frac{1}{4}$ wrinkled

$\frac{3}{4}$ of these will be yellow = 3/16 wrinkled yellow

$\frac{1}{4}$ of these will be green = 1/16 wrinkled green

Algebraic method: The above phenotypic ratio among F_2 progeny can be obtained algebraically based on product rule of probability of two independent events. The gametic combination of one parent is multiplied with that of other parent as:

$$(\tfrac{3}{4} + \tfrac{1}{4})(\tfrac{3}{4} + \tfrac{1}{4}) = (\tfrac{3}{4} + \tfrac{1}{4})^2$$
$$= (\tfrac{3}{4})^2 + 2[(3/4)(1/4)] + (\tfrac{1}{4})^2$$
$$= 9/16 + 3/16 + 3/16 + 1/16.$$

Significance of independent assortment: Two new combinations of genes in a dihybrid cross are produced. Likewise, the F_2 progeny of trihybrid cross show six new combinations and so on it goes on increasing with the polyhybrid crosses. This knowledge of producing new combinations due to independent assortment of genes helps the breeder to produce and breed new combinations of useful characters.

The independent assortment of genes is not universally applied.

2.3 REASONS FOR REMAINING UNNOTICED THE MENDEL'S LAW FOR 35 YEARS

Mendel's discoveries remained unnoticed for 35 years during 1866 to 1900. There had been certain special reasons for long delay in recognition to Mendel's work.

(1) Mendel had chosen discontinuous characters and many biologists were concerned with the problems of evolution to study the gradual changes in continuous variation. The concept of inheritance of continuous variation was not proved with Mendel's law.

(2) Mendel's law could not be proved with other organism by him and Nageli, the famous botanist who worked with hawkweed, *Hieracium* and failed to achieve a demonstration of segregation. It was because the hybrids between two different varieties in this organism arise directly from diploid tissue in

the ovary without fertilization of gametes and thus some of the embryonic tissues will have exactly the same genetic constitution as its maternal parent and will not thus segregate for different characters among its offspring.

(3) Mendel's findings indicated that there was constancy of hereditary factors among generations *i.e.* the factors do not change but expressed themselves in new and different combinations among the offspring. The constancy of hereditary factors was unacceptable to those biologists who were interested for a source of variability in evolution.

(4) Mendel's preoccupation with probability events and mathematical ratios was an unfamiliar approach to biology.

(5) The physical basis of heredity (in terms of nuclear and chromosomal division) was not known.

(6) Mendel's results were published in an obscure periodical of limited circulation.

(7) Mendel was by then having trouble with the administration of monastery of which he had become the abbot (head) and hence could not work further.

2.4 BASIC MONOHYBRID CROSSES

The cross between two purebreds (homozygote) is called a *parental cross* or the P_1 *cross*. The progeny produced from parental cross is heterozygous and called as the first *filial* or F_1 *generation*. The progeny produced from mating of two F_1 individuals are called the F_2 *generation*.

An individual heterozygous for one pair of gene (having two different alleles at a locus, Aa) is called a *monohybrid*. Here, the mono refers one pair of gene and the hybrid means heterozygous or heterozygote having different genes. Thus, the two alleles at a locus are different in a monohybrid individual. A cross of two individuals which are heterozygous for one pair of gene is called a *monohybrid cross*.

There are *six types of monohybrid crosses:* These six crosses can be classified as:

(i) *Crossing of homozygote:* Three crosses involve the crossing of homozygote of either type (AA x AA, aa x aa and AA x aa) and each of the three crosses produces only one kind of progeny *viz.* the *first cross* (AA x AA) will produce all progenies with genotype AA, the *second cross* (aa x aa) will produce all the progenies with genotype aa, and *third cross* (AA x aa) will produce all progenies with genotype Aa. It can thus be concluded that a cross between any two homozygote (as the above three crosses) will produce all the progenies of only one kind having the same genotype.

(ii) *Crossing of homozygote with heterozygote*: This involves two crosses (Aa x AA and Aa x aa, say *cross number 4 and 5*) which are similar in that one of parent is heterozygote (Aa) and other parent is homozygote (either any of the two homozygote, AA or aa) and both the crosses will produce two types of progeny (heterozygous and homozygous) in equal proportions. [The cross Aa x AA will produce ½ Aa : ½ AA whereas the other cross (Aa x aa) will produce ½ Aa : ½ aa]. Moreover, the genotypes of two kinds of progenies in both the crosses are identical to those of the parents. The second type of cross (say lat cross number 6) involves the crossing of both the heterozygous parents (Aa x Aa). The last cross (heterozygous mating) produces three types of progeny in the ratio of ¼ AA: ½ Aa: ¼ aa.

(iii) *Heterozygous mating*: This last cross of heterozygous mating *(cross no. 6)* is special with respect to illustrate the segregation of genes in gametes during meiosis (the two members of each pair of homologous chromosomes are distributed into a separate gamete) and recombination of genes in the zygote in all possible combinations as a matter of chance. This is called the *law of segregation and recombination* which is the first law of genetics. This law was discovered by Mendel and hence the subject of the expression of phenotypic ratios due to segregation and recombination of genes is referred to as *Mendelian genetics*.

Chapter 3

Multiple Alleles

Mndels' fundamental experiment on garden pea contributed to basic knowledge about the unit nature of inheritance of characters which are inherited without change or dilution through generations. Mendel found F_2 ratio of 3:1 for monohybrid crosses and 9:3:3:1 for dihybrid crosses as a result of segregation of two different alleles for one and two pairs of genes, respectively among which one allele was dominant and the other was recessive. These F_2 ratios were called as Mendelian ratios. Mendelians' laws of inheritance were confirmed following the rediscovery of Mendels' work in 1900 with other organisms.

3.1 CAUSES OF MODIFICATION IN MENDELIAN RATIOS

Post Mendelian findings did not show the appearance of the phenotypic ratios as per Mendelian rules (Mendelian ratios) but some modified ratios were observed without following Mendelian rules of heredity and it was observed for both the monohybrid as well as the dihybrid crosses. These modified phenotypic ratios could not be explained by the presence of only two alleles for a gene pair among the population and by law of dominance at one locus. When the normal phenotypic ratios based on law of inheritance (Mendelian ratios) were not obtained it was but natural to think of other possible mechanism of inheritance to explain these modified phenotypic ratios.

The *phenotype is the expression of genes (genotype)* in such a way that the gene expression can be observed or measured. The genes are found to express themselves phenotypically in different ways. There are a number of *causes of variation in gene expression.* The gene expressions, based on modified phenotypic ratios, have been explained on the basis of the followings:

- *Multiple allele system:* Presence of more than two alleles for a gene pair.
- *Gene interaction:* Gene action which includes additive and non-additive gene action (interaction),

- *Lethal gene* has effect on modification of phenotypic ratios,
- *Environmental dependence of genotype:* Some genes produce their effects by interacting with *environmental agent,*
- *Linkage of genes* present adjacently on the same chromosome,
- *Sex controlled inheritance*

This chapter deals with the multiple allelic system; in the next chapter (number 4) have been discussed the different ways of modification of phenotypic expression of genes due to different types of gene action controlling the characters affected by one or more pairs of genes, the effect of lethal genes in modifying the phenotypic ratios and environmental dependence of genotype. The effect of linkage phenomenon of genes on modification of phenotypic ratios has been discussed in the subsequent chapter number 5 and the sex controlled inheritance has been given in chapter 6.

3.2 MULTIPLE ALLELES

Before a discussion on multiple alleles, it is necessary to understand the meaning of the two terms *viz.* locus and alleles.

Loci: The location (position or site) of a gene on a chromosome is called the *locus* (loci- plural). Thus, every gene has a locus on the chromosome. This was first described by Mendel that the factors (now called as genes) are present in pairs. Mendel called the genes as the factors. Therefore, an individual has only two genes at a locus and called as *diploid*. Therefore, maximum two genes are present at corresponding positions on homologous chromosomes of diploid individual.

Alleles: The two genes present at a locus of homologous pair of chromosomes may be identical in their structure or they may not be identical but both the genes are present at the same locus affecting the same character. These different forms of the same gene (in terms of different structure) found at the same locus on homologous chromosomes are called *alleles*. There are thus three points to consider a gene as an allele *viz.* the two genes are different in structure, present on the same locus on homologous chromosome and affect the same character.

Consider an example of a particular gene present at the same locus (called as locus A) affecting a trait. There will be present two genes at this locus in an individual. These two genes in an individual may be identical in structure or not. If both the genes at this locus A are identical in their structure, the two genes may be written as AA or aa and the individual will be called as *homozygote* to have homozygous genotype (AA or aa). The homozygote produces only one kind of gamete *viz.* the individual with genotype AA (homozygote) will produce two gametes, each having the same allele (A) and the second homozygous individual with genotype *aa* will produce two gametes, each having the same allele (*a*). If the two genes at this locus are different in structure, one of the two genes will be written as A and the other gene as *a* and the individual will be called as *heterozygote* to have heterozygous genotype (Aa). These two genes (A and a) are called as alleles of each other, because they are the alternate forms of the same gene (being differing in structure) but they are present on the same homologous locus on the homologous chromosome affecting the same character. The heterozygote (Aa) produces two gametes of different types having different alleles, one gamete will have the allele A and the second gamete will have the allele *a*.

Suppose the character is the presence or absence of horns (called as polled condition). One of the two genes present on the same locus will develop the horns in a cow while its allele which differ in structure (alternate form of that gene which cause a cow to develop horn) will cause another cow not to develop horns and hence the cow will not have horns (called as polled). It has been observed that the alleles causing the absence of horn in a cow is dominant, designated by the capital letter P whereas its alleles which causes the horn to develop is recessive, designated by the small letter *p*.

Concept of multiple alleles: During post Mendelian era, it became known that more than two alternate forms of a gene (alleles) occurred in population. The normal allele is called as wild allele while the newly formed allele is called as the *mutant*. The newly arisen allele (mutant) is produced from normal gene as a result of change in chemical structure of normal gene. This change is called *mutation* which is very rare. The mutation may occur several times in a gene and so several forms of the same gene may be produced. A large numbers of alleles at a locus are thus expected to be present but in different individuals of the population. Only two alleles can exist in a single individual because there are only two locations on a pair of homologous chromosomes. Therefore, the genes may exist in different allelomorphic states having two or more forms (alleles) but more than two forms of a gene (alleles) will be present in different individuals of the population. *This situation of existing more than two allelic form of a gene at a given locus in the population is called a multiple allelic system or series and the alleles are called as the multiple alleles.* When more that two alleles control a character, it is said that the character is controlled by multiple alleles.

Symbols for multiple alleles: The mutant trait that deviates from the ancestral type is usually chosen the basis for the symbol of gene. The multiple allelic system /series is indicated (written) by the dominance hierarchy of alleles. The dominant allele to all others in the series is distinguished by capital letter, the recessive to all others in the same series is designated by the small letter and the intermediate alleles between these two extremes will be designated by some superscript with a small letter. Thus, the multiple allele will be written in dominance hierarchy as C > C^{ch} > C^h > c.

3.3 CHARACTERISTICS OF MULTIPLE ALLELE

The multiple alleles of a series occupy the same locus on the homologous chromosome. In diploid individuals only two alleles of the multiple series are present in the cell of an individual. The crossing over does not occur within the alleles of a same multiple allele series because only two forms of a gene are present at homologous locus in an individual. They influence the same character but each allele has its different manifestation. Thus, they carry the same function but having different degree of efficiency.

The wild type allele (normal gene) is nearly always dominant over all the mutant alleles while the other mutant alleles in the series may show dominance or codominance among themselves or there may be an alternate phenotypic effect. The crossing of mutant with mutant produces mutant type and not the wild type. The two members of an allelic series are governed by the principle of segregation and recombination.

3.4 THEORY OF MULTIPLE ALLELES

The mutation theory of the origin of allelomorph has been widely accepted. However, two other theories have been put forth. These are the theory of pseudo-allelism (or close linkage among genes) and the heterochromatin theory of allelism.

Under the concept of multiple alleles, it has been postulated on the basis of mutation theory of allelomorphs that no crossing over occurs between alleles of a multiple series. But the other possibility is that some gene loci may be so closely located and linked to each other that they are inherited together with minimum chance of crossing over. Such genes, undoubtedly different genetically, may behave like the alleles of the same gene and by mistake may be taken as multiple alleles. Such genes occupying two separate loci but having close linkage and hence very rare chance of crossing over, often inherit together are called pseudo-alleles (Lewis, 1951). Thus, *the pseudo-alleles are two closely linked genes and affect the expression of normal gene showing position effect*. Lewis conducted experiment on inheritance pattern of a sex linked character, the eye colour, in Drosophila. The normal eye colour is red which is wild type but with other shades of eye colour like, pink, apricot, cherry, buff, pearl, eosin, white. These were considered to form a multiple allelic series. The experiment of Lewis showed that red flies are cross over and therefore he concluded that apricot and white are not actually the alleles but they are pseudo-alleles occupying adjacent loci.

On the other hand, Pontecorvo (1958) had considered the *pseudo-alleles* as the different mutational and recombination sites within the same gene and thus the gene was regarded as sub-divisible defined by functional properties. Recombination through crossing over may also occur within such a gene rather than only between the genes. This functional gene has been termed as cistron (Benzer, 1957). The pseudo-alleles are functionally related and closely linked to inherit together most often. The mutant loci are so close to the normal allele that crossing over between them is not possible. The red, apricot and white colour which appear to form multiple allelic series are actual examples of (pseudo-alleles) close linkage because red hybrids are produced when crossing over is greatly increased. In this context it needed to change the classical concept of gene that a classical gene may consist of more than one functional unit or cistron.

The alleles which produce the same phenotypic effect in homozygous state, nevertheless differ, are called as iso-alleles. Some genes behave as though it was a compound gene. Stadler had found that these genes are not exactly alike and called as sub-alleles.

There is another theory of allelism, known by the heterochromatin theory. This theory states that the heterochromatin is brought next to genes by the chromosomal breakage and rearrangement and suppresses the expression of genes. There is evidence that position effects in maize are sometimes due to the transposition of very minute invisible fragments of heterochromatin.

3.5 EXAMPLES OF MULTIPLE ALLELES

The presence of multiple alleles has been found on autosomes as well as on sex-chromosomes.

Coat colour in Cattle: Three colour patterns in cattle are governed by 3 alleles at a locus in a dominance hierarchy of alleles as $S^h > S > s$. These three colour patterns are as:

- $S^h S^h$ = Hereford pattern (white-face, white underside of body and top of neck),
- S S = Angus breed (solid colour, with no white spot),
- s s = Holstein breed of cattle (white spotting).

A fourth allele at this locus (S^c) has also been reported. This fourth allele produces the colour-sided pattern in Pinzgauer breed of cattle (Dutch). The colour-sided pattern contains the presence of coloured hairs on head, neck and sides of body, and a white strip extends from withers down the neck and behind the animal, continuing along the underline to the brisket. The gene, S^C is co-dominant over allele S^h, thereby the heterozygotes ($S^h S^c$) show both patterns (Hereford and colour-sided pattern) and this allele is completely dominant over other two alleles (S and s).

The four alleles have been fixed in these different four breeds in homozygous state, as mentioned above. The crossbreeding results (crossing of four breeds in different combinations) on colour of purebred animals of these different four breeds, their F_1 and F_2 progenies have indicated that these 4 alleles producing different colours are located on the same locus and hence these four alleles are the multiple alleles. The combinations of these 4 alleles produce the following phenotypes with 10 genotypes as indicated below-

Homozygous	Heterozygous	Phenotypes
$S^h S^h$	$S^h S$, $S^h s$	Hereford pattern (white-face)
S S	Ss	Solid colour (Angus)
$S^c S^c$	$S^c S$, $S^c s$	Colour-sided (Dutch cattle, Pinzgauer)
-	$S^h S^c$	Colour-sided and white-face
s s	-	Holstein spotting (white)

Coat colour in Horses: The coat colour patterns in horses are controlled by genes at several loci (see next chapter, under modifying genes). One colour series is controlled by at least three or four alleles. The 4 alleles with dominance hierarchy are: A'> A > a'>a. The 4 phenotypic classes (colour patterns) produced are:

- Wild type (bay colour with no markings, controlled by allele, A');

- *Bay pattern* (black mane, tail and lower parts of legs with rest of body usually of brown or reddish brown in colour, controlled by the allele, A);
- *Seal brown* which is similar to black except that lighter brown or tan hair over the muzzle and ears, and the flank area (controlled by the allele, a'); and
- *Black* hair on whole body (controlled by the most recessive allele in homozygous state, aa).

The four colour patterns with genotypes are shown here –

Phenotypes	Genotypes
Wild type	A'A', A'A, A'a', A'a
Bay pattern	AA, Aa', Aa
Seal Brown	a'a', a'a
Black (Solid colour)	a a

Some reports had shown that the wild allele (A') is not present at this locus and hence only three alleles (A, a' and a) control the three colour patterns.

Coat colour in Mice: *Agouti locus with lethal effect*: The agouti locus has 5 alleles (A^y, A^W, A, A^t and a) in the series. The gene for yellow (A^y) is lethal in homozygous condition ($A^y A^y$) in embryonic stage and hence the yellow colour is produced from heterozygous state with one no-yellow allele. The progenies produced from mating of yellow mice are 2/3 yellow: 1/3 non-yellow. Thus, the phenotypic ratio is modified. The allele A^y does not show complete dominance to other four alleles in this series which show dominance hierarchi as: $A^W > A > a^t > a$. Some times, the dominance hierarchy among alleles is considered as: $A^y > A^W > A > a^t > a$. These alleles in this series produce yellow, white belly agouti, solid agouti (full colour), black with tan belly, and non-agouti (black), respectively.

The second type of coat colour in mice is produced by the *albino* locus, having 5 alleles, present on different chromosome. The albino allele is mutant of the gene (C) controlling the production of melanin in the hair, skin and eyes. The 5 alleles with dominance hierarchy are $C > C^C > C^h > C^e > c$. These 5 alleles produce full colour, Chinchilla, Himalayan, extreme dilute colour and albino phenotypes, respectively, when in homozygous state as well as in heterozygous state with other alleles (recessive alleles) of dominance hierarchy.

Coat colour in Rabbits ((rodents): One type of coat colour in rabbit is called albino series and governed by 6 alleles at a locus. The alleles are designated in order of dominance hierarchy as $C > C^d > C^{ch} > C^l > C^h > c$. The 6 alleles produce 21 genotypes but only 6 phenotypic classes (6 types of skin colour) as under:

Full colour or *agouti*	CC, C C^d, CC^{ch}, C C^l, CC^h, Cc
Dark chinchilla	$C^d C^d$, $C^d C^{ch}$, $C^d C^l$, $C^d C^h$, C^d c
Chinchilla (Silver gray)	$C^{ch} C^{ch}$, $C^{ch} C^l$, $C^{ch} c^h$, C^{ch} c
Light chinchilla	$C^l C^l$, $C^l C^{ch}$, $C^l c$
Himalayan white	$c^h c^h$, $c^h c$
Albino (pure white)	c c

The coat colour of wild type rabbit is known as *agouti or full colour* and the allele for agouti is usually dominant, represented by capital letter C. These animals have banded hairs, being gray near the skin, followed by a yellow band and finally a black band or brown tip.

The *chinchilla* rabbits appear silver grey (diluted agouti) due to the optical effect of black and gray hairs. They lack yellow pigment. The allele for chinchilla shade is represented by C^{ch}. There are dark as well as light shades of chinchilla.

The allele for *Himalayan* (Russian) is represented by c^h. This coat colour pattern is white except for black extremes of ears, tail, feet and nose.

There is no pigmentation in *albino* rabbits and their eyes remain pink due to lack of pigment in iris of eyes. The albino allele is represented by the small letter c.

Coat colour pigments in Dogs: It is controlled by 3 alleles at a locus with dominance hierarchy of $A^s > A^y > a^t$ and produces 3 colours with the genotypes as:

Dark pigment	$A^s A^s$, $A^s A^y$, $A^s a^t$
Tan colour	$A^y A^y$, $A^y a^t$
Spotted pattern like	$a^t a^t$
(tan & black, tan & brown)	

Coat colour in foxes *with lethal effect:* There are three alleles controlling the coat colour in domestic foxes which show co-dominance and the two alleles are lethal in embryonic stage. The three alleles are represented as: W, w and p. The alleles w and p have lethal effects both in homozygous (ww and pp) and in heterozygous state (wp), resulting death at embryonic stage or at the most within a few hours after birth, in few cases. Three coat colours are produced. These are silver fur (WW), white-faced silver (Ww), and platinum colour (Wp). It is important to note that white-faced silver and platinum colours are produced from heterozygous genotypes. The lethal effect of w and p alleles modifies the phenotypic ratios resulting from certain mating *viz.* Ww x Ww, and Wp x Wp.

Eye colour in Drosophila: The normal eye in Drosophila is red with its variant colours. There are probably 12 alleles present at a locus on X chromosome for this system and produces several phenotypes. The white eye color was probably one of the first mutant allele known in this insect and subsequently different alleles producing different colour shades are designated as below:

Phenotype	Genotype	Phenotype	Genotype
Wild type Or Red	W	Honey	w^h
Coral	w^{co}	Buff	w^{bf}
Blood	w^{bl}	Tinged	w^t
Eosin	w^e	pearl	w^p
Cherry	w^{ch}	Ivory	w^i
Apricot	w^a	White	w

The wild type (red) colour is completely dominant and white is completely recessive to all other alleles in the series. Each allele in the series except w produces pigment but with less intensity in the series as indicated above between the eye colour of parental homozygote.

Wing size in Drosophila: The Character wing size in Drosophila is controlled by 3 multiple alleles *viz.* L^+, L^{vg} and L^a with dominancy hierarchy as $L^+ > L^{vg} > L^a$. The normal wing size or wild type is long represented by the allele L^+ and dominant to other two alleles. The allele L^{vg} in homozygous condition produces vestigial wings and the allele L^a produces antlered wings.

In Drosophila, there are some wing abnormalities ranging in size from no wing to normal wings. These abnormalities producing gradations in the amount of wing are also controlled by multiple alleles when associated with certain gene combinations *viz.* normal wings being produced by the allele Vg^+ whereas no wing condition is produced by the allele Vg^{nw} in homozygous condition. The others are small stump (Vg), narrow strap like wing (Vg^{st}), notched (Vg^{no}), and nicked (Vg^{ni}). The wild type allele (Vg^+) is dominant over other members of the series with dominance hierarchy as $Vg^+ > (Vg^{st}=Vg^{no}=Vg^{ni}=Vg^{nw}=Vg)$.

Shell colour of land snail: The colour pattern is due to 3 alleles at a locus showing dominance hierarchy as Brown > pink > yellow. Thus the yellow colour is most recessive. The 3 alleles can be represented as B, P, Y, respectively for the three colours in descending dominance. The three phenotypic classes of colours with their genotypes will be produced as:

Brown	BB, BP, BY
Pink	PP, PY
Yellow	YY

Plumage colour in mallard Duck: The trait is dependent on 3 alleles with their dominance hierarchy as $M^r > M > m$ and produces the following 3 phenotypes with their genotypes as:

Restricted mallard pattern	$M^r M^r$, $M^r M$, $M^r m$
Mallard	MM, Mm
Dusty mallard	m m

Blood group in Man: *Dominance and codominance:* Before a discussion on A, B, O blood group system of man, controlled by multiple alleles, it is better to understand about the antigen- antibody reaction. There are two main components of blood *viz.* cells and plasma. The plasma is liquid and composed of clotting protein fibrinogen and serum. The studies have shown that two kinds of substances are present in the blood *viz.* antigens (or agglutinogens) and the antibodies (or agglutinins). The RBC's are generally antigenic carrying antigens (or agglutinogens) and the blood plasma contains the contrasting antibodies (agglutinins).

The clumping of R.B.C., known as agglutination, takes place in some cases when blood of a man is mixed with the serum of other man but in some other cases the clumping does not occurs. The clumping takes place due to the antigen-antibody reaction (discovered by Landstenler, 1900).

The blood of man and other higher vertebrates has the property to react with foreign substance so as to eliminate or neutralize or immune that foreign substance. The foreign substance is called *antigen* (or agglutinogen) which may be plant or animal protein or may be the bacterial or viral toxin. The blood responds to the entry of such an antigen in the form of producing another protein molecule which combine or interact with the foreign body (antigen) in some way. The protein produced in the blood to immune the antigen is known as *anti body* (or immune body or agglutinins). The anti body production is the result of the modification of molecules of gamma globulin proteins which are synthesized by plasma cells. As a result of the interaction between antigen and anti body, the antigen is changed in another form in some way so that the antigen is destroyed, inactivated, phagocytized or eliminated from blood circulation.

The antibodies are specific for a particular antigen. These antibodies may be acquired ones which are produced by the plasma cells during entry of foreign antigen in the blood stream. Such antibodies are called *acquired antibodies*. On the other, some antibodies are produced naturally and normally by the blood, even in the absence of antigen and such antibodies are called as *natural antibodies*. The antibodies involved in A, B, AB, O blood groups of man are the natural antibodies.

The well known example of multiple alleles is the ABO blood group system in humans controlled by 3 different alleles at a locus. These alleles are A, B, and O. The group O is taken as normal whereas the alleles A and B represent two dominant mutations which have occurred on the same locus and represent co-dominant alleles. These blood groups are inherited in simple Mendelian fashion. The heterozygous mating between blood group A and B (*i.e.* AO x BO) can produce children of all the blood groups. There are 3 alleles (A, B, O) where there is co-dominance between A and B, and both these are dominant over O. This is the mixture of co-dominance and dominant alleles. The 3 alleles form 6 genotypes *viz.* AA, AO; BB, BO; AB; and OO in the population. Thus following 4 blood groups (phenotypes) are produced as:

Blood groups	A	B	AB	O
Genotypes	AA, AO	BB, BO	AB	OO
Reaction with: Antigen A	+	-	+	-
Antigen B -	–	+	+	-
Anti body (in serum)	B	A	none	A&B
Agglutinates	A & AB cells	A & AB cells	none	A, B, AB cells
or serum agglutinin	Anti-B	Anti-A	none	Anti-A & Anti-B.
Antigen pre sent on RBC	A	B	AB	None
	----	------	Universal recipient	Universal donor
Agglutinated by serum of: % population	O, B 43	O, A 14	O, A, B 6	None 37

The allele A produces antigen-A and allele B produces antigen B while the allele O does not produce any antigen. A person having antigen in his R.B.C's his plasma will have antibodies against the other antigen *viz.* if a person has antigen **A** in R.B.C's, it has antibody **B** in his plasma. The agglutination occurs on account of the reaction between antigen and antibody. The antigens are not proteins but muco-polysaccharides (sugar + amino-acids). The four blood groups occur depending upon the presence and absence of antigens and antibodies.

The R.B.C's of a man with blood group O, having no antigen, are not clumped by the serum of any blood group and hence this person can donate blood to all but he can take blood only from a man of his own group and hence known as *universal donor.* The serum of AB blood group man does not clump the R.B.C's of any blood group and hence they are *universal recipients* but can donate blood only to the person of their own blood group. The Rh factor has changed the concept of universal donor and recipient. The persons of groups A and B can take and donate their blood to their own groups but they can donate blood to people of blood group AB and receive blood from O group.

Use of ABO blood groups: The discovery of ABO blood group system has been proved very useful to mankind in medico-legal aspects. It helps in solving the legal problems of disputed parentage, in solving the problem of illegitimacy and the disputed cases of claimants to estates or in criminal proceedings

Testing ABO blood groups: The coagulation test to ascertain the blood group is conducted. Antisera-A and antisera-B are take either on slide or in test tube and adding a drop of given blood in each tube or on each slide. Now thorough mixing is done and observations are made as under:

(i) Coagulation of blood in tube A with antisera –A indicates that the blood group is A.

(ii) Coagulation of blood in tube B with antisera –B indicates that the blood group is B.

(iii) Coagulation in both the tubes indicates that the blood is of AB group.

(iv) No coagulation in either of the tube indicates the blood group is O.

Haemoglobin polymorphism in human: The normal hemoglobin is determined by the allele A, sickle cell haemoglobin by the allele S (which causes the RBC to be sickle under reduced oxygen tension and causes haemolytic anaemia becoming fatal before the age of 20). The individuals with a third haemoglobin C in homozygous condition produce mild anemia with hemolysis. All the 3 haemoglobins are produced by 3 alleles present at a locus and all the 3 alleles have co-dominance relationship. Thus 6 genotypes produced resulting 6 phenotypes as:

AA, SS, CC, AS, AC and SC.

The RBC acid phosphatase enzyme in man is controlled by 3 alleles at a locus with co-dominance among all the 3 alleles, designated as A, B and C producing 6 genotypes and 6 phenotypes.

There are some multiple allelic systems in farm animals affecting blood groups.

Sex-linked multiple alleles: The multiple alleles are also present on sex-chromosomes. The feather colour of pigeons is an example of sex – linked multiple alleles. Three alleles have been reported at a locus on Z chromosome. The three alleles can be designated as A, B, and b, producing ash red, blue and chocolate feather colours, respectively. The three alleles have the dominance hierarchy as: A > B > b. In pigeon, the males are homogametic (ZZ) and the females are heterogametic (ZW). The genotypes in two sexes and feather colours can be shown as:

Genotypes		
Males (ZZ)	Females (ZW)	Phenotypes (feather colours)
AA, AB, Aa	A (W)	Ash red
BB, Bb	B (W)	Blue
Aa	a (W)	Chocolate

Chapter 4

Gene Expression (Action)

The existence of genes, to play their role in controlling a character, is inferable only from the expression of genotype in the form of a phenotype (character). The expression of a gene is different in the presence or absence of another gene present on the same or different locus. This is the basis of variation in gene expression (action or effect) which is of different types. The gene expression to produce phenotype is of two type *viz.* additive gene action and non-additive gene action (gene interaction).

4.1 ADDITIVE GENE ACTION (AGA)

The *additive effect of genes* means the individual effect of genes which is independent of the presence or absence of another gene(s) and hence it is produced without being interacting with another gene(s) of the same gene pair or of another gene pair. This implies that the effect of a gene is not affected by the presence or absence of another gene. The important aspect of additive gene action is that each gene is contributing gene, adding some value to the character and the effect of each allele is considered to be arithmetically equivalent and additive or cumulative, for which these genes are called as additive or cumulative.

The term additive gene action is used for explaining the effect of polygenes or minor genes affecting the economic characters in farm animals. This type of gene action leads to the population mean to be changed by replacement of a gene with another gene. The additive gene action merely implies an increasing or decreasing effect of a gene on the phenotypic value of a trait. Thus, the gene effect is expressed as the change in population mean due to the introduction of a gene. The amount of change in population mean attributable to gene substitution is called the average effect of gene. The gene influencing a trait by adding some amount to the phenotypic value of the trait is called favourable or positive allele while the gene which decreases the phenotypic value is called the unfavourable or negative allele or neutral gene. This is the actual meaning of additive gene action that the presence of each allele either adds or deducts the genetic value by one unit.

In additive gene action, there is no sharp distinction between genotypes, but many gradations are present in between the two extreme phenotypes. The skin colour inheritance in human may be taken as an example of the additive gene action (Davenport's theory). In this type of inheritance some genes are called the *contributing genes* which make their contribution in darkening of the skin, genes A, B, etc. whereas the genes *a* and *b*, etc make no contribution in affecting the skin colour and hence are called as neutral genes for their no contribution to skin colour. Therefore, the genotype *aabb* is called as the *residual genotype* which has no contributing gene. In this type of gene action, no gene is dominant or recessive and hence this indicates *lack of dominance* or *lack of epistasis*.

Additive type of gene action is most important in influencing most of the characters of farm animals which are of quantitative in nature (measured in metric units). The examples are production of milk, wool, egg, meat, growth rates and body weight, quality traits of milk, meat, fleece, egg. These traits are polygenic in nature.

4.2 NON-ADDITIVE GENE ACTION (NAGA)

The *non-additive gene effect* means such an effect of gene which is produced in the presence of another gene of the same locus or of different loci and hence the gene produces its effect after being interacting with another gene. Thus, the non-additive gene effect is dependent on presence of another gene and hence a gene produces its effect when combined and interact with another gene. In this type of gene action, the addition of a gene to the genotype does not add an equal amount to the phenotype. Thus, it is called as non-additive gene action (NAGA) which is the nonlinear expression of genes, also called as the *gene interaction* or gene combination effect which is of the following two types.

4.2.1 Allelic Interaction

The first is the allelic interaction when the genes may interact with their own allele. This is also called as intra-allelic interaction. The example is the *dominance* which may be complete, incomplete or partial and overdominance. The allelic interaction depends on the expression of heterozygous genotype. The phenotypic expression of heterozygous genotype is variable. It may be either similar or different to varying degree to the phenotypic expression of either homozygote. The example of similar phenotypic expression of heterozygote to either homozygote is the dominance whereas the examples of phenotypic expression of heterozygote different to homozygote to varying degrees are the incomplete or partial dominance and over-dominance.

(1) Dominance or complete dominance: The dominance arises when the phenotypic expression of heterozygous genotype is similar to either of the homozygous genotype. The allele of a gene pair is called as *dominant* if it expresses itself in the heterozygote by inhibiting the expression of another allele. The allele whose expression is masked / inhibited is called *recessive allele*. This phenomenon is called as *dominance*.

When the dominance is complete, identical phenotypes are produced by the heterozygous individuals and by dominant homozygous individuals as:

Genotypes	A_2A_2		A_1A_2 A_1A_1
Phenotypic Value Scale	0	1.0	2.0

Fig. 4.1: Dominance Model

The phenotypic ratios of monohybrid crosses among F_2 progenies with complete dominance are 3:1, *i.e.* [3 (dominants):1 (recessives)]. The examples of traits showing complete dominance, in different species are:

Species	Characters	Dominant phenotype	Recessive phenotype
Cattle	Horn	Polled (No horn PP, Pp)	Horns (pp)
	Ear notch (Ayrshire)	Notch (NN, Nn)	Normal (nn)
	Body Colour	White face	Coloured (solid)
	-do-	Solid colour	Irregular white spotting
	-do-	Black hair coat	Red hair coat
	-do-	Red	Yellow (light red)
	Hoof type	Cloven hooves	Mule feet
Sheep	Coat colour (wool)	White wool (Bb, Bb)	Black wool (bb)
	Fleece type	Hairy fleece	Wooly fleece
	Eye colour	Brown	Blue
Swine	Coat colour	White	Black
	-do-	Black hair	Red hair (Hampshire)
	Foot condition	Mule footed (MM, Mm)	Cloven hoof (mm)
	Ear	Erect	Drooping
Poultry	Plumage colour	White	Coloured
	Comb shape	Rose comb (RR)	Single comb (rr)
	-do-	Pea comb	Single comb
	Feathers on legs	Feathered (FF, Ff)	Clean legs (ff)
	-do-	Normal feathers	Silky feathers
	Feathers colour	Coloured	White
	-do-	Black feathers	Red feathers
	Skin colour	White skin	Yellow skin
Horse	Hair colour	Black hair coat	Chestnut or sorrel
	-do-	Bay	Black (non-bay)
	-do-	Smooth hairs	Curly hairs
	Movement	Trotting	Pacing

Contd...

Contd...

Species	Characters	Dominant phenotype	Recessive phenotype
Turkey	Coat colour	Bronze (RR, Rr)	Red (rr)
	Feather condition	Normal (HH, Hh)	Hairy (hh)
Rabbit	Hair length	Short (LL, Ll)	Long (ll)
	Hair colour	Black hair (BB, Bb)	white hair (bb)
	Hair morphology	Normal	Wuzzy- unkempt
Drosophila	Body colour	Gray (b'b', b'b)	Black-ebony (bb)
	Eye shape	Lobe shaped (LL, LL')	Oval (L'L')
	Eye colour	Red (S'S', S'S)	Sepia (SS)
	Wing size	Normal (vg'vg', vg'vg)	Vestigial (vg vg)
Guinea pig	Coat colour	Black (BB, Bb)	White (bb)
	Hair length	Short (LL, Ll)	Long (ll)
	Coat texture	Rough (RR, Rr)	Smooth (rr)
Fox	Coat colour	Red (BB, Bb)	Silver black (bb)
	Body Size	Normal (DwDw, Dwdw)	Dwarf (dw dw)
	Behaviour	Normal (VV, Vv)	Waltzing (vv)
Dog	Body colour	Dark (solid)	Albino (White spotting)
	-do-	Red hair	Yellow hair
	Hair size	Short	Long
	Hair texture	Wire haired (WW, Ww)	Smooth hair (ww)
	Deafness	Normal (DD, Dd)	Deaf (dd)
	Ear	Erect (EE, Ee)	Drooping (ee)
	Barking	Barkness (BB, Bb)	Silence (bb)
	Tail	Stumpy	Normal
Cat	Hair length	Short	Long hairs
	Hair colour	Black	Brown hairs
	Skin colour	Tabby	Black or blue
	-do-	Colour	White
	-do-	Agouti	Non-agouti

(2) Codominance: This is also called as lack of dominance. The co-dominance occurs when the two alleles at a locus do not show dominance-recessive relation in their expression but both the alleles are capable to express them phenotypically in the heterozygous condition and the phenotype is a mixture of the phenotypic traits produced by either of the alleles in homozygous condition. The co-dominance differs from additive gene action in the sense that the effect of co-dominant alleles is not the additive though both alleles express themselves in heterozygous condition.

The co-dominance of alleles modifies the F_2 phenotypic ratios of monohybrid crosses from 3:1 (with complete dominance) to 1:2:1, *i.e.* [1(dominant homozygote):2

(heterozygote):1 (recessive homozygote)]. The examples showing co-dominance for qualitative traits are as under:

Coat colour in Shorthorn cattle: The character coat colour in Shorthorn breed of cattle is controlled by one pair of gene which shows co-dominance to produce hair colours on the body. The two alleles are denoted as R and W which produce three different coat colours *viz.* Red (RR), White (WW) and Roan (RW). The heterozygous animals (RW) had a different colour of the body called as *roan* which is admixture of red and white hairs on the body. The heterozygous mating produce red, roan and white offspring in 1:2:1 ratio and thus the phenotypic and genotypic ratios are equal and similar. Both the alleles (R and W) have their effect in heterozygous condition and hence called as co-domonant alleles which do not prevent the effect of each other. Thus, the dominance was lacking or absent.

Body size of Hereford cattle: The body size of Hereford breed of cattle is controlled by one pair of gene denoted as N and C. These two alleles produce animals of three different sizes *viz.* normal size (NN), dwarf size (CC) and comprest dwarf size (NC) which is intermediate. The mating of comprest dwarf animals produce the animals in the ratio of 1 Normal (NN): 2 Comprest (NC): 1 dwarf (CC). The mortality rate is high among dwarfs (CC) and it is impossible to mate them with normal to produce comprest as the *dwarfs* do not survive to maturity, having lethal effect.

The examples of genes expressing like co-dominance in other species are as under:

Species	Characters	Homozygote 1	Heterozygote	Homozygote 2
Sheep	Wool quality	Normal(N'N')	Medulated (N'N)	Hairy (NN)
Guineapig -	Coat colour	Yellow(YY)	Cream (YW)	White(WW)
Poultry -	Feather morphology	Normal (NN)	Mild Frizzle(NF)	Frizzle(FF)
	Feather colour	Black	Blue	White
Drosophila -	No. eye facets	800 (B'B') Normal	250-500 (B'B)	60 (BB)
	Eye colour	Red eye	Pteridine (reduced pigment)	Sepia colour
Goat –	Ear size	Long ears	Medium	Short
Swine -	Body colour	Black	Black & Red spots	Red

The blue Andalusion fowls, arise from the combination of alleles for white and black feathers, is not a true breeding variety. The blue Andalusian fowl, exhibiting codominance between black and white feather colour alleles, is in real sense a fine mosaic of black and white areas that appear to be blue.

The *coat colour in horse* show lack of dominance. The *palomino* horses (copper or golden colour) never breed true due to heterozygosity for colour genes (Cc). The palomino are produced from mating of *Chestnut* (full colour, CC) or Sorrels (which are homozygous for one of the alleles) with pseudo albino or white *Cremello*, the light cream colour (cc) which are homozygous for another allele (Castle, 1948).

The *sickle cell disease in humans* is characterized by severe chronic anaemia, retarded physical development and pains in joints, muscle and abdomen. This disease is the result of a gene, S, when homozygous (SS) and is almost fatal without exception. The homozygous individuals (SS) had all the RBC in their blood of grossly abnormal form (sickle shape), instead of the normal round shape, due to deprivation of access of oxygen to the blood. The heterozygous individuals are healthy but having some RBC as sickle shape and all of the cells can be sickled by oxygen deprivation though a greater reduction in oxygen tension will be required than in persons of SS genotype. In normal persons (AA) only haemoglobin A is found in the RBC. The genotype AS produce both types of haemoglobin (A and S). In this case neither gene affects the expression of other gene. The heterozygote has seldom anaemia (mild). The SS individuals seldom survive to reproduce and usually die of anaemia. The heterozygote (AS) are more resistant to malaria than normal AA persons.

The co-dominance is illustrated in *human blood groups* (MN system and ABO blood systems). The genes M and N are co-dominant for MN group whereas the genes A and B are co-dominant for ABO system.

(3) Incomplete or partial Dominance: This is the invariable expression of co-dominant alleles, when the dominant allele does not completely mask the expression of its recessive allele. In this case, the phenotypic value of heterozygote is different from either of the two homozygotes but some where in between to them.

The partial dominance in quantitative terms indicates that heterozygote is not exactly at the mid point of the phenotypic value of two homozygotes, whereas the incomplete dominance indicates that one allele is more dominant than the other.

Genotypes		A_2A_2		A_1A_2	A_1A_1
Phenotypic Scale	Value	0	1.0	1.5	2.0

Fig. 4.2: Incomplete Dominance model of gene action

However, irrespective to the term used for this type of allelic interaction, it is important to note that the expected phenotypic ratios of monohybrid crosses should be 1:2:1, similar to co-dominance, instead of 3:1 expected with complete dominance.

(4) Over-dominance: The over dominance is also allelic type interaction and is indicated when the phenotypic value of heterozygote is superior to either of the two homozygote.

Genotypes	A_2A_2		A_1A_1	A_1A_2
Phenotypic Value Scale	0	1.0	2.0	2.5

Fig. 4.3: Over dominance Model

This is a type of gene action as a result of which the progeny is superior to either of the parent. The gene is a complex organic molecule. Therefore, a chemical reaction between two alleles (A and a) may be different and greater than the reaction between AA or aa. Thus, genes in heterozygous state enter into some kind of interaction whereby a greater net result is produced than the genes in either of the homozygous condition and this makes the heterozygous individual to be superior to either of the homozygous.

. As a result of this type of gene action, the heterozygous individuals pass through unfavourable environmental conditions more successfully than the homozygote because "A" allele may respond more favourably to one type of environment and "a" allele to the different environment. Thus "Aa" condition produces more favourable results than AA or aa condition.

The phenomenon of overdominance is illustrated in the inheritance of a particular *blood type in rabbit (RBC antigen)* where 2 alleles are responsible: One allele produces one antigen and the second allele produce second antigen, and a different third antigen (which was not present in either homozygous) was produced in the homozygous individuals.

The *human haptoglobin (binding haemoglobin)* is another example. The two homozygotes produce 2 haptoglobins which are proteins and the heterozygote produce a third (not present in either of the 2 homozygotes) plus these 2 haptoglobins which are produced in the 2 homozygous individuals.

The phenomenon of overdominance, besides the allelic interaction, may probably involved many pairs of genes (non-allelic) affecting the same character. In case of polygenic inheritance, the effect of any one pair of genes may be rather small but the combined effect of many pairs result in a considerable advantage to the heterozygote and the combined effect of many loci is greater than the sum of the total effect of the additive effect of genes. The expression of hybrid vigour by crossing certain lines, strains, or breeds could be a result of over-dominance together with dominance and epistasis. Heterozygote are at advantage because of the combined effect of many pairs of genes brought together rather than the small effect of any one pair of gene.

These types of gene action (allelic interaction) are all varied expressions of dominance and do not differ too much except in overall effects and the inter-relationship among them may well be illustrated by taking a hypothetical example as under. The value of *A* is taken as 5 units and of *a* = 2 units.

Type of gene action	AA	Aa	aa
Lack /absence of dominance (Additive)	10	7	4
Complete dominance	10	10	4
Partial /incomplete dominance	10	8	4
Overdominance	10	12	4

(5) Lethal and Sub-Lethal Genes

It has been recognized that most of the mutations are recessive which are detrimental (deleterious) to the organism. There are many detrimental genes in farm animals. However, some mutation are beneficial like polled trait in cattle is beneficial to man but such beneficial mutations are few.

The detrimental genes have been defined and classified depending on the time and extent of their effect (drastic effect) on the organism. The mutant allele may affect the vigour or vitality of the individual at any age of the individual. Accordingly, *the detrimental genes* are classified into three types *viz.* detrimental or non-lethal (called as sub-vital genes), semi-lethal (or sub-lethal) and lethal genes.

(i) Sub-vital genes or *non-lethal:* This group of detrimental genes express themselves to some extent only in reducing the vigour and vitality of the individual, and thus they can escape to be detected.

(ii) Lethal genes: Second group of detrimental genes (Lethal genes) cause death of the individual. The effect of lethal genes starts from gamete formation resulting in *prenatal death* (before birth) at zygotic stage or embryonic (foetus) stage during pregnancy or *peri-natal death* at the time of birth or soon after birth within few hours or *post-natal death* in early life (young age). However, a number of genetic and non-genetic (environmental) factors may be responsible for death. The genetic cause of death is due to genes which are incompatible with development and these genes are called lethal genes.

(iii) Semi lethal or sub-lethal: Third group of detrimental genes causes death of individuals but sometime at later age of life and called as *semi-lethal* or *sub-lethal genes.* The semi-lethal or sub-lethal genes also cause some death losses in animals *e.g. dwarf condition* in Hereford cattle produced by mating of comprest with comprest. The dwarf animals are born alive but die before sexual maturity.

Abnormalities and lethal genes are mostly recessive in inheritance caused by one or more loci. However, in some cases the recessiveness is not complete but the gene has its phenotypic effect in heterozygous state. The heterozygote are sometimes more desirable than homozygous normal *e.g.* Dexter breed of cattle, infertility in Swedish dairy cattle due to *gonadal hypoplasia* is hereditary but is correlated with high milk production with higher fat content.

The mode of expression of lethal genes is that they being mostly recessive or partial dominant must be in homozygous condition to have their full effect. The frequency of recessive genes which are either detrimental or sub-lethal is very low in the population and they appear in homozygous state only through inbreeding. This is one of the reasons that inbreeding is avoided. However, when the frequency of lethal or detrimental genes is high in the population, they are produced without inbreeding.

Increased rate of inbreeding results the recessive lethal to come into play and thus inbreeding leads to higher death losses in the young ones from conception to birth or up to weaning age. Inbreeding is thus very important to discover the recessive mutations particularly for the mutations resulting in defective development in which the heterozygote does not have selective advantage.

Sickle cell disease in humans is caused by a gene in homozygous state (ss). This gene has its drastic effect to result in death. Such genes are called lethal genes. The fully dominant lethal allele arises occasionally by mutation from a normal allele and majority of mutations are recessive in their effect. However, the dominant lethal have more effect in a population because they are fatal to both dominant homozygous as well as to heterozygous individuals. The completely recessive lethal gene have their effect only in homozygous condition, but some of the recessive lethal alleles also exhibit a distinctive phenotype without lethal effect when present in heterozygous state *i.e.* in a co-dominant system. Such genes result in the modification of Mendelian ratios if they kill the organism before birth, or before hatching in poultry or before germination in plants.

(*i*) **Dominant detrimental / lethal genes:** The examples of dominant detrimental/ lethal characters are as under:

The *creeper condition in fowl* is due to a dominant gene (Cc) which has an effect on the vitality when homozygous (CC). The creeper birds are always heterozygous (Cc) because the cross of creeper with creeper produce 2:1 ratio of creeper to normal. The normal birds are recessive homozygous (cc), the creepers with short and deformed legs and wings are heterozygous (Cc) and the dominant homozygous (CC) birds die before birth.

Hairless condition in Mexican dogs is produced by the heterozygous condition (Hh). The homozygous recessive dogs (hh) are normal whereas puppies homozygous for H allele (HH) are usually born dead with abnormalities of mouth and absence of ears. Thus, heterozygous creeper (Cc), normal (cc), and homozygous dominant (CC) is lethal.

The *coat colour in mice* is controlled by genes located at about 24 or more loci and some of these genes also affect skeletal growth and development of certain body tissues. The heterozygous yellow mice (Yy) survive but the dominant yellow mice (YY) die during pregnancy. The yellow colour gene is dominant (Y) and lethal over non-yellow. Thus, heterozygous yellow (Yy), non-yellow (yy), and homozygous yellow (YY) is lethal.

The gene for *coat colour (platinum) in foxes* is dominant and lethal. The homozygous dominant foxes with platinum colour (PP) die before birth and the mating of heterozygous platinum foxes (Pp) produce platinum to silver progeny in 2:1 ratio instead of 3:1. The silvers are homozygote recessive (pp).

The *gray colour in sheep* is due to a partially dominant gene. The homozygous gray sheep (GG) have defect of digestive tract (abnormal abomasum) and die during embryonic stage or within few months after birth. The heterozygous gray sheep (Gg) survive. The black colour sheep are homozygous recessive (gg).

Pelger anomaly of rabbit (abnormal WBC number segmentation) is expressed in heterozygous state of two alleles (Pp) and the normal individuals are homozygous recessive (pp). The homozygous dominant genotype (PP) causes grossly deformed skeleton causing death before or very soon after birth.

Polydactyly in cattle (extra toes on one or all feet) is most probably controlled by dominant gene. Such animals show lameness and hence undesirable. The gene is thus detrimental.

The *umbilical hernia in cattle* (Holstein-Friesian) appears at the age of 8-20 days, persists up to 7 months age, is limited to male calves and controlled by a dominant gene.

Brachyphalangy (shortening of middle joint of one or more fingers) also called *brachydactyly* and the *thalesemia* in man are dominant characters which cause lethality before sexual maturity in heterozygous individuals. The homozygous individuals for thalessemia have severe anemia (thalessemia major) and die before maturity, heterozygote have mild anemia (thalessemia minor).

Huntington's chorea in human is a disease, caused by a dominant gene and results in progressive degeneration of the nervous system. This causes death after about 30 years of age. This disease is caused by the dominant latent lethal (sub lethal) gene, having its effect after sexual maturity, getting a chance to reproduce.

(ii) **Recessive detrimental / lethal genes:** There are some alleles which are lethal in homozygous recessive state. The followings are some of the examples of recessive detrimental / lethal characters:

Amputated condition (absence of legs) in *Swedish Friesian cattle* is caused due to a recessive lethal gene. The calves are still-born or die soon after birth. The mating of normal with normal produces an amputated calf which is usually dead at birth.

The *parrot beak in Shorthorn cattle* (abnormal lower jaw with impacted molar teeth): The gene in heterozygous condition shows no detectable sign of gene's presence and the cattle are normal. The gene in homozygous recessive condition result the still-born calves or they die after birth within one weak and they have parrot beak.

Hairless and semi-hairless cattle are the two characters which are due to recessive gene. Another character *snorter dwarfism in Angus and Hereford cattle* is semi-lethal. The dwarfs have difficulty in breathing for which the term *snorter* is used. The gene is actually semi-dominant and has some effect in heterozygous condition.

Bull dog in cattle is recessive lethal (semi-dominant lethal). The heterozygous (Dd) are Dexter type showing the effect of lethal gene having short leg. The homozygous dominant (DD) cattle are known as Kerry type having long size legs. The homozygous recessive (hh) produces a lethal bull dog calf *(achondroplasia)*.

Syndactylism (having one toe in place of two, on one or more feet) has been reported in Holstein-Friesian cattle and is probably controlled by a recessive gene.

White heifer disease has been reported in *Shorthorn* heifer with constricted hymen, missing of anterior vagina and cervix and rudimentary uterine body is probably a sex limited recessive character.

Short vertebral column is due to a recessive gene. The calves are still-born or die soon after birth.

Prolonged gestation of 310-315 days in cattle seems to be due to a lethal recessive gene.

Parrot mouth dwarfs has been reported in *Southdown sheep*. This is due to semi-lethal recessive gene, killing all lambs within a month of birth.

Amaurotic idiocy is a recessive hereditary abnormality causing death within the first few years of life in homozygous recessive state. This condition is characterized by mental and physical retardation, abnormal lipid metabolism and total blindness. The heterozygous individuals are normal, indicated that the allele causing the defect is recessive.

4.2.2 Non-allelic interaction or *Epistasis*

The epistasis is similar to the dominance but it occurs when the character is influenced by many pairs of genes and these genes interact with their non-alleles (other pairs of genes) present on other loci of the same chromosome or of other chromosome. This is also called as inter allelic interaction and generally known as *epistasis* which is of several types.

Definition of Epistasis: When one pair of gene influences the expression of another pair of gene, it is known as *epistasis*. The epistasis involves the interaction between genes that are not alleles (interaction between genes present on separate loci). Thus, the genes at one locus interact with genes at another locus to produce a character that can not be produced by either locus acting independently. In other words, the genes at one locus modify the expression of genes at another locus. In this type of inheritance, the effect of one gene does not add directly to the effect of another gene to influence the phenotype but the phenotype is the result of interaction of genes at two or more loci.

Thus, an *epistatic factor (gene)* is one which prevents a gene other than its allele from exhibiting its normal effect on the development of the character and the gene which is being prevented from exhibiting its normal effect in the development of the character due to the epistatic gene is known as *hypostatic gene*.

It is not known to certain that in what manner the genes may act to influence the quantitative traits but some kind of epistasis have been observed and reported for qualitative characters.

Types of Epistasis: The different types of gene interaction (epistasis) for two gene pairs, when each gene pair assorting independently, have been given here.

Classical ratio for two gene pairs: As the epistasis involves the interaction between two pairs of genes, it is essential to have the knowledge of dihybrid ratios, so as to know whether the epistasis is present or not and that too which kind of epistasis is present. This needs to know the classical dihybrid ratio expected from dihybrid crosses, based on complete dominance between alleles at both the loci. The classical dihybrid F_2 ratio was observed in garden pea for yellow-green and round-wrinkled characters. The F_2 ratio depends on the dominance recessive relationship between alleles at two loci. The F_2 phenotypic classical ratio with complete dominance at each of two loci was 9:3:3:1 (yellow round: yellow wrinkled: green round: green wrinkled):

	AA BB	AA bb	aa BB	aa bb
	2 AA Bb	2 Aa bb	2 aa Bb	
	2 Aa BB			
	4 Aa Bb			
Total	*9 (A - B -)*	*3 (A – bb)*	*3 (aaB -)*	*1 (aabb)*

The different types of epistasis may or may not modify the above phenotypic classical ratios of 9:3:3:1. The numbers of phenotypes in classical ratios for dihybrid cross are four whereas these may be four or less than four in case of epistasis. The appearance of less than four phenotypic classes is the evidence of epistasis. Two or more phenotypic classes of the classic ratios are combined together as a result of epistasis, if the numbers of phenotypic classes are less than four. Broadly speaking, the epistasis which modify the classical ratios (9:3:3:1) can be grouped in two categories *viz.* simple statistics and duplicate epistatis.

The simple epistasis is when the genes at one locus modify the expression of genes present at second locus. Thus epistasis is only in one direction and the genes at second locus have no effect on those at first locus. The examples of simple epistasis are dominant epistasis and recessive epistasis. In simple epistasis only two genotypic classes of the classical ratios are combined, producing three phenotypic classes.

The duplicate epistasis act in both the directions, the genes at both loci interact simultaneously and the phenotypic classes are reduced to two except in one case of duplicate interaction (incompletely duplicate epistasis) in which there are three phenotypic classes.

The phenotypic ratios as a result of epistasis of different types between two gene pairs are modified in different ways as shown below:

Simple epistasis: This is one way epistasis, being of two types, *viz.* recessive and dominant epistasis-

(1) Recessive epistasis (9:3:4): There is complete dominance at both gene pairs. *One gene in homozygous recessive state (cc) is epistatic to both the alleles of other gene pair (B and b).* This is *supplementary gene action* or recessive epistasis. The combination of both the genes in dominant condition (A- B-) produces another phenotype. The F_2 ratio becomes as 9:3:4.

Many species of mammals have albino gene which exhibit recessive epistasis. The dominant gene C controls the production of pigment melanin, but the recessive gene in homozygous condition (cc) fails to produce enzyme that produce melanin and hence causes albinism. The recessive homozygous genotype (cc) at the C locus masks the expression of the alleles at another locus (B and b).

The example of this type of gene action is *coat colour in rabbit*. The black hair colour is dominant over brown. The inter se mating of F_1 dihybrid progeny (BbCc) produced 3 coat colours among the F_2 progeny as:

9 Black (B - C -) : 3 Brown (bbC -) : 4 Albino (B - cc and bbcc)

Another example of recessive epistasis is *coat colour in rat*. On crossing black mice with albino mice, the F_1 dihybrid produced are Agouti (Bb Cc). On inter se mating of F_1 dihybrids, the F_2 individuals are 9 agauti: 3 black: 4 albino as -

F_2 progeny 9 (B – C -) : 3 (bbC -) : 4 [(B - cc) + (bb cc)]

= 9 Agouti : 3 Black   : 4 [3 Albino + 1 Albino]

The agouti colour pattern has colour bands in a way that the hairs are of grey colour near the skin, followed by yellow band and finally black or brown at the tip of hairs. The agouti colour gene is A whose effect is masked by recessive genotype at other locus (bb). The mice with bb genotype mask the effect of colour gene A- and produces black coat.

The dominant allele B in the absence of C produces albino mice (BB cc and Bb cc). The dominant allele C in the presence of dominant allele B produces agouti colour. Thus two independent pair of alleles interact to produce the coat colour in a way that one dominant (B-) produces its effect (albino) in the absence of C (cc-homozygous recessive) but the gene C can produce its effect (to produce agouti) only in presence of other gene (B-). Therefore, these two pairs of genes are called as supplementary genes.

(2) Dominant epistasis (12:3:1): There is complete dominance at both gene pairs. The dominant epistasis occurs when a dominant gene at one locus interacts to suppress or modify the expression of genes at second locus. Therefore, the *dominant gene of one pair (A) is epistatic to both the alleles of other pair (B and b).* When A is epistatic to B and b, the F_2 ratio becomes as 12:3:1. In this case, the two phenotypic classes (A –B – and A – bb) of F $_2$ progeny expressed similarly.

The examples are *coat colour in dogs* as well as *in cats* which appeared as 12 white : 3 black : 1 brown and *colour of summer squash* appeared as 12 white : 3 yellow : 1 green.

The dominant gene B produces black colour. The gene A produces white colour and masks the effect of colour gene (B). Thus, B produces black colour but in the absence of A because the gene A is epistatic to B and b. White dogs carry colour gene (B) that are inhibited from forming colour in the hair due to the presence of gene, A. The individuals in F_2 generation appear in 12:3:1 ratio as under-

P   AA BB (White)   x   aa bb (Brown)

F_1 Aa Bb (white)
F_2 12 [9 (A – B -) + 3 (A – bb)] : 3 (aaB -) : 1(aabb)
         = 12 [9 white + 3 white]   : 3 Black   : 1 Brown

The brown dogs (aa bb), lack the gene for black colour (B), therefore, brown in colour, because they also lack the inhibitor (epistatic) gene A. Such genes are often called as *inhibitors.*

Duplicate epistasis: These are both ways epistasis. Thus, the epistasis acts in both directions. These are of the following four types –

(3) Duplicate dominant epistasis (15:1): *The dominant gene at one locus (A) is epistatic over dominant gene of other locus (B and b) and vice versa (B is epistatic to A and a).* The dominant allele at one locus (A) is epistatic to its non allele (B and b) whereas the dominant allele at other locus (B) is epistatic to its non allele (A and a). Thus, the dominant alleles of both the loci produce the identical phenotype by masking the manifestation of genes of other pair. The ratio appeared as 15:1.

The examples are the presence and absence of *feathers on the shanks of legs in poultry* (15 featered: 1 clean shank with genotype aabb*); bursa seed capsule shape* (15 triangular: 1 oval shape). The black Langshan breed of chicken has feathered shanks whereas most of other breeds have no feathers on shanks (clean shanks).

The crossing of feathered shank poultry (AA BB) of Black Langshan breed with clean shank poultry (aa bb) of Buff-rock breed (Plymouth Rocks) produced all feathered F_1. The inter-se mating of F_1 produced the following results:

15 [9 (A – B -) + 3 (A –bb) + 3 (aaB -)]   : 1 (aabb)

= 15 [9 Feathered + 3 Feathered + 3 Feathered] : 1 clean

The presence of dominant genes at either locus as well as at both loci causes the feathers to grow on the legs whereas the double recessive genotype (aabb) results absence of feathers on the legs.

(4) Duplicate recessive epistasis (9:7): In this type of epistasis, the recessive gene at both loci in homozygous condition masks the effect of dominant combination of both loci. Thus, *aa is epistatic to its non alleles (B and b) and bb is epistatic to its non alleles (A and a)*. In this case, the last three phenotypic classes of classical ratio expressed similarly and hence show similar phenotype. The F_2 ratio appeared as *9:7. The genotypes having recessive homozygous genotype for either gene locus as well as for both gene loci produce identical phenotypes*.

Fowl colour: There exists a locus controlling white feathers in poultry. The recessive genotype at this locus (aa) produces white feathers while A gene produces colour. Homozygous white Silkie fowls mated to homozygous white Plymouth fowls produce F_1 birds which are all coloured, and a ratio of 9 coloured: 7 white is obtained among F_2 offspring. The F_1 individuals, Aa Cc, have both the genes, A and C, necessary for the production of colour while the parents with genotype AA cc and aa CC, have only one gene, either A or C. Therefore, the presence of recessive genes for either locus in homozygous condition produces white.

Deafness in human also shows this type of gene action because recessive genes of both pairs (aa and bb) mask the manifestation of dominant genes of other pair *i.e.* aa is epistatic to B and b whereas bb is epistatic to A and a. Both the deaf parents have only one gene either A or B but the F_1 getting both the genes A and B have normal hearing capacity. Thus, dominant alleles of both pairs are required for normal hearing. Among F_2 offspring, 9 normal: 7 deaf individuals were produced.

9 (A –B -) : 7 [3(A – bb) + 3 (aaB-) + 1 (aabb)]

Human hearing = 9 Normal : 7 deaf

Fowl colour = 9 coloured : 7 white

(5) Dominant and recessive epistasis (13:3): In this type of epistasis one dominant combination at one locus is epistatic over one recessive combination at other locus. The *dominant allele at one locus (A) is epistatic to both alleles of other locus (B and b) whereas the homozygous recessive genotype of other locus (bb) is epistatic to both the alleles at other locus (A and a)*. In other words, the dominant gene at one locus and the recessive gene at second locus are mutually epistatic. Thus, *the dominant allele of one gene pair in homozygous (AA) and heterozygous condition (Aa) as well as the recessive homozygote of another gene pair (bb) produce the same phenotype*. In this case, the 1st, 2nd and last phenotypic classes of classical ratios expressed similarly and hence show similar phenotype. The ratio become as 13:3.

Feather colour in chickens is an example of this type of epistasis. Bateson and Punnet (1908) crossed White feathered Leghorn (IICC) with White Plymouth Rock (iicc) having white feathers and obtained white feathered F_1 progeny. Among F_2 progeny, the white and coloured individuals appeared in a 13 white: 3 coloured individuals. The F_1 birds having Ii Cc genotype are white in colour with coloured tips. The F_2 birds presented the usual dihybrid genotypic ratios with phenotypic ratios as:

13 [9 (A –B -) + 3 (A – bb) + 1 (aabb)] : 3 aaB -

or 13 [9 (I – C -) + 3 (I - cc) + 1 (iicc)] : 3 ii C-

    = 13 [9 white + 3 white + 1 white] : 3 coloured

All the F_2 individuals having inhibitor gene in its dominant allelic form (II, Ii) or cc were white whereas the individuals lacking I allele but having CC or Cc genotype were having coloured feathers. The gene I is the inhibitor and C is the colour gene whose expression is inhibited by I. Thus, the gene C fails to develop colour (melanin) because the gene I modifies (inhibits) the expression of gene C. The coloured gene (C) in its homozygous recessive state (cc) does not allow the melanin to be produced and hence produce white. The genotype cc is epistatic to any genes for colour. This type of epistasis is termed as "Dominant and recessive epistasis" because dominant gene I is epistatic to C, c and the recessive gene (c) of other pair is epistatic to I, i. This is also called as the *inhibitory factor inheritance.*

(6) **Incompletly duplicate epistasis (9:6:1):** There is complete dominance at both gene pairs. The interaction between both dominants produces new phenotype. This is also called as *duplicate genes with cumulative effect.* In this type of epistasis, the recessive genotype for either pair of gene masks the manifestation of dominant genes of other pair to a certain extent. There is incomplete masking of recessive combination of both loci on dominant combination of both loci. The genotype aa is partially epistatic to its both the non alleles (B and b) and bb is epistatic to its both the non alleles (A and a). Thus this type of interaction occurs *when the dominant condition at both locus produce one phenotype (= 9/16); the dominant condition at either locus (but not at both) viz. A-bb (= 3/16) and aaB- genotypes (= 3/16) produce identical (same) phenotype;* and the absence of any dominant gene at either locus (aabb = 1/16) produce a separate phenotype. In this case, the two middle classes expressed identical phenotype. The ratio appeared as 9: 6:1.

The example is coat colour pattern in Duroc swine (9 red: 6 sandy: 1 white). There is one sandy coloured race of Duroc swine having red swine (AA bb) and white swine (aa BB). On crossing them produce F_1 (Aa Bb) with red colour. The F_2 offspring having at least one A and one B gene are red, lacking both A and b are white, and having at least one A with 2 b's or at least one B with 2 a's are sandy.

F_2 offspring   9 (A – B -) : 6 [(3 A – bb) + 3(aaB -)]   : 1 (aabb)

                = 9 Red : 6 [3 Sandy + 3 Sandy)]   : 1 white

In this type of epistasis, the dominant gene at either gene pair produce sandy colour, dominant gene at both loci produce red colour and absence of dominant gene at both loci produces white colour.

4.3 NO MODIFICATION OF RATIOS BY EPISTASIS

Complementary gene action: Above have been given the examples of characters controlled by two gene pairs showing epistasis and modified the dihybrid classical ratios of 9:3:3:1 among F_2 (with dominance recessive relation between alleles of both gene pairs). However, there are some examples of gene interaction (epistasis) wherein the classical dihybrid ratios among F_2 are not modified. Such examples of epistasis of two pairs of genes influencing the same character of comb shape in chickens and coat colour in rats can be cited. The two gene pairs interact but do not modify the phenotypic ratio of 9:3:3:1.

Certain factors can not express themselves except in the presence of some other factor or the modification of their expression may be due to the presence of some other factor.

(i) *Comb shape in chicken*: Domestic breed of chicken have different comb shapes. The birds of Wyandotte have *rose comb;* Brahmans and Cornish breeds have *pea comb;* Leghorn, Langshan and white Plymouth Rock breeds have *single comb.*

When rose comb fowls (RR pp) were mated to pea shape comb fowls (rr PP), the F_1 was produced with another type of comb, Walnut shape (Rr Pp). When the Walnuts (F_1) were mated *inter se,* the four types of comb shape were produced among the F_2 progenies. The F_2 results were as under:

RR PP	RR pp	rr PP	rr pp
2 RR Pp	2 Rr pp	2 rr Pp	
2 Rr PP			
4 Rr Pp			
= 9 Walnut :	3 Rose :	3 Pea :	1 Single

Neither single nor walnut comb was present in the original parental lines. The F_2 results indicated the followings-

- The normal F_2 ratio (9:3:3:1) was obtained which suggested that two pairs of genes are involved and that dominance recessive relation exists between each of two pairs of alleles.

- The ratio also suggested that F_1 walnut comb shape chickens were dihybrids (RrPp).

- Two more phenotypes (Walnut and Single) appeared in F_2 which are the result of interaction of two gene pairs. The phenotypic class representing a ratio of 1:16 indicated that the single comb type is the result of the combined effects of two pairs of recessive genes (rr pp). The phenotypic class representing a ratio of 9:16 indicated that walnut shape is the result of at least one dominant gene at each of two loci (R - P -).

- It is clearly an interaction between non-alleles, because none of the comb type can be explained based on one pair of genes. The two genes R and P are non-allelic but they were dominant on their alleles (R over r and P over p). These two dominant genes responsible for rose comb and pea comb interacted to produce walnut comb, similar to the co-dominant alleles and the absence of both dominant genes (rr pp) resulted in a different phenotype as single comb. Therefore, non-allelic genes acting in this manner are called *co-epistatic genes or complementary*.

Comb shape in poultry is thus governed by two pairs of genes with complete dominance at both gene pairs. New phenotypes are produced by interaction of dominants at both gene pairs (walnut shape) as well as by interaction of both homozygous recessives (single shape) in contrast to supplementary gene action. Both rose (Wyandott) and pea (Brahman) comb shape is dominant to single (White Leghorn) shape comb.

The homozygous Walnut may be produced by inbreeding and selection, because R and P are apparently located on different chromosome pairs. But if R and p (and like manner r and P) had been located on the same chromosome (not allowing the crossing over) it would have been impossible to obtain a homozygous strain of Walnut comb.

(ii) Coat colour in rats: The yellow rats were mated to black rats, all the F_1 progeny were gray in colour. The *inter-se* mating of F_1 gray, produced the F_2 progeny of four colours in a ratio of 9 (gray): 3 (black): 3 (yellow); 1 (cream). The F_2 results in this case are explained on the same logic as was used to explain the epistasis involved in the inheritance pattern of four comb shapes in poultry above.

The complementary action of genes is another expression of gene interaction and the characters of economic importance are affected by this type of gene action, and usually many gene pairs are involved, if the characters are of a quantitative nature and in that case the expression is less clearly marked than in the above example.

(iii) Umbilical hernia in swine: This trait is at least stimulated by 2 independently inherited factors. Three inbred lines A, B and C in swine have no umbilical hernia. The crossing of inbred line A either with inbred line B or C produce all the crossbred progeny having umbilical hernia. The F_1 from both crosses were culled for umbilical hernia and inbred, and in both the cases the hernias disappeared from the lines in 2 generations. The conclusion was that line A carrying a factor potent for hernia which expressed itself when introduced in the genetic environments provided by B and C lines. The cross between line B and C did not develop hernias.

4.4 MODIFYING GENES

In addition to the different types of epistasis mentioned above, there are other types of gene action. The inheritance of coat colour in some species of domestic animals suggests that the inheritance of some characters is not as simple as explained above. A mutant gene expresses itself in a background of normal alleles at all other loci. All genes have a dual effect - one is that they are necessary for the well being of the individual and if a portion of these genes is lost, it leads either in improper development of the individual or results in its death.

The *modifying genes* are those that modify the effects of genes present on other locus controlling a character. For example, take the inheritance of coat colours in mammals. The pigment melanin is the basic requirement of all colours in animals.

This melanin is of two types *viz*. black (*Eumelanin*) and red melanin (*phaeomelanin*). These two pigments are modified by genes at several loci. These genes affect the synthesis of pigments and associated enzymes as well as the distribution and location of pigment granules in the skin, and hair to produce various shades of colours.

(i) Eye colour in Drosophila: It is due to a single factor (gene pair), but in addition to this, seven other factors (genes) that cause modifying effects have been identified. The gene *w* for white eye prevents the colour development in homozygous condition in females (w/w) and in hemizygous condition in males (w/y). The other genes are necessary, though not for the production of white eye but for the normal development of fly in which the white eye can be expressed. The interruption at white locus of the *w* mutant inhibits the development of pigment which results in a white eye.

(ii) Coat colours in mammals: Regarding coat colour inheritance, the wild colour is known as agouti which is a familiar colour in wild rats, mice and rabbits. The agouti is produced by a dominant gene (A) whereas its recessive condition (aa) produces non-agouti which is usually black unless modified by other genes. The agouti series has other alleles in different species. The mechanism of production of melanin pigment from the amino-acid tyrosine involving the catalytic action of an enzyme (tyrosinase) under the control of a gene (C) and the absence of the production of enzyme (Tyrosinase), causing the deficiency of melanin under the control of a recessive allele in homozygous condition (cc) resulting in albinism has been explained in chapter 10 under the article "*How genes function*".

The 3 basic genes for colour in wild animals are A, B and C. The C is dominant gene which produces melanin pigment in the hair and skin. Its recessive allelomorph c in homozygous condition (cc) results in albinism. The gene B is another dominant gene which produces black pigment with C. The gene B's allelomorph in homozygous state (bb) changes the black colour to brown, which is less completely oxidized melanin pigment. The gene A is dominant gene which modifies the expression of B and C so as to produce agouti or wild coat pattern. In the domestic horses, it modifies the action of B and C to produce bay colour.

At albino locus A, there are some other alleles which act as the production of variable amount of tyrosinase and result in production of modified phenotypes. The two coat colour phenotypes in rabbits and rats may be cited here. These are the *chinchilla* (gray colour which is diluted agouti colour) and *Himalayan* (white with black extremities of the body *viz*. tail, feet, neck).

At locus B, the genes decide the type of melanin (*Eumelanin* - Black or brown; or *phaeomelanin* - red or yellow) depending on other modifying genes present at locus E. The modifying effect cause the series of colours from black (dominant extreme) to red or yellow (recessive end of series).

At locus D, the genes cause clumping of pigment granules which leads to a decrease in light absorption and dilution of colour. The dilution gene may be dominant or recessive.

The examples of coat colour inheritance in different mammalian species are given below–

(1) Coat colour in **cattle**: The coat colour in HF cattle (white spotting) is due to a single pair of gene but the coat colour varies from almost white to almost black having black spots of varying size. The indication that the modifying genes are responsible for the expression of the genes for spotting comes from the successful selection for spot size in either direction. The variation in white spotting in Holstein cattle occurs to the extent that some cattle have very little amount of white that they are almost all black while others are almost white. All these indicated that white spotting is modified by genes present at several loci.

Yellow colour in Guernsey cattle is due to a recessive gene when homozygous (cc) which dilutes the red colour but does not affect black. The black (B -) is dominant over red (bb). The Angus cattle are black and do not carry the dilution gene (BB CC). Hereford cattle are of genotype bb CC whereas the Guernsey cattle are of the genotype bb cc.

(2) *Coat colours in horses:* There are a number of loci which control and modify the coat colours in horses. The effect of genes at locus A has already been discussed in chapter three on multiple alleles. This locus (A) determines the variability in expression of black pigment (aa) to its full to remain limited to only some parts (black mane, tail, legs with red body, called bay). The *black (aa)* is recessive to bay (A -). The *bay like* pattern (brow or reddish body with black markings on mane, neck and foot) has also been reported by some workers to be due to the presence of another allele, A^t which is dominant over A and *a* alleles. The *seal brown*, which is modified black with light brown areas on muzzle, eyes, legs and flanks, is said to be controlled by another allele at this locus (A^l) which is recessive to A but dominant to the allele *a*.

The shade of melanin (black or brown) is determined by the genes at locus B. The dominant allele (B) produces black, except as restricted by A - whereas the homozygous state of recessive (bb) produces *chocolate brown*.

The gene at locus C is the only allele in horses (c^{cr}) known as cream dilution gene which is codominant and partially epistatic to ee. The term *cremello* is used for this cream dilution. The homozygous dominant (CC) expresses the colours as determined by genes at other loci (A, B and E). The heterozygous state *(Cc)* results in *palomino* due to partial dilution (modification) of red or yellow pigment but does not affect the black melanin. The *chestnuts* (reddish colour) and *sorrel* (light red) are diluted to the *palominos* (Copper or golden colour, with lighter mane and tail than rest of body).

The genes at locus E further influence the type of melanin (black or brown) which needs the presence of gene E whereas its recessive homozygous state *(ee)* restricts the expression of black and brown so that red or yellow (phaeomelanin) is produced on the whole body (including mane, tail and legs) and causes the reddish colour (light or dark shade), called as *chestnut*. The light shade is called *sorrel*.

The gene at locus *D*, causes dilution of colour in horses (called as dilution gene which reduces the intensity of all pigmentation). The gene *D* in the homozygous condition *(DD)* combined with bb results in type A albino and the DDBb genetic state results in type B albino. Neither of the above one are, however, true albino;

the true albino condition can result only from the presence of cc genes. The buck skin or dun colour is produced by the *DdBb* genetic state and the *Palomino* colour (copper or golden) from the Ddbb genetic state. These colour patterns can be shown as:

Colour patterns	Genotypes
Albino type A	DD bb
Albino type B	DD Bb
Dun colour	Dd Bb
Palomino	Ddbb

The genotypes for five coat colours in horses can be written as:

Coat colours	Genotypes
Black	aa B – E -
Chocolate brown	aa bb E -
Bay with black points	A – B – E –
Bay with brown points	A – bb E –
Chestnut - any genotypes having ee:	A – B – ee,
	A – bb ee,
	Aa B – ee, and
	aa bb ee.

The gray and roan patterns are also exhibited in horses with inter spread of white hairs with coloured hairs. The proportion of white hairs in genetically gray horses is progressively increased with increase in age, some breeds of horses have no coloured hair in older age. The gray is dominant (G -) to non-gray (gg). The situation of the presence of white hairs in roan horses is different than gray pattern, in that the proportion of white hairs in roan horses is not changed with progressive increase in age, as in case of gray horses. The roan pattern is usually absent on the head and feet compared to body. Some reports have indicated that roan gene (R) is partially dominant or co-dominant with lethal effect in embryonic stage when homozygous (rr).

(3) Coat colour in cats: The coat colour in cats is of *tabby pattern* which consist black stripping on a yellowish-gray background (similar to agouti which is full colour). The agouti pattern (full colour) is due to dominant gene at the A locus (A -) whereas the homozygous genotype at this locus (aa) produces black. There are three different tabby patterns *viz. Mackerel tabby* which is like wild pattern, found in mongrel cats; *blotched tabby* which consists irregular stripping than mackerel tabby; and third tabby is that which has no stripping except in small amounts on the leg and tail, found in Abyssinian breed. Three alleles in a dominance hierarchy of $T^a > T > t$ are responsible for three kinds of tabby. The allele T^a controls the tabby of Abyssinian breed, T controls Mackerel tabby and t controls blotched tabby (most recessive trait to other two tabby).

A dominant gene at B locus (B -) is necessary for the expression of black (aa). The recessive gene at B locus in homozygous state (bb) produces brown or chocolate. The recessive gene at B locus (b) is epistatic to aa. There are references indicating that a third allele which is recessive to b allele produces milk chocolate colour which is lighter brown.

Two alleles at C locus produce two popular colour patterns. The allele C^b in homozygous condition (C^b C^b) produces a brownish body colour due to reduced intensity of black with darker extremities than body and the eyes being sometimes blue. This is found in Burmese cats. The second allele (C^S) in homozygous state as C^S C^S produces almost white body with dark extremities as in case of Himalayan rabbits and this colour pattern is found in Siamese cats. Further, these two alleles are co-dominant and hence the heterozygote (C^b C^S) produce an intermediate phenotype. However, both these alleles are recessive to C (responsible for full colour expression determined by genes at other loci).

There is absence of true albino colour in cats like horses and dogs, though white cats are produced by a dominant gene (W -) at another locus, in which the genotype ww produces non-white phenotype. The eyes of these cats may be blue, yellow or odd-coloured. This gene is an example of pleiotropic effect on hearing ability, as half of the white cats are deaf. The white hairs are the result of the failure of the production of melanin due to a dominant inhibitor gene (I).

A dominant gene (S -) at another locus seems to control the irregular white spotting in cats, known as piebald spotting pattern. The piebald is a dominant trait.

Another locus in cats is found responsible for dilution of colour. The full colour is produced by D – whereas the recessive dilution gene (d) when homozygous (dd) dilutes the colours. The black is diluted to blue and yellow to cream colour.

A *sex-linked locus* with co-dominant alleles exists in cats. This locus produces black, yellow and tortoise coat colour, as explained in chapter number 6.

(4) *Coat colour in Dogs:* About 10 gene loci have been reported to affect the coat colours in dogs.

The locus A (agouti series) has four alleles with dominance hierarchy as A^y > a^w > a^t > a. All these four allele, are observed in German Shepherds and respectively, produce cream or gray-coloured body with a brown or light brown saddle (A^y); cream or gray-coloured body with a black saddle (A^w); black and tan colour (called as bicolor, A^t), the black varying from a black saddle to black on whole body except the points as seen in many Dobermans, the tan colour is an intense (reddish tan rather than the light tan or cream) with genotype (A^y – or A^w -) and the last solid black in homozygous state (aa).

The black or brown eumelanin (Chocolate) depends on the genes at E locus. The black is determined bybaaB- and brown by aabb. The recessive gene for brown (b) is epistatic to *aa*. The dominant gene at E locus produces *eumelamin* (Black or brown) whereas the recessive homozygous genotype (ee) causes the *phaeomelanin* (red or yellow) to be produced and ee is epistatic to aa, B -, and bb. These results the following body colours in dogs-

Solid black	aa B – E –
Solid brown (Chocolate)	aa bb E –
Red or reddish yellow with dark nose and lips	aa B – ee
Yellow or golden with pink nose and lips	aa bb ee

The genes at C and D locus result in dilution of colours. The dominant gene C produces full colour. Another gene at C locus (*Chinchilla*, c^{ch}) modifies the *phaeomelanin* that changes tan to cream colour, and red and yellow to light yellow and golden colours. But the allele c^{ch} does not affect the *eumelanin* like the cream dilution gene producing *palomino* in horses.

The gene at locus D in homozygous recessive state (dd) affects the production of melanin of both types. The black is diluted to blue and the effect of dilution of yellow is similar to chinchilla dilution, described for C locus. The recessive gene d has an effect like *dun* dilution in horses, though in horses it has dominant effect.

The *piebald pattern* (irregular spotting) also exists in dogs, produced by recessive state (ss) and the dominant gene S produces an effect of absence of spotting. The amount of spotting varies from very little white to the complete white. The recessive condition of a gene at W locus (ww) results in all-white dogs with pigmented eyes.

The progressive graying with age has also been observed in dogs, similar to that in horses and controlled by a dominant gene (G).

Two other loci (T and M) have also been found associated with white. The small pigmented spots (ticks) on white background appear due to dominant gene T. The amount of white in the coat is increased by a co-dominant gene M in heterozygous state (Mm), producing a pattern of dilute and intense patches, called *merle* whereas the homozygous dominant condition (MM) produces almost all-white coat with pale blue eyes and pleiotropic effects including blindness, deafness and some structural abnormalities of the eyes.

(5) **Coat colour in mice:** In mice, white spotted pattern behaves as a single recessive to self coloured coat but spotted mice range from nearly self coloured to nearly all white. The variations are attributed, for the most part, to modifying genes which affect the white spotting.

4.5 VARIABILITY IN GENE EXPRESSION

Besides above types of gene action, some other types of gene expression like penetrance and variability of genes, have also been found in resulting modification of Mendelian original phenotypic ratios.

(1) **Penetrance:** This is the ability of a gene to be expressed phenotypically in only some individuals of the population. The penetrance of a gene is the frequency (percentage) of expression of a trait in the population. Thus, *the degree (percentage) of phenotypic expression of genes in the population is known as penetrance.* The expression of the gene depends on environmental conditions as well as on genetic background. When a gene produces a constant phenotypic effect, it is said to have 100% penetrance (complete penetrance). The dominant genes have 100% penetrance in heterozygous individuals while the recessive gene could properly be said to have

zero penetrance. Thus, penetrance may vary from zero to 100%. The penetrance may thus be complete or incomplete.

The *polydactyly* (presence of extra to normal number of fingers and or toes) in men is controlled by the presence of a dominant gene (P) whereas its recessive allele (p) in homozygous condition (pp) governs the normal number of fingers on each limb. The presence of dominant allele in all heterozygote (Pp) should produce polydactyly but it does not hold true in all cases and some of the individuals are not polydactylous, indicating that the allele is penetrant but to a lesser degree.

The other examples of penetrance are *Blue scleroties, stiff little finger, diabetes insipidus and multiple exotoses* (cartilaginous tumors of the long bones) in human which are governed by single dominant genes but with reduced penetrance.

The *blue sclera* anomaly had a penetrance of 90% and it is expressed by a blue discolouration of the white (sclera) of the eye and by bristleness of the bones.

In several of the pedigrees the *diabetes* skip (an individual whose father or mother had the condition and some of whose children are also affected, but one individual remains unaffected). The diabetes mellitus develops only in those individuals of the twins who take more carbohydrate foods (starch and sugars) in their diet.

The penetrance of dominant gene for *multiple exotosis* is lower than 75% in women but considerably higher in men. The penetrance may be equal or unequal in the two sexes.

The obstruction of the orifice of the stomach known as *pyloric stenosis* is due to a recessive gene with some reduction in penetrance in boys but much more reduction in girls. The trait becomes sex-limited when the penetrance is completely reduced in one sex.

The expression of the D allele in horses for coat colour is remarkably constant with few exceptions where the D allele fails to produce the usual phenotypic effect (Palomino – a dilute colour in the heterozygous condition Dd or in the presence of at least one B). A mare with genotype aa Bb Dd was found black and not Palomino by Castle and King (1959) and when it was mated with a chestnut stallion (a- bb dd), it produced a palomino foal. Thus, the black mare had gene D but failed to produce any phenotypic expression (palomino) which indicated that the D allele in this case has low penetrance or no penetrance.

(2) **Expressivity:** The degree of effect produced by a penetrant genotype is known as expressivity. It represents merely an *expansion of penetrance* in the visible range. When a penetrant character (controlled by a gene) show individual variation, it means the gene is having variable expressivity. Thus, the gene in one individual may produce its effect to a lesser degree but to a greater degree in another individual. *The expressivity is the variable degree of gene expression among different individuals carrying this gene.* Thus, a character may be penetrant but it may be quite variable in its phenotypic expression in different individuals.

Variation in the expression of genes for tail length in sheep has been observed. The *length of the tail* varies from almost normal to a condition where a part of the spinal column may be missing.

Another example is the *"snorder' dwarfism in beef cattle* which vary from very small animals that die soon after birth to those which attain normal size surviving for several years of age.

In human, the well known example is of variable expressivity is *juvenile cataract of the eye* (Lutman and Neel, 1945) due to a dominant gene whose expression may vary from a *slight milkiness of the lens to dense opacity.*

A dominant gene is associated with an "allergic diathesis" with various manifestations in different individuals as asthma, vasomotor rhinitis, atopic dermatitis or hay fever (Schwartz, 1953).

Further observations regarding the *polydactyly in human* indicated that this condition is not present in all the four limbs, but it may be observed either in one hand and not in the other or it may be in the hands and not in the feet. Thus, although the gene (P) which controls the trait is present in the individual and have shown its effect (penetrance) but with varied expression within the individual itself. *This variable effect of a penetrant genotype within the individual is also called the variable expressivity.*

Viability is another example of variable expressivity. The sub lethal alleles in homozygous recessive condition also varied in their phenotypic expression and thus they are not fatal to all the individuals before sexual maturity. Some lethal genes have their variable effect from complete lethality to sterility if the genotype survived to sexual maturity. Thus, the viability is another ontogenetic manifestation of a gene. Some genes do not affect the life span, some others result into sterility if the individual survive to sexual maturity, others definitely reduce the chances of living and still others are lethal in homozygous state.

The expressivity of a gene (genotype) is influenced by the environment. The identical twins raised on different diets attain different height. The exposure to childhood disease to an individual of the twin also affects the expression of adulthood characters.

The variable expressivity, like partial dominance, can be explained by modifying genes and or by environmental influences.

4.6 ENVIRONMENT AND GENE EXPRESSION

There had been controversies in early days regarding the relative importance of heredity and environment. It is now clear that both are important. The potentiality of an individual is determined by heredity (genetic constitution) called as genetic potentiality which is achieved by the proper environment. The control mechanism for differentiation, organization, and growth in living beings responsible for the development and carrying out the living activities are dependent on both the genetic and environmental factors. The development of living system is so complex to understand all the possible steps involved in the expression of any one characteristic.

Some genes do have their consistent effect to produce the phenotypic effects under all sort of environmental diversities but other genes respond differently and the phenotype appears as a result of close-knit interaction or cooperative result of many factors. For the development of a character neither genes nor environmental influences can be solely responsible but it is their cooperative result and one or the other factor may have negligible effect. Thus, both heredity and environmental influences work together for the development and expression of a character via many individual steps involving interaction between the genotype and environment.

Here, the *environmental dependence of genotypes* has been covered mainly for the modification of expression of major genes in relation to change in environment.

A number of characters can be enlisted with the relative influence of variability in hereditary or in environmental factors acting upon them. In some cases the environment may be more effective and the heredity, being similar in all the individuals, may not have any effect in the variation of the character. The example of *measles* fits beautifully in this category, in which every one inherits susceptibility to it but its occurrence depends entirely on the environmental contact of the individual with the causative agent.

There is a second category of characters which are more affected by the heredity and no environment of sufficient potency can alter the expression of genes present and responsible for. The *ABO blood groups* in human are remarkably stable in most environments to which the human beings are subjected. The gene for antigen A always produce antigen A in the cells regardless of environmental influences and the antigen A is not developed in the absence of gene producing it. There are only a very few environmental conditions, such as the *disease leukemia*, which change the blood type of a particular person.

There is third category of characters in which the variations are found due to both the heredity and the environment. The body weight and other *quantitative characters* have genetic component but are more influenced by the diet of the individual and other environmental factors.

Susceptibility to *diphtheria* is dependent on a simple gene and immunity by its dominant allele. But for the contraction of the disease, infection with the causative agent is necessary. The individuals with inherited immunity (resistance) to diphtheria would not have the disease upon infection to the bacillus but with inherited susceptibility would develop it. Therefore, the heredity has the variable effect in this case. In another situation, the individuals with inherited susceptibility to disease will have the disease on exposure to the pathogen but unexposed individuals will not have. This is the case of variable effect of environment.

This leads to the confusion because in one case the diphtheria was the result of hereditary influence but in another it may be concluded that environment was responsible. Further, an example may be taken when a person is made artificially immune to diphtheria (by means of toxin antitoxin) in spite of the inherited susceptibility. However, it does neither mean that he does not inherit the susceptibility nor that he will not transmit the susceptibility. The assumption of the heredity and the environmental effect for a condition can not be invalidated by the fact that a causative agent is known for a condition nor that the condition can be changed or cured. On the other hand, it does not mean that a genetically determined abnormality can not be controlled by the requisite environmental manipulation. Thus, all the characters are limited by hereditary and environmental effects realizing that the heredity is a relative term, when the variation in the character is caused mainly by differences in the genetic make up of the individual.

The different behavior of similar genotypes in different environment is the basic fact of gene action. However, this never means that the gene might have been changed but it may be that the reactions controlled by the gene are influenced by the environment. High temperature promotes the chemical reactions more rapidly than a low temperature. There may be cited a number of examples of conspicuous inherited characteristics which are influenced by the environmental conditions.

4.6.1 Environmental factors causing variation in expression of genes

The environmental factors controlling the qualitative characters may be external environmental factors which affect the population as a whole and the internal environmental factors which affect at individual level.

(I) External environmental factors

1. Temperature: The *temperature* can play a very significant role in gene expression. Chemical reaction proceeds more rapidly at high temperature than at low temperature. The basic developmental effect of many genes is to control the rate of specific reaction and a change in temperature can be expected to have an effect on development. It is not necessary that the genes act during the temperature effective period but the gene may have its action earlier and the period may coincide with an embryological process at the time of utilization of gene product. It may also be possible that a substance required for the gene action is produced during the temperature effective period.

The temperature influences the *coat colour* in Himalayan rabbit and Siamese cats which are born white but later develop black pigment at the extremities like paw (feet), nose, ears and tail. The animals develop no colour at birth because they have the body temperature of the mother in *utero*. The dark colour is developed, as the animals grow, at the extremities where the temperature is somewhat lower than body temperature. The black pigment (melanin) responsible to produce colour in skin is genetically controlled. The gene responsible for black colour causes the production of an enzyme which is necessary for the formation of a black pigment. However, this enzyme is produced at slightly low body temperature than at normal temperature. The enzyme is inactivated at high temperature. Therefore, the black pigment is formed only on the extremities of the body (nose, ears, feet and tail) where the temperature is relatively lower than main parts of the body which are white in colour.

A strong temperature effect has been shown on the formation of melanin pigment responsible for development of black colour. The experiment with Himalayan rabbit indicated that shaving off the fur from the animal and keeping the animal under changed temperature results in change of colour. If the environment (temperature) is changed, after removing the dark fur, by keeping the animal under warm condition, the white fur grow instead of dark fur. On the other hand, if the white fur is removed and the animal is kept under cold condition, the fur grown is black. Thus, the phenotypic expression of the genes (for pigment formation) depends on the temperature at the time of fur growth.

The *development of wings* in Drosophila is also influence by temperature. The gene affects the organs of Drosophila to develop into wings in 35 % individuals raised at high temperature (25°C) and it cases a reduced penetrance from 35 to only 1 % if the flies are raised at low temperature (17°C).

The *number of facets in Bar-eye mutation* is another character in Drosophila affected by temperature. The rearing of Drosophila at high temperature has been found to develop *Bar eye* compared to that reared at low temperature. The number of facets decreases in Bar-mutation with the increase in temperature from 15°C to 31°C, while for an allele of Bar-mutation, known as *Infrabar,* the number of facets increases with the increase in temperature. Temperature changes during the period from 90 to 97 hours after fertilization influences the Bar eye in Drosophila. The eyes of Drosophila have about 50 facets at 35°C during this period and about 150 facets at 25°C compared with 700 to 800 facets for wild type.

The *Bristle formation* in Drosophila is also influenced with the changes in temperature at any time throughout larval and pupal stages.

The small water insect, Daphnia, commonly found in ponds, show temperature sensitivity to their *survival.* They survive at 20°C and die when the temperature goes to 27°C. On the other hand, a mutant type of this water flea survives and thrives in warmer temperature (25°C - 30°C) and dies at low temperature (20°C).

2. Sunlight: The sunlight has also been found to change the expression of genes. The ability of genes to express themselves in relation to sunlight is more evident in plants. The light provides energy and is essential for growth and development of all plants. The plants grown in dark do not develop chlorophyll and therefore appear as albino, although they have genes for chlorophyll.

Hypersensitivity to sunlight is heritable trait in Southdown sheep. The liver of the affected individuals fails to function properly and the byproduct of metabolism (Phyllaerythrin) is not excreted but starts accumulating in bloodstream and skin whereby activated by sunlight, causing the *eczema to develop on face and ears* which may become fatal if the animals are exposed to sunlight. But the animals are free from symptoms if kept indoors away from sunlight and allowed to graze during night hours.

Cancer eyes of Hereford cattle is believed to be inherited and provoked (frequently developed) in areas of intense sunshine, developed in younger animals maintained on high plane of nutrition and more incidence in white eyed than in pigmented eyelid cattle.

The *face freckling* (spotting) occurs among the people of certain genotypes exposed to sun. The potentiality to produce freckles was acted upon by sunlight to produce different phenotypic extremities. Few freckles were developed on the face of a girl who worked indoors as compared to many freckles developed on the face of its identical twin girl who worked out in the sun.

3. Nutrition and culture conditions: The nutrition (food) has many functions providing energy and material necessary for structural functions. Any living individual can not show its full genetic potential in lack of nutrition.

The change in diet leads to change in the expression of genotypes. The appearance of *yellow fat in rabbit* depends on two factors – one is the recessive gene in homozygous condition (*yy*) and second is inclusion of green vegetable in the diet (xanthophylls). The yellow fat disappears if the diet is devoid of green fodder. An enzyme found in the liver control the process of breakdown of certain yellow compound (Xanthophyll) of the green diet. The breakdown of Xanthophyll to colourless derivatives does not occur when the rabbit lacks the enzyme and the Xanthophyll is stored as such in the animal's fat giving it yellow colour. Thus, the difference between white and yellow fat rabbits is caused due to the presence or absence of enzyme. The allelic alternatives Y and y determines the presence or absence of the enzymatic function. A single dose of Y allele is sufficient to produce enzyme needed to breakdown all the Xanthphyll eaten by the rabbit. The genotype *yy* produce white fat on a Xanthophyll free diet. Therefore, the genes determine the potentialities which ate exploited by the environment under which the genes perform their functions.

Abnormal abdomen in Drosophila (irregular abnormal segments) is affected by environment (culture condition). This abnormality is caused by a gene but abnormality is developed in the flies which hatched first when the culture bottles are moist and it is not expressed in the flies which hatched later or raised in bottles having less moisture, even the flies carry the gene. Thus, the environment is the determining factor in the expression of gene to develop the character (abnormality, in this case).

The *crooked calf* (deformities of front legs) are developed by feeding lupine and high level of lead during pregnancy or feeding of diet low in manganese.

The *Cyclopean malformed lambs* (single eye centrally located) can be produced by feeding the weed veratum californicum during early pregnancy.

(II) INTERNAL ENVIRONMENTAL FACTORS

1. Hormones: The hormones are also important in this respect. The genes for growth can not express themselves due to lack of growth hormone produced by anterior pituitary gland. A gene has been reported to cause *dwarfism in mice* through failure of the production of growth hormone.

The phenotypic differences between two sexes are observed in reproductive structures, secondary sexual characteristics, growth pattern, and behavior of each sex. These sex differences are associated either with different sex chromosomes present in two sexes (sex-linked traits) or due to different sex hormones that results the expression of genes differently in two sexes (sex limited traits and sex influenced traits). The detail discussion on these characters has been given in chapter 6.

2. Age of animal: It is well known that gene expression is age specific. The genotype of every living being is determined and fixed at zygotic stage, remaining unchanged throughout the life. However, the phenotypic changes in many characters occur in all organisms according to their age. Some genes express themselves early in embryonic life, some exert their effects at birth, some a few weeks after birth, some other express themselves at different ages of later life (younger age/sexual maturity, old age) of the organism.

The following are the examples of age specific gene expression:

(i) *Short tail in sheep* is an inherited condition for which the genes express early in embryonic life at the time of bone formation.

(ii) The genes for the formation of blood group antigens express in early embryonic life.

(iii) The effect of eye colour genes in humans starts a few weeks after birth and not at birth.

(iv) In humans, alkaptonuria (darkening of urine-an abnormality) starts at birth.

(v) Infantile amaurotic idiocy at 4–6 months of age.

(vi) Vitamin D – resistant rickets at 1 year of age.

(vii) Muscular dystrophy at 2–5 years of age.

(viii) Hereditary baldness at 25–35 years of age.

(ix) Genes for diabetes mellitus and Huntington's chorea (nervous disorder – a progressive mental deterioration) had their effect over a wide range of ages (30–60 years).

The genes for semen production, appearance of moustaches and beards in males; breast/udder development, ova production, milk production and egg production in females; growth of puberal hair in both males and females starts expressing with sexual maturity, etc. However, in all these cases, the genes responsible for the expression of these characters are present at zygotic stage but their expression (visible effects) depends on the age of organism.

3. Chemical environment: To speak of gene activity without specifying a particular environment has no meaning. There are a number of situations /examples of modification of gene expression in the absence of proper substrate /chemical environment of genes.

The *diabetes mellitus* is inherited in humans but it is not manifested unless the person consumes a large amount of carbohydrate over a long period of time and the pancreas is overloaded by sugar accumulation. The sugar level is controlled by the pancreatic hormone, *insulin* in normal condition but in diabetes the sugar level is not properly regulated, causing high blood sugar level and excretion of sugar in the urine leading to a change in body metabolism from the use of sugar for energy to the excessive use of fatty acids. The consequence is the coma and death of the diabetic patient.

Scrotal hernia in Swine though heritable but it is influenced by the maternal effect.

4.6.2 Phenotypic Stability (Canalization) and Environment

The stability means a constant state and hence the genetic stability refers to imply a constant state for a population genome. The phenotypic stability is the consequence of stabilizing selection (selection of individuals near the mean of population). The tendencies for a constant state in development are of fundamental importance to maintain a uniform phenotype with genetic variability in secrete form (*hidden genes, known as cryptic genetic variability*). Some buffered ontogenetic sequences of events in development are responsible for the steady state.

The phenotype is an end product of development via diverse genetic pathways. Each species has the phenotypic uniformity and adaptive norm. An individual has the capacity to develop it self, through the internally buffered sequences of processes, and counteract the upsetting tendencies of environmental stresses. This capacity of counteracting the enviroenmental stress is called the *development buffering or flexibility*. The developmental buffering makes the individual react to the environmental change by switching from its normal developmental path to an alternate path which in turn helps the individual to develop unimpeded by environmental stresses. The stability for population genome is promoted by stabilizing selective mechanisms and genetic interactions. The basis of stabilizing selection is the superiority of heterozygote due to the gene interaction. The heterozygote are well buffered to withstand environmental variability and hence able to live successfully in a wider range of environmental conditions than homozygote. Therefore, the heterozygote has the advantage to show superiority even in the stress condition. A biochemical basis for superiority of heterozygote was proposed by Haldane (1954) suggesting that if homozygote controls only a single protein form (allozyme, globin, etc.), the heretozygote allows two forms which could function in more diverse environmental conditions. Thus, heterozygote shows grater stability having wider range of enzyme activity over temperatures compared to that of either homozygote at that temperature. Therefore, hereozygote have more homeostatic (buffering) adjustments than homozygote.

Waddington (1957) coined the term *canalization* to the developmental buffering of genotype-environment interaction during growth or to a balance between a flexible ontogenetic process and inflexible final phenotype, so that there is maximum fitness in a wide variety of environment. This makes the phenotype to develop unimpeded (without obstruction) by environmental stress or by underlying genetic variability. He described the canalization as one of the products of stabilizing selection which eliminates the homozygotes which are sensitive to the potentially disturbing effects of environmental stresses. This leads to fixation of alleles in homozygous balanced state and reduction of genetic variance but the genetic variability is present in a balanced state.

The high degree of uniformity is shown by various characters particularly which are associated with fitness *e.g.* wing shape and body proportions in insects have usual very small variability and are important for survival in nature. The absence of phenotypic variability can be taken in the absence of genetic variability but this is not true. The reason for uniformity lies in the fact that the development of phenotype for canalized characters (having uniformity and developmental buffering) is caused by interactions between many genes so that a minor change in one gene usually has little or no effect on the phenotype. The development system is therefore said to be buffered or canalized.

Very rare events like mutation or heat strock (environmental stress) upset the canalization (buffering) and cause a change in the normal phenotype. The environmental stress disrupts the development of the normal phenotype by exposing the hidden polygenes to selection. Thus the polygenes (with minor effect), which exist but can not express their effects on the phenotype (called as hidden genes causing cryptic genetic variability), start showing their effect due to environmental stress and cause a change in the normal phenotypes and a new phenotype is produced.

The presence of canalization (buffering capacity against variability – secret/ hidden/cryptic genetic variability, which comes into expression under environmental stress) can thus be detected by giving environmental shock which disrupt the phenotype during development and by selecting such genotype which differ from normal phenotype. If the normal phenotype is upset by environmental stress or in the presence of a mutant gene and the new phenotype start showing inherited differences by exposing the underlying genetic variability (which already exists as hidden) it then indicates the existence of canalization (buffering) for the phenotype. This canalization (buffering) of phenotype is upset by environmental stress or in the presence of mutant gene by exposing the expression of hidden polygene to environmental stress.

Selection and canalization: Waddington (1953) produced crossveinless wings (like sex linked mutant trait) in Drosophila melanogastator flies by subjecting them to environmental shock. The continued environmental shock treatment each generation to the crossveinless flies and breeding such flies separately produced the new phenotypes (crossveinless) in the selected lines in high frequency with the phenocopy treatment (environmental stress). More interestingly, the new character (crossveinless wings) appeared in some flies without treatment. This indicated that new phenotype is caused by numerous genes whose effects were exposed by phenocopy treatment. Thus, selection after phenocopy treatment appears to change the primary expression of the genotype and also changes the zone of canalization (buffering) of phenotype. Milkman (1961) also obtained the similar effect of heat treatment on crosveinless character.

The genotypes for their sensitivity to particular environment can be detected by use of environmental stress or by incorporating major gene mutation (like bar eye or scute). The underlying genetic variability which can modify a character can be exploited by selection up to selection limit. The selected phenotype may be retained as an *"assimilated"* genotype even after removing the environmental stress or removing major gene (Waddington, 1961).

The selection limit (plateau) is attained with continued selection leaving no genetic variability and the homozygosity is reached for all the genes controlling the trait under selection, though reverse selection is effective to the control level. Similar is the effect of inbreeding to reduce heterozygosity to a certain level beyond which further decrease in heterozygosity does not occur. Therefore, genetic variability is maintained in a population in spite of the effects of all the forces acting to reduce the genetic variability. Lerner (1954) termed this phenomenon as the *genetic homeostasis*. This can be defined as the capacity of the population to maintain genetic equilibrium in variable environmental conditions by favouring heterozygotes in natural selection. In other words, it is the ability of a genotype to withstand (oppose) environmental fluctuations. The genes which appear in homozygous state due to selection may have their pleiotropic effect on fitness to reduce viability. This resist to a further increase in homozygosity and hence the population maintains genetic variability. This is the *population homeostasis.*

The genetic homeostasis involves both individual resistance and population resistance (buffering) which results in consistent performance of a population and may not be associated with heterozygosity or homogeneity. The overall action of the genetic system under environmental conditions decides the maintenance of a state of equilibrium and the integrated physiological processes during the course of development decides the mechanism of genetic homeostasis. The genotypes which have the capacity to make physiological adjustments for stability of development in various environments are called as *homeostatic genotypes.* Thus, the stability of a genotype in various environments is the property of homeostatic genotypes. *Homeostasis* thus denotes the power of an organism to hold the physiological processes within normal limits in spite of varying external conditions.

4.6.3 Phenocopy: Environmentally induced phenotype

The environment has its effect to the extent that one of the environments permits the expression of a genetic character while other environment does not permit the expression of the same character. Response of the genotype to the environmental conditions gives rise to the characters of organisms and sometimes the characters are changed due to changed environmental condition in the same way as they are changed when a gene mutates. Thus, change in environment sometimes induces non-hereditary phenotypic change which is similar to that caused by mutation. This non-heritable environmentally induced modification in phenotype resembling the phenotype of a known mutant gene is called *"phenocopy".* The phenocopy is the production of same phenotype by two genotypes under changed environment.

Such induced change may persist for the life of the individual and it may also persist in the progeny or may disappear in next generation. The production of phenocopy depends on the strength of environmental change to modify the effects of genes. There are some examples of G-E interaction in qualitative characters to illustrate the concept.

It is assumed that that the environmental agents sometimes affect certain chemical reactions which are similar to mutations and results in similar end products. To determine the inheritance of a phenotypic change, appropriate crosses are made and the difference between mutation and phenocopy is tested.

The term phenocopy was coined by Goldschmidt (1935). He subjected the larvae of Drosophila for short interval at different periods of development cycle to higher than normal temperature (35°C). This resulted into several phenocopies of wing abnormalities indistinguishable to genetic mutants of vestigial (vg) series.

Waddington (1953) subjected the Drosophila to environmental shock which produced flies *lacking crossveins wing*, similar to that caused by a sex-linked mutation, but this was non inherited phenocopy.

There are some phenocopy producing agents, drugs and treatments. The *diabetic symptoms* in genetically normal people appear due to specific environmental changes like infection, injury or removal of pancreas in humans. This is the phenocopy of diabetic individual of genetic origin. Similarly, the phenocopies of normal individuals can be produced by treating with insulin the diabetic patient that inherits diabetes. The diabetes in humans and animals seems to be a hereditary disease. A patient is unable to metabolize the glucose properly. Consequently, the glucose level in the blood stream becomes high to the extent of its excretion with the urine, finally causing death. The maintenance of blood sugar level is under the control of insulin produced by the pancreatic gland. The insulin prevents the occurrence of diabetes. The diabetes develops only when the insulin production is inadequate for a long time by the over consumption of carbohydrate. The insulin treatment does not permit the diabetes to develop.

The diabetes in animals may also be induced by introducing antibodies against an individual's own insulin, by pituitary and thyroid hormone and by chemicals such as alloxan; and in all these conditions the effects are not inherited but are only environmentally induced.

The normal *body colour in Drosophila melanogaster* is brown with its variant yellow colour. If the normal larvae with normal genotype for brown body are raised on diet containing silver salt (silver nitrate), they develop yellow colour like the yellow mutant but give rise to brown in normal environment.

The Himalayan *albino rabbits* have black colour on feet, ears, nose, and tail but the remaining body is white. If the rabbits are shifted to cold place after cutting of white hairs (from white parts), the black hairs appear instead of white hairs.

The phenocopy of *creeper trait* (shortened limbs) caused by genes in poultry can be produced with phenocopy treatment by injecting boric acid into chick eggs.

Landauer (1948) produced phenocopies of several mutations in the fowl with *skeletal effects*. He injected eggs of normal genotype with insulin before incubation or during development. This resulted in rumpless phenocopies if treatment was given in early stages upto 72 hours, shortening of limb (micromelia) if injected on 5th and 6th day, and other abnormalities of skull and upper beak similar to that of several mutant genes. The insulin effects are accompanied by hypoglycemia (a symptom of carbohydrate metabolism) after the establishment of circulation. An injection of nicotine amide counteracts the effect of insulin by way of acting as a coenzyme in cell respiration. The rumples phenocopies were also produced in chickens with normal genotype by mechanically shaking the eggs at a critical period and also by treating the embryos at 48 hours of age with sodium methyl arsenite. These chemicals produce shortened limbs similar to a dominant mutation called "creeper". These are two mutations in chicks causing rumpless abnormality characterized by the absence of caudal vertebral and tail structures including muscles, feathers and oil secreting gland which supply oily secretion to the feathers.

Both the mutations have the same end product but differ in their nature of inheritance – one being dominant and the other recessive with low penetrance, location (*i.e.* they are not alleles and also not closely linked), and also differ in structural details, chemical composition of bones and embryonic development.

The mouse kept at reduced atmospheric pressure during pregnancy produce the offspring lacking regions of the urogenital system similar to that of a mutant gene (Danforth short tail).

Human *limb abnormalities* including complete absence of hands and feet (acheiropody) are rare and inherited in some families. A serious effect of sleep inducing drug (thalidomide) by pregnant ladies result in limb defects and the eye defects following measles during pregnancy, resemble to hereditary abnormalities of limbs and eyes. The sleep inducing drugs produced their effect on skeletal formation of the foetus at the time of formation of limb buds and the babies are born with shortened limbs giving a flipper like appearance called phocomelia. Thalidomide has similar effect in mice, monkeys and rabbits. The dietary changes and atomic bomb testing were assigned as the cause of these sudden increases in phocomelia in early seventies of 20th century in German. Irradiation induced several skeletal and neurological anomalies resembling in genetically produced phenocopies.

Chapter 5

Linkage and Crossing Over

After rediscovery of Mendel's law in 1900 by 3 Scientists, the law of independent assortment was tested by Bateson and Punnet (1906) by making a dihybrid cross between two varieties of sweet pea. They found a different ratio than reported by Mendel among the F_2 progeny. The proportions of parental type progenies were higher and new combinations (recombinant types) were lesser than expected on the basis of independent assortment of genes. Such deviated results are logically possible if the gene pairs failure to segregate and assort (distribute) independently, and also if more than one pair of genes are present on a chromosome with an association or link between different gene pairs.

The presence of more than one pair of genes on the same chromosome pair is possible if the number of genes exceed to the number of chromosomes in the cell of an organism. This was made clear first time by Sutton (1903) who proposed the theory that chromosomes are the bearers of the hereditary units (genes). According to Sutton-Bovery hypothesis, the number of chromosomes in any organism is considerably less than the number of genes (Mendelian factors). Under this situation, large number of genes will have to be located on one pair of chromosome. Therefore, it is likely that all the genes present on each pair of chromosome may transmit in groups rather than singly.

5.1 LINKAGE

The phenomenon of existence of numerous pairs of genes on the same pair of chromosome and their inheritance together is called linkage. The two or more pairs of genes residing in the same chromosome and transmitting together are said to be *linked* and the phenomenon is said to be *linkage*. Such linked genes present on a chromosome form a *linkage group*. The number of linkage group corresponds to the chromosome number in haploid individuals and to the number of chromosome pairs in diploid organisms.

This chapter deals with the process of linkage and crossingover, effect of linkage on normal gene segregation and that how the linkage is important in modifying the normal phenotypic ratios among progenies.

Dihybrid individual produces two types of games *viz.* parental and recombinant types, in equal proportion, when there is independent assortment of two pairs of genes. The linkage and crossing over result in deviation of the production in the proportion of parental and recombinant gametes. The Mendelian dihybrid ratios of F_2 cross and of testcross are modified. This has been explained here -

5.1.1 DEVIATION IN DIHYBRID F_2 CROSS

Bateson and Punnet (1906) worked on the inheritance of two characters (flower colour and shape of pollen grains) of sweet pea. They crossed a variety of sweet pea having purple flowers and long pollen grains (PP LL) with another variety that had red flowers and round pollen grains (pp ll). They found that flower colour and pollen shape are controlled by separate gene pairs with dominance of purple flower and long pollen grain over the alleles for red colour and round pollen. They made a dihybrid cross of purple-long and red-round. The F_1 hybrids were heterozygous for purple-long (Pp Ll). These F_1 were selfed to produce F_2 progeny. The results of F_2 generation were clearly a deviation of normal classical F_2 dihybrid phenotypic ratio of 9:3:3:1 proposed by Mendel.

Another cross was made by them in confirmation of the deviated normal classical F_2 dihybrid phenotypic ratios, with different combination of characters. The purple-round plants (PP ll) were crossed with those having red flowers and long pollens (rr LL). The resulting dihybrid F_1 were again had purple flower and long pollen. On crossing the F_1s, it was again observed the deviation of normal classical F_2 dihybrid phenotypic ratio.

In both the crosses, they found that the F_2 progeny with parental combination were in higher proportions and recombinants occurred in lesser proportion than expected on the basis of independent assortment of two pairs of genes. They developed a theory of *"Coupling and repulsion"* to explain their results, depending upon the dominant and recessive genes coming from one or from two parents.

The cross in which the dominant genes at both loci come from the same parent (one chromosome) was said to be in *coupling phase* and the cross that include two dominant and recessive genes from different parents was said to be in *repulsion phase*. Thus, linkage relationship is of two types *i.e.* coupling phase and repulsion phase. Accordingly, purple-long x red-round cross illustrates the coupling phase whereas the purple-round x red-long cross illustrates repulsion phase. These two terms (*coupling and repulsion*) are used to distinguish the two possible arrangements of two pairs of genes on a single pair of homologous chromosome.

5.1.2 Deviation in Dihybrid Test cross

Morgan (1910) worked on the inheritance of two characters (eye colour and wing length) of Drosophila. He selected the flies differing in two different characters controlled by two pairs of genes, one pair affecting the eye colour (Purple, pr and red, pr′) and the other pair affecting the wing length (vestigial, vg and long, vg′). The red eye colour and long length of wing were normal and hence taken as wild type. The red eye colour is dominant over purple colour and the long wing gene is dominant over vestigial gene.

The first cross was that of coupling phase. *The second cross* was that of the repulsion phase when each of the two parent was homozygous for only one of the two dominant gene pairs. The resulting dihybrid F_1 females were *test crossed* with double recessive homozygous males, in both the crosses. The results obtained were drastically different than Mendelian prediction of 1:1:1:1 test cross phenotypic ratio.

5.1.3 Conclusion of the results of F_2 cross and test cross

It was evident from the results obtained from F_2 and test cross that the coupling as well as the repulsion phase relationship of genes gave similar trend of results in F_2 and test cross data. The similarity in both the crosses *viz.* F_2 *cross and test cross* (both coupling and repulsion relationship) was that *parental type progenies were in higher proportions than recombinant types.* Based on the deviation of results, Morgan replaced the Bateson theory of *coupling and repulsion* with *linkage and crossing over*. Morgan found that coupling and repulsion are two aspects of a single phenomenon to which he called as *linkage*.

The term *linkage* was defined as the tendency of genes present on the same chromosome (linkage group) to enter the gametes in parental combinations (tendency of genes to remain together during the process of inheritance) whereas the tendency of genes, residing on the same chromosome, to enter the gametes in the combinations other than those of parents (recombinants *i.e.* new combinations) was defined as *crossing over.*

The theory of linkage states that:

- The tendency of genes to remain in their parental combination is due to their presence on the same chromosome,
- The parental combinations of linked genes remain unchanged during the process of inheritance, and
- The genes on a chromosome are in a linear arrangement. The distance between linked genes on a chromosome is variable and decides the strength of linkage.

The deviation from the test cross ratio (1:1:1:1) and the F_2 ratio (9:3:3:1) based on independent assortment of genes indicated that two pairs of genes do not assort independently and are not located on different pairs of chromosomes but must be present on the same pair of chromosome. Morgan explained that the higher frequency of parental type progeny in either test cross or F_2 generation might be due to the presence of two pairs of genes on the same pair of chromosome, because the different pairs of genes being located on the same pair of chromosome are always transmitted together having come from the same parent (linked genes). This explains partly why the parental gene combinations remain together and occurred in higher proportions. When the two pairs of genes are located on the same pair of chromosome and are inherited together (linked group), there should have been no recombinants. These results neither follow the independent assortment law nor show linkage phenomenon but fall between the two. In other experiments also the recombinants occurred in almost all the cases, with few exceptions. Now, the matter is to explain the production of recombinants that how they occur (the process by which the recombinants occur) when there is a linkage between the genes. This means that some process is taking place in the formation of recombinant gametes during meiosis so as to produce the intermediate results between linkage and independent assortment of genes. The process which forms the recombinant gametes and produces the intermediate results is the *crossing over*.

5.2 CROSSING OVER

Morgan postulated that recombination of linked genes is accompanied through a process by which the homologous chromosomes exchange parts (Johanssen, 1909). This process of physical exchange of parts between homologous chromosomes is called crossing over. The event of crossingover occurs between nonsister chromatids of homologous chromosomes at the time of synapsis during late prophase and metaphase of meiotic cell division. This involves actual exchange of chromosome material. Each of the four chromatids goes into separate gamete to form four gametes. The two gametes carry the gene combinations identical to the parents and other two gametes are produced by crossingover and hence called as crossover gametes or *recombinants*. These two recombinant gametes on fertilization produce two recombinant phenotypes which are different from parental phenotypes. These recombinant phenotypes are the result of crossing over. The percentage of recombinant is also referred to as the percentage of crossover which is simply the number of recombinant phenotypes expressed as a percentage of the total number of progeny.

The *cytological basis of the crossing over* was described by Johanssen (1909) through the formation of chiasma at the tetrad stage during spermatogenesis in Salamonders. He noticed that during meiosis at the time of pairing (*Synapsis*) of duplicated homologous chromosomes (tetrads or four-strands), two sister chromatids showed a cross-shape structure with each other while the other two do not. This *cross-shape structure* of tetrads is called as *Chiasma* (plural- chiasmata). Thus chiasma is probably the visible manifestation of crossing over. Thus the crossing over occurs between chromatids and not between unduplicated chromosomes.

Conditions of the recombinants to occur: The recombinants can result either due to independent assortment when the two gene pairs are present on different chromosomes or due to crossing over when the genes are linked on the same chromosome. Thus, there are two types of recombinants:

(*i*) **Inter chromosomal recombination** which arises from the genes located on different pairs of chromosomes assorting independently. The frequency of recombinants in this case is always ½ or 50 % of the total gametes produced in a test cross.

(*ii*) **Intra chromosomal recombination** which arises from the genes located on the same chromosome by the process of crossing over during meiosis. The frequency of recombinants in this case is less than 50 %. This frequency provides very useful genetic information to provide genetic variability and to know the relative position of two or more genes located on the same chromosome (linkage or genetic map).

5.2.1 ILLUSTRATION OF CROSSING OVER

The phenomenon of crossing over can be illustrated by genetic studies involving crosses and by cytological proof.

(i) **Genetic studies:** The genetic crossing over was first considered a consequence of the exchange of parts between two homologous chromosomes. The phenomenon of crossing over provided the most reasonable physical mechanism for recombination of genes to occur between the members of a linkage group. The assumption of crossing over was based on the fact that the recombinants are actually produced by the exchange of chromosome segments between homologous chromosomes for the genes located on the same chromosome. The results of genetic studies showed crossing over (exchange of parts between homologous chromosomes) because it was the only possibility of recombinants being produced. Therefore, in order to produce recombinants between two different allelic pairs located on the same chromosome pair, crossing over must occur between the loci of genes involved.

(ii) **Cytological proof:** The cytological proof of crossing over was demonstrated by Stern (1931) in Drosophila and McClintock and Creighton (1931) in Mize. There were faced some difficulties to prove crossing over cytologically. The main difficulty was to demonstrate the physical exchange of parts of homologous chromosomes because the members of a pair of homologous chromosome are not visible clearly from each other due to their identical appearance. Secondly, the four chromatids are identically coiled around each other during crossing over. Moreover, the crossing over can not be seen in living cells.

Stern Experiment: Stern recognized the value of cytological marker for demonstrating crossing over. He obtained female Drosophila having each of the two X-chromosome structural abnormalities. He found an aberrant X-chromosome to which a portion of Y-chromosome was attached and also found another distinguishing feature of the other X-chromosome which was composed of two equal fragments, each with its centromere (the X- chromosome being broken into two pieces). These abnormalities lead to the recognition of the two X-chromosome from each other as well as from normal X- chromosome. The female Drosophila carrying these abnormal chromosomes and heterozygous for carnation (eye colour recessive mutant to red eye colour) and for bar eye (small eye, a dominant eye shape over normal wild type) were crossed with males of carnation colour and normal eye shape. The two X- chromosomes from the female with distinct cytological markers (attached –X which was extra long piece on one side of centromere and broken piece) and normal X-chromosome female with no marker were visibly distinguishable and could be identified in the progeny from the cross microscopically. The results were recorded on genetic aspect (linkage and crossing over) as well as cytological aspect (chromosome appearance). Examination of the chromosomes of the progeny whose phenotypes showed genetic recombination indicated that genetic recombination was followed as a consequence of appropriate exchange of identifiable chromosome segment. The two abnormalities (broken piece and with long piece, wild type) present on two different X-chromosomes can only come together on one member of a pair of chromosome if crossing over takes place. He found that the parental types determined genetically (Bar, carnation and + +; wild type) always had parental chromosome configuration (Bar, carnation flies) having bar, carnation and + + (wild type) flies. On the other hand, the chromosome of the recombinant progeny had new chromosome arrangements, with Bar + and + carnation. Carnation, is a recessive mutant eye colour (c) to wild type red colour (C) whereas the bar shape eye (small or narrow, B) is dominant over wild type normal eye shape (b). Thus, the genetic, physical and cytological crossing over was established.

To demonstrate the crossing over with chromosome exchange, it is thus important to have chromosomes which are cytologically different in which the crossing over occurs. The above experiment showed relationship between genetic and cytological crossing over, and that cross over was a result of physical exchange.

5.2.2 When crossing over occurs

The crossing over occurs during the prophase of meiosis at the tetrad stage (four strands) of homologous chromosomes. The homologous chromosomes are paired during meiosis and this pairing is called as *synapsis*. If crossing over takes place during pairing of homologous chromosomes it may occur either when only two strands are present or after two strands have splitted (duplicated) to make four strands, known as tetrad stage. The ability to identify the four products of meiosis (two parental types and two recombinants) in all the concerned breeding experiments are the genetic evidence of crossing over giving strong proof that crossing over occurs at four-strand or tetrad stage called as chromatid stage. The four different products of meiosis are possible if crossing over occurs at tetrad stage, because had the crossing over occur at two strand stage, there would have never been more than two different products of a given meiosis.

The four strand crossing over can be illustrated by making use of attached X-chromosome in Drosophila and tetrad analysis in moulds (fungi) and other lower organisms which are haploid.

5.2.3 Kinds of crossing over

The Crossing over may be single, double or multiple depending upon the number of chiasmata formed. When only one chiasma is formed it results in single crossing over forming a single crossover gamete. This type is most frequent. Double crossing over results due to the formation of two chiasmata and it is less frequent type. The multiple crossing over take place when more than two chiasmata are formed and it does not occur frequently.

5.2.4 Frequency of crossing over

The frequency of crossing over between two loci is expressed in terms of percentage of crossing over which is directly proportional to the distance between two loci. Thus the distance between two loci on a chromosome is indicated by the percentage of cross over. When two genes are located on a chromosome very close to each other, the frequency of their crossing over is less to produce cross over progeny. Some times the more closely located genes do not allow crossing over. In this case only parental types of progeny are produced due to linkage of genes and thus the frequency of crossing over is zero because no cross over progeny is produced. The percentage of crossing over is obtained from the data and helps to work out the relative distance between various loci on a chromosome which in turn determines the exact location of the genes on the chromosome. Thus it is of great significance because it helps in the construction of chromosome maps. The unit of crossing over is morgan according to J.B.S.Haldane.

5.2.5 Factors affecting frequency of crossing over

The strength of linkage or the frequency of crossing over is influenced by genetic, physiological and environmental factors. These include intra-species variation (genotypes, chromosome structure and position of genes to centromere), differential effect of sex, temperature, chemicals and irradiation etc.

The gene mutation may affect the frequency of crossing over. In *Drosophila melanogaster*, a gene on 3rd chromosome (c 3 G) completely suppresses crossing over when homozygous in females but enhances recombination when heterozygous. Change in chromosome structure, *i.e.* order of genes on a chromosome (chromosomal aberration) in one pair of chromosome was found to increase the frequency of crossing over in other normal non-homologous chromosome. The inversion of chromosome segments greatly reduced recombination in heterozygote.

The genes located near the centromere show reduced crossing over and the position of genes relative to centromere of the chromosome can be changed by structural chromosomal aberrations. The crossing over is less frequent near centromere, at the tips of chromosomes and near another crossover. The crossing over is less likely to occur between the genes that are located relatively closer together on the chromosome and so recombinant gametes are less likely to be produced. On the other hand, a greater number of crossover gametes will be expected when linked genes are located relatively far apart on the chromosomes.

The crossing over in Drosophila males is negligible. Likewise, males of mammals also show little crossing over. But the crossing over does not occur in females of silkworm

The increase in age of the female of Drosophila is associated with decrease in the rate of crossing over to a certain age and thereafter it increases again in later age. Deviation in temperature from 22°C in both the direction has been shown to increase the rate of crossing over in Drosophila.

The chemical composition of food, starvation, antibiotics and X-ray irradiation increase the crossing over.

5.3 EXPECTED RATIOS UNDER INDEPENDENT ASSORTMENT OF GENES AND LINKAGE

The test-cross data as well as the F_2 data indicated the following situations:

(i) Independent assortment of two gene pairs: This results in equal proportion of parental combinations and new combinations (recombinants), with further subdivision of both types of combinations in equal proportion resulting 1:1:1:1 ratio.

(ii) Linkage of genes: The linkage of genes may be complete or incomplete, depending on the extent of crossing over:

- *Complete linkage between two gene pairs (No crossing over):* This results in only the parental types with both parental types in equal ratio.

- Incomplete linkage (Some *crossing over between two gene pairs):* This results in parental as well as in new combinations in unequal proportions, the parental combinations being higher than new combinations. This is the case of incomplete linkage or crossing over which results in 4 phenotypic ratios, two of each parental and new combination.

(1) Complete Linkage: *It means no crossing over.*

This is a rare phenomenon. The linkage is caused by linked genes being located on the same chromosome. Thus, the two or more genes located on the same chromosome should remain together in all cases to produce only the parental combinations and no new recombinants should be produced. This type of linkage is called as complete linkage which can be explained under the condition of no crossing over. This is only possible that the genes are so closely located in the same chromosome that they are always inherited together.

The example of complete linkage of genes was found for the genes present on 4th chromosome of *D. melanogaster:* the mutant carrying genes controlling bent wings and shaven bristles (bt svn / bt svn genotype). When these flies are crossed to normal flies (bt' svn' / bt' svn'), the F_1 produced are heterozygous for the mutant genes. The test cross progeny are either bent shaven or normal and thus only the original parental types progenies are produced without appearing any cross over (recombinant) types. This strong linkage among two genes was evident in both coupling and repulsion phase.

Studies have shown that crossing over is either absent or rare in male Drosophila and in female silkworm moth which are both hemizygous.

(a) Male Drosophila: Morgan (1910) crossed the F_1 heterozygous female (for eye colour and wing size) with double recessive male (purple eye-vestigial wings, pr vg / pr vg). He obtained the parental as well as recombinants as obtained earlier. But when he changed the sex in the test cross *i.e.* F_1 heterozygous males were test crossed with double recessive females, the complete linkage was observed because no recombinants were produced. Thus the *linkage strength is different in the two sexes in Drosophila,* crossing over rarely or never occurs in the male.

Another example of complete linkage in this insect (Drosophila) can be cited. The gene for gray body colour (B) is dominant over its allele producing black or ebony body (b) and another gene for wing size resulting in long wing (V) dominant over its allele *(v)* producing vestigial wings. Morgan (1919) crossed gray bodied vestigial winged flies (Bv /Bv) with black bodied long winged (bV /bV). The F_1 were heterozygous having gray body long wing (Bv /bV). The test cross taking F_1 as male with double recessive female produced only two kinds of progeny which were of parental types with no cross over type. Thus the linkage was complete. On repeating this experiment in a way that the F_1's were female in test cross and the double recessive parent was male (black-vestigial), the test cross progeny produced had both parental as well as cross over types, the parental types were about 83

% and rest 17 % were cross over types. Thus, there is perceptible distance between the genes for black and vestigial which separates them. The cytological experiments of spermatogenesis in male have shown that homologous chromosomes undergo pairing

(b) *Female silkworm moth*: A Japanese geneticist, Tanaka observed that there was no crossing over in female silkworm (*Bombyx mori*).

It was generalized based on above experiments that there was no crossing over in *hemizygous* sex but this notion failed because in most other organisms the crossing over has been found to occur in both sexes.

(2) **Incomplete Linkage**: *It means that some crossing over takes place.*

The complete linkage is very rare and the experimental results fall between complete linkage and independent assortment which means that some crossing over occurs that leads to incomplete linkage. The following examples can be cited for incomplete linkage:

Drosophila: Two pairs of genes on 3rd chromosome, hairy (h) which is a bristle mutant and approximated (app) which is a wing vein mutant, show linkage. Both genes are recessive to wild type. The test cross showed that there were 90% parental types and 10 % recombinants. Similar results were obtained in coupling and repulsion phase.

Linkage in certain sex-linked characters in Drosophila have been observed *viz.* between white eyed and miniature wings; white eyes and yellow body colour; yellow body colour and cut wings. These genes are present on X- chromosome. The white eye and miniature wings are recessive mutants to normal eye (red) and normal wings size. Morgan (1911) reported 37.6 % cross over between these two loci for eye colour and wing size. The yellow body colour and white eye colour are both mutants to normal or wild type (gray bodied and red eye). The test cross data resulted in 1.5 % recombinants in both coupling and repulsion phase of linkage. The crosses involving yellow body colour and cut wings (a recessive mutant on X-chromosome) produced 20 % recombinants.

Poultry: The genes controlling the *Comb shape* and *length of legs* (short or normal) are reported to be linked. Rose comb (R) is dominant over single comb (r) whereas creeper (short legs) gene (C) is dominant over normal leg length (c). However, these two genes do not maintain their linkage completely, as the test cross data indicated that some recombinants were produced that were in lesser frequency (8.3 %) than the parental combinations (91.7%).

Two other pairs of genes affecting *feather colour* (I, white and i, coloured, black) and *feather texture* (F, frizzle, and f, normal or smooth) are also linked (Hutt, 1933). The gene I was dominant inhibiting the expression of melanin pigment and the individual was white feathered. The test cross data indicated that the genes showed incomplete linkage both in coupling and repulsion phase. The recombinants produced were 18.2 % in coupling phase and 19.7% in repulsion phase. A third loci affecting *head shape* (crested and noncrested) is linked with other two loci affecting the feather colour (black and white), and texture (frizzle and smooth).

The two loci which are sex-linked affecting *rate of feathering* (normal or slow and rapid feathering), and *plumage colour* (Silver and gold) are linked together, being present on sex chromosome. Rapid feathering and gold feathering are controlled by recessive genes in homozygous condition, respectively. A third gene locus affecting colour pattern (barred and black) is also present on sex chromosome. The barred (white bars on top of head, feathers are banded with bars of black on white background) is dominant (B) over black (b).

Mice: The genes for coat colour in mice (Black and Himalayan) and hair texture (freezy and normal smooth) are linked on the same pair of chromosome. Himalayan colour is recessive to full colour (black) and the frizzy hair coat is recessive to normal smooth hair.

Pigs: The genes for skin colour (Black and red) and for ear condition (erect and droopy ears) are linked together. The black is controlled by dominant gene while its recessive allele causes white hogs. The gene for erect ears is dominant over droopy ears. The Hampshire hogs are known to be homozygous for dominant genes at both these loci whereas the Duroc hogs are homozygous recessive for both pairs of genes.

5.4 LINKAGE DETECTION

There are two methods to prove/detect the linkage, based on above results. One is the *test cross* (back cross of F_1 with double recessive parent) and the second method is F_2 *data* obtained from *inter se mating of F_1.*

The ratios of different phenotypic classes obtained either from test cross or F_2 data are compared to the ratios expected on the basis of independent assortment. The significance of departure in the ratios obtained from the test cross or F_2 with the normal expectation (based on I.A.) is tested by *chi-square test* (test of independence for linkage). Among these two tests of detecting linkage, the *test cross* is the most convenient method and used:

- For determining the linkage and
- To know relative position of the locus within the chromosome.

Indication of linkage: In a test cross (Aa x aa), a dominant gene segregates from its recessive allele in the ratio of 1:1. Similar will be for segregation of second gene pair (Bb x bb). If the two gene pairs segregating independently (taking together) in the ratio r_1:1 and r_2:1, their joint segregation will be in the ratio $r_1 r_2$: r_1: r_2: 1, producing 4 phenotypic classes [four combinations in equal ratio of 1:1:1:1 as a result of (1:1) (1:1)]. Thus, there will be 4 phenotypic classes in equal numbers in case of I.A. of two pairs of gnes. The complete linkage is indicated if all the testcross progeny are of parental types only, whereas the linkage is incomplete if recombinants are also observed but in lesser frequencies than of parental types. The testcross ratios having fewer than 50% recombinants suggest linkage.

The F_2 ratio for single gene pair (with complete dominance, A and a) is 3:1 and for two gene pairs (A,a and B,b) segregating independently will be 9:3:3:1 obtained from (3:1) (3:1). The deviation, from these F_2 ratios (9:3:3:1) based on independent segregation, will be due to linkage of genes. When the deviation in these ratios occurs due to linkage (either in a test cross or in a F_2 cross), the parental type progenies are always in greater numbers and the numbers of recombinants are fewer, depending on the strength of linkage.

Test of linkage: The presence of linkage is tested by χ^2 and square root method: (i) Chi-square test: The *linkage is tested* by *chi-square (χ^2) test* based on the observed numbers of 4 phenotypic classes (AB, Ab, aB and ab) considering their observed numbers (O) as: a_1, a_2, a_3 and a_4. The expected numbers (E) for joint segregation of two loci in the ratio 9: 3: 3: 1 will be, respectively as: (9/16) N; (3/16) N; (3/16) N and (1/16) N. Where,

$$N = a_1 + a_2 + a_3 + a_4.$$
$$a_1 = AB; a_2 = Ab; a_3 = aB; a_4 = ab$$

$$16 = \frac{9}{16} + \frac{3}{16} + \frac{3}{16} + \frac{1}{16}$$

$$\chi^2_{(Total)} = \sum \frac{(O-E)^2}{E} \text{ tested at 3 degree of freedom.}$$

The above information on the observed numbers of all the 4 phenotypic classes can better be understood if arranged in a tabular form as given under –

A locus:

		A	*a*	*Total*
B locus:	B	AB = a_1	aB = a_3	$R_1 = a_1 + a_3$
	b	Ab = a_2	ab = a_4	$R_2 = a_2 + a_4$
	Total	$C_1 = (a_1 + a_2)$	$C_2 = a_3 + a_4$	$N = (C_1 + C_2)$
				$= (R_1 + R_2)$

The total chi-square is partitioned into chi-square due to A locus, due to B locus and due to linkage, as under and each is tested at 1 degree of freedom.

$$\chi^2_{(A\ locus)} = \frac{(C_1 - 3\,C_2)^2}{3N} = \frac{(a_1 + a_2 - 3\,a_3 - 3\,a_4)^2}{3N}$$

$$\chi^2_{(B\ locus)} = \frac{(R_1 - 3\,R_2)^2}{3N} = \frac{(a_1 + a_3 - 3\,a_2 - 3\,a_4)^2}{3N}$$

$$\chi^2_{(Linkage)} = \frac{(a_1 - 3\,a_2 - 3\,a_3 + 9\,a_4)^2}{9N}$$

All the above 3 chi-square (χ^2) values are tested at 1 degree of freedom. The sum of all the 3 chi-square values will give the total chi-square. [*See solved example 5.1*]

(ii) *Square root method:* The linkage may also be detected by *square root method.* The double recessive phenotypes among the F_2 progeny can be distinguished but the crossover percentage may not be known. In this case, the frequency of double recessive phenotype among F_2 progeny can be used to estimate the frequency of non-crossover gametes when the F_1 is in coupling phase and to estimate the frequency of crossover gametes when the F_1 is in repulsion phase.

Coupling phase: In coupling phase, the double recessives (aabb) will be one of the two parental types (AABB, aabb) and their (aabb) frequency will be half of the total parental types because other half will be double dominants (AABB). Let the frequency of 4 types of gametes be p for AB, q for Ab, r for aB and s for ab gametes.

The probability of uniting two ab gametes to form double recessive (ab/ab), which are parental type, will be s^2. Therefore, *percentage of parental gametes*

$$= 2\sqrt{\text{frequency of double recessive}}$$

For example, suppose the crossover percentage is 20, then parental gametes are expected be 80 percent (half of which, 40 % will be ab and rest half 40 % will be AB). The probability of uniting two **ab** gametes to form the double recessive genotype (ab/ab) will be 0.40 x 0.40 = $(0.40)^2$ = 0.16 = 16 percent. This is the frequency of double recessive. Thus, the percentage of parental gametes

$$= 2\sqrt{\text{frequency of double recessive}}$$

= 2 √ 0.16 = 2 x 0.4 = 0.80 or 80 percent. Now if 80 percent are parental types, the other 20 percent must be recombinants.

Repulsion phase: In repulsion phase, the double crossover (aabb) will be one of the two recombinants (AABB, aabb) and its frequency will be half of the total recombinants. With the same logic, the probability of uniting two ab gametes to form double recessive, which are recombinants in repulsion phase, will be s^2.

Therefore, the *percentage of crossover gametes* $= 2\sqrt{\text{frequency of double recessive}}$.

For example, with 20 percent crossover, 10 percent of the gametes are expected to be ab. Therefore, the probability of double recessive phenotype (ab/ab) = 0.10 x 0.10 = $(0.10)^2$ = 0.01 = 1.0 percent. Thus, if 1.0 percent are double recessive, the *percentage of crossover gametes*

$$= 2\sqrt{\text{frequency of double recessive}} .$$

= 2√0.01 = 2 x 0.10 = 0.20 = 20 percent

.The test cross over the F_2 has a special advantage that this cross provides simple ratio of 1:1:1:1 for comparison between the observed (from experiment) and expected (based on the hypothesis of I.A. of genes). But it is difficult to conduct a test cross than to get F_2 *progeny* by selfing of F_1 in some self-pollinating species, *e.g.* wheat. This is because the test cross requires emasculation in self –pollinating plants and cross fertilization from pollen of other plant (homozygous recessive). On the other hand, the F_2 plants are obtained easily from self fertilization of F_1 which requires only protecting the F_1 plants from foreign pollen. Therefore, when test cross is difficult to perform, it is easy to produce F_2 by selfing of F_1, but the F_2 ratios (9:3:3:1) are more cumbersome to handle than test cross ratio of 1:1:1:1.

In a dihybrid test cross data, the frequency of cross over directly estimate the percentage of crossing over and hence the strength of linkage. The percentage of recombinants equals the percentage of cross over and it is also taken as the distance between two gene loci.

5.5 IMPORTANCE OF CROSSING OVER AND LINKAGE

The crossing over creates as many combinations as there are different gene pairs. It thus increases the likelihood of variation among population and act as a potent force for generating variability. The variation is a vital tool to evolutionary development. The numerous combinations caused by crossing over can be acted by natural selection. It is thus possible that favourable combination is created which may increase the economic and adaptive value of the individual.

Secondly, the crossing over has shown that genes occur in organized linkage groups and in linear order. The frequency of cross over is constant and characteristic for any pair of linked genes under standard conditions. Thus it helps to locate the position of more and more genes and to form the genetic or chromosome maps which represent the relative positions and distance of genes in each linkage group.

The plant breeders are always trying to produce new and useful varieties of plants of economic value by recombining genes through crossing of existing varieties. But this information is rarely utilized because some better gens are linked with bad genes (inferior genes) and therefore they are transmitted together to the progeny. On the other hand, linkage has its own practical advantage in making a linkage group of some qualitative characters with quantitative characters. This association of qualitative characters which are easily recognizable with quantitative helps the breeder as an index of the linked quantitative characters for selection purpose, sine the entire linked group is inherited together. These genes are called as marker genes.

The linkage association in farm animals between a qualitative and quantitative characters is not possible because so many genes are involved in affecting the quantitative characters and the expression of any single gene pair is very much weak (minor). Secondly, linkage relationship will have to be determined for each animal which vary among individuals. Moreover, it is difficult to differentiate the association among characters whether it is due to linkage or due to pleiotropic effect of the same gene or through a common physiological pathway affecting the two characters.

5.6 EFFECT OF LINKAGE

The linkage of genes has the following effects:

(i) Linkage and classical F_2 ratio of dihybrid cross (9:3:3:1): The two gene loci either in coupling phase (AABB and aabb) or in repulsion phase (AAbb and aaBB) will produce dihybrid F_2 ratio as 9: 3: 3: 1 without linkage. The percentage of recombinants in F_2 generation of a dihybrid cross is less with linkage than expected without linkage. The parental type F_2 progeny in coupling phase is expected to be 62.5 % based on [(9 + 1)/16] and the recombinants to be 37.5 % based on [(3 +3)/ 16], without linkage whereas in repulsion phase the proportion of two types of F_2 progeny is reverse, *i.e.* 62.5 % are recombinants and rest 37.5 % are parental types. The linkage of genes disturbs this ratio of parental and recombinants among F_2 progeny. The proportion of recombinants with linkage is reduced among F_2 progeny than expected ratio.

Similar tendency of lesser proportion of recombinants than expected (50: 50) is observed in test cross when the two gene pairs are linked.

(ii) Linkage and epistatic ratios: The classical F_2 ratio of 9:3:3:1 and the test cross ratio of (1:1:1:1) is modified due to epistasis. For example, the epistasis

changed the test cross ratio into 1:1:2 combining two phenotypic classes without linkage. When the genes showing linkage are linked, the parental combinations are observed in greater numbers than recombinants. This indicates the linkage. This can be illustrated by an example of coat colour in rabbit as follows:

The black colour in rabbit is dominant over brown. The gene, B, responsible for black hair is dominant over its allele (b) which produces brown colour of hair whereas the albinism is a trait expressed by homozygous recessive genotype (cc) and the dominant allele (C) produces coloured hairs. The genotype CC and Cc produces pigment while cc produces albino. The cross of homozygous black rabbit (BBCC) with albino genotype (bbcc) produces hybrids (BbCc) with coloured hairs. The test cross made by crossing of F₁ heterozygote (BbCc) with homozygous recessive (bbcc) having albino colour will result into the following phenotypes with expectation as:

CcBb	:	Ccbb	:	ccBb and ccbb
1	:	1	:	2
(Black)	:	(Brown)	:	(Albino) with no linkage

The ratio of 1:1:2 instead of 1:1:1:1 is caused by epistasis and hence the epistasis has modified the expression of the colour gene. The colour gene in homozygous state (bb) has modified the effect of the allele C and produced brown colour of the genotype Ccbb and the same allele in homozygous state (cc) has modified the effect of brown gene (B) in the genotype ccBb producing albino.

Now if the two loci are linked, the parental phenotype (black) appeared in greater number than the recombinants (brown). This indicates the linkage. The third phenotype (albino) can not be properly classified as parental phenotype because the expression of colour gene is modified due to linkage. Thus, only the black and brown phenotypes are used for calculations to estimate the crossover percent.

(iii) Linkage and sex linked genes: The dihybrid, in a test cross, for sex linked gene, does not exist in homogametic sex as for autosomal linkage. This is the major difference produced by linkage.

(iv) Linkage and pleiotropy: The close or complete linkage produced the pleiotropic like effect. The complete linkage does not produce any recombinants whereas very close linkage produce some recombinants. It is required high number of test cross progeny to prodce one each of the two recombinants in the cross to reveal close linkage. The probability of crossing over between very closely linked loci is increased with higher numbers of progeny produced. Therefore, linkage is not important in farm animals compared to that in fruit fly. This is one of the problems associated with linkage studies of genes closely located.

5.7 CHROMOSOME MAPPING

The genes are located on chromosomes in a linear arrangement having their fixed relative position on the chromosome. The genetic map (or chromosome map), known as chromosome mapping, is the graphical representation of genes on the chromosome indicating distance of genes between loci and their relative position from each other in each linkage group. The distance of genes indicates the *map distance* that how far the two genes are located on the chromosome while the relative position of genes indicates the *gene order* (sequence of genes) on the chromosome.

The *map distance* (distance between genes) is represented in percentage of cross over (recombination) between loci along the chromosome. This is because the amount of crossing over, (percentage of crossover progeny produced), is proportional to the distance between two loci on a chromosome. This is based on the hypothesis formulated by Sturtevant (1913) that each gene has a definite position (locus) on its chromosome and the place for occurrence of chiasma (synapsis or pairing of chromosome for occurrence of crossing over) is a matter of chance and hence the frequency of crossover is proportional to the linear distance between genes on the chromosome. He thus suggested that the percentage of crossover can be used as a quantitative index of the gene distance on the chromosome. This helped in chromosome mapping. Thus, map distance is measured from the frequency of crossovers obtained either from test cross or F_2 data. The map distance is measured in map units (anti Morgan). One percent recombination is equivalent to one map unit of length on the chromosome. The percentage of crossing over is constant between any two loci but it is different between different genes. The maximum percent of crossover can not exceed to 50 percent and thus the relative length of any chromosome is 50 map units. Therefore, one map unit is 1/50 of the length of chromosome.

The map distance in terms of map units is not helpful in deciding the actual location of different genes on the same pair of chromosome. This information on actual location of different genes requires the relative distance between each of the two genes as well as other genes located on the same chromosome. The test cross involving three gene loci is required to know the actual position of different genes (sequence of genes on the chromosome).

The *gene order* (sequence of genes on the chromosome) is linear and detected from *three-point cross* (trihybrid cross involving three genes in a cross at a time). In three point cross, three types of crossovers are possible *viz.* two single crossovers (crossover between first two genes, region I; between second two genes, region II) and combined or double crossing over simultaneously between first two and second two loci, such as taking 3 loci A, B and C. The single crossover may occur between AB region (region I) in some chromosomes, between BC region (region II) in some chromosomes and double crossovers in some other chromosomes. The double crossovers results from the simultaneous crossing over in both the regions. This is referred to as the multiple crossingover. The test cross involving three points cross helps to reveal the occurrence of double crossovers. The comparison of double-crossover gametes to the parental gametes determines the linear order of genes on the chromosome. The genes in the double crossover gametes which have been transposed compared to the parental gametes are in the middle.

Assuming linkage, the parental type progeny are expected to be more frequent (more than 50 %) than any of the crossover types. Such results indicate an alternative to independent assortment of genes (*i.e.* linkage between genes).

Methods to estimate the gene order: The gene order may be estimated by two methods which are the elimination method and the double crossover method.

(i) Elimination method: This requires the computation of the frequencies of recombinants, taking two pairs of genes at a time and ignoring the third pair of gene.This can be illustrated citing the example of a trihybrid cross of fruit fly, performed by Bridges and Olbrycht (1926). They crossed the flies having *crossveinless wing* (cv) flies (males) with rough eyes (echinus gene, *ec*) and cut wing margin (ct*).* All these three characters are sex-linked and recessive to their normal (wild) type characters. Therefore, the F_1 flies were heterozygous normal (wild) for 3 characters. The F_1 female should have produced 8 types of gametes in equal numbers but it was observed on backcrossing them with hemizygous recessive males (ec,cv,ct) that 8 types of F_2 progeny were not in equal proportions as shown here -

Phenotypes	Genotypes	Observed numbers	Types
Crossveinless	+ Cv +	2207	Parental
Echinus cut	Ec + Ct	2125	"
Echinus crossveiless	Ec Cv +	273	S.C.O. (I)
Cut	+ + Ct	265	"
Echinus	Ec + +	217	S.C.O. (II)
Crossveinless cut	+ Cv Ct	223	"
Scute Crossveinless cut	Ec Cv Ct	3	D.C.O.
Wild type	+ + +	5	"

The frequencies of recombinants are computed taking two genes at a time as:

Steps	Loci taken	Ignored locus	Parental types	Recombinants	Freq. recomb.
I	Ec and Cv	Ct	2207+2125+217+223	273+265+5+3	$\dfrac{46}{5318}$
			(4772)	(546)	(0.102)
II	Cv and Ct	Ec	2207+2125+273+265	217+223+5+3	$\dfrac{48}{5318}$
			(4870)	(448)	(0.084)
III	Ec and Ct	Cv	2207+2125 +3+5	273+265+217+233	$\dfrac{978}{5318}$
			(4340)	(978)	(0.184)

The recombinant frequencies between two gene loci are the map distance between the respective two gene loci. The above results have indicated that crossing over has resulted the formation of 10.2 % gametes between Ec and Cv gene loci, 8.4 % gametes between Cv and Ct gene loci and 18.4 % gametes between Ec and Ct ene loci. Therefore, the gene distance between two respective loci was 10.2, 8.4 and 18.4 units. In addition to these, some gametes may be produced which contain both crossovers at the same time, called as *double crossovers* (D.C.O.) and this phenomenon is called *double crossing over*. However, D.C.O. progeny may be produced in few cases, the DCO gametes being least frequent than single crossover in two regions or parental combinations.

(ii) Double crossover Method: The double crossover method depends upon identifying the double crossovers whose frequencies are expected to be least and that which of the three possible arrangements (sequences) of loci on the chromosome could produce the double crossover classes. It is important to note that production of *DCO type gametes depend on the gene order and the map distance between loci.* The method of DCO depends on:

- The estimation of DCO which are of least frequent type
- To know that which of the linear order of gene can produce the DCO genotypes.

The fundamental experiments on three point cross or trihybrid cross taking three sex-linked genes of fruit fly were conducted by Sturtevant (1921), Bridges and Olbrycht (1926), and others.

Two experiments on three – point crosses were conducted by Bridges and Olbrycht (1926) on Drosophila. *In one experiment* male flies (parental generation) having rough eyes (echinus, ec.) were crossed with female flies having scute bristles (absence of bristles, sc.) and crossveinless wings (cv). All these three characters were recessive to their normal or wild type characters so as the F_1 females were heterozygous normal for the three characters and hence the F_1 females produced 8 types of gametes in equal numbers.

In the second experiment, the F_1 flies were produced by crossing original parents (parental generation, P_1) having crossveinless flies (males) with female flies having echinus eyes (rough eyes) and cut wing margin (ct, a sex linked recessive mutant gene). In both these experiments of three points cross, the original parents (P_1) were heterozygous.

Sturtevant (1921) conducted an experiment on Drosophila by crossing yellow (body colour, y) carmine (eye, c) females with forked (a bristle mutant gene, f) males and the F_1 females were mated to yellow carmine forked males (y c f). Thus the original parents were y c + / y c + (female) and + + f / y (male parent) and hence heterozygous.

On the basis of the findings, the following conclusions were emerged-

(i) Parental combinations were more than 50 % of the total progeny (hence with highest frequency), the double crossover classes were least frequent (with lowest frequency) and single crossovers each at both the regions (I and II) were intermediate in frequency of the parental and double crossover classes.

(ii) In estimating the single crossover frequency at the two regions, for the two loci respectively, the double crossover progeny are added in the number of progeny of single crossovers of both the regions separately.

(iii) Sum of the frequencies of the single crossover classes at the two regions excluding the double crossover class, constitute the distance for the loci which are at the two extreme sides.

The above findings can be generalized more clearly, taking 3 loci as A, B and C in linear order and designating the numbers of 4 types of test cross progeny (ignoring the reciprocal classes *i.e.* taking two reciprocal classes as one class) as: N_1 for parental classes, N_2 for recombinants at region I (between A and B gene), N_3 for recombinants at region II (between gene B and C), and N_4 for double crossover classes as under:

Classes	Phenotypes and genotypes	Types of progeny	No. of progeny
1		Parental	N_1
2		"	
3		Single crossover	N_2
4		at region I (A-B loci)	
5		Single crossover	N_3
6		at region II (B-C loci)	
7		Double crossover	N_4
8		"	
Total			N

The *map distance* between two loci is calculated as under;

$$\text{A-B loci} = \frac{N_2 + N_4}{N}$$

$$\text{B-C loci} = \frac{N_3 + N_4}{N}$$

$$\text{A-C loci} = \frac{N_2 + N_3}{N}$$

The $\dfrac{N_2 + N_3}{N}$ will give the maximum distance indicating that A and C loci are located farthest apart at the two extremes and B locus is located in between the two loci.

Interference and Coincidence: It is fairly characteristic that double crossovers do not occur as often as the single crossover in individual regions and also in some experiments the double crossovers do not appear at all. It may thus be concluded that once crossing over occurs, it reduces the probability of another crossing over in an adjacent region of chromosome. Therefore, the occurrence of crossing over in one part of the chromosome is not independent of its occurrence in other part of the same chromosome. This means that the formation of one chiasma reduces the probability of forming another chiasma in an adjacent region of the same chromosome. This reduction in chiasma formation might be due to physical inability of the chromatids to bend back upon themselves within certain minimum distance. Therefore crossing over in one region decreases the likely hood of crossing over in nearby other region. In other words, the crossing over in one region interferes (inhibits) the second crossing over in the adjacent region of close vicinity. Consequently, fewer than expected double crossover types are observed. This phenomenon of decreased probability of second crossing over in adjacent region of same chromosome due to the occurrence of crossing over at one point is called *interference* (Muller, 1916). Interference is thus the effect of a crossing over in one mathematical term, the *coefficient of coincidence*, which is simply the ratio of observed to the expected double crossovers. Therefore, *Coincidence* = $\dfrac{\% \text{ observed D.C.O.}}{\% \text{ expected D.C.O.}}$

The *frequency of double crossovers* (D.C.O.) is obtained by multiplying the frequencies of single crossovers at two regions according to the expectation of the law of probability. Therefore, the *expected double crossovers* = p (crossing over at I region) x p (crossing over at II region). The coincidence is complement of interference and their sum is equal to unity. Thus, *interference (I)* = 1- coefficient of coincidence = 1- (observed DCO) / (expected DCO). In general, the interference is more for crossing over to occur over the short distances of loci in the chromosome so that within a certain less distance there is no double crossover (resulting I = 1.0, full interference) and the coincidence is zero. On the contrary, for the farthest apart loci, the interference becomes less and with increasing distance between loci, the interference disappear entirely resulting coincidence equal to unity. The positive interference indicates that the chromosomes show rigidity (stiffness or obstinacy) and do not coil tightly to each other during meiosis.

$$\text{Coincidence} = \frac{\text{Freq. of observed D.C.O.}}{\text{Freq. of expected D.C.O}}$$

Frequency of expected D.C.O. = (Freq. of C.O. at region I) (Freq. of C.O. at region II)
Frequency of observed D.C.O. = Coincidence x Frequency of expected D.C.O.
Frequency of single C.O. (I) = Map distance (I region) – DCO %
" " (II) = Map distance (II region) – DCO %
" parental type = 100 % - (SCO at both regions + DCO)

The parental, single crossover and double crossover are each of two types and hence the frequency obtained of any type of gametes is divided into two parts equally of each type.

The data of four experiments, given under numerical examples 5.4, may be analyzed in terms of the interference and coincidence of D.C.O. simultaneously, as under:

Experiment numbers	I	II	III	IV
Gene order	sc ec cv	ec cv ct	y c f	cv ct v
Frequency of S.C.O.:				
Region I	0.076	0.102	0.339	0.064
Region II	0.097	0.084	0.121	0.132
Double C.O. (Observed)	0.0000	0.0015	0.0258	0.0055
Double C.O. (Expected)	0.0074	0.0086	0.0411	0.0084
Coincidence	0.0000	0.1744	0.6277	0.6548
Interference (1-Coincidence)	1.000	0.8256	0.3723	0.3452
Total no. of progeny	1980	5318	2555	1448

Use of genetic maps: The chromosomal basis of genes was demonstrated by parallelism of actual crossing over between homologous chromosomes (cytological evidence) and genetic crossing over (based on results of crossing of individuals). Thus, the association of genes with chromosome was provided by the fact that there is an interchange of genes by crossing over during meiosis when some material between two homologous chromosomes is interchanged.

(1) The chromosome maps based on linkage and crossing over have already established the validity of the following facts:

- Genes are present on the chromosomes,
- Genes are located at specific loci in different chromosomes,
- Genes are arranged in a linear order on chromosomes,
- Specific genes are present in specific chromosomes.

(2) The available information on genetic maps can be used to predict the results in offspring generation produced from either dihybrid or trihybrid cross. With the help of given gene distance between two loci in a dihybrid cross, the frequency of different types of progeny expected to be produced can be predicted. Similarly, in case of trihybrid cross, the available information on genetic distance between three loci, *viz.* A-B, B-C and A-C, can be used to predict the frequency of different types of progeny expected to be produced *viz.* parental types, single crossover types at two regions and double cross over types.

(i) **Dihybrid cross:** The map distance between two loci can be used to find out the expected results from any type of mating with the help of gametic checker board. In a dihybrid cross, two parental types and two recombinants (crossovers) are produced. Further, each of two parental types is expected to be produced with equal frequencies and same is for the two recombinant types which will also be produced n equal frequencies.

(See solved example 5.2)

(ii) **Trihybrid Cross:** In a three point test cross experiment (3 linked loci), 4 classes are produced and each of which is further of 2 types *viz.* two parental types, two crossover type for region I (between A and B locus), two crossover type for region II (between B and C loci), and two double crossover types, each of two types within all the 4 classes will occur with equal frequency. The results of the offspring generation can be predicted, provided map distances, type of mating and coincidence or interference are known.

(See example 5.3)

Chapter 6

Sex Controlled Inheritance

The inheritance of some characters is controlled by sex in one or the other way. The expression of these characters is either under the control of genes present on sex chromosomes or under the control of sex hormones. The sex chromosome carry genes which control sexual characters (masculine and feminine characters) as well as other characters not related to sex of the individual. The sex controlled inheritance may be categorized as sex influenced characters, sex limited characters and sex linked characters.

6.1 SEX INFLUENCED CHARACTERS

These characters are controlled by such genes whose phenotypic expression in homozygous condition (either dominant or recessive genotype) is similar in two sexes but *the gene expression in heterozygous state is different in two sexes*, in such a way that one allele is dominant in males but it is recessive in females. Thus, the nature of interaction between alleles (expression of heterozygous genotype) depends on the sex of the individual. The allele recessive in males is usually dominant in females. The presence of sex hormone has an influence on gene interaction and hence on the development of the character in the individual. Such genes are present on autosomes, so both the sexes are alike in genetic constitution (same genotypes) and give rise to a sexual dimorphism. These characters are expressed in both the sexes.

In general, secondary sex characters *e.g.* body size, appearance, temperament are the examples. The best known sex influenced characters are the inheritance of horns in sheep, Mohogany colour pattern in Ayreshire cattle and beard in goat. The other characters are white forelock, absence of upper lateral incisor teeth, one type of skin abnormality (ichthyosis), shuttering, baldness, short index finger in human. These characters occur more frequently and more severely among men than women. The inheritance pattern of some of these sex influenced characters is as follows-

(i) **Horns in Sheep**: Dorset breed of sheep are always horned in both sexes whereas Suffolk, Shropshire and Southdown breeds of sheep never bear horns in either sex. The horn condition is controlled by dominant gene, H. The crossing of these two breeds irrespective of the way the cross was made (reciprocal cross) produces F_1 with horns if male but without horns (hornless or polled) if female. Thus, the F_2 ratio is 3 horned: 1 polled in males whereas 1 horned: 3 polled among females. This is because the gene for horn is dominant in males but recessive in female. The heterozygous genotype in the two sexes behaved differently in their phenotypic expression by producing horn in males but no horn in female. Further, it was observed that the horns of F_1 heterozygous males are much shorter not only to rams of horned breed but even shorter than the ewes of Dorset breed. This indicated that polled condition is dominant but incompletely. The crosses can be represented as:

Parents		Dorset (horned)	x	Suffolk (polled)	
		HH	↓	hh	
F_1		Hh (Horned if male, Polled if female)			
F_2	Genotypes	Sex	HH	Hh	hh
	Phenotypes	Males	Horned	Horned	Polled
		Females	Horned	Polled	Polled

(ii) **Coat colour in Cattle**: The coat colour in Ayrshire cattle is also a sex influenced character. The Ayrshire cattle have red and mahogany spotting on a white background. The mahogany is a very deep red, almost maroon and found in Maine-Anjou and Ayrshire cattle. The red colour is observed in Hereford and Shorthorns. The white spotting is almost always associated with red or mahogany colours, though the inheritance of spotting is independent of red or mahogany.

The Ayrshire cattle of both the sexes with MM genotype are mahogany for coat colour while with *mm* genotype are red, but the heterozygote with Mm genotype behaved differently in the two sexes, the males are of mahogany spotting while the females are of red spotting. The gene for mahogany colour is completely dominant in males but completely recessive in females. On the contrary, the gene for red colour is dominant in females but recessive in males. This has been shown below:

Parents		Mahogany	x	Red	
		(MM)	↓	(mm)	
F_1		Mm (Mahogany if male, Red if female)			
F_2	Genotypes	Sex	MM	Mm	mm
	Phenotypes	Males	Mahogany	Mahogany	Red
		Females	Mahogany	Red	Red

(iii) **Baldness in men**: The pattern baldness in men is a familiar example of sex influenced character. The baldness may be genetic or environmental. It may

occur due to some disease, thyroid defects or radiation and also it is found to be an inherited character. The condition of getting hair thinner progressively on the top of head until little or none is left is due to a single mutant gene, B. This gene for pattern baldness (B) exhibits dominance in men but acts recessive in women. The pattern baldness develops in men when the gene is in the homozygous (BB) or heterozygous (Bb) condition and in women it will develop only when homozygous (BB). Thus, the individuals of both sexes with genotype BB are bald and with genotype bb are non-bald (normal hair) whereas the heterozygous men (Bb) are bald but heterozygous women are normal (non-bald). The cross will be as:

Parents	Bald		x	Non bald	
	(BB)		↓	(bb)	
F₁	Bb (Bald if male, Non bald if female)				
F₂	Genotypes	Sex	BB	Bb	bb
	Phenotypes	Males	Bald	Bald	Non-bald
		Females	Bald	Non-bald	Non-bald

The expression of the gene as a character is related to the concentration of male hormone when the expression of the heterozygous class is altered by the cellular environment (hormone concentration), this means that the gene B gets its effect in association with male hormone. This is supported by the evidence that the women after developing a tumor in the adrenal cortex became bald, developed a beard condition and also a deep voice because adrenal cortex normally produces a small amount of male sex hormone, bur after surgical removal of tumor the hairs on head appear, beard condition disappear and female voice resumed. Likewise, under male hormone therapy, the women suffer thinning of the hair on the head and growth of hair on the upper lip appears. The normal baldness in women with genotype BB may be due to small amount of male sex hormone present in the blood stream of women and this amount is sufficient to cause baldness in BB women but insufficient to cause baldness in Bb women.

(iv) **Index finger in men:** The gene for short finger (S) is dominant over long (s) in men but the dominance of gene is changed in women. The expression of the character (finger size) is such that:

SS genotype = Short index finger in men and women,

Ss genotype = Short index finger in men but long finger in women,

ss genotype = long index finger in both men and women

(v) **White forelock in men:** The gene W producing white forelock acts as dominant in males but recessive in females.

WW genotype = white forelock in men and women

Ww genotype = white forelock in men but normal in women

ww genotype = normal forelock in men and women

6.2 SEX LIMITED INHERITANCE

Some characters are an extreme type of sex-influenced characters. However, *these characters are expressed only in one sex* in contrast to sex influenced characters and hence called as sex limited characters. The genes controlling these characters are present on autosomes and hence both the sexes are alike in genetic constitution (same genotype) but their phenotypic expression is different in two sexes, being limited to only one of the two sexes, depending on the presence or absence of one of the sex-hormones. These characters give rise to a sexual dimorphism. These genes control the expression of primary and secondary sexual characters and their action is clearly related to the sex hormone.

The genes, which express themselves in the presence of certain sex hormones, are considered to be sex-limited. For example, the genes for *beard, size and shape of the penis* are present in both the sexes but the genes are expressed only in males. Likewise, the genes for *breast development* in human beings and *udder development* in farm animals are present in both the sexes but their expression is limited in females only. *Milk production in mammals* and the *egg production in birds* etc. fit beautifully in this category as the expression is limited to females of the species in spite of the fact that milk production and egg production genes are carried on chromosomes of bulls and cocks also. The expression of these characters is dependent on the development of certain anatomical structure which provides the particular product (semen, milk, egg). Men have *beard* but women normally do not, though some abnormalities in hormone secretion may cause appearance of beard in women and breast development in men.

The other *secondary sex characters* include not only physical traits (penis, breast, udder, ovary) but differential response of the nervous system depend upon gene whose expression is sex limited.

Creamy Puff Plumage colour in fowl: Morejohn (1953) had demonstrated that the adult plumage of red jungle fowl, the ancestor of domestic chicken, showed sex limited pattern of inheritance for plumage colour. The adult females of recessive genotype **(ss)** develop a *creamy buff plumage*, whereas the males of ss genotype develop the normal plumage (wild type) indistinguishable from that of the males which have SS or Ss genotype. The *creamy buff plumage* is limited to only in females. The female sex hormone is necessary for creamy puff plumage gene to express.

Genotypes	Sex	SS	Ss	ss
Phenotypes	Males	Wild	Wild	Wild
	Females	Wild	Wild	Creamy buff plumage

Plumage type (cock feathering) in fowl: The *cock feathering* is sex limited character controlled by single pair of autosomal genes. The plumage type (feathering) of the two sexes in most breeds of poultry is strikingly different. The feathering pattern is hormone dependent. The cock feathering is being a showy plumage represents larger comb and wattles and long, pointed curved, fringed feathers on neck, tail (sickle feathers), is expressed only in males and produced by recessive allele h. The *hen feathering* (short and round feathers without a fringe) is expressed in both the sexes and produced by dominant gene H. Thus, the males are of two types (hen feathered and cock feathered) but all females are hen feathered. The hen feathering gene (H) is dominant over cock feathering gene (h). Therefore, hen feathering genotypes can be HH or Hh. The chicks of the genotypes HH and Hh of either sex are hen feathered, whereas the homozygous recessive (hh) are hen feathered if females but cock feathered if males. The *cock feathering is produced by* *homozygous recessive genotype (hh) limited to the male sex* and also requires the presence of male hormone. Thus, the male hormone is apparently necessary for the cock feathering gene to express. This indicated that the poultry breeds having different plumage in the two sexes (Hamberg, Campines) are homozygous recessive (hh) while other breeds (Sebright) having hen feathering in both sexes are homozygous dominant (HH). The hybrids of these two types of birds will express their feathering pattern according to the genotype and sex (presence or absence of male sex hormone). In these crosses, all the females are hen-feathered irrespective of the genotype but the males are hen-feathered if carry H gene and cock-feathered if carry hh genotype. *In other words*, the gene for hen feathering (H) is dominant over cock feathering gene (h) in the presence of any sex hormone and thus the allele H causes hen feathering. The cock feathering is limited only in males and thus the recessive gene (h) in homozygous condition (hh) produces cock feathering in the presence of male sex hormone but produces hen feathering in the presence of female sex hormone.

Genotypes	Sex	HH	Hh	hh
Phenotypes	Males	Hen feathering	Hen feathering	Cock feathering
	Females	Hen feathering	Hen feathering	Hen feathering

The expression of the gene H and h depends on sex hormone. The experimental evidences indicated that removal of testes of male with hen feathering or the ovary in female results in the production of cock feathering (Donn, 1929). Thus cock feathering occurs even in the presence of H gene and in the absence of sex hormone either due to removal of gonads or destruction by disease. This indicated that the hen feathering gene H appears to be dominant over cock feathering in the presence of either sex hormone. The cock feathering is produced by the genotype hh only in males or in ovariectomised females. The allele h thus does not prevent cock feathering but it can not express itself in the presence of sex hormone. The hh genotype leads to cock feathering in castrated chickens of either sex, and if the castrated chickens are treated with male hormone the cock feathering is developed while treatment with female sex hormone develop the hen feathering. This indicates that feathering pattern is hormone dependent.

Colour pattern in Butterfly: The white colour in butterfly is a sex limited character, expressed only in females (Gerould, 1911). The males are yellow but females are of two types – yellow and white. The white colour appeared in the female only and depends on a dominant gene (W) so that the females of the genotype WW, Ww are white and of ww are yellow, but the males of all the 3 genotypes are yellow. Thus, males are always yellow. The *white colour is a sex limited character* which depends upon the genotype (WW, Ww) and also requires the presence of female sex hormone. The mating of yellow male (WW) with yellow female (ww) produces yellow F_1 male but white F_1 females, and the F_2 ratio with all yellow males but 3 white: 1 yellow female as:

Genotypes	Sex	WW	Ww	ww
Phenotypes	Males	Yellow	Yellow	Yellow
	Females	White	White	Yellow

Similarities and differences in sex influenced and sex limited characters

The similarity lies in that the genes for both kinds of characters are present on the autosomes and hence the genetic constitution (genotype) is same for both the sexes. The difference between two types of inheritance is that the dominance of gene is sex dependent for sex influenced traits in a way that the expression of gene in heterozygous genotype is determined by the sex hormone. On the contrary, in case of sex-limited inheritance, the gene dominant in the male is also dominant in female. The effect of sex hormone is different in two types of inheritance, in a way that sex hormone determines the gene interaction (dominance or recessive alleles) for sex influenced inheritance while the sex hormone, in case of sex limited inheritance, determine expression of the trait (*i.e.* whether the trait will or will not be expressed). Thus, the expression of any genotype for the sex limited trait is determined by the sex hormone.

6.3 SEX-LINKED INHERITANCE

The *genes present on the sex chromosome* (X chromosome in mammals and Drosophila or Z chromosome in birds and moth) are said to be *sex-linked genes*. The inheritance of these genes is called as the *sex-linked inheritance*. The characters controlled by the sex-linked genes are called as *sex linked-characters*.

Inheritance pattern of sex-linked characters: The genes present on the X chromosome in mammals are represented twice in females (XX) but once in males (XY). Therefore, recessive characters of this type are expected to express more often in males than in females. This is because that the recessive character can express itself in a female if the recessive gene is present on both the X chromosomes (double dose of gene is required) while the character in males can be expressed with single dose of gene. A female parent transmits its genes to both of her male as well as female offspring but the male parent transmits its genes only to his female progeny and not to his male offspring because the Y chromosome, which has no gene, is transmitted to male progeny by male parent. *These characters are expressed in both the sexes but the expression of the character depends on the sex of the parent* in which the character was introduced in to the cross. In case of the characters controlled by autosomes, the sex of the parent has no effect on the appearance of the character in the progeny. As a consequence, the reciprocal crosses for autosomal genes give similar results but *the results of reciprocal crosses are different for sex linked traits.* This means that a given character brought in by the male or female parent produces different results for sex linked characters. This is because the two members of sex chromosomes are not homologous either morphologically or genetically and so there is no crossing over between sex chromosomes (X and Y).

Complete vs incomplete sex-linked genes: The linkage of genes may be complete or incomplete, depending upon the presence of genes on the homologous part or non-homologous part of X and Y chromosomes. Some part of X and Y chromosome are identical (homologous) having homologous genes and called as homologous region whereas other portion of X and Y chromosomes are non-homologous.

The synapsis and crossing over takes place between homologous parts of the two chromosomes (X and Y) and therefore, genes present on homologous parts of the two chromosomes do not always inherit along with other genes of their respective chromosomes. Such genes are called as *partially sex linked*. In contrast to this, the genes located on non-homologous parts of X chromosome and inherit together due to no crossing over between non-homologous parts are called *completely sex-linked genes*. The genes for haulandric inheritance are completely sex-linked because of their presence on differential parts of chromosome having no homologous genes. Thus, the Y-linked genes inherit along Y chromosome and hence expressed only in males. Therefore, the Y linked (haulandric) genes and the sex-linked (X-linked) genes present on non-homologous part (of X chromosome) are known as *sex-linked genes* and their inheritance is called as *sex linkage*.

Characteristics of sex-linked recessive and dominant characters

The sex linked genes show dominance-recessive relationship and the co-dominance among the alleles affecting a character. The dominance of alleles changes the inheritance pattern of the sex-linked characters.

A sex linked recessive gene (character) expresses itself in the following way:

(i) The incidence of sex linked recessive trait is *more in males (q)* than in females (q^2). The proportion of male recessive to female recessive is 1:q if the recessive gene is more. The red-green colour blindness in humans is due to sex linked recessive gene. When the sex linked recessive gene is very rare (low frequency), its frequency in females (q^2) will be too low as there will be hardly any homozygous recessive female at all and the recessive gene of the female will remain hidden in the heterozygous condition. The hemophilia is a sex linked recessive trait and it is extremely rare in women for two reasons- one is the low frequency of the gene and second is that homozygous females are not viable.

(ii) The sex linked recessive traits express themselves showing the *crisscross inheritance* (To *skip a generation*). The process of skipping of generation is called as *crisscross inheritance or skip- generation inheritance*. The skip generation inheritance occurs when the female is homozygous. But when the female is carrier (heterozygous) the recessive trait appears in father as well as in sons.

(iii) The females do not express the trait unless it was expressed in the paternal parent.

A dominant sex linked gene and hence sex linked dominant character expresses in the following ways:

(i) The carrier female will express the trait. The *female dominant are always more numerous* than male dominants. Thus the dominant trait is mostly present in females than in males. This is in the contrary to the sex link recessive characters that the women have twice as many chances as men of receiving the gene for the reason of having two X chromosomes.

(ii) There is *no skipping of generation* for sex linked dominant traits. It is expressed in all female progeny of a male which express the trait. The ocular albinism in man is controlled by partially dominant gene.

(iii) If the female parent does not express the trait, it is not expressed in male progeny from the mother.

Morgan and his associates gave the concept of sex-linked inheritance, working on the inheritance pattern of eye colour in Drosophila, concluding their experimental findings that gene for eye colour is located on the X-chromosome and hence sex linked. The best known and earliest example of sex-linked character in men is the *haemophilia* known as bleeding disease or bleeder's disease, reported only in males in the royal family of Spain. The examples of sex-linked genes in various species of animals with their details have been given as under:

Sex linkage in Drosophila: Morgan (1910) first time noted that reciprocal crosses in Drosophila produced different results in F_1 and F_2 progeny. He found that white eye was recessive to red eye. He made the *first cross* by crossing the red eye female (wild type - dominant) to white eye (recessive) male and observed that all the F_1 individuals of both sexes were red eyed. But F_2 data did not fit a 3:1 ratio of red: white eyed. Interestingly all the F_2 white eyed flies were males whereas among the F_2 red eyed progeny both the males and females were observed. There was a deficiency of recessive white eyed males among F_2 due to their low survivability. All the F_2 females were red eyed with no white eye whereas the F_2 males were red as well as white in a 1:1 ratio.

The *reciprocal cross* made by changing the sex of parents exhibiting the character (*i.e.* red males were crossed with white females) gave quite different result. Among the F_1 progeny, all the females were red eyed and all the males were white eyed. When these F_1 were crossed among themselves (*inter se* mating of F_1), the F_2 progeny had red and white eyed individuals in about equal numbers in both the sexes. Thus, the results of reciprocal for F_1 as well as for F_2 generation were different to the first cross. The results of two crosses have been shown digrametically as:

I. Original cross	II. Reciprocal cross

Parents X^+X^+ x X^wY X^wX^w x X^+Y

Red female   ↓   white male   white female   ↓   red male

F_1 X^+X^w : X^+Y X^+X^w : X^wY

       Red female   red male   Red female   white male

F_2 X^+X^+ : X^+X^w : X^+Y : X^wY X^+X^w : X^wX^w : X^wY : X^+Y

     Red females   red   white   red   white   white   red

              male   male   female   female   male   male

        2459   1011   782   ¼   ¼   ¼   ¼

            3470

Morgan postulated that Y chromosome is inert and therefore the genes on the single chromosome of a male are expressed even if they are recessive. On the other hand, female have two X chromosome and hence the females express the recessive trait only when homozygous for recessive gene. The conclusions drawn were:

(i) The white gene (w) is present on the X chromosome (Sex linked character) and Y chromosome had no allele, hence white gene will be transmitted from the male parent to only females of F_1 and the males will receive the Y chromosome,

(ii) The white eyed females had gene for white eye on both its chromosome (X^wX^w),

(iii) The white eyed males receive X chromosome with *w* gene from mother and Y chromosome (no gene) from father, producing white eye in all the males,

(iv) The female progeny received one X chromosome with w gene from mother and the other X chromosome with dominant W^+ gene from father,

(v) One X chromosome having red allele will come from the red female parent. Thus, the F_1 female will be heterozygous. This heterozygous female will transmit X chromosome to F_2 males which will result 50 % of F_2 males as white eye.

These results clearly indicated the linkage of genes expressing two colours (presence of two colour genes on X chromosome). Moreover, the sex linked genes are transmitted from father to grandson through the daughters and never directly through sons. Thus, the *crisscross* pattern of inheritance (from male parent to F_1 females and then to the F_2 males) of recessive X linked gene was evident. Morgan called this phenomenon as sex linkage. The sex linkage is due to genes located on X chromosome and the character (eye colour) controlled by sex linked genes has no association with sex of the individual.

In *Drosophila, some other sex linked characters* controlled by sex-linked genes (completely) have also been reported. The followings are the examples:

Dominant gene (B) producing bar eye (narrow reduced eye);

Recessive gene (*v*) producing vermillion eye colour;

Recessive gene (ec) producing echinus (rough eye);

Recessive gene (y) producing yellow body colour;

Recessive gene (sc) producing scute bristles (abscences of some bristles);

Recessive gene (f) producing forked bristles;

Incomplete recessive gene (bb) producing bobbed bristles (short);

Recessive gene (cv) producing crossveinless wings;

Recessive gene (ct) producing cut wings margin and

Recessive gene (m) producing miniature wings

Sex linkage in humans: About 120 inherited characters are controlled by sex linked genes. Some of them are *red- green colour blindness, hemophilia (bleeder's disease), agammaglobulinemia, deafness, white forelock, toothlessness, oscillation of the eye ball, optic atrophy, near sightedness (myopia), defective iris, muscular dystrophy* *(pseudohypertrophic), double eye lashes (distichiasis), etc*. The inheritance pattern is similar to that for white eye colour in Drosophila.

(i) Hemophilia in humans is completely sex linked recessive character in human. It occurs only in men. The hemophilic people in the population are nearly 0.01 percent which means that hardly one person in 10,000 is suffered with this disease and taking the same frequency of homozygous females will be around 0.0001 percent which is negligible which means that one in one crore women will be homozygous. The chances of homozygous girl to be born are very rare and further a girl with severe bleeding die before adolescence. The hemophilic condition stops the blood to clot on exposing to the air. It has been seen that a small cut or injury cause continuous bleeding which may lead to death due to loss of blood as a result of excessive bleeding.

The hemophilia is under the control of a sex-linked recessive allele present on X chromosome and hence the single dose of the allele causes this ailment. On the other hand, a woman requires two genes to cause the problem. A person who has normal gene for the sex linked recessive trait can not transmit the disease (defect) caused by sex linked recessive gene. Thus, a man whose blood clots normally can not transmit the allele for hemophilia to his children, irrespective of high incidence of the disease in his family. The reason is that a man has only one X chromosome on which the allele for hemophilia may be located and therefore a normal man (whose blood clots normally) will not have gene for hemophilia. If the recessive allele is present on the X chromosome of a man, then he will bleed, even with small injury. The hemophilic girls \surviving to marriageable age, can produce normal daughters, though they will be carrier for the hemophilic gene. But theses carrier daughters on marriage with normal man transmit the hemophilic gene to half of their sons which are grand sons of the hemophilic man. But a carrier daughter on marriage to a hemophilic man can produce a hemophilic girl.

(ii) Colour blindness in human is another example of completely sex linked recessive character. It is characterized as the inability of a person to distinguish between some colours. Among various types of colour blindness, the red-green colour blindness is most common described by Horner (1876). It is caused by sex linked recessive allele which fails to produce colour sensitive cells in the ratina of eye and hence the eyes results in colour blindness.

The condition of colour blindness is more common in men than women. This is because two dose of recessive gene are required for expression of recessive character (colour blindness) in females while only one dose of recessive gene is required to express the recessive character in males. Men are not carrier. Colour blind woman always have colour blind father, and if this woman marry with colour blid man, will always produce colour blind sons as well as colour blind daughters only. Normal woman whose father is colour blind produce sons having both colour visions in equal ratio.

The *inheritance of colour blindness* (genotypes and phenotypes of parents and the progeny) can be studied from all possible marriages among the couples of different visions. There will be 6 possible combinations of marriages among women of three vision types (normal XX; carrier XX^c; colour blind $X^c X^c$) and men of two vision types (normal XY; colour blind X^cY). The progeny (sons and daughters) from marriage of persons with similar vision types (normal x normal and colour blind x colour blind) will have vision of the parental types. Now, the remaining couples can marry in 4 possible combinations. These 4 possible marriage combinations will produce the sons and daughters with genotypes and the vision types as shown below:

Parents	$XX \times X^cY$	$X^c X^c \times XY$	$XX^c \times XY$	$XX^c \times X^cY$
	↓	↓	↓	↓
Sons	XY	X^cY	XY X^cY	XY X^cY
Daughters	XX^c	X^cX	XX X^cX	XX^c X^cX^c

(iii) Some **incompletely sex-linked genes** in man are present on that portion of the X chromosome which is homologue with a portion of Y chromosome. These are the genes causing total *colour blindness, Xeroderma pigmentation, progressive degeneration of retina and nephritis (kidney disease).* These genes segregate like autosomal genes but in association with sex of individual and represent in the X and Y chromosome as allelic pairs. Therefore, the incompletely sex linked characters do not show crisscross inheritance.

(iv) The **sex linked dominant character** in man is the *defective enamel of teeth.* It occurs more frequently in women compared to men because the women have two X chromosomes but men have only one X chromosome.

(v) **Sex-linked lethal:** The gene causing *muscular dystrophy* is recessive sex-linked lethal gene. This is the condition when the muscles wear up in early age, leading to death in early age.

Sex linkage in birds: The female is heterogametic (having one X chromosome) in birds (fowl, pigeon, duck), moths, butterflies whereas the males are homogametic (having two X chromosomes, XX), in contrast to mammals and Drosophila.

(i) The **barred feathers in poultry** (white bars on top of head, feathers are banded with bars of black on white background), is a colour pattern characteristic of Plymouth Rock breed of fowl and is a *sex linked dominant* colour pattern over black or red unbarred plumage. This forms the best example. The barred is normal and dominant (B) over black (b). The crossing of barred with black male produces only barred males and black females in F_1 generation. The *inter se mating* produces barred and black individuals in equal ratio among males as well as among females. The *reciprocal cross* of black female with barred male produced all the F_1 progeny as barred (males as well as females). In F_2 generation, barred and black individuals were found in a ratio of 1:1 among females but all F_2 males were barred.

(ii) The **slow and rapid feathering in domestic chicken** (Leghorn) is another example. The difference is more apparent from 8-10 days after hatching. The slow feather chicks have no tail feathers and only very small wing feathers whereas a rapid feathering chick has tail feathers of about one inch long and wing feathers extended to the tail. The slow normal feathering (K gene) is dominant over rapid feathering (k gene), the recessive gene causes an increase in feathering rate. The rapid feathering is thus a recessive sex-linked trait.

(iii) **Silver coloured plumage in chickens** is produced by dominant gene (S) whereas recessive gene (s) produces gold coloured plumage.

Sex linked characters in other species of animals: The *hemophilia* in dogs is a sex linked recessive trait. The hemophilia is a genetic disease, caused by a deficiency of some factor that influences blood clotting reaction. The mating of normal males (AY) to carrier female (Aa) can only produce hemophilic pups. From this mating, only half of the pups produced (aY) will be hemophilic and other half will be normal (AY). On the other hand, all the females produced from this mating will be either AA or Aa and hence will be normal phenotypically.

Tortoiseshell or **calico pattern in cats:** The sex-linked locus controlling coat colour in cat has been reported. This locus controls the production of the black and yellow pigment in the coat of the cats. This is due to a pair of co-dominant alleles (B, Y) present on X chromosome. The black coat colour is due to the allele represented by B in homozygous state (BB) whereas the yellow (or orange) colour is due to the allele Y in homozygous condition (YY). The two alleles are co-dominant and hence produce another colour in heterozygous state (BY), called as tortoise pattern which is due to the presence of both black and yellow hairs, as the roan colour pattern occurs in cattle due to the presence of red and white hairs. As the locus is sex-linked, only the females are heterozygous having both the alleles (B and Y) and hence produce the tortoiseshell colour pattern. The males are hemizygous and hence can not produce tortoiseshell or calico pattern. The females on the other hand, have black (BB), tortoise (BY) and yellow (YY) coat colours.

Sex linkage is also observed in moth, similar to sex linkage in fowl. Normal dark *wing colour in currant moth* (Abraxas grossulariata) is dominant (L allele) over lacti colour (l, allele), a pale milky coloured variety.

About 20 sex linked genes have been reported in mouse.

Contribution of sex linkage: The *sex-linked inheritance* has made it clear that homogametic sex receives equal inheritance from both the parents for sex linked genes but the hemizygous sex receives only from homozygous parent. Thus the two parents for sex linked genes do not contribute equally to their offspring and hence the reciprocal crosses yield different results. The female progeny receives equal inheritance from both the parents for sex linked characters but the male progeny receives only from their mother.

The sex-linked recessive traits *skip a generation* remaining unexpressed in the F_1 generation. The genes are transmitted from a parent of one sex to the progeny of other sex in subsequent generation. A male transmit sex linked traits to his grand sons through his daughters and not through his sons and likewise the female parent transmit the sex linked character to her grand daughter through her sons and not through her daughters. This *crisscross inheritance* has the importance in establishing the relationship between gene and the sex chromosome, and to study the transmission of the sex linked disorders casing the disease.

Use of the knowledge of sex linkage: The knowledge of sex linkage is helpful to predict about the children of affected individuals which is evident from hemophilia in men. The principle of sex linkage in poultry has been practically used to determine the sex of the baby chicks. Sexing is done on the basis of marker genes for colour pattern such as barred feathers and this type of sexing is called *auto sexing*. There is another marker gene which is used for auto sexing without involving different colour in poultry. The growth of feathers is influenced by this gene. When male homozygous for the slow feathering gene are crossed with females having normal allele for rapid feathering, the females to be produced from this cross will be slow feathering because they receive marker gene (K) from their father and can be distinguished from the heterozygous males (Kk) with in few hours of hatching because males will show rapid feathering.

6.4. Y-CHROSOME HEREDITY

Though Y- chromosome is genetically inactive or inert, but some genes have been reported to be located on Y chromosome and such genes are called as haulandric genes. The word haulandric is derived from two words *viz. haul* means whole and *andros* means male. This is because of their presence only in males. In Drosophila, one gene, b, has been found on both X and Y chromosomes. The homologous recessive genotype (bb) in either sex produced shorter and slender bristles (bobbed). The traits controlled by the Y linked genes have a very simple pattern of inheritance. All the male offspring express the trait of the father whereas none of the female progeny expresses the trait. The inheritance controlled by Y linked genes is called *haulandric inheritance* and it is sex limited.

Morgan (1942) mentioned at least 5 gene loci which have alleles on X and Y chromosomes in human. A dense growth of long hair on the outer rim of ears in man (hairy pinna) is one of the characters transmitted from fathers to all of their sons and to none of their daughters (Dronamraju, 1965). The gene controlling the *hairy pinna (hypertrichosis of ears)* seems to be present on non-homologous part of the Y chromosome. This gene has high penetrance but not 100% for an occasion it may skip a generation. The same phenotypic effect like Y linked gene has another gene which is autosomal. Therefore, it requires nearly complete pedigree for positive identification.

Another example of haulandric inheritance is a type of *ichthyosis (scaly skin)* called as porcupine man. However, there are some evidences of doubt on its being Y linked (Penrose and Stern, 1958).

Chapter 7

Animal Cell and Cell Division

The *genes*, the hereditary units transmitted from one generation to next, reside in the nucleus of a cell and hence it is most essential to study the cell structure and its division into two cells. The discovery of living cells has been made possible by the invention of compound microscope by Zacharias Jansen of Holland in 1590.

The term cell has been taken from Greak word, cella which means hollow space. Robert Hooke (1665) from England used the term cell to the cavities (hollow space) observed by him in the sections of cork under microscope. The first systematic study of cell structure was published by Marcello Malpighi, ten years later, in 1675. The cell is very small, individual unit and the constituent of the body of all the plants and animals. It is therefore, the cell is known as the smallest structural unit of life, and act as the microscopic building blocks of animal body. The cell is very minute structure ranging in diameter from about 10 to 100 micra ((1 micra = 1/ 1000 milimeter).

The idea of cell theory was first given by Dutrochet, H.J., Frenchman, and others in the beginning of 19th century (1802 – 26) but Schleiden, M.L., German botanist and Schwann, Theodor, German zoologist in 1839 outlined the basic features of cell theory to the discovery of earlier workers. Negeli (1846) added an additional feature that new cell is always formed by the division of pre-existing cell. Virchow, R. (1858) supported Negeli and worked on cell division. This additional feature was essential to reject the belief of spontaneous creation of life and Louis Pasteur of France also supported this. The cell theory has now two main features: the living beings are composed of cells and that all the cells arise from pre-existing cells. This had helped to define the living being as the one which is composed of cells capable of reproducing. The cell theory is universally applicable. It holds true from the single unicellular structure of bacteria and protozoa to the higher multi cellular organisms (man, animal and plants).

7.1 TYPES OF CELLS

The living beings have either single cell in their body or composed of many cells, called as multi cellular organisms. The unicellular and multi cellular living beings are classified as prokaryotes and eukaryotes depending on the type of cell they have. The cells are of two types, mainly depending on the presence of nuclear membrane. These are the prokaryotic and eukaryotic cells.

(*i*) The **prokaryotic** cells are composed of cytoplasm and the nuclear material (DNA and RNA) remaining in the cytoplasm without the nuclear membrane. Thus, there is no well organized nucleus in these cells. The cells are nuclear equivalent and hence called as prokaryon. The karyon is a synonymous term to nucleus. There are no mitochondria, endoplasmic reticulum, Golgi body, lysosomes, centrioles, etc. in the cytoplasm. The organisms having such cells are called prokaryotes, *e.g.* bacteria, blue-green algae. The bacteria are single celled living organisms.

The bacteria are unicellular, prokaryotes, rod-like or spherical or spiral shapes, living forms found every where in abiotic (air, water, earth) and biotic (animals and plants) ecological habitats of earth. The prokaryotic bacterial cell has cytoplasm enclosed in an outer covering with three layers. The outer most is the *slimy layer* of polysaccharides, middle strong rigid *cell wall* containing lipids, proteins, carbohydrates phosphorus and some mineral salts, and an inner *plasma membrane* of lipoproteins. The cytoplasm is dense and colloidal having granules of glycogen, proteins and fat but does not have mitochondria, endoplasmic reticulum. The ribosomes occur freely in cytoplasm.

The bacterial cell has no true nucleus (devoid of nuclear envelope). There is nucleus like zone in cytoplasm which contains a single, large, circular chromosome with double-stranded DNA molecule (bacterial chromosome, called nucleoid) but has no protein and RNA around the DNA molecule like eukaryotes. The bacteria are autotrophic, saprotrophic and parasitic to plants and animals, for their nourishment.

The DNA molecule is the genetic material of all bacteria. The DNA occurs in the ribosome-free cytoplasmic regions of the bacterial cell. The hereditary material (DNA) is a single large circular chromosome (100 µ long) made up of a circular molecule of double strand DNA and called as bacterial chromosome or nucleoid. It has no protein around the DNA molecule as in case of eukaryotes. Some RNA is present. The DNA molecule is super coiled which is due to nucleoid proteins and RNA. The circular chromosome of *Escherichia coli* has about 50 or more highly twisted or super coiled loops with about 4 million nucleotide pairs and molecular weight of 2.8×10^9. The DNA is tightly coiled up in nucleiod in such a way that most of the cytoplasmic particles are excluded but without surrounding the nucleoid by a membrane.

Each bacterial cell also contains much smaller circular duplex (1 to 20 in number), DNA molecules, called as plasmids, besides the single large circular chromosome. They are similar to viral DNA in size and replicates in fixed numbers along with the bacterial chromosome. The functions of plasmids vary from being incorporated into the host cell chromosome and called episomes, to being sometimes transferred from one bacterial cell to another during conjugation.

(ii) The *eukaryotic cells* have true nucleus which means that the nuclear materials (DNA containing stainable and thread-like chromosomes, RNA rich nucleolus and nuleoplasm) are enclosed in a nuclear membrane which separates the cytoplasm and the nuclear substances. Such organisms are called as eukaryotes. The eukaryotic cells are found in most of the animals (from protozoa to mammals) and plants (algae to angiosperms). Secondly, the eukaryotes have more amount of genetic material (DNA molecule) than prokaryotes and virus. Thus, the DNA molecule of eukaryotes contains many units of genetic information in the form of chromosomes which occur as a single unit (single chromosome) in prokaryotes and virus.

(iii) The *virus* is an exception to the cell theory. There are also some other organisms which do not fit in cell theory. For example, the living substance in Paramecium (protozoa), Rhizopus (fungus) and Vaucheria (alga) is enclosed in a wall and there is no organization like that of the cell.

The virus was discovered in 1899 by Beijrnick. The virus has been taken from Latin word, venom or poisonous fluid. The *virus* is a unicellular, ultramicroscopic and infectious particle (pathogenic agent), transmitted by infection and replicated in the host cell, and causes a number of diseases in man, animals and plants, besides bacterial virus known as *bacteriophage* which parasitizes the bacterial cell. The virus is the non-living inert particles outside the host cell, regarded as a living chemical, though the biologists consider the virus as individual cells.

The virus, the simplest forms of life, has only one type of nucleic acid either DNA or RNA, called as *viral chromosome*, and enclosed in a coat of protein molecule (capsid). The viral chromosome may be single stranded or double stranded. The nucleic acid of virus and bacteriophage occurs as a single molecule of DNA or RNA which is either linear or circular formed by 1000 to 250,000 nucleotides. The viruses exist as replicating nucleic acid molecule without protein coat inside the host cell and can multiply only in the living cells of the host. However, the viruses have no plasma membrane, cytoplasm and nucleus as well as no energy yielding and no synthetic enzyme system, and their metabolic activities depend on host cell, for which a virus can not live a free mode of life but lead a parasitic mode of existence, living as intra-cellular parasites of specific host like animals, plants and bacteria and causes damage to the cellular structure of the host. Accordingly, the viruses have been given the names after the host cell to which they parasitize viz. plant viruses (parasitizing the plant cell), animal viruses, fungal viruses, bacteriophage (parasitizing bacteria), and cynophages (parasite blue green algae).

The *bacteriophages* had been the best experimental materials for molecular genetics, such as discovery of DNA as the carrier of genetic information, discovery of genetic code and protein synthesis mechanism.

7.2 STRUCTURE OF ANIMAL CELL

The animal cell has two major components, namely *cytoplasm* and *nucleus*. The cytoplasm is a major part enclosed in the cell membrane whereas the nucleus, a dark staining body, resides within the cytoplasm. The cell membrane is the outer portion of the cell. The nuclear material is separated from the cytoplasm by a thin membrane called as the nuclear membrane. A typical eukaryotic cell has thus an outer plasma membrane, cytoplasm and nucleus.

7.2.1 PLASMA MEMBRANE

The outer most layer of animal cell is called cell membrane or plasma membrane or plasmalemma. The plasma membrane is a thin between 75 and 100 Å in width and permeable membrane to liberate the cellular secretions.

The plasmalemma in plant cell is lined and surrounded by a thick rigid outer additional wall of cellulose called as cell wall which protects and provides additional strength and rigidity to the plasma membrane. The cell wall is absent in animal cell and this is the most important difference between animal and plant cell.

The plasma layer seems to be of three layers and is composed of a thin double layer of lipid molecules and a layer of protein is absorbed on both surfaces of the membrane. The plasma membrane has a number of pores lined with protein molecules. The ER and Golgi body of the cytoplasm have continuity with the pores of plasma membrane. The plasma membrane controls cellular semi permeability, absorption, excretion and secretion. It also helps in transport of substances in and outside the cell and thus controls the cell form and cell activity. The materials are transported into the cell by the process of *phagocytosis* or pinocytosis which involves an invagination of plasma membrane to form a vacuole (food vacuole or phagocytic vacuole). Likewise, the materials are thrown out by a reverse process through plasma membrane. The transport or movement of individual molecules or ions by plasma membrane is by active and passive transport. The plasma membrane also provides an interaction of cell with its surroundings and also acts as a protective function. There has been observed a relationship of plasma membrane with cell organnels like endoplasmic reticulum (ER), Golgi bodies and nuclear envelop.

The plasma membrane contains proteins, lipids and small portion of about 1-5% oligosaccharides. The relative proportion of protein and lipid in plasma membrane greatly differ in different cells, the lipids vary from 28 to 78% depending on the tissue and the organism. Proteins are the major components of all biological membranes, to provide mechanical support and to help in transportation of material in and out of the cell or an organelle. There are also other types of proteins, other than structural proteins, to function as enzymes, antigens and receptor molecules. Some proteins remain associated with membrane surface which can be easily removed from membranes and called as peripheral proteins or extrinsic proteins whereas other type of proteins make an integral part of membrane and can not be easily removed and called as integral proteins or intrinsic proteins.

7.2.2. CYTOPLASM

The portion of the cell between plasma membrane and the nucleus is called as the cytoplasm or cytoplasmic matrix. The cytoplasm is a colloidal (granular) substance, enclosed in plasma membrane. The colloidal nature of cytoplasm is helpful in viscosity changes, intracellular motions, spindle formation and cell cleavage. This is a complex mixture of different inorganic elements, and many organic molecules comprising carbohydrates, proteins, lipids, nucleic acids and the building blocks of organic substances like nucleotides, nucleic acids, fatty acids, etc. The cytoplasm is the most important part of the cell. It contains soluble protein and enzyme which forms 20-30 per cent of the total protein content of the cell. The glycolysis and activation of amino-acids for protein synthesis are performed by the enzymes present in cytoplasm. The cytoplasm also contains transfer RNA (*t*RNA), microfilaments and micro tubules. The micro tubules form the main components of mitotic spindle, cilia and flagila.

The *cytoplasmic matrix contains various cellular elements* called *"organneles"* which play important role in many vital functions of the cell. The various cell organneles are endoplasmic reticulum, ribosomes, Golgi complex, lysosomes, centrosomes, mitochondria, plastids (plant cell) etc. Someone do not consider to the mitochondria and plastids as cell organnels for the reason that they are self-replicating like nucleus and hence independent instead of organnels. The DNA is also found in mitochondria and plastids. The mitochondrial DNA and plastids DNA function similar to that of the nuclear DNA.

(i) **Endoplasmic reticulum (ER):** The ER is a double walled membrane formed of a series of membrane bounded tubules, vesicles and cisternae usually folded in layers and forms a hollow structure of a continuous system connecting the nuclear membrane and cell membrane. The double membranes of ER are separated by a narrow lumen between them and this narrow lumen is continuous into the lumen of the nuclear membrane which gives a look like ER as an outgrowth of the nuclear membrane. The ER forms the nuclear envelop around the chromosomes (nucleus) as well as it also form the new plasma membrane, after each cell division. The double membrane of ER on its outer surface is rough bearing numerous small dense, granular and rounded particles called *ribosome*. The presence of ribosomes on the outer surface of E.R. makes it granular and called as rough ER while those areas of ER which are devoid of ribosomes are called as smooth ER. The ribosomes free from ER are also found in the cytoplasm. The cells involved in protein synthesis have well developed rough ER, particularly the pancreatic and liver cells where the secretary proteins are synthesized on the attached ribosomes. The major sites of protein synthesis within the cell are the ribosomes. The smooth ER is found in glycogen rich regions.

The large amount of ER provides *mechanical support* by forming the skeleton of the cell developing a network of membranes in cytoplasm and dividing it into a number of chambers. Secondly, it also helps in *exchange and translocation* of substances from cytoplasm to outside the cell and vice versa. Thirdly, the rough ER plays a role in *protein synthesis,* due to ribosomes. The proteins, which are exported outside the cell to the place of their utilization, are synthesized by the ER. For example, the rough ER having attached ribosomes carries synthesis of secretary proteins on these ribosomes and it exports them. The protein molecules synthesized on attached ribosomes are discharged and enter into cavity of ER to be stored there or exported outside. The transport is accomplished by interacting the ER, Golgi membrane and plasma membrane. The smooth ER is involved in *synthesis of lipids* with the help of Golgi complex and in the *synthesis of glycogen*. The vacuoles located through the cells as part of ER or Golgi complex may act as reservoirs for essential secretary materials (waste products) of the cells.

(ii) **Ribosomes:** These are mostly found in cytoplasm and are the sites of protein synthesis. The ribosomes are granular, dense and round particles attached on certain areas of the outer layer of ER. The ribosomes are composed of protein and ribosomal RNA. They have several (more than 50) proteins, their formation may involve more than 50 genes and their biosynthesis involves several regions of the cell. The ribosomes are not self-replicating particles. The proteins and RNA are two major components of ribosomes in equal proportions in both small as well as in big units. A ribosome contains one small subunit (30s) and a big subunit (50s). The **s** indicates the sedimentation rate. Most of the subunits of ribosome are processed in the nucleolus and transported in the cytoplasm. There is association and dissociation of subunits of ribosomes depending on Mg^{++} ions concentration. There is positive relation between Mg++ ions concentration and association of subunits. The ribosomes are found in attached and free forms. They arrange themselves and attach on the outer surface of ER during its growth phase and they are also found free in the cytoplasm. Mostly the ribosomes are found in cluster for which these are called as polyribosomes or *polysomes* or sometimes as *ergosomes*. These polyribosomes are held together by strands of RNA, a string known as mRNA and perform two important functions to bind with mRNA carrying the coded message and secondly to combine with tRNA which brings amino-acids to the ribosomal assembly unit. The ribosome contains no lipid. The 60% RNA of ribosome is helicle (double stranded) containing paired bases.

The rRNA is synthesized by the nuclear genes. The rRNA is formed together with formation of different ribosomal proteins that are synthesized in cytoplasm. The ribosomes are the sites where the amino-acids are\assembled for protein synthesis. It is well established that function of ribosome is the *protein synthesis* but the relative role of rRNA and ribosomal protein is doubtful.

(iii) **Golgi complex:** This name has been given after its discoverer, Camillo Golgi (1822 – 85) in the cytoplasm of nerve cells. This is smooth lined membrane tubules and cav and more commonly found in cells involved in secretion. Like ER, the Golgi complex forms an extensive membrane system in the cell having a continuous lumen. This forms a distinct channel which is separated from the cytoplasm and the nucleoplasm of the cell and forms a place where material may be synthesized, modified and transported.

The form of Golgi complex varies from compact discrete granule to the dispersed filamentous reticulum, a series of double membranes. The Golgi complex may be located any where in the cell and derived from the nucleus envelop in the same manner as ER. This mainly consists of lipoproteins, low level of DNA, RNA, polysaccharides and enzymes.

The Golgi complex has a function of storage of synthetic products (proteins) of ribosomes and ER as well as secretion of certain carbohydrates. It is the primary site of synthesis of large carbohydrate molecules. It also produces secretary granules and formation of acrosome. The main function of Golgi complex is the packaging and transporting the material from the central control site to the cytoplasm, being connected with the nucleus. The reverse pinocytosis is involved in performing this function of exporting the materials (mucus secretions, thyroxin compounds, pigments, lactoprotein of mammary gland) from the cell across plasma membrane. However, its rupture may cause cell destruction.

(iv) **Lysosome:** The lysosomes are tiny, dense, spheroid and membrane bounded vesicles. The lysosome is a lytic body containing active hydrolytic enzymes which are capable of lysing (break down) of all the major constituents of living organisms and hence called as lysosome. The enzyme contained by lysosome is for intra-cellular digestion of foreign substance (bacteria, etc.), in which the material is taken into the cell, broken down and excreted. The membrane surrounding lysosome is lipoproteinic. The lysosomes are formed from Golgi vesicles. The lysosome contains various hydrolytic enzymes for the digestion of proteins, polysaccharides and glycosides.

There are four types of lysosomes depending on the function they perform *viz.* primary lysosomes, secondary lysosomes, residual bodies and autophagic vacuole (autophagosome). The lysosomes, being lytic in nature are involved in digestion of external particles (phagocytosis, which means engulf), digestion of intracellular particles (digestion of the substance of same cell), digestion of extracellular particles (reverse to phagocytosis), and cellular digestion (digestion of dead cell).

(v) **Centrosome:** These are circumscribed region of Cytoplasm. These are capable of replication. There are two portion of each centrosome and called as centrioles or diplosomes. These centrioles are attached to each other at right angles such that there is a cylinder like look of each centriole. There are 9 triplet fibers of the wall of each centriole arranging around a central axis. Each triplet fiber has three secondary fibres or tubules arranged in a line. The centrosome is formed of the centrioles and the centrosphere, the later refers to the cytoplasm at the poles of the spindle. The centriole contains DNA and RNA. The centrioles are semi-autonomous and each centriole produces daughter centiole, thus self replicating. A centrosome is originated through replication of centrioles.

The centrosome has its role by forming poles in division of animal cell while the plant cells are devoid of centrosomes. These are primarily involved in the formation of spindle fibers, cilia, etc. and found to form the meiotic apparatus and to direct the migration of the separating chromosomes during nuclear division. The centrioles separate and moves to opposite poles of the nucleus at the time of cell division, spindle is formed between them arranging the chromosomes at the equator. The chromosomes are dragged towards the pole due to the contraction of spindle fibers.

(vi) **Vacuoles:** These are the store depots of waste products, soluble pigments, excess water, etc.

7.2.3. MITOCHONDRIA

These structures now known as mitochondria were first noted by Kallikar (1880). Fleming (1882) and Altman (1890) discovered thread like and granular structures in the cell. The term mitochondria was given by Benda (1887) and derived from mitos (which means thread) and chondrion (which means granule). There are different views about the origin of mitochondria like origin from promitochondrian, by division or budding, autonomous replication and nuclear origin.

The mitochondria are small rod shape bodies bounded with a double membrane envelope. The two membranes are separate by a space of 6 -8 nm. These structures have the enzymes for respiratory metabolism and a circular DNA molecule and ribosome.

The mitochondria have 65-70 % proteins, 25-30 % lipid and a small amount of DNA and RNA. The mitochondrial DNA is not related to the nuclear DNA and involved in extra chromosomal mitochondrial inheritance. Thus, mitochondria have their own protein synthesizing mechanism independent of nuclear control and for this the mitochondria are said to be semi-autonomous. The mitochondrial DNA controls the characters like, *petite* colony in yeast, cytoplasmic male sterility in maize, etc.

The mitochondria are the sites of chemical reactions that supply energy to the living cells. These are generally known as the ***power house*** of the cell and are the source of all energy required for life. The mitochondria supply energy to the cell by the oxidation of substrate (oxidation reaction to supply electrons) and conversion of released energy (through transferring electrons that synthesize ATP) in the form of bond energy of ATP (adinosine triphosphate, energy rich compound) and catalyzing synthetic reactions to all energy consuming process of the cell.

7.2.4. NUCLEUS

The nucleus was discovered by Robert Brown (1833). A synonymous term, karyon is used for nucleus. The nucleus is an oval-shaped or roughly spherical body almost in the centre of the cell, with a diameter of 5 – 7 micra. The nucleus is known as the controlling centre of the cell for the reason of containing the chromosomes (DNA which is hereditary or genetic material) on which are located the genes which control all the vital activities of the cell *viz*. growth and reproduction of the cell and responsible for genetic differences and similarities among individuals. The nucleus may thus be considered as the heart and brain of the cell.

Generally there is one nucleus in the cell whereas its shape and size varies. Some specialized cells of eukaryotes *e.g.* red blood cells of mature mammals are devoid of nucleus, for which the word corpuscles is used in place of cells. The size of nucleus depends on the activity of the cell, being large in active cells than in a resting cell. The nucleus has three components *viz*. nuclear membrane, nucleolus and chromosomes. There is a fluid of nucleus called as nucleoplasm which is enclosed in the nuclear membrane, and contains the nucleolus and chromosomes.

(i) **Nuclear Envelope:** The nucleus is bounded by double thin membrane called as nuclear envelope or nuclear membrane, the outer membrane is continuous with ER. This separates the nucleus from cytoplasm. The nucleus is connected with the cytoplasm by pores of the diameter of 400 to 800 A°, covering about 10 % of its surface area and these pores allow thw passage of protein molecules. The nuclear envelop is only a mechanical boundary between nucleus and cytoplasm. On the

basis of the presence or absence of nuclear envelop, the organism are classified as prokaryotes and eukaryotes.

This nuclear membrane contains two lipoproteinous membranes. The outer membrane is continuous with the ER. The outer line of nuclear membrane is smooth and interrupted by pores (perforated) where the outer and inner membranes are joined and the nuclear membrane acts as a semi-permeable membrane between nuclear and cytoplasmic materials. There is a direct contact between cytoplasm and nucleus through these pores. However, the free transfer of material is not allowed by a diaphragm across each pore and hence there is selective continuity. The diaphragm is not present in the plant cell. The double membrane appears to be in active contact with the ER and cell membrane.

This nuclear membrane usually breaks down at the end of prophase in both mitosis and meiosis cell division and again formed after cell division. The nuclear membrane is broken down into vesicles which disappeared in cytoplasm and again give rise to the nuclear envelope either by the process of gathering the old vesicles together or by forming *de novo*.

There seems to be no functional significance of nuclear envelope for the reason that there is no difference between cytoplasm and nucleoplasm, there are pores in the envelope, nuclear envelope disappears during cell cycle and the *m*RNA which is synthesized in the nucleus escaped through the opening and closing of pores and reaches to the cytoplasm for protein synthesis.

(ii) **Nucleoplasm**: There is a colloidal substance filled up in nucleus and it is called as nuceoplasm. The nucleoplasm contains materials for building of DNA (phosphoric acid, both the sugars which are ribose and deoxyribose), RNA, various amino-acids and enzymes.

(iii) **Nucleolus:** The cell nucleus has a spherical, colloidal, darkly stained body which remains associated to a specific nucleolar organizing chromosome and the region of chromosome where the nucleolus is associated is called as nucleolar organizing region. More than one nucleolus may sometimes be present and called as nucleoli. The nucleolus mainly has RNA and some phospholipids. The nucleolus disappears during prophase cell division (mitosis and meiosis) and appears again during telophase. The most important function of the nucleolus is the synthesis of ribosomal RNA (*r*RNA).

(iv) **Chromosome**s: The thread-like structures in the nucleus were found and discovered by Strasburger (1875) and these structures were called as chromosome by Waldeyer (1888). The chromosomes are known as the *vehicles of heredity*. The details of chromosomes have been given in chapter 8.

7.3 CELL DIVISION

The cell of plant and animal is divided in to two cells called as daughter cells. In most of animals and plants, there are two types of cell division *viz.* mitosis and meiosis which are both dynamic processes. The mitosis cell division is a prerequisite for growth and maintenance of life whereas the meiosis is the pre-requisite for reproduction.

7.3.1 MITOSIS CELL DIVISION

This cell division occurs in somatic (body) cells and produces the daughter cells possessing the chromosome complement same to that of the mother cell. The mitosis results in multiplication of cell numbers and is responsible for development of an individual from zygote to adult stage. There is life-cycle of a cell which is completed in two phases *viz.* the interphase (resting phase) and the phase of cell division.

The *interphase* is a stage in between the division of cell when the cell is not undergoing any type of cell division either mitosis or meiosis. The interphase is thus the interval period between two successive cell divisions. It is also known as the preparatory phase when the nucleus and cytoplasm are metabolically and synthetically active. The chromosomal DNA molecule is duplicated during this phase. At the beginning of either type of cell division, each DNA molecules replicates making its exact copy. As a result of copying, two identical functional strands of chromosomes called chromatids are formed and both the chromatids remain attached with a common centromere. The period of interphase is divided into 3 phases *viz.*

(i) Initial growth period (G_1) - a resting phase for growth of young daughter cells in size;

(ii) DNA synthesis period (S) during which DNA synthesis (replication) takes place; and

(iii) Second growth period (G_2) - also a resting phase. The nuclear volume is increased in G_2 stage during which rRNA and mRNA are synthesized. The duration of different phases vary in different tissues and different organisms.

The *phase of cell division (mitosis)* is completed by two processes, *viz.* katyokinesis and cytokinesis. During mitosis, an exact copy of each chromosome is formed (duplication or replication of chromosomes), separated into two chromatids and moved to opposite poles, forming two daughter nuclei and distributed an identical set of chromosomes to each of the two daughter cells followed by constriction of cytoplasm (formation of daughter cells by division of cytoplasm). The replication and distribution of chromosomes (nuclear division) is called Karyokinesis whereas the division of cytoplasm to form two daughter cells is called cytokinesis.

(A) *Karyokinesis:* The karyokinesis is completed in the four following stages:

1. Prophase: This phase is of longest duration. The DNA synthesis is already completed during ineterphase. The chromosomes are thin, filamentous uncoiled structures in the beginning of prophase but become visible with two chromatids and one centromere in the nucleus as a result of their shortening and thickening due to coiling and addition of protein matrix on nucleo-histone fibers. Thus, the chromosomes becomes coiled, condensed (shortened) and distinct as thread like structure during prophase. Secondly, each chromosome is splitted longitudinally in to two sister chromatids and these sister chromatids are attached at the position of centromere (or Kinetochore) to the spindle tube. The nuclear membrane started breaking down and the nucleolus disappears by the end of prophase. The nuclear membrane also disappears, being converted into ER and the spindle fibers form.

2. Metaphase: Each chromosome becomes compact, has two chromatids and moves to the centre of the cell, known as equatorial or metaphase plate. The spindle tubules start appearing either at late prophase or in the beginning of metaphase and attached to chromosomes at the centromere. One of the two centrioles of the centrosome moves towards and reached to the opposite pole to form mitotic spindle by microtubules. The spindle is completely formed. The chromosomes move and arranged on a plate (Metaphase plate or equatorial plate) midway of the two poles of the spindle. The arrangement of the chromosome is such that smaller one remains towards the interior and larger ones to the periphery. The chromosomes are clearly visible at this stage and karyotype can be made, if required for chromosomal studies.

3. Anaphase: This is the shortest phase. This phase begins with the splitting of chromosome at the centromere with each half going to its respective mitotic centre (centriole) and move towards the opposite pole of the spindle. Each chromosome takes its shape depending on the position of centromere by dragging the arms of chromosome behind their centromeres, by the contraction of chromosomal fibres.

4. Telophase: This is the mitotic phase when identical set of chromosomes is assembled at each pole of the cell. During this phase, the chromosome starts uncloiling, mitotic spindle disappears, the ER forms new nuclear membrane enclosing each chromosome sets forming one nucleus at each pole, and the nucleolus reformed in each daughter nucleus.

(B) Cytokinesis: As soon as the karyokinesis (division of nucleus by replication and distribution of chromosomes) is complete, the cytokinesis (division of cytoplasm and separation into two daughter cells) begins to form two cells. The cleavage furrow in the cytoplasm is formed by the cyclosis movement of cytoplasm. This furrow deepens and pinches the cell into daughter cells. Thus, the cytoplasm divides into two equal parts (cytokinesis) across the equatorial plate, producing two daughter cells which are identical to the original mother cell.

The time taken to complete mitosis depends on the species and the environment, like temperature and nutrition, but generally takes about 20 hours. The prophase and telophase take longer time whereas anaphase takes shortest time to complete.

The mitosis cell division is of great *significance*. It is responsible for growth and development of an organism. It distributes the chromosomes equally from a parent nucleus to the daughter nuclei. Thus, it maintains an equal amount of genetic material (DNA content) in the newly formed cell. The mitosis cell division besides being responsible for growth by producing new cells, it also replaces the old and damaged tissue by new cells.

7.2.2.MEIOSIS CELL DIVISION

This is the second type of cell division during which the number of chromosomes is reduced to half. It occurs in sexually reproductive species in their sex cells (reproductive cells) to produce gametes (haploid sex-cells). Thus, it occurs prior to gamete formation. Higher plants and animals (multicellular) develop from a single cell, *zygote* which is a result of fusion of pre-existing cells of opposite sex, called as gametes. The gametes are produced by cell division in both the sexes. The cell division producing the gametes is called as the meiosis and the formation of gametes is known as gametogenesis (spermatogenesis in males and oogenesis in females). The gametes produced in the male sex, are called as male gametes whereas those produced in female sex are called as female gametes.

The process of meiosis is completed in two successive divisions called as meiosis I and meiosis II. A diploid cell undergoing meiosis is called as meiocyte. Each meiocyte is in the interphase during which the DNA is duplicated, before initiation of meiosis I. The first meiotic division is the reduction division (resulting in the reduction of the chromosome number without any division of the chromosomes) and produces two haploid cells from diploid parental cell whereas meiotic II division separates the sister chromatids of the haploid cells (produced in meiosis I division). The reduced number of chromosome obtained during meiotic I division remains constant (haploid) during the meiotic II division. Thus, the meiotic II division, is like mitotic division, occurs in both the haploid sister cells (produced from Meiotic I division) which are again divided and produce four daughter cells, each with haploid chromosome number.

There is an interphase before meiosis, like that in mitosis. The interphase also consists 3 phase *viz.* G_1, S and G_2 phases. The meiotic division starts as soon as the DNA synthesis is complete with very short or absence of G_2 phase.

The detail events or stages of meiotic I and II cell division are as under:-

MEIOSIS I DIVISION: This meiotic I division is more important and is accomplished in four phases *viz.* Prophase, metaphase, anaphase, and telophase.

Prophase I: This phase of cell division is of very longer duration because synapsis and crossing over occur during this phase. This stage has been subdivided into following 5 sub stages:

(i) **Leptotene:** This means thin thread and this stage starts after interphase. During this phase, the chromosomes look long, thin (thread-like) and uncoiled threads with bead-like structures (chromomeres) due to dense, bead-like swellings (coiling) in them all along their length.

(ii) **Zygotene:** The zygotene means yolked thread. The chromosomes move due to attraction of homologue of a pair and the *synapsis* (pairing of chromosomes) takes place during this phase. The paired chromosomes are called as bivalent. The pairing starts at any position (centromere, chromosome ends or other place) between homologous portions.

(iii) **Pachytene:** The word pachytene means thick thread. The chromosomes are shortened, coiled and thickened appearing thread-like or thick rod structures. Each pair of chromosome (bivalent) splits longitudinally into two sister chromatids (which are haploid in number) and thus, a bivalent really has four chromatids in the nucleus lying parallel to each other, called *tetrad*. The crossing over (exchange of genetic material) between non-sister chromatids takes place during this tetrad stage. The crossing over occurs due to break of non-sister chromatids caused by *endonuclease* enzyme and inter change of broken chromatid segments between two non-sister chromatids of the same tetrad. The union of broken chromatid segment

with non-sister chromatid occurs by the enzyme called *ligase*. This point of interchange (crossing over) is called *chiasma* which is seen as a X- shape figure. The chiasmata are the result of crossing over rather than a cause. The formation of one chiasma restricts and interfers the formation of other chiasma in the adjacent region. It is important here to mention that the pachytene chromosomes are longer than chromosomes of mitotic metaphase and hence useful for morphological studies.

(iv) **Diplotene:** The diplotene means double thread. The chromosome further become thick and short and homologous chromosomes start separating and this starts at centromere towards the ends (called as terminalization). The homologous chromosome are held together only at certain points along the length which are called as chiasmata. These chiasmata are the consequence of crossing over. The nucleolus starts disappearing and the nuclear membrane disappears.

Diakenesis: The only difference between diplotene ad diakinesis is that bivalent becomes more contracted at this stage. The nucleolus disappears totally. The chromosomes can be counted easily at this stage. The bivalent tend to be near the periphery.

Metaphase I: The chromosomes have more clear outlines. The spindle apparatus appears and the bivalent appears in the form of an equatorial plate.

Anaphase I: The first anaphase is the movement of chromosomes of bivalents from equatorial plate to the poles. The sister chromatids do not separate while going to the same pole. The chromosome number is reduced and each pole has a haploid number of chromosomes after anaphase I. Thus, meiosis is called as a reduction division.

Telophase I: This starts with the formation of nuclear membrane around the chromosomes at both the poles. Now there is a small interphase before start of second meiotic division.

MEIOSIS II DIVISION: There may or may not be interphase between meiotic I and II division. This division is really a mitotic division and for that it may be referred as meiotic mitosis.

The chromosomes have two sister chromatids with one functional centromere during second prophase while there is arrangement of chromosomes at metaphase plate during second metaphase. During second anaphase, the centromere splits and the two chromatids (which are now chromosomes) pass to the two poles after which the second telophase starts as well as the cytokinesis. The chromosomes assume their reticulate configuration and nuclear membrane is formed. The cytokinesis separates the two nucleus from each other with which the meiotic nuclear division is completed.

The meiosis gives rise to four daughter cells which resemble to each other for chromosome number but they differ due to exchange of chromosome portions during crossing over. The four daughter cells are haploid.

7.4 GAMETOGENESIS

The process of formation of gametes in sexually reproducing organisms is called as gametogenesis. The spermatogenesis is basically the meiosis of sex cells of gonads in animals. The production of mature germ cells begins with the sexual maturity of an individual. There is a rapid growth period after which these mature germ cells are called as *meocytes* **or** *axocytes*. These meocytes are called as *spermatocytes* in males and *oocytes* in females. This process produces 4 daughter cells from each meocytes.

1. Oogenesis: The oogenesis is completed in three stages *viz.* proliferation of primary oocyte, prolonged growth period prior to maturation division for formation of large amount of yolk and third is the maturation stage for nuclear changes to occur. There are two maturation divisions with a different nature. The first division gives rise to one large cell (secondary oocyte) and one minute cell called as I polar body. The II division divides the large secondary oocyte again into one large cell and one very small cell, the II polar body. During this time, I polar body may or may not divide into two equal cells. Thus, in all 4 cells are produced out of which the single large cell is the mature ovum and rest smaller cells (3) are the polar bodies incapable of fertilization and regenerate.

2. Spermatogenesis: The process of spermatogenesis is also completed in 3 stages equivalent to those in the formation of ovum. The spermatozoa arise from the epithelial cells that leave the tubules of the testis. All the stages of spermatogenesis can be seen from cross section of normal mature testis. There are two distinct types of cells *viz. Sertoli cells (Nurse cells)* which supply nutrition to the maturing spermatids and second are *male germ cells.*

The spermatogonia start multiplying after sexual maturity. There are two nuclear divisions after growth and enlargement of primary spermatocytes. Each spermatocytes give rise to two secondary spermatocytes after first nuclear division. These secondary spermatocytes give rise to two spermatids from every secondary spermatocyte. Thus 4 spermatids are produced from every primary spermatocyte. The reduction division occurs in the primary spermatocytes while forming the secondary spermatocytes and hence secondary spermatocytes are haploids. The secondary spermatocyte has a second division which results in the formation of spermatids. There are then morphological changes in spermatids being elongated and packing of chromatin material in the head, followed by a slender mid piece and a tail.

Spermiogenesis is the final maturation of spermatids into spermatozoa. This process of maturation begins in the seminiferous tubules and completed in the epididymis. Here it is important to note that the nuclear envelope of spermatid is double with localization of entire nucleus in the anterior area of the cell. Golgi complex, which is formed of numerous small vacuoles help in the formation of a cap called as *acrosome.* The tail is formed from the centriole.

The sperms remain for 5-7 days in the seminiferous tubules after spermatogenesis and remain attached to the Sertoli cells and some biochemical changes takes place. *Spermeation* then takes place which is the process of movement of spermatozoa to epididymis after detaching from Sertoli cells.

Significance of meiosis: The meiosis is the prime and essential requirement of sexual reproduction. The plants and animals of higher order are diploid (have chromosome numbers in pair).

The meiosis divides a diploid cell (called as *meiocyte,* a diploid cell undergoing meiosis) into 4 daughter cells (gametes) with half number of chromosomes represented by **n** and hence the gametes are haploid sex cells. The meiosis is also known as reduction cell division for reducing the genetic material to equal half. The reduction is essentially required to keep the chromosome number constant in sexual reproduction which is accompanied by fusion of sex cells, forming the zygote which is diploid having **2n** chromosomes. Without reduction of chromosome number to half in sex cells, the chromosome numbers will be doubled after fusion of sex cells in each generation, and the resulting situation will then become difficult. The chromosome numbers remain constant after fusion of sex cells (sexual reproduction) from generation to generation and it is possible only when the chromosome numbers are reduced through meiosis to half in sex cells (gametes) before their fusion.

Thus, meiosis is essential to produce sex cells with reduced chromosome number, to restore the chromosome number as characteristic of a species and it provides new genetic combinations through synapsis (crossing over) by producing the gametes with new genetic combination.

Differences between mitosis and meiosis

1. Mitosis occurs in body cells whereas the meiosis occurs in reproductive cells.

2. Mitosis requires only one sequence whereas meiosis requires two successive divisions.

3. The prophase of mitosis is of short duration without any sub-stage whereas the prophase of meiosis is of longer duration completed in 5 sub-stases.

4. The chromosomes split longitudinally into two sister chromatids during prophase of mitosis whereas in meiosis the chromosomes remain single.

5. There is no pairing of chromosome and no crossing over in mitosis while in meiosis pairing occurs, doubling of chromosome (forming tetrads) occurs in pachytene stage and crossing over takes place.

6. In metaphase stage, the chromosomes consists two chromatids and centromere of each bivalent divides in mitosis but in meiosis the chromosomes are tetrads, the homologous chromosomes are separated and the centromeres do not divide.

7. The anaphase chromosomes of mitosis are single, longer and thin whereas the meiotic chromosomes of anaphase are short, thick and bivalent having two chromatids.

8. Mitosis produces two diploid cells which are all alike and similar to parent cells while the meiosis produces four haploid cells which are all not alike and similar to parent cells due to crossing over and half number of chromosomes.

Chapter 8

Chromosomes

Karl Negali (1842) observed as rod shape bodies in plant cells and Hofmeister (1848) as darkly stained bodies in pollen mother cells. Strasburger (1875) mentioned that the thread like structures, occur as chromatin network in the nucleus, looks like beads (beaded structure) during leptotene stage of meiotic prophase. All the body cells of all the living beings contain some threadlike structures in the beaded form.

8.1 WHAT ARE CHROMOSOMES?

The term chromosome was coined by Waldeyer (1988) to the structures present in the body cells which are rod shape, thread-like and darkly stained due to their great affinity to basic dyes. The Chromosome is a Greek word, *chrom* means colour and *some* means body. Thus, chromosome means "coloured body" inside the nucleus. The chromosomes have two areas regarding their staining property, some parts stain very darkly (Heterochromatin) while other parts stain relatively lightly (Euchromatin).

The chromosome present in prokaryotic cells (bacteria) is single, large and circular with double- stranded DNA and remains lying in the cell cytoplasm as there is no nucleus in prokaryotes. The chromosomes present in eukaryotic cells remain isolated from the cell cytoplasm being enclosed in the *nucleus.* The chromosome contains two long strands, made up of DNA, being twisted forming a spiral-like structure (helix).

The chromosomes occur in all living beings with their definite number, size, shape and morphology. They maintain their morphological and physiological properties during transmission to successive generations, have special staining regions, special organization (chemical and molecular), and special functions (capability of self-reproduction and protein synthesis).

The number, shape, size, morphology, types, staining properties, chemical and molecular structure (organization) of chromosomes have been given in this chapter (8); the changes that occur in chromosome numbers and structure (chromosomal aberrations) have been given in chapter 9; the evidence that DNA is genetic material and its (DNA) molecular structure, the classical and molecular concept of genes and gene functions have been given in chapter 10.

8.2 CHROMOSOMAL THEORY OF HEREDITY

The chromosomes are considered as the physical bases of heredity. They contain the units of inheritance which are now called as *genes* to which were earlier called as Mendelian factors or determinants. The chromosomes (*genome*) are capable of transmitting the hereditary material (genes - DNA) to the next generation. Walter S. Sutton (1902) proposed *'Chromosomal Theory of Heredity'*. The salient feature of this theory are-

(i) Each diploid cell contains two identical sets of chromosomes. One set of chromosome is maternal inherited through ovum and the other set is paternal inherited through sperm. Two chromosomes of equal size and shape, one from mother and one from father, constitute a pair called as *homologous pair* of chromosome.

(ii) The chromosomes are able to retain their structural uniqueness and individuality.

(iii) The chromosome behavior at the time of gamete formation during meiosis, in terms of the mechanism of segregation of characteristics, gave the evidence that Mendelian factors or determinants or genes are located on chromosomes.

(iv) The genes (Mendelian factors) located on each chromosome play their significant role in the development of an organism from a zygote.

VEHICLES OF HEREDITY

The chromosomes have DNA (as their main chemical constituent) which is universally accepted as the genetic material to carry genetic information from parent to offspring. The chromosomes, being the carriers of genetic material (genes), are known as vehicles of heredity as they transmit the genetic material from one generation to the next. This is accomplished from the evidences that the chromosomes:

- Have the property of duplicating in a precise manner and equally dividing by mitosis so as full complement of chromosome is provided to each daughter cell,

- Behave as per expectation of heredity to make equal contribution of both the parents as a result of meiosis during which there is random mixing and crossing over providing ample variability among individuals, and

- The chromosomes and the changes in their numbers and structure (chromosome aberrations) are associated with the transmission of specific characters.

8.3 PHYSICAL FEATURES OF CHROMOSOMES

1. Length (size) of chromosomes: The chromosomes vary in their size in different organisms and in different tissues. Different chromosome pairs differ in their size (length) within the nucleus of the same cell. The size of the chromosomes is measured at meiotic metaphase when they can be stained and observed. The chromosome appear doubled (sister chromatids) connected at the centromere. The length of chromosome vary in different species, being as short as 0.25 μ in birds and fungi, 3 μ in fruit fly (Drosophila), 5 μ in man, about 10 μ in maize to 30 μ in other plants. [1 μ = 0.001mm, 1 nm = 10^{-6} mm.]. The width of chromosome at anaphase may vary from 0.22μ to 2μ.

Change in chromosome size during cell cycles: The chromosomes (*genome*), which are the nucleoprotein complex, are invisible and form an interwoven network of fine twisted but uncoiled threads of *chromatin,* during resting phase of nondividing eukaryotic cells. This chromatin is shapeless randomly dispersed in the nuclear matrix.

During cell division, the chromatin threads start condensing in compact structures by helical coiling. The chromosomes appear distinct threads during prophase stage of cell division. These chromosomes become short and compact taking definite size and shape (V, L or J shape) during metaphase and anaphase. The chromosomes appear in to a species – specific numbers. At this stage (metaphase) the chromosomes are suitable for conducting studies. In last stage of cell division (Telophase) the chromosomes again form the chromatin net after uncoiling.

2. Shape of chromosomes: The shape of chromosome is observed at anaphase stage. One of the prominent features of most chromosomes is the primary constriction (centromere). The centromere means the "central part" but it is not always in the center of the chromosome but occurs at different places on the chromosome. Thus, the position of centromere on the chromosome determines the chromosome shape. The position of the centromere and the length of chromosome are important to distinguish one chromosome from the other.

Depending on the position of centromere, the shape of the chromosome may be rod shape when the centromere is present on terminal (end) point of chromosome (*acrocentric* chromosome or *telocentric* when the centromere is located at the very end of a chromosome). It becomes difficult sometimes to differentiate between acrocentric and telocentric chromosome.

The sub terminal position of centromere produces two unequal arms of chromosome giving rise to *J shape* to the chromosome (*submetacentric*) and a medium centromere (located in the center) produces two arms of the chromosome of equal size giving *V shape* to the chromosome (*metacentric chromosome*).

3. *Number of chromosomes:* The nucleus of eukaryotes contains definite number of chromosomes with definite size and shape. The number of chromosomes in the nucleus is constant among the animals of the same species. Thus, the chromosome number has its significance in the identification of species

The *chromosomes exist in pairs.* The animals and plants having chromosomes in pairs in their somatic cells (body cells) are called as *diploid* whereas the sex cells (gametes - sperms and eggs) are haploid having only one member of each pair of chromosome. The chromosomes numbers are written as 2n (diploid number) or n (haploid number). The n indicates the haploid or gametic chromosome number (when the chromosomes are not paired e.g. the number present in the gametes or sex cells) while 2n indicates the diploid number (when the chromosomes are paired) or somatic chromosome number in an individual (the total number of chromosomes present in pairs in somatic cells, body cell). Thus, the number of chromosomes in somatic cell is called the diploid number and so the diploid nucleus contains the chromosomes in pairs. The sex cells (gametes) are haploid because only one member from each of homologous chromosome pairs is found in the nucleus of a gamete).

The two members of each pair are of equal size (length) and shape, having their centromeres located in the same position with exception of sex chromosomes. The two members of a pair of chromosomes being similar in shape, length and position of their centromere are called as *homologous chromosomes.* The *homo* is a Greek word, meaning equal or the same and *logous* means proportion. These two members of a pair of chromosome come in contact at zygotene stage and they form pair length wise throughout their length.

Sex chromosomes versus autosomes: One pair of chromosome in each somatic cell of any mammal including man, and Drosophila is known as sex chromosome, because this pair is involved in sex determination. The two members of sex chromosome may or may not be homologous (depending on the sex of the individual) and called as X and Y chromosome. The X chromosome is much larger than Y chromosome. The body cells of male contain one X and one Y chromosome (which are not homologous), whereas those of female contains two X chromosomes (which are homologous). The chromosomes other than this pair of sex chromosome are called as *autosomes.*

The eukaryotes are mostly diploid which means that each somatic (body) cell contains two sets of chromosome, one each set coming from maternal and paternal parent and these sets are comparable (in length and shape depending on the position of centromere) and called homologous chromosomes.. The chromosome numbers vary from 2n = 4 in *Haplopappus gracilis* to 2n = > 1200 in some pteridophytes.

The diploid numbers of chromosomes present in the nuclei of different species of animals have been given here as under:-

Table 8.1: Chromosome numbers (2n) in different animal species

Species	Common name	Diploid No. (2n)	Species	Common name	Diploid No. (2n)
Homo sapiens	Man	46	*Canis familiari*	Dog	78
Bos indicus & *B.taurus*	Cattle	60	*Felis domestica*	Cat	38
Bos bubalis	Buffaloe	50	*Mus musculus*	Mouse	40
Ovis aries	Sheep	54	*Drosophila*	Fruit fly	8
Capra ibex	Goat	60	*Blata germanica*	Cockroach	23
Equus caballus	Horse	64	"	female	24
Equus asinus	Donkey	62	*Felis domesticus*	Cat	38
Camelus bacterianus	Camel	74	*Cavia cobaya*	Guinea pig	64
Oryctolagus cunuculus	Rabbit	44	*Rattus rattus*	Rat	42
Gallus domesticus	Poultry	78	Rana	–	26
Chimpanzee	--	48	Grasshopper	--	24
Columbia livia	Pigeon	80	Ascaris	–	2
Sus domestica	Pig	38	*Culex pipens*	Mosquito	6
Macaca mulatt	Monkey	42			

Polyploids: There are also organisms having more than two sets of homologous chromosomes in nuclei of their cells. Such organisms are called as *polyploids*. These are mostly the plant species. In such cases, the chromosome numbers are represented by the base number designated by x. The common wheat is hexaploid which means that it contains 6 sets of homologous chromosomes with total number of chromosomes as 42. Therefore, the numbers of chromosomes are represented as:

2n = 42 or   n = 21 and

2n = 6x *or* x = 7.

4. Morphology of chromosome: The *inner part of each chromosome* contains a long double helicle structure called as deoxyribonucleic acid molecule (DNA). The DNA looks like a ladder being twisted in opposite ways at the ends. The thickness of a chromosome is generally hundred times than of DNA but the length of DNA is several hundred times the length of chromosomes. Watson and Crick gave the double helicle structure of DNA. There is string (DNA chain) on beads rather than beads on string.

The thickness of the chromosomes may be equal throughout its length or there may be some constriction (narrow part) at some place. There are thus different regions in a chromosome depending on the constriction.

(i) Primary constriction: There is a comparatively narrow part of chromosome at some place than the remaining chromosome and this narrow part is called as

the *primary constriction*. This primary constriction has a constant number and position for a specific chromosome of all the cells and all the individuals of a species, forming a feature of identification. In the middle of primary constriction, there is a spherical area called the *centromere or kinetochore*. The *centromere* produces the primary constriction in chromosome.

The centromere is either cup-like or plate-like disc about 0.20 to 0.25µ in diameter formed of non-chromatin material. The centromere has 3 layers. The first layer is the *dense layer* which is 30 – 40 nm thick with convex outer surface to which are attached the microtubules of spindle fibres to penetrate through it upto the chromatin fibre. The centriole thus helps in the attachment of microtubules of the chromosomal spindle fibers and chromosomal movement to the two poles during cell division. The second layer of centromere is the inner *less dense layer* of 15 – 30 nm thick; and the third is the *outer fibrillar material* which is on the convex face of centromere.

Another important function of centromere, besides that it helps in the chromosomal movement during cell division, is that it may help in the formation of microtubules through polymerization of tubulin (a protein used in formation of tubules) and hence the formation of spindle fibre

The number and position of centromere is variable for different chromosomes and species. The position of primary constriction gives shape to the chromosome by dividing the chromosome in two arms (see above para - shape of chromosome). The number of centromere are generally two, one on each chromatid but the number may vary from none to many.

(ii) Secondary constriction: Sometimes, *secondary constriction* is also observed at any point of chromosome. It is a lightly stained constricted region on the chromosome and associated with nucleolus during interphase and takes part in formation of nucleolus and hence known as *nucleolar organizer region*. This contains genes coding for 18S and 28S ribosomal RNA. The location of secondary constriction is constant for a given chromosome.

When the secondary constriction is present in the distal region of an arm (sub-terminal constriction region), the chromosome region beyond secondary constriction is called *satellite* which remains attached to rest of the body by a thread of chromatin. The satellite is thus the terminal part of a chromosome beyond secondary constriction and may appear as a round or elongated *knob*. Such chromosomes having satellite are *marker chromosomes* and called *SAT-chromosomes*.

(iii) Tertiary constriction: The tertiary constriction is also present on the chromosomes and this help to distinguish one chromosome from others.

(iv) Telomers: The terminal regions (tips) on either side of chromosome are called *telomers*. These are the ends of chromosomes which are round and sealed providing stability to the chromosomes for protecting their individuality. The telomers are usually repellant in action, being not attracted to the broken ends of the chromosome breakage. The chromosome ends thus can not be associated with other parts of the homologous or non-homologous chromosomes, but the broken ends may join.

(v) **Chromatids:** A chromosome has two long spiral threads, *chromatids* at metaphase stage. These two chromatids are joined at the centromere. One chromatid may have more than one *chromonema* (pl - chromonemata). Small granules are observed on chromonema and called as *chromomeres*. At the time of division of centromere in the beginning of anaphase, each of the two chromatids gets one chromatid and changes in to a chromosome.

8.4. KARYOTYPE AND IDIOGRAM

The different species of animals and plants have certain constant features regarding the chromosome number, their size and shape of individual chromosomes. The chromosomes of a species are arranged according to these features (shape, size and structure) and this arrangement is called *karyotype* of the species whereas the representation of karyotype in the form of diagram is called *idiogram*. Therefore, the group of characteristics that identifies a particular chromosome set is known as *karyotype* and represented by a diagram called *idiogram*, in which haploid set of chromosome of an organism are arranged in a series of decreasing size. These ideograms may be symmetric and asymmetric which is used as an indication of the primitive and advanced feature of an organism. The asymmetric karyotype (advanced feature) shows large differences in smallest and largest chromosome of the set and having fewer metacentric chromosomes.

8.5. STAINING REGIONS OF CHROMOSOMES

Some stains like feulgen or acetocarmine are used to stain the chromosome at prophase. On staining with basic dyes, two different regions on chromosome appear. The part which is stained is called *euchromaic region* and other which do not stain is called as heterochromatic region. The *euchromatin* contains genetic elements and forms the major portion of chromosome. It is thus genetically active portion of chromosome because of its DNA molecule synthesizes the RNA molecules.

.The *heterochromatin* is in the condensed state containing 2-3 times more DNA than euchromatin. The genes in heterochromatic region though present but are inactive because it does not direct the synthesis of RNA (transcription) and proteins. The time of replication of heterochromatin is different than rest of the DNA. The heterochromatin is of 3 type *viz. constitutive, facultative and condensed heterochromatin.*

8.6. TYPES OF CHROMOSOMES

The chromosomes are of different types depending on the type of cell they occur, their involvement in sex determination, number and position of centromere (narrow part) in the chromosome, and also depending on their bigger size (enlarge).

(i) The chromosomes present in different types of cells are called as *eukaryotic chromosomes, prokaryotic chromosomes* and *viral chromosomes.*

(ii) According to the involvement of chromosomes in sex determination, the eukaryotic chromosomes are of two type viz. *sex chromosomes* and *autosomes.*

(iii) Depending on the number of centromere, the chromosomes are called as *monocentric* with one centromere, *dicentric* with two centromeres (one in each

chromatid), *polycentric* with more than two centromeres and *accentric* without centromere.

(iv) Based on the position of centromere, the chromosomes are categorized as: *Acrocentric* which are rod shape when the centrome is present at terminal position resulting one arm very long and the other being absent; *Telocentric* which are also rod shape with subterminal centromere; *Submetcentric* which are S shaped with centromeres slightly away from centre of chromosome and *Metacentric* which are V shaped due the central position of the centromere.

8.7 SPECIAL TYPES OF CHROMOSOMES

Depending on the special structure (enlarge size) of chromosomes in some special tissues of eukaryotes, these are of two types *viz. Lampbrush* chromosomes and *Polytene chromosomes*. These are special types of chromosomes. These special chromosomes are present in some body cells of eukaryotes or during some particular stage of their life cycle.

(a) *Lampbrush chromosomes:* Flemming (1882) first observed these chromosomes in amphibian oocytes and later studied by Ruckert, J. (1892) in the oocytes of Sharks. These chromosomes are found in primary oocyte nuclei of invertebrates and vertebrates (insects, amphibians, reptiles, sharks and birds) which produce large and yolky eggs. They are also found in plants. These chromosomes are found also during the prolonged diplotene stage (prophase) of first meiotic division and in spermatocyte nuclei of *Drosophila*. The prophase stage of first meiotic is extremely extended in those animals having large yolky eggs. The oocytes during prolonged stage grow and synthesize nutrients for the future embryo. This results an enlargement of chromosome taking unusual configuration.

The characteristic feature of these chromosomes is the remarkable change in structure like loops projecting in pairs from chromomeres. A number of loops are projected out from the chromatid axis and may vary from one to nine from a single chromomere giving lampbrush appearance for which these chromosomes are called as lampbrush chromosomes. These chromosomes are bivalent each having two chromatids. There is an enormous increase in length of these chromosomes, even larger than polytene giant chromosome of salivary gland. The largest chromosome may attain a length of 1 mm as observed in Urodele amphibian and may attain length of about 5900 µ in Salamander oocyte. These chromosomes can be seen under light microscope.

The size of loops varies from 9.5µ in frog to 200 µ in newt. These chromosomes are formed during active synthesis of mRNA molecule for future use by the egg during cleavage when no synthesis of mRNA is possible because the chromosomes are actively engaged in the meiotic cell division. The loops produce the mRNA molecules of different kinds.

(b) *Polytene or salivary gland chromosomes:* These chromosomes were first observed by E.G.Balbiani (1881) in the salivary gland of dipteran species. These are inter-phase chromosomes which are so large to be visible with naked eye. The nuclei of the salivary gland of the larvae of dipterans (like, Drosophila) have unusually long and wide chromosomes which are about 100 to 200 times in size of the mitosis and meiosis chromosomes. They are also found in the cells of fat bodies of larval stage of some Diptera.

There is no cell division in the cells of salivary gland after these are formed but there is replication of chromosomes several times resulting in exceptionally giant size, called *polytene*. These chromosomes are thus multistranded chromosomes composed of a large number of chromonemata. The process of replication of chromosomes several times is called *endomitosis*. Thus, the chromatin replicates without cell division, called as *endo-reduplication* which keeps on increasing the number of chromonemata, which may be upto 2000 or even more chromonemata per chromosome. These chromosomes are formed during early larval development at the end of mitotic cell division, by continuous DNA replication for about 10 times.

These chromosomes may attain a length of 2000 μ in D. melanogaster. A polytene chromosome consists of dark coloured bands alternating with the light coloured interbands. The bands are rich in DNA and therefore darkly stained. In certain regions, the structure of chromosome is modified resulting in enlargement of bands into lateral loops. These enlargements or puffs are called as chromosomal puffs or Balbiani rings which form large rings around chromosome which are probably formed by unwinding or uncoiling of chromonemata resulting into loops which increase the thickness of chromosome providing fuzzy appearance. Some specific genes control the puff formation. These contain large amount of RNA besides DNA and protein and hence may synthesis the RNA.

8.8 CHEMICAL STRUCTURE OF CHROMOSOME

The eukaryotic chromosomes are made up of a complex substance called *nucleoprotein (nucleic acid and protein)*. The nucleic acid is an organic acid found in the nucleus of the cell (hence called nucleic acid) whereas the protein is histone and or protamine. Some amount of calcium is also present.

The nucleic acid contains *deoxyribonucleic acid* (DNA) and *ribonucleic acid* (RNA). The nucleic acid *deoxyribo-nucleic acid (DNA)* is hereditary material present in almost all living beings except in some plant and animal viruses in which RNA is the genetic material.

The chromatin contains more proteins (60 %) than DNA (35 %) and RNA (5 %). The DNA is permanent component of chromosome and is the genetic material which plays central role in controlling heredity whereas both kinds of proteins are important in regulating the gene activity in eukaryotes.

The DNA in the largest chromosome is about 4 cm long with molecular weight of 80×10^9 while in case of human chromosome the DNA varies from 1.7 to 8.5 cm long in uncoiled condition..The histones plays structural as well as regulatory role.

The nucleic acid (DNA) carries the genetic information which is expressed in terms of protein synthesis. This is because the synthetic and degradation capacities of a cell are determined by enzymes which are proteins. The enzyme is capable to change the behavior and function of any cell. The DNA thus directs the synthesis of protein.

The details of chemical structure of genetic material (the nucleic acids - DNA and RNA) have been given in chapter 10.

8.9 MOLECULAR STRUCTURE (ORGANIZATION) OF CHROMOSOMES

Each chromatid of eukaryotic chromosome contains DNA fiber which is single greatly elongated and highly folded and hence unistranded with 100 A^0 thickness. This nucleoprotein fiber is composed of a single, linear, double-stranded DNA molecule wrapped in equal amount of a variety of DNA binding proteins (histone and non-histone proteins) as well as variable amount of different kinds of RNA.

Structural proteins or packaging proteins: The DNA binding proteins are structural proteins or packaging proteins. These are bound to DNA along the length to help in packaging. The main structural proteins are the histones. The histones are of 2 types.

Histones: The first type is the histones present in eukaryotic cells. The histones are small proteins (low molecular weight) having basic amino acids (arginine and lysine) with positive charge so as to bind to DNA and responsible for packing (coiling) of long DNA molecule into nucleosome. There are 5 histones placed in two groups.

(a) Nucleosome histones: The first category is of 4 nucleosome histones (H2A, H2B, H 3 and H 4), each of which has about 102 – 135 amino acids forming the inner core of nucleosome. All the nucleosome histones are similar in all the species. One *histone octamer* or core particle is about 11nm in diameter and 6 nm in height and formed by uniting two of each of 4 types of nucleosome histones, being present every 200 base pairs.

(b) H1 histone: The second category of histones is the *H1 histones* which are tissue specific and large having about 200 amino acids. The H1 histones are responsible for packing of nucleosome into 30nm fibre and these are present once per 200 base pairs having loose association with DNA.

The histones have two functions – as structural element helping in coiling and packing of long DNA molecule and as preventing the transcription to the specific segment of DNA unless there is dissolution of histones by some molecular signals.

Nucleosomes: The *nucleosomes* are fundamental packing units of chromatin giving to the chromatin the *'beads on a string'* appearance. The nucleosome is disc shape forming about 10 nm of diameter. This nucleoprotein fiber of 10 nm forms the first level organization of chromatin observed in interphase nucleus.

Each nucleosome contains a core particle (histone octomer) and small linker DNA. The core particle has octamer of histones with two copies of each nucleosome histone (H2A, H2B, H3 and H4). The DNA strand of 146 base pairs is tightly wrapped around the core particle forming two circles. The core particle is of about 11 nm diameter and 6 nm in height. The *linker DNA* is a small part of DNA having only 4 base pairs which may vary in different species. The length of linker DNA in sea-urchin sperm may vary from 1 to 80 base pairs. One unit of histone H1 is associated with spacer DNA.

A chromatin fiber or nucleoprotein fiber of 30 nm diameter is formed by packing of nucleosomes upon one another. Now this gives an appearance "like beaded structure". The beads are nucleosomes and connected by small linker DNA. In inter phase nucleus the thin chromatin fiber of 10 nm in diameter becomes 30 nm in diameter by spiral coiling of thin chromatin with 6 – 7 nucleosomes per turn. This has solenoid type of ultra-structure of 6 – 7 nucleosomes per turn. The solenoid is packed in to another helix, the super-solenoid having 400 nm in diameter in mitotic chromosomes. There is further condensing of super-solenoid to give final shape to metaphase or anaphase chromosome.

Chapter 9

Chromosomal Aberration

The constancy of genetic material is assumed for most of the time in interpreting the breeding results based on laws of heredity. However, it is very much evident that the genetic changes take place due to a number of reasons and these changes create variation among the individuals of a population. The change may be in the chromosome or in the gene. The genetic changes, in general, are called *mutations* but more commonly used for gene mutation. The mutation can be classified further based on the cytological visibility of the change in genetic material. The change in chromosome can be cytologically visible in the nucleus whereas the change in the gene is not visible cytologically. The change in chromosome(s) either in their numbers or in their structure is called as *chromosomal aberrations*. The change in gene is called as gene *mutation* or point mutation.

9.1 CHROMOSOMAL ABERRATIONS

The chromosomal changes are of two types, *viz.* numerical changes and structural changes in chromosomes.

9.1.1. Numerical Changes (Change in chromosome numbers)

The changes in chromosome number (heteroploidy) are of two types, depending on the change in one set (pair) of chromosome (aneuploidy) or in whole set of chromosomes (euploidy).

(1) Aneuploidy: It involves only the change in single chromosome within a pair and occurs when there is a loss or addition of one (or both) chromosome of a pair, and known as hypoploidy or hyperploidy, respectively. The aneuploidy arises due to the nondisjunction during first meiotic division. This represents the absence or presence of extra chromosome number different from the multiple of the basic chromosome number and denoted as: 2n-1 or 2n+1, etc. The aneuploidy is a Greek word (*aneu* means uneven and *ploidy* means unit), indicating uneven number. The different types of aneuploidy are as:

(a) Monosomy (2n-1): This represents the loss of one or more chromosome from the complete diploid set. Such diploid organisms are called as monosomics and produces two kinds of gametes *viz.* n and n-1.The n-1 gametes in diploids are either non functional (lethal) or have reduced fertility or high mortality and hence the monosomics in diploids can not be tolerated. But in polyploids which have several chromosomes of the same type, the monosomics can be tolerated. The monosomics can be double monosomics (2n-1-1= 2n-2) or triple monosomics (2n-1-1-1= 2n-3) and can be produced in polyploids such as wheat.

(b) Nullisomy (2n-2): This represents the loss of both the chromosomes of a homologous pair of chromosomes. Thus there is a loss of single pair of chromosome. The nullisomic diploids show reduced survival, fertility and vigour but in polyploids the nullisomics survive to maturity.

(c) Trisomy (2n+1): Trisomy is the presence of one chromosome extra to the diploid chromosome. Thus, one chromosome of a pair is represented three times and this extra chromosome may be any one of the chromosome of a haploid set. The numbers of possible trisomics to be present are equal to the haploid chromosome numbers. The trisomy had been reported to change the morphology (size and shape) of Datura. The trisomy has also been observed in man producing morphological abnormalities *viz. Mongolism (Downes' syndrome)* caused in children (mentally retarded due to defective development of central nervous system and malformed with short and broad neck, short hand and feet) having 47 chromosomes instead of 46. In man, there is an extra 21 chromosome (trisomic) or due to an extra portion of chromosome 21 as a result of translocation. Trisomy for 18th chromosome has been observed with malformed ears, short hands and defective nervous system, children usually die upto one year of age. Trisomy for 13th chromosome (Patau's syndrome) has been observed with cleft-palate and many other abnormalities resulting death soon after birth. The trisomics are used for locating genes on specific chromosomes. The segregation of genes in trisomics do not follow Mendelian ratio.

(d) Tetrasomy (2n+2): The tetrasomic individuals have a particular chromosome represented in four doses in a diploid chromosome complement (thus 2 chromosomes extra in a pair of chromosome). Thus, one chromosome of diploid is present in quadruplicate. This results during meiosis as a result of formation of a quadrivalent by extra chromosome.

Abnormalities in sex-chromosomes: Klinefelter's syndrome involves the sex chromosome and found in XXY, XXXY or even XXXXY individuals which are males but they possess female characteristics and are mentally retarded.

Turner's syndrome also involves sex chromosomes with XO chromosome composition in women. Such women have some abnormalities like webbing of the neck, low-set ears, wide-set eyes, underdeveloped breasts and sterility. .

(2) Euploidy: The euploidy is the change involving entire sets of chromosomes. The diploid individual either losses or gains a complete set of chromosome. The euploidy is a Greek word (eu means even; ploid means unit, or sets of chromosomes or number of chromosome sets). The euploidy may be monoploidy or polyploidy.

(a) Monoploidy: There is difference between haploids and monoploids. The individuals having only one set of chromosomes (n) in their body cells are called monoploids while the individuals having one set of chromosomes in gametes (germ cells) are called as haploids. Thus monoploids have only one set or single basic set of chromosomes (n) whereas the haploids have half the chromosome number found in normal individual. Among diploid (2n) organisms, the monoploids and haploids are identical whereas among polyploid organisms, the haploid individuals of the tetraploids (4n) or hexaploids (6n) will have 2n or 3n chromosomes and their monoploids will contain only one set (n) of chromosomes. The haploid individuals are classified as euhaploids and aneuhaploids. The haploid individuals of polyploids are known as euhaploids while the haploid individuals of aneuploids are known as aneuhaploids.

The chromosome number of haploids can be doubled by *colchicines* treatment and the so produced individual will be completely homozygous for all genes. The homozygous diploids are produced from haploids or monoploids. This produces the experimental material. This practice has been commercially exploited in the production of tomato variety (Marglobe) and a barley variety (Mingo). The haploids are also utilized to know the role of individual chromosomes.

The haploid individuals are small in size than diploids and also produce leaves, flowers, fruits and seeds of smaller size. The haploids also have smaller pollen diameter, small stomata and small nucleus as compared to the diploids. The haploids are also comparatively weak or less vigorous, sterile for the reason of having no regular pairing partner of chromosomes during meiosis and susceptible.

The haploids arise parthenogenetically due to unfertilized eggs of diploids particularly in insects and some flowering plants *viz.* cotton and tomato. The haploids can also be artificially produced by some treatments *viz.* X-ray treatment, colchicines treatment, temperature shocks, delayed pollination, inter specific hybridization and pollen culture among which distinct hybridization and pollen culture are most important to produce haploids. The haploids have been produced in large numbers in potato, barley, tobacco, rice.

(b) Polyploidy: The organisms having more than two sets of chromosomes are called as polyploids. These individuals have one or more haploid sets of chromosomes extra in addition to the original diploid sets of chromosome. These are called as triploids (3n), tetraploids (4n), pentaploids (5n), hexaploids (6n) and so on depending on the presence of extra number of the chromosome sets over diploids.

Among animals the polyploids are observed in Drosophila, Bombyx mori, bees, wasps, Ascaris, hamaster, etc. Various kinds of polyploidy have also been reported in different body tissues of man. However, the polyploidy in animals is rare bur more common among plants, like peanuts, oats, sorghum, clover, berseem, barley, wheat, rye, corn, cotton, apple, banana, grapes, potato, plum, sugarcane, sweet potato, strawberry, coffee, tobacco, rose plant, chrysanthemum, genus Solanum.

The polyploids, depending on their source of additional chromosome sets, are of two kinds, *viz.* autoploids and alloploids.

Autopolyploidy: The auto means self and hence the autopolyploidy indicates that phenomenon which involves the multiplication of homologous chromosomes *i.e.* when the same basic sets of chromosomes are multiplied, like AAA (autotriploid), AAAA (autotetraploids), if the diploid organism is represented as AA. The autopolyploids can be produced artificially or they also occur naturally. Doob grass grows in North India is natural autopolyploid.

The polyploids are normally larger due to increased cell size and more vigorous. The leaves, flowers, fruits and seeds of polyploids are large, though the seeds are few and flowering is delayed, the cell division is slow and hence growth rate is slow. These are not resistant to frost because of increase in water content with increase in cell size. Some abnormalities, like wrinkled leaves and dwarfing have been observed in polyploids.

Allopolyploids: In this type of polyploidy, the non homologous sets of chromosomes are involved. These non homologous chromosomes come from different species. The prefix '*allo*' indicates different or non homologous.

The polyploidy often results in sterility because of genetically unbalanced recombination to the germ cells. However, the segregation of chromosomes is complete in autopolyploids and hence they are fertile. The tetraploid plants have large size pollen grains, cells of leaves, stomata, xylum etc compared to a diploid plant and hence more vigorous bearing large size fruits, seeds and flowers which are economically more important. The polyploidy helps to evolve new species. The allopolyploidy is more important in this respect because it accumulate diverse genome in the new evolved species and hence provides better adaptability and offer the chances for natural selection.

9.1.2. STRUCTURAL CHANGES IN CHROMOSOME

The changes in chromosome structure, in addition to the numerical changes, are also possible. The presence and definite linear order arrangement of many genes on each single chromosome facilitates the structural changes. The changes in the chromosome structure may be either in the total number of genes (gene loci) on a chromosome or rearrangement of genes on the chromosome. These changes are called as *chromosomal rearrangement* or *chromosomal aberrations* or *chromosomal mutations*. The structural changes occur through breaks in the chromosome, or in its subunit (the chromatid) and it results in a loss, gain, or rearrangement of particular part of the chromosome(s).

The structural changes in chromosomes depend on the number of breaks, their location and their joining pattern. There are two ends due to each break and their fate may be any one of the following three: First is that the two broken ends may reunite immediately and hence there is no change. Secondly, there is a loss of chromosomal segment which may not have centromere and the ends remain un-united. Third fate of the broken part may be that one or both ends of break may join those produced by a different break, causing an exchange.

The chromosomal aberrations may be classified in two ways. First classification is the change in the homologous chromosomes or non-homologous chromosomes. These are *intra-chromosomal aberrations* which remain confined to a single chromosome of a homologous pair and these include deficiencies (deletions), duplications (additions) and inversions, and *inter-chromosomal aberrations* when the breaks are interchanged by the non homologous chromosomes and include translocation. The second classification is the changes involved either in the number or arrangement of gene loci. The loss (deficiency or deletion) and duplication (additions) are the changes involving in the *number of gene loci* whereas the inversion and translocation are the changes involved in the *arrangement of gene loci*. Brief description of all the four types of chromosomal mutations is given here as under:-

(i) **Deficiency or deletion:** The deficiency refers the loss of a part of chromosome causing loss of gene(s). The gene loss may be of a single gene or a group of genes (block of genes) depending on the length of chromosome part to be lost. A number of factors or agents like drugs, chemicals, radiations may cause a break in chromosome. The break may occur either in one chromatid or both the chromatids of a chromosome at any time during cell division in somatic as well as in germ cells. Further, the deletion may be terminal occurring at either end of chromosome or it may be intercalary causing loss of intercalary part of chromosome and the terminal parts are reunited. Thus the intercalary deficiency is the result of two breaks followed by reunion of terminal parts.

Loss of dominant allele in heterozygous individual causes the locus to become hemizygous and the recessive allele present on homologous chromosome express phenotypically after deletion and this is called as pseudo-dominance. The deletions disturb the pairing of chromosomes and hence crossing over is impossible in a chromosome part near deficiency. This type of chromosome aberration occurs in both animals and plants. The chromosome with deficiency in plants is sterile and hence not transmitted except in rare cases. However, the deleterious effect of deficiency depends on the amount of genetic material (part of chromosome) deleted, *e.g.* small deficiency in maize is retained / transmitted. In man the deficiency is caused in long arm of 22nd chromosome and it is called *Philadelphia* chromosome. Another deletion in man occurred in short arm of 5th chromosome which is called as the *cat cry syndrome.* The deletion of large size part causes detrimental effect. The individuals having deficiency usually die before attaining maturity.

(ii) **Duplication or addition**: The duplication is the presence of same part of gene more than once in a haploid complement and known as the repeat. Thus, the duplication occurs when there is replication of one part of chromosome two or more times in a pair of homologous chromosome. The duplication may arise either by unequal crossing over in the homologous chromosome or by the attachment of a deleted part of one chromosome to the other. The repeat may be the result of attachment of the deleted part from one chromosome to the other chromosome of the pair. The bar eye (narrow eye) in Drosophila is an example of duplication. There are other examples of duplication in Drosophila without any lethal effect.

The duplications change the structure and function but do not affect the viability. These are thus less deleterious, though more frequent. The duplications increase the number of genes in the chromosome and hence more important in evolution. The additional genes are free to mutate to a new form but without having immediate deleterious effect to the organism. However, some meiotic complications may arise due to large duplication which may result in reduction of fertility and hence reduces their own probability of survival. The duplication may result in a changed phenotype which is called the *position effect*. The bar effect (narrowing of eyes) in Drosophila is greater with each added part of the chromosome more times if the duplicated parts are on the same chromosome, *i.e.* each added part narrows the eye still farther and the effect is further intensified when additional duplications occur in one chromosome. This is the position effect.

(iii) **Inversion:** A part of chromosome is inverted at 180° as a result of two breaks in a chromosome and reuniting the intercalary part in reverse order. Thus, the chromosomal parts rotate at 180 degree, *i.e.* the broken parts are reinserted in the chromosome from one chromosome to the other. The insertions are of two types depending on the involvement (inclusion or exclusion) of centromere in the inversion and called as peri-centric (inverted parts having centromere) and para-centric (inverted part having no centromere). The crossing over within the inverted part of a para-centric inversion, results a dicentric chromosome (having two centromeres) and an acentric chromosome.

The inversion results in different phenotype due to position effect since the gene changes its position or locus on the chromosome. The inversion may results in origin of new species. There is suppression of crossing over as a result of difficulty in pairing due to inversion. Thus, the inversions are called as *crossover suppressors*. This may help in maintaining the heterozygosity.

(iv) **Translocation:** The translocation indicates transfer of chromosome parts from one chromosome to another. These translocations are inter-chromosomal or heterosomal chromosomal aberrations which occur between non-homologous chromosomes. The more important are the reciprocal translocations or segmental exchange involving mutual exchange of chromosomal parts between non-homologous chromosomes. The translocation can be divided based on the part(s) of non-homologous chromosomes which is detached and reunited. A small part of a chromosome may be attached to the end of other non-homologous chromosome and called as simple translocation and caused by single break. There may be exchange of parts between non-homologous chromosomes and known as reciprocal translocation.

There may be change in morphology (appearance) of chromosome. This happened due to centric fusion and may bring a change. Secondly, the genetic polymorphism is introduced in the population as a result of translocation and important in development of new varieties. The translocation has position effect too.

9.2 MUTATION

The mutation covers all genetic changes including gene mutation and chromosome mutation. The *gene mutation* is the chemical change in gene which is called as point mutation. The *chromosome mutation* is the change in either structure of the chromosome (inversions, translocations, deletions, duplications of nucleotide sequence) or in the number of chromosomes (heteroploidy). However, the term mutation is restricted to a change at a particular locus (called point mutation or gene mutation) and thus differs from chromosome aberration.

1. Definition and Concept: Hugo De Vries coined the term mutation to a sudden and permanent change in the genome observed phenotypically.

The mutation is a change at gene level rather than at chromosome level. Mutation is a sudden and spontaneous heritable change produced by a change in the base pair of nucleotide of the DNA segment (gene) so that a new allele produced is different from the original gene in its base sequence and has an effect on the expression of a particular character.

The mutant allele has an effect different to that of the normal allele and produces an alternate phenotype of the trait. For example, in *Drosophila melanogaster*, the normal colour of the eyes is red. A white eyed male fly appeared in the normal population of red eyes. It was later on found that white eye colour is recessive to the normal red eye colour. The appearance of white eye due to recessive allele was referred to as the gene mutation. Thus, mutation provides new genetic material on which selection can operate.

A gene can change into another form which may either exist in the population or it may be a new allele that does not exist in the population. The different allelic forms of one or the other gene (A, a) arise only as a result of mutation. It is taken that one of the two alleles at a locus arises by mutation from the other.

2. Molecular pathway of mutation to occur: Some copy error may occur during replication of DNA and this copy error leads to a change in DNA structure, resulting in a new gene (DNA).

The change in DNA is due to a change in nucleotide sequence (codon) either by deletion, insertion or substitution. This change in nucleotide sequence results in a different amino-acid in a particular protein, *e.g.* in sickle cell anemia (Hb S). A codon changes from GAA (glutamic acid) to GUA (valine) and thus valine is incorporated instead of a glutamic acid residue, resulting in production of another protein (abnormal protein, Hb **S)** and another phenotypic expression of the character.

Illustration of mutation: A single change in a single base pair (nucleotide) may results in a change of gene product, *e.g.* production of abnormal human hemoglobin S, in place of Hb A and Hb B at its beta chain. There are about 140 amino-acid residues in each of alpha and beta chain of normal hemoglobin A. The sequence of amino-acids in the beta chain of Hb A is:

Position:	1	2	3	4	5	6	7	8
Aminoacid:	Valine	histidine	leucin	theorine	proline	glutamic acid	glutamic acid	lysine etc

The triplet codes for three amino-acids *viz.* glutamic acid, valine and lysine are:

Glutamic acid	Valine	Lysine
GAA		*AAA*
GAG	GUU	AAG
	GUC	
	GUA	
	GUG	

The glutamic acid at 6th position of beta chain of Hb A is replaced by valine and abnormal hemoglobin (Hb S) is produced but when the glutamic acid at 6th position of Hb A is replaced by lysine, it also produces another type of abnormal hemoglobin called as hemoglobin C which is lethal.

The replacement of first A base (adenine) of glutamic acid (GAA) by U changes the codon as GUA and translates into valine. Likewise, the G base (guanine) in codon for glutamic acid (GAA) is replaced by A (adenine base) and produces the codon as AAA which codes for lysine. Thus, a change in the single nitrogen base (nucleotide, A to U or G to A) of the gene Hb A produces the change in amino-acid (glutamic acid, GAA into valine, GUA and the glutamic acid, GAA into lysine, AAA) of the chain and hence the gene product (protein, hemoglobin) is changed which has phenotypic effect other than the original gene product.

The mutations are random, permanent and rare. However, the mutation rate varies from one strain to others, from one individual to another and from one locus to another in the same strain. There are some genes whose effect is only to increase the mutation rate of other genes (loci) and such genes are called "*mutator genes*". For example, in corn a dominant gene (Dt) on 9th chromosome is a mutator gene which mutates the colourless gene (a) present on 3rd chromosome into coloured gene responsible for producing dots in maize grain.

The mutations are recessive, lethal or semi-lethal and may occur simultaneously at two or more loci. The mutations, being deleterious to the organism, are at low frequencies in the population. This is because the individuals showing mutant trait are less adapted and hence the natural selection acts against mutations keeping their frequencies less than expected. The action of natural selection against mutant types is indicated by the observed deviations from expected ratios of phenotypes.

3. Types of mutations: The mutation may be classified in a number of ways depending upon the cell type of the body in which mutation occurs, nature or origin of cause (known or unknown causes), size of the part of chromosome affected by mutation (single nucleotide to entire gene or chromosome), different phenotypic effect, recurrence of mutation (once or repeatedly), direction of change (forward or reverse mutation) and change in nucleotide content or in location of gene (basis of mutation).

(i) Cell types of body in which mutation occur: The mutation can be somatic and gametic. When the mutation occurs in somatic cells (non reproductive cells) it is called as somatic mutations but if the mutation occurs in germ cells or sex cells, it is called as gametic or germinal mutation.

The somatic mutations are not heritable and remain limited to the organism in which they occur. Thus, somatic mutations are not genetically important. However, the Delicious apple and the Novel orange have resulted from somatic mutation. The progeny from grafts and buds have perpetuated the original mutation. Somatic mutations also occur in animals. The occurrence of hair spot in red portion of the coat of Hereford cattle is an example of somatic cell mutation.

The gametic or germ cell mutations produce permanent heritable change. These mutations create genetic variations, producing numerous varieties of plants and animals. colours in farm animals. The other examples of germ cell mutations in animals have been mentioned elsewhere in this book dealing with multiple alleles and other topics.

(ii) Origin or cause of mutation: The mutation may occur spontaneously due to unknown cause. These are called as *background mutations or spontaneous mutations.* The spontaneous mutation may result of natural radiations (gamma rays and cosmic rays produced from radium and other radio active substances), natural gases, sudden change in temperature and metabolic activities as a result of formation of certain chemical compounds. The mutation rate is increased with the increase in in age of animals and plants. The plants grown from old seeds (5-6 years) have more mutations than grown from fresh seeds.

Secondly, the mutations can be induced artificially (Muller, 1927) and hence called as *induced mutation.* The mutations are induced by giving treatment to the individuals, like radiations and chemicals called mutagenic agents. Some genes mutate under the influence of some mutator genes.

(iii) Size of mutation: A very small part of DNA *viz.* single nucleotide or nucleotide pair may be involved and called as *point mutation.* The mutation may involve more than one nucleotide pair, *gross mutation,* to the entire chromosome or set of chromosome, chromosomal aberration.

(iv) Extent of effect on phenotype: The mutated alleles may produce slightly modified phenotype or clearly identified phenotype in homozygous or in heterozygous combination with other alleles. They can only be detected by special techniques. Such mutated alleles are called as *iso-alleles.*

Some mutant alleles have greater effect on phenotype affecting the viability of organism to different extent. They are called as *lethal* which kill all individuals before adult age; *semi- lethal* which kill most of the individuals (say up to 90 %); and *sub-vital (detrimental)* which have relatively greater viability.

(v) Recurrence of mutation: The newly arisen mutant allele may be due to a single mutational event and called as *single mutant allele* or the mutational event recurred regularly called as *recurrent mutation.*

A newly mutant allele may increase or decrease in its frequency or may be lost from population by chance elimination or fixed in the population. The fate of a new single mutation depends on the population size, selective advantage and the distribution of progenies in families in which the mutant gene is arises.

The recurrent mutation results to an increase in frequency in a large population and can not be lost by sampling process but they are established in the population. The frequency of this mutant gene depends on mutation rate per generation which varies for different loci but seems to be reasonably constant for a particular locus under constant environmental conditions.

(vi) Direction of mutation: The mutation may be in a *forward* or *reverse direction.* When the wild type phenotype is changed into an abnormal phenotype due to mutation is called as *forward mutation.* Likewise, the abnormal phenotype is changed into wild type due to the reason that the mutant allele mutate back to the original form (wild type). This is called as *reverse or back mutation.* The rate of forward and reverse mutation is different.

(vii) Basis of mutation: The mutation may bring change in the nucleotide content of the gene called as *structural mutation* or there may be a change in the

location of the gene within the genome leading to position effect called as *rearrangement mutation.*

(a) *Structural mutation:* These are of two type *viz. substitution and frame shift.*

The substitution mutation involves the substitution of one nucleotide (base pair) for another in DNA. These substitution mutations are of two types depending on the type of base pair to be substituted. The substitution may be within purine bases (one purine base is substituted with another purine) or it may be within pyrimidine bases (one pyrimidine base with another pyrimidine). Such substitution is called the *transition* mutation and it occurs through deamination and base analogue. Secondly, the substitution may be between purine and pyrimidine bases called as *transversion mutation.*

The *frame shit mutation* occur when base pairs are added or deleted leading to addition of one or more extra nucleotide in a gene, called as *insertion mutation,* or the change may lead to loss of some portion of a gene, called *deletion mutation.*

(b) *Rearrangement mutation:* The location of gene is changed in the genome. The mutation may occur in the *cis or transposition* of the same gene and they may produce different effects. Secondly, the numbers of gene replicates are non equivalent on the homologous chromosome and produce different phenotypic effect. Thirdly, there may be movement of the gene locus either within the same chromosome (*inversion*) or to a non-homologous chromosome (*translocation)* and these mutation produce new phenotype when the gene is reallocated near heterochromatin.

4. Significance of mutation: The mutation is said to be a necessary evil. This statement can be supported from the fact that the mutations are important in causing genetic variation for selection to work and for evolution to occur but on the other hand, the mutations are harmful. However, the mutations due to their low rate play a relatively minor role as a factor to disturb the genetic equilibrium.

(1) Mutation is necessary: The mutations are of biological importance as they lead to the *biological evolution.* In the absence of mutation the new genetic variation will not arise, the members of a species will remain alike and the new species would have not come up without mutation.

Moreover, the mutations are essential for *biological existence* which depends on change and not by maintaining the *status quo* under the changed environmental conditions. The organism would cease to exist or would have remained in very simple form in the changing environment. The absence of mutation would eliminate the genetic variability which is essentially required for adaptation of any species in a changing environment. Thus, in the absence of mutation the possibility of evolution through natural selection favouring hereditary alternative forms is eliminated.

There is a reference in Holy Bible about a man of great stature born to the giant having polydactyly (6 fingers) on each hand and foot. This character is dominant originating from a mutation affecting finger numbers. In cattle mutation occurred in gene controlling horn (p) and the gene mutated to its allele of polled (P, without horn). The polled trait is dominant over horn. Short leg sheep of Ancon breed is also the result of mutation in gene controlling normal leg size. Likewise, different colours in farm animals have been originated as a result of mutation in gene for original colour.

As a result of several mutations in the same gene, various alternative forms of a gene (multiple alleles) occur in the population and produce different phenotypic effects. It is generally agreed that more than two alternate forms of a gene (multiple alleles) are present in the population for a number of loci. These multiple alleles are created by mutation. The creation of a new allele causes change in the existing genetic structure of the population by increasing the number of genotypes.

The mutation is the sole and primary and ultimate source of genetic variation of a population while genetic recombination, migration and selection are of the secondary level. The mutation leads to genetic differences and *provides the raw material* for selection and evolution. The occurrence of mutation consistently in every generation over a long period of time gradually changes the gene frequency in a population. All the local populations of a species would have been identical without mutation. The role of secondary level factors to give rise to genetic differences starts only if there are allelic differences at a number of gene loci which arise due to mutation. This is because without the existence of mutational variability at individual locus there is nothing different to recombine. Likewise, migration is only effective to create genetic differences in a population only when there is genetic variability between migrants and native population. Thus, the mutant conditions have allowed the present evolutionary stage. The allele in a population is either got fixed or lost irrespective of the size of population though the rate of fixation and loss of allele from a large population is very small. The allele once lost can be regained through mutation. Thus mutation provides a means of genetic variability and plays a special role as the *spoiler of genetic fixation.*

(2) Mutation is harmful: The mutation exerts an adverse effect on the genetic structure of population. The mutation induced changes (mutant alleles) are mostly unfavourable because most of the mutant alleles are harmful and deleterious. The change caused by mutation depends on its degree of harmfulness and the rate at which the mutant alleles arise (mutation rate). The mutant alleles can do harm to the extent that they cause death of the individual and these mutant alleles are called as *lethal* or *semi-lethal genes.* The process of mutation and reduced viability brings down the frequency of the gene under mutation.

Higher mutation rate are harmful and dangerous for the existence of any species due to the fact that most of the mutant forms of alleles are deleterious. Therefore, if there would have been too much mutation, the species could have not continued to exist because of accumulation of so many deleterious effects of mutant alleles. Thus, the level of mutation must be less. The direction of mutation is random and unpredictable for which the mutations are usually deleterious.

However, the mutations are rare and hence their direct effect on gene frequencies is very small. It requires many generations to bring a change in gene frequencies due to mutation. The mutation rates are not constant for any locus being influenced by temperature, radiation hazards, chemical treatments etc. The mutation rates also vary between loci and also between races or species. The mutation rates have been indicated ranging between one gamete in one lakh gametes to one gamete in hundred million gametes (10^{-5} to 10^{-8}) contain a mutation at a locus. Thus, mutation alone can produce only very slow changes in gene frequency and hence they are not important. However, the effect of mutation which is seemingly minute become important and may play a substantial role in the process. Mutation at separate locus is slow but each locus is capable of mutation and hence a number of mutations are expected in every gamete each generation because the entire genome contains thousands of loci. A number of loci determine the general functions (viability, fertility, disease resistance, vigour etc.) and hence the mutations at these loci may accumulate resulting in the total genetic variation to affect these functions to a greater extent than individual locus mutants.

Chapter 10

Molecular Genetics

This chapter has been devoted to other branches of genetics *viz*. Molecular genetics, Biochemical genetics, Developmental genetics and Bacterial genetics.

10.1 MOLECULAR GENETICS

The study of the molecular structure of genetic material (gene, a part of DNA) and its functions in the phenotypic expression of characters is called the molecular genetics.

10.1.1 Genetic material - evidence and structure

The chromosomes, known as hereditary vehicles, are chemically composed mainly of proteins and nucleic acids (DNA and RNA), although other chemical constituents are present. In the beginning, it was thought that one of the two component parts of nucleoproteins (protein and nucleic acids) should be the genetic material. Thus, there had been controversy about the proteins or the nucleic acids as the genetic material.

Evidence of genetic material: Molecular biologists conducted experiments to get an evidence of genetic material. F.Griffith (1928), bacteriologist in England, conducted an experiment injecting a mixture of two strains of bacteria (pathogenic, which lost the infectivity due to their heat killing, designated as S-III versus non-pathogenic which were live designated as R-II) of pneumococcus (pneumonia causing bacteria) into mice and noted death of mice. This showed that heat killed virulent bacteria (S-III) has caused transformation of non-pathogenic bacteria (R-II) into pathogenic bacteria. This phenomenon of transferring the character (virulence) of the virulent but heat killed strain (S-III) into another strain (non-virulent, R-II) by using a DNA extract of virulent strain (S-III) was called as *bacterial transformation* or *Griffith effect*. However, Griffith could not explain the cause of bacterial transformation. Later on, Avery *et al.* (1944) repeated the Griffith's experiment and found that DNA is the transforming substance which could bring about transformation of avirulent bacteria into virulent form.

Another experiment on bacteriophage infection (infection of bacteria by bacteriophage) carried out by Harshey and Chase (1952) also indicated that DNA is the genetic material of certain bacterial virus.

The experiment was conducted by H. Fraenkel in California on tobacco mosaic virus (TMV), which contains only RNA and no DNA. The infectivity test of both RNA and protein fractions of TMV had shown that RNA alone and not the protein fraction of TMV was a cause of infection. This gave the evidence that RNA also carry genetic information. Thus, the *nucleic acids actually constitute the genetic material.* However, *DNA is always the genetic material* and RNA is non-genetic in most of the cases except where DNA is absent (virus).

Structure of genetic material (nucleic acid): Friedrich Miescher (1844-1895) isolated the nucleic acid in 1868 from the nuclei of RBC of pus and named as nuclein. The nucleic acids are complex molecules, bigger than proteins, containing carbon, oxygen, hydrogen, nitrogen and large amount of phosphorus but no sulfur in contrast to proteins.

There are two kinds of nucleic acids depending upon the type of sugar present *viz.* ribose and deoxyribose (lacking one oxygen atom). These are accordingly called as *ribonucleic acid* (RNA) in which the sugar is ribose and the *deoxyribonucleic acid* (DNA) in which the sugar is deoxyribose. The RNA is commonly found in cytoplasm as well as nucleus. The DNA is found mainly in the nucleus except in some cases (plant virus- in which RNA is genetic material and DNA is absent). The DNA in eukaryotic cells is long double-stranded spirally coiled, in virus there is a single molecule of DNA (single strand) which is enclosed in protein coat whereas in bacteria, mitochondria and plastids of eukaryotic cells it is circular lying naked in the cytoplasm.

The nucleic acid is found to associate with various proteins in combinations called nucleoproteins. Levene (1869-1940), after separation of nucleic acid from protein, found that the nucleic acid is formed of thousands of smaller parts called *nucleotides.* Thus, the nucleic acid is a long chain polymer of nucleotides (polynucleotide). *The nucleotides* contain a sugar, a nitrogenous base and a phosphate (phosphoric acid).

. The *sugar* of nucleotide is a 5- carbons sugar (pentose sugar). There are two types of sugars in nucleic acids which are named as *ribose* and *deoxyribose*. In case the sugar is lacking one oxygen atom, it is called as deoxyribose. Both these sugars are not ordinarily found at the same time in any particular nucleic acid.

The *phosphate group* of each nucleotide is attached to the sugar at its number 5 carbon position. The nucleotides in the DNA molecule are joined by linkage of chemical connection between the sugar of one nucleotide to the phosphate of next nucleotide. The sugar and phosphate groups are the constant components of all nucleotides in a nucleic acid.

A *nitrogen containing group* of nucleotide is present in association with each sugar at its number 1 carbon position. The nitrogen containing group is more variable having either one or two carbon-nitrogen rings and function as a base called as nitrogen base (hydrogen-ion acceptor) whereas the phosphate group has its acidic nature. The nitrogen bases having one carbon-nitrogen (C-N) ring are called *pyrimidines* whereas having two C-N rings are called *purines.* The two pyrimidines are *cytosine* (C) and *thymine* (T) while two purines are *adenine* (A) and *guanine* (G), found in DNA. The RNA has also four nitrogen bases which are also two pyrimidines (*cytosine* and *uracil,* in place of *thymine*) and two purines (*adenine* and *guanine*).

Nucleotides: The DNA contains four deoxyribonucleotides whereas the RNA contains four ribonuleotides and these are named according to the nitrogen base attached. The term nucleoside is used for a combination of a nitrogen base and a sugar without the phosphate group. The four nitrogen bases, nucleosides and nucleotides of DNA molecule are as under:

Nitrogen base	Deoxyribonucleoside (Base + Sugar)	Deoxyribonucleotide (Nucleoside + Phosphate)
Adenine (A)	Deoxyadenosine	Deoxyadenylic acid
Guanine (G)	Deoxyguanosine	Deoxyguanylic acid
Cytosine (C)	Deoxycytidine	Deoxycytidylic acid
Thymine (T)	Thymidine	Thymidylic acid

The nucleotides, in addition to be present in DNA, are also found in nucleoplasm and cytoplasm but with the difference that the phosphate present in DNA is in monophosphate form whereas that present in nucleoplasm and cytoplasm is in its triphosphate form. The triphosphate form is essential because during DNA replication, the DNA polymerase enzyme can act only on triphosphate of deoxyribonucleotides.

(i) Structure of DNA: The DNA molecule is present in the nucleus of the cell, extends the length of the chromosome, more or less in its center. Watson and Crick (1953) proposed the *"double helical"* structure of DNA. They mentioned that DNA molecule in eukaryotic cells is a long two-stranded structure of polynucleotide chains running in opposite directions. The two strands of DNA are twisted (coiled like a rope) to form a helix (spiral-like structure). Such a twisted structure like a rope is called helix.

Each of the two strands of DNA contains thousands of units of *nucleotides.* Thus, each strand of DNA is called a polymer of nucleotides, for the reason of being composed of many repeated units of *nucleotides.* The polymer means many parts (*poly* means many and *mer* means parts). Each nucleotide contains a special nitrogen base, the sugar (deoxyribose) and phosphoric acid, as mentioned above. In each nucleotide, the nitrogenous base is linked to a sugar and the sugars are linked together by a phosphoric acid molecule. The nucleotides in each strand of DNA are linked together by a chemical connection between the sugar (deoxyribose) of one nucleotide and the phosphoric acid of the next.

The two strands (sides) of DNA molecule are held together by relatively weak connections of hydrogen bonds between two nitrogen bases of two strands (base pair or complementary base pair). The nature of hydrogen bonding is such that adenine and thymine are attracted to each other and likewise is the cytosine to guanine. This is the *base pairing principle*. Further, the adenine can bond only to thymine by two hydrogen bonds (A to G) whereas guanine can bond only to cytosine by three hydrogen bonds (G to C), without any exception.

The sequence of bases in one polynucleotide should determine the sequence of bases in the other polynucleotide of double helix. These two strands are thus called as complementary because the base pairs are complementary. The base pairing of purines with pyrimidines results to a constant diameter of 20°A of the DNA molecule. This is because the two pyrimidines (single C-N ring) require lesser space whereas the two purines (two C-N rings) require more space to pair. The distance between two base pairs (one step) in the nucleotide chain is 3.4°A with each turn of about 34°A long and hence there are 10 base pairs in each turn. The bases of double stranded DNA has a quantitative relationship in which A = T and G = C, or A+G=C+T, or A+G / C+T = 1.0. However, it is not necessary that A+T equal G+C.

The two complementary strands of DNA are coiled in a manner so as to be separated from one another only by uncoiling allowing their ends to revolve freely. The coiling is helical (spiral) structure like a circular stair case maintaining same diameter and same width of the steps with a connecting railing on either side. The railing (backbone) of each strand (side) is composed of the phosphate-sugar linkage which are continuously repeated without change.

The *gene is a part of DNA molecule*. On an average, a gene (sometimes called a *cistron*) probably contains about 600 consecutive base pairs but this number vary from gene to gene. Thus, the gene is a structural part of a chromosome. Hundreds and possibly thousands of genes are present in each chromosome and each gene has a fix or especial position called a *locus*. Each gene has at least two alternate forms called as *allelomorphs* which have contrasting or different effects on the trait.

This model of DNA structure has helped to explain that how the changes in heredity (mutation) may occur. The error in base pairing may happen occasionally due to some reason (environmental effects, *viz.* radiation etc.) and this may result the substitution of different nucleotide during replication in place of the nucleotide usually present at that position. Thus, all descendant molecules will have the new sequence of nucleotides in the DNA.

(ii) **Structure of *RNA and its types*:** The RNA is a single-stranded structure of polynucleotide chain. The polynucleotides are arranged in a linear sequence connected together by 3'-5' phosphodiester bonds. In some viruses, the RNA is double-stranded. There are two main differences between RNA and DNA. The first is that RNA contains the sugar ribose in place of deoxyribose of DNA, thus the nucleotides in RNA are called *ribonucleotides* whereas in DNA the nucleotides are the *deoxyribonucleotides*. The second difference is that RNA molecule contains the nitrogen base *uracil* in place of *thymine* of DNA.

The purines are not necessarily equal to pyrimidine in RNA. The RNA is of different types specially for performing different functions in protein synthesis. However, the RNA is also the genetic material in most of the plant viruses and some animal viruses. The RNA is found in cytoplasm and in nucleolus. The RNA occurs freely in cytoplasm as well as in the ribosomes.

Non-genetic RNA: In case of eukaryotes in which the DNA is the genetic material, there are 3 major forms of RNA molecules which are non-genetic. These three forms are the messenger RNA (*m*RNA), ribosomal RNA (*r*RNA) and transfer RNA (*t*RNA) or soluble RNA (sRNA). Each kind of RNA molecule differs among them, each is transcribed from different genes and each has a unique function in the cells but all the three forms are likely to be produced in the nucleus. The rRNA and tRNA live longer than *m*RNA. This is because after a short time, RNA hydrolyzing enzyme (RNAases) cause the mRNA to break down after protein synthesis takes place.

(a) Messenger RNA: The mRNA is synthesized in the nucleus as a complementary strand to one of the two strands of the DNA. Its main function is to carry the genetic information from DNA to the cytoplasm for protein synthesis and constitutes about 10 % of the total RNA present in the cell. The mRNA contains the same information as coded in that part of DNA. It immediately diffuses out of the nucleus into the cytoplasm after its synthesis and deposited on some ribosomes. The synthesis of mRNA from DNA is called as *transcription*.

The specificity for a particular amino-acid sequence is determined by mRNA. The transcription of eukaryotic DNA to produce mRNA begins with the synthesis of long precursor molecule by RNA polymerase II from the template strand of DNA. This enzyme functions by catalyzing formation of 5'end towards 3' phosphodiester of RNA "backbone" by "reading" the DNA template in the 3'to 5' direction. The developing mRNA molecule is anti- parallel and its nucleotides are complementary to those of the DNA template strand. The *m*RNA chain growth is rapid from 15 to 100 nucleotides per second in vitro. The immediate product of transcription in eukaryotes is a molecule of many more ribonucleotides that comprising the ultimate functional *m*RNA. This primary transcript in eukaryotes is generally referred to as heterogeneous nuclear RNA (hn RNA). The molecules of hnRNA range from 500 to 50,000 nucleotides long. They are very rapidly processed in to much smaller molecules. Some estimates predict that as little as 10% of most original transcript eventually reaches the cytoplasm as *m*RNA. Following its transcription in eukaryotes, hnRNA is processed into functional *m*RNA in 3 steps.

The mRNA is of short duration in prokaryotes. Its lifespan is of about 2 minutes in bacteria. However, in eukaryotes, the mRNA remains for more times, say from a number of hours to even for days.

(b) Transfer RNA: It is also called as soluble RNA (sRNA) or sometimes as adapter RNA. It is of about 10-15 % of total RNA of the cell. This molecule serves as adapter molecules to bring amino-acids to the site of protein synthesis. There is at least one tRNA that is specific for each of the 20 amino-acids. There are more than one tRNA molecules for each amino-acid. The tRNA functions as an adapter because there is a group of three bases known as anticodon. The *anticodons* are the nucleotide triplets in the tRNA molecule. During protein synthesis, codon on the mRNA attracts the proper tRNA by specific interaction between codons of mRNA and antcodons of tRNA. The amino-acid phenylalanine is coded by UUU, the anticodon on one type of tRNA to accept phenylalanine is AAA. The tRNA is transcribed from several particular sites on template DNA and comprises about 15% of the RNA present at any one time.

(c) *Ribosomal RNA:* It is bulk of the cellular RNA up to about 80 % of the total RNA of the cell and occurs in ribosomes. It is relatively rich in cytosine and guanine. The rRNA is formed inside the nucleus. The set of polypeptides synthesis is the ribose which is a small peptide that averaged about 175 x 225 A°, composed of protein and rRNA. The ribosomes that function in polypeptide synthesis occur in linear groups connected by mRNA and such groups of ribosomes are called *polysomes.*

10.1.2 GENE CONCEPT

After the discovery of Mendel's law of inheritance, the gene concept came into existence. The different concept of gene have been given in different ways as -

(i) Classical concept (structure) of gene is that a gene is the hereditary or genetic unit based on the following criteria-

A gene is a unit of physical or physiological function, present at a definite location on the chromosome and produce a specific phenotypic expression (character). Secondly, a gene is a unit of structure which not be divided by crossing over and hence a unit of segregation or transmission. Thirdly, a gene is a unit of change by mutation.

(ii) Molecular concept (structure) of gene is given in relation to the parallel behavior of DNA with chromosomes. Molecular studies (structure and function of DNA in protein synthesis) had shown that chemically, a gene is a part of DNA as one or a sequence of nucleotides. The molecular structure of gene be given in the following terms (Benzer, S.):

Cistron: The cistron is a part of DNA containing a number of nucleotides which can synthesize a single polypeptide chain of an enzyme. It is the unit of physiological function and in real a gene. It is synonymous with the gene function. A polypeptide chain is formed of a number of amino-acids and each amino-acid molecule contains 3 nucleotides. Therefore, the hemoglobin for its globin protein fraction would need two cistron, one each for α and β chains.

Muton: The muton is the smallest unit of DNA which may undergo a change resulting mutation and produce a phenotypic effect. The mutation occurs as a result of change in any nucleotide. There are thus many positions or sites within a cistron as per the number of nucleotides where the mutation can occur. In this way, the muton may thus be a single nucleotide or some part of nucleotide, as the smallest unit of mutation to occur and hence to qualify the term gene. This is because any change in a nucleotide (base pair of triplet) modifies the message carried by the codon.

Recon: It is the smallest unit of DNA capable of crossing over and hence recombination between two adjacent nucleotides.. The crossing over may take place anywhere in DNA. A recon may be as small as one nucleotide pair in DNA.

Replicon: It is the unit of replication.

Operon: The combination of operator gene and sequence of structural gener acting together as unit is called operon. Thus it is several ghas.

Complon: Two or more polypeptide chains which complements to each other through their active groups. this unit of complementation is called complon and used to replace cistron.

(iii) **Modern concept of gene** in terms of its essential features is that-

"The genes are part of DNA, located on chromosomes, at a specific position (locus), arranged in linear order like the beads on a string and tansmitted from parents to the offspring. A gene may occur in many alternate forms (multiple alleles) due to mutation and hence a gene is capable of mutation. The genes duplicate themselves very accurately through DNA replication, express themselves by producing enzymes (proteins) through genetic code and finally determine the various characteristics of organisms".

Split Genes: Some genes ae called as split genes that are splitted into several distinct units separated by regions of non coding DNA. These intervening non coding sequences are called as *introns* where as those having biological information are called *exons.* Those split genes are also called as *discontinuous genes* or *mosaic genes.* A gene may have no introns or many introns.

10.1.3 Gene functions

The gene is a portion of DNA molecule and acts mainly by producing the enzymes. These enzymes are involved in different steps of metabolism. The enzymes are proteins formed of polypeptide chain of amino-acids. The type of protein produced is determined by the arrangement of amino-acids. The sequence of amino-acids in a polypeptide chain is determined by the genes.

The genes perform the following two major functions:

- The gene are self-replicating *(duplication of chromatin, DNA)* and
- The genes synthesize the *proteins* through production of RNA molecules (mRNA) which help the ribosomes in protein synthesis.

Before the process of protein synthesis is described, it is essential to know about the constituents of proteins (amino-acids), the arrangement of different amino-acids in sequential order to form different proteins and the linear arrangement of nucleotide bases in DNA to determine the sequence of amino-acids in RNA (genetic code).

What are proteins: The proteins are the polymers of amino-acids. However, the proteins instead of being build up by one kind of *monomer,* they are built from 20 different *amino-acids.* The amino-acids are organic molecules that are linked together linearly by the covalent peptide bond between carboxyl group (COOH, organic acid) of one amino-acid and the amino group (NH_2) of other amino-acid. Thus, the amino-acids in protein molecule are held together (bind or link) by *covalent bonds.* This binding forms a long polypeptide chain. The process of linking is called *polymerization.* There are 20 amino-acids which form polypeptide chains of different proteins. This bonding involves elimination of a hydroxyl from the carboxyl group of one amino-acid and hydrogen from another. The polymerization (binding of amino-acids) to form protein needs catalytic agents. The catalytic agents in the living body are enzymes which are proteins too.

Each amino-acid has three parts *viz. carboxylic group* (COOH) of organic acid, the basic *amino group* (NH_2). These two parts of amino-acids are common to all the

20 amino-acids. The third part of amino-acids is different for different amino-acids. This third part may be *aliphatic (CH), alcoholic (OH), carboxylic (COOH), aromatic (benzene), phenolic* (one H of benzene is replaced by OH), or *basic.*

The different arrangements of these amino-acids form the various types of proteins and a change of a single amino-acid in a polypeptide chain can change the function of a protein.

The 20 amino-acids are: Alanine, Arginine, Asparagine, Aspartic acid, Cystein, Glutamine, Glutamic acid, Glycine, Histidine, Isoleucine, Leucine, Lysine, Methionine, Phenylalanine, Proline, Serine, Threonine, Tryptophane, Tyrosine and Valine.

These 20 amino-acids, necessary to form a protein molecule, are dispersed in the cell (cytoplasm). Some of the amino-acids are synthesized in the cell of a man but other amino-acids are supplied from diet and called as essential amino-acids.

Genetic code: All the proteins are synthesized by the cell. The synthesis of different proteins requires the arrangement of a different number of amino-acids in a different sequential order. The linear arrangement of nitrogen bases in DNA determines the sequence of amino-acids in a protein molecule. *The nucleotide sequence in mRNA which specifies one specific amino-acid is a code word* or *codon. The sequence of triplet nitrogenous bases in mRNA molecule* (that contains the information for protein synthesis) is the *genetic code.* One specific amino-acid of polypeptide chain is coded by a sequence of 3 nitrogenous bases (triplet) in DNA. This is known as triplet code which means a three letter code. Such triplet of bases (codon) is capable of selection and inserting a particular amino-acid in a protein molecule.

The order of arrangement of nitrogen bases is important, because the change in sequence order of bases make a change in formation of a different amino-acid. For example, the sequence of UUC codes for the amino-acid *phenylalanine* whereas the sequence of CUU codes for the amino-acid *leucine* and another sequence of UCU codes for the amino-acid *serine.*

Each code consists of three nitrogenous bases. This is because single letter code consisting only one nucleotide can provide only 4 codons or words *i.e.* one word for each nitrogenous base (A, G, C. and U are nitrogen bases in *m*RNA) which are not sufficient to code for 20 amino-acids. Similarly, the combination of two nitrogen bases (double letter code) could provide only 4 x 4 = 16 codons or words which would have also not been sufficient to code for 20 amino-acids. Three letter codons of 4 nitrogen bases (4 x 4 x 4 = 64 words) are sufficient to accommodate the code words for 20 amino-acids.

The information are though coded in the form of nitrogenous base sequence in DNA molecule but the code letters of RNA nitrogen bases are used for genetic code. This is due to the message from DNA is passed on to the cytoplasm by *m*RNA and the code on *m*RNA is translated into the sequence of amino-acids.

The nitrogen bases in DNA molecule are A, T, C, G and the corresponding (complementary) bases in *m*RNA transcription are U, A, G, C. This implies that if the nitrogen base in DNA is A (adenine), the complementary nitrogen base in *m*RNA will be U (uracil); if the base in DNA is T (thymine), the complementary base in mRNA transcription will be A; the base C (cytosine) of DNA will have its complementary base G (guanine) in mRNA and likewise the complementary base of G (guanine) of DNA will be the C (cytosine) in *mRNA* transcription.

Essential features of genetic code: The genetic code has the following characteristics-

- The code is triplet, having three nitrogenous bases of mRNA in a specific sequence.
- The code is commaless. This means no punctuation is needed between any two words.
- The code is non-overlapping. Thus, same letter is not used for two different codons.
- The code is degenerate. This means that more than one word can be used for a particular amino-acid or say that most of the amino-acids except two (methionine and have 6 codons, etc. Amino acid can be coded by more than one codon which is multiple system of coding providing a protection to organism against many harmful mutations.
- The code is non-ambiguous- a particular codon always code for the same amino-acid.
- The code is universal - same genetic code is applied in all kinds of living organisms.

Table 10.1: Triplet codes of *m*RNA for 20 amino-acids

First base letter	Second base letter								Third base letter
	U		C		A		G		
U	UUU	Phe	UCU	Ser	UAU	Tyr	UGU	cys	U
	UUC	"	UCC	"	UAC	"	UGC	"	C
	UUA	leu	UCA	"	UAA**	"	UGA**	"	A
	UUG	"	UCG	"	UAG**	"	UGG	Tryp	G
C	CUU	"	CCU	Pro	CAU	His	CGU	Arg	U
	CUC	"	CCC	"	CAC	"	CGC	"	C
	CUA	"	CCA	"	CAA	glun	CGA	"	A
	CUG	"	CCG	"	CAG	"	CGG	"	G
A	AUU	Ileu	ACU	Thr	AAU	Aspn	AGU	Ser	U
	AUC	"	ACC	"	AAC	"	AGC	"	C
	AUA	"	ACA	"	AAA	Lys	AGA	Arg	A
	AUG *	Met	ACG	"	AAG	"	AGG	"	G
G	GUU	Val	GCU	Ala	GAU	Asp	GGU	Gly	U
	GUC	"	GCC	"	GAC	"	GGC	"	C
	GUA	"	GCA	"	GAA	Glu	GGA	"	A
	GUG *	"	GCG	"	GAG	"	GGG	"	G

*Chain initiation codon ** Chain termination codon

The two functions of genes *viz.* duplication of DNA molecule and protein synthesis have been explained here-

1. DNA duplication: It is well known that the chromosomes are duplicated during cell division (both mitosis and meiosis). Each chromosome manufactures a new chromosome identical to itself. This is known as DNA replication. The DNA replication is semiconservative. This is because each daughter DNA molecule is a hybrid conserving one parental polynucleotide chain and the other is the newly synthesized strand.

The first step in duplication process is the untwisting (uncoiling or straightening) of the double-stranded DNA molecule and the two strands are separated into single strands. Each of the original single strand manufactures its new strand by attracting the appropriate nucleotides according to base pairing principle. The new strands of DNA are formed on RNA primers in 5′–3′ direction from 3′–5′ template DNA by the addition of deoxyribonuleotidies to the 3′ end of primers RNA. The addition is by DNA polymerase III in the presence of ATP. The RNA primer is the primer strands of RNA which are synthesized by DNA directed RNA polymerase. The DNA synthesis proceeds simultaneously on both the strands as small segments of 1000-2000 nucleotides. The enzyme polymerase ligase helps to join them and in completing the formation of polynucleotide chain. The enzyme RNA polymerase initiates the replication of DNA or RNA strand and regonizes 5′ end of DNA.The uncoiling of original DNA strand and synthesis of fresh DNA stands go simultaneously.

This process results into the production of two new double-strand DNA molecule wherein the bases are united as per base pairing principle *i.e.* the nitrogen base A pairs with T and G pairs with C. These two newly produced double-stranded DNA molecules are exactly the same to the original double-stranded DNA molecule, except few copy errors (gene mutations).

2. Protein Synthesis: The DNA performs another important function of protein synthesis. The information carried in the genes present in the nucleus is transferred to the site of protein synthesis in the cytoplasm. This process (transfer of information) is aided by another nucleic acid called RNA. The DNA in the nucleus produces RNA molecules (mRNA) which moves to the cytoplasm and assists the ribosomes (in the cytoplasm) for protein synthesis.

There is one way flow of information in the procoess of synthesis of protein and called as *central dogma*. This states that genetic information flows from nucleic acids to protein. The transcription is the first step of central dogma and does not involve a change of code due to complementary between DNA and *m*RNA. The second step (translation) involve a change of code from nucleotide sequences to amino acid sequences. The flow of information is as shown here:

```
            Transcription                    Translation
DNA----------------------→RNA ----------------------→        Protein
```

Thus, protein synthesis involves two steps *viz.* mRNA production (*transcription*) together with its movement in the cytoplasm and protein production (*translation*). The two steps involved in protein synthesis have been discussed here.

(i) Transcription (RNA production): The genetic information (message) contained in DNA (genes) present in the nucleus of the cell are transferred (transmitted) to the site of protein synthesis in the cytoplasm. This process is called as *transcription*. The transcription requires the help of another nucleic acid (mRNA). This kind of RNA, called messenger RNA (mRNA) is synthesized in the nucleus by the coding strand of DNA. This process of synthesizing mRNA strand is complementary to one DNA strand. One strand of DNA acts as a template in the synthesis of a molecule of mRNA. The nucleotide sequence of mRNA is complementary to that of the template DNA strand, except that of uracil replaces thymine.

The mRNA is synthesized with an enzyme called RNA polymerase in the nucleus of the cell, along one strand of double stranded DNA. The RNA *polymerase* opens up a section of DNA strands to pair with complementary bases and thus provides a message of the same sequence (transcription). The RNA *polymerase* moves along the DNA template and the growing RNA strand separates allowing the hydrogen bonds between two DNA strands to rejoin.

The mRNA after it synthesizes in the nucleus, moves out of the nucleus to the cytoplasm and becomes associated with ribosome to form a complex called as polysome or polyribosome. The ribosomes are also composed of one type of RNA called as ribosomal RNA (rRNA) where the protein molecules are synthesized.

The *basic function of mRNA* is to carry the genetic message of DNA into cytoplasm where it is responsible for directing the synthesis of polypeptide chains.

The ribosomes in the cell are the sites of protein synthesis. The ribosomes are composed of proteins and rRNA. The ribosomes involved in protein synthesis form a group of ribosomes (called as polysomes or polyribosomes) by holding together by delicate strands of RNA.

The polyribosomes perform two functions:

- To bind with mRNA which carries the coded message from DNA and
- To combine with tRNA which transfer the amino-acids to the ribosomal assembly plant.

The RNAs of all types are transcribed from a template strand of DNA, catalyzed by several enzymes collectively referred to as DNA-dependent RNA polymerase or simply *RNA polymerase*. Eukaryotes produce the following 3 RNA polymerase, each with a distinctive function:

- *Polymerase I*, located in the nucleolus, synthesizes only rRNA;
- *Polymerase II*, found in the nucleoplasm, synthesizes Hn RNA (heterogenous nuclear RNA) which is a precursor of mRNA; and
- *Polymerase III*, also found in nucleoplasm, synthesizes only transfer RNA (tRNA) and 5s RNA.

During the process of transcription, there is a localized unwinding of the double helix, with transcription proceeding in the 5' to 3' direction from the single stranded region of the DNA template. This localized unwinding moves along the molecule followed by recoiling of the helix behind the newly synthesized RNA. The DNA molecule splits and a single strand of this molecule serve as a template for assembling of single strand of RNA with each base in the RNA strand being complementary to the bases on the DNA molecule strand. During formation of RNA strand, its each base takes up its proper position on the DNA strand having the base combination of A to U, U to A, C to G and G to C. The free ends of the bases in the RNA strand are connected with a new ribose, phosphate, base sequence. The single strand of RNA so produced is separated from the DNA template and passes into the cytoplasm.

The strand of DNA used as the template for RNA synthesis is called the sense strand, the other strand is the anti-sense strand. The same strand is always not the sense strand throughout the entire DNA molecule or chromosome. The region of DNA actually transcribed in to RNA is called the coding region. However, there are sequences preceding the actual coding region that are required for transcription. These sites, called *promoters*, are the sites of recognition and interaction between RNA polymerase and DNA that signal transcription to begin.

(ii) **Translation (Protein synthesis):** The in-going information from DNA to RNA, the language (nucleotide sequence) remained the same, while the in-going information from RNA to protein, the language is changed from a nucleotide sequence to an amino-acid sequence. The language of DNA available in the form of mRNA is translated into the language of protein. This process of using information contained in mRNA to make a protein is called *translation.* Thus, the sequence of nucleotides in mRNA is changed (translated) into the sequence of amino-acids of a polypeptide chain. The sequence of triplets (codons) in mRNA determines the sequence of amino-acids in the enzyme or other protein being synthesized on a particular ribosome.

The mRNA transmits the code, received from DNA for a certain protein, to the ribosomes in the cytoplasm. Amino acids are present in the cytoplasm but these are not in active form as such and hence they do not take part in protein synthesis. These amino-acids need activation which is done through ATP. The amino-acids of the cytoplasm are selected and their sequence to form a protein molecule is encoded in the base sequence of *m*RNA. This requires the help of another type of RNA, called *t*RNA. The *t*RNA found in the cytoplasm identifies amino-acid through anticodon corresponding to the codon in the *m*RNA molecule and transfers it to the ribosomes. The nucleotide triplets in the *t*RNA molecules are called *anticodones.* In other words, the amino-acids are brought to the ribosome by their appropriate tRNA to be incorporated in to protein. The *t*RNA identifies a particular amino-acid by means of anticodon which corresponding the codon of the *m*RNA. The *t*RNA carries this amino-acid to the ribosome along the *m*RNA molecule. The different amino-acids are bound together to form molecules in the ribosomes as per the code sent by *m*RNA. Different kinds of *t*RNA are present in the cytoplasm and recognize certain amino-acids, assemble them in the sequence specified by the DNA molecule (gene). The different codes are sent by DNA to form different proteins in the ribosomes.

The *first step* is incorporating an amino-acid in to a protein which involves the amino-acid attachment to its correct *t*RNA. The nucleotide sequence of *m*RNA determines the type and sequence of amino-acids in a protein chain, but the amino-acid themselves do not interact with the *m*RNA. Secondly, the *t*RNA (adapter or accepter of soluble RNA) take up proper activated amino-acids from cytoplasm and transfer them to the *m*RNA. One end of *t*RNA molecule contains anticodon and the other end makes an attachment with one of the appropriate amino-acid to involve in protein synthesis. With the movement of molecules of *t*RNA with attached amino-acid along the molecule of *m*RNA, the amino- acids are placed in position to be connected together to form a chain of protein. After attaching the amino-acid to each other, the *t*RNA molecules are free from *m*RNA and from amino-acids.

The interaction between *m*RNA and *t*RNA takes place only on the ribosomes. The ATP provides the energy required for the reaction involved between amino-acid and *t*RNA. The amino-acid is first activated by ATP to form *amino-acid adenylate* after which reacts with *t*RNA to form *amino-acyl tRNA*. When the ribosome reaches the *m*RNA, the translation of the nucleotide code into amino-acid sequence is complete. Thus, it is a two steps process *viz. amino-acid activation* and *transfer of activated amino-acids to tRNA*.

(a) Amino acid activation: The activation is controlled by specific enzymes. Each amino-acid has its own specific enzyme, in general. At least 20 different *t*RNA molecules and 20 different enzymes are required for 20 amino-acids. The enzyme, aminoacyl-*t*RNA-synthetase, catalyses the reaction of a specific amino-acid with ATP and forms the aminoacyl-adenosine monophosphatase (AMP) and pyrophosphatase.

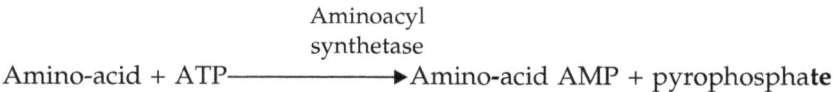

Aminoacyl
synthetase

Amino-acid + ATP————————➤Amino-acid AMP + pyrophospha**te**

The amino-acid + AMP are referred to as an activated amino-acid which is linked to adenosine by a phosphatase ester bond and is now raised to an energy level whereby it can react with its tRNA.

(b) Transfer of activated amino-acid to tRNA- The enzyme bound activated amino-acids aminoacyl adenylates become attached with their tRNA molecule.

Amino acid AMP + tRNA ————————➤Amino acid tRNA + AMP

Following amino-acid activation and binding to tRNA, the charge tRNA diffuses to the ribosome, where actual assembly in to polypeptide chains takes place. The process of the sequential incorporation of amino-acid in to proteins by peptide bond formation is completed.

10.1.4. HOW GENES FUNCTION

Several chemical reactions are involved in carrying out the functions of a cell. All these chemical reactions need specific enzymes to work as catalysts for performing the reaction. The ultimate result of gene expression is the development of a character in terms of its phenotype. The main function of genes to produce phenotype is through protein synthesis which involves a series of the chemical reactions controlled by enzymes. The production of enzymes is controlled by genes. The chemical reaction has two requirements of the presence of a catalyst and the energy. The enzymes act as catalysts in initiating the chemical reaction and the energy acts as the power to perform and complete the chemical reaction. The specific reaction requires specific enzyme.

The enzymes are proteins composed of amino-acid subunits. Each enzyme is a specific protein and is composed of a particular number and sequence of amino-acids. The number and sequence of amino-acids are different for different enzymes. It is the enzyme that make possible for an individual that which type of reaction is needed to proceed for performing a particular function.

The adenosine-triphosphate (ATP) is a single energy level compound, in all living organisms, which react with innumerable compounds and causes them to take part in reaction. The ATP molecule is like a nucleotide molecule in the sense that ATP consists of adenine, ribose sugar but three phosphate groups instead of one phosphate group as in case of nucleotide. The second and third phosphate groups are bound by a high energy bond. The ATP has one low and two high energy phosphate bonds, depending on the amount of energy liberation. An individual utilizes high energy bond between the second and third phosphate group in chemical reaction.

The example of starch formation from glucose may be cited to understand the process. The starch is a polymer made up of monomer units of glucose molecules. The glucose is first converted into glucose phosphate by ATP and it is reduced to adenosine-diphosphate (ADP). The glucose phosphate molecule takes part in reaction with the help of catalyzing agent (enzyme) and starch is formed. Likewise, the ATP also takes part in other reaction.

Another example of gene action that can be cited is the production of a pigment, melanin. One of the common components of many proteins is the phenylalanine, an amino-acid found in blood and it is metabolized by converting into another amino-acid, called tyrosine. The conversion of phenylalanine into tyrosine requires the presence of an enzyme (phenylalanine hydroxylase) which is controlled by a gene (P). In the absence of gene P, the enzyme will not be produced and hence the phenylalanine will not be converted into tyrosine. It will result improper metabolism of phenylalanine and it causes a disease called phenylketonuria (PKU) which is associated with mental retardation.

The amino-acid tyrosine is the major precursor of melanin. The tyrosine is metabolized into *melanin* (pigment) under the control of an enzyme, *tyrosinase*, which is under the control of a gene (C). Thus, gene C in the form of genotype CC, synthesizes the enzyme tyrosinase so that melanin pigment is produced. The melanin is produced in part due to the enzymatic action of gene C. In the absence of this dominant gene (C), the recessive genotype (cc) is produced at this locus and the enzyme tyrosinase is not produced and hence melanin is not produced. Consequently no pigment (melanin) is produced. The deficiency of melanin results *albinism* (no colour).

(a) **Normal metabolism**

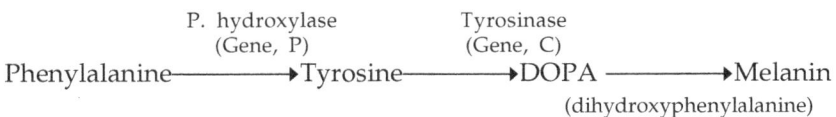

P. hydroxylase   Tyrosinase
(Gene, P)   (Gene, C)
Phenylalanine————→Tyrosine————→DOPA ————→Melanin
                                   (dihydroxyphenylalanine)

(b) **Abnormal metabolism**

Phenylalanine....No gene P, no enzyme (no tyrosine) = PKU;

Tyrosine .... no gene C, no enzyme (tyrosinase) = Albinism (no melanin).

The melanin is the basic pigment and source to produce various colours of skin, hair, and eyes of animals. After the production of melanin, requires the production of specific enzyme to produce specific colour of farm animals. The synthesis of different enzymes to produce different colours requires different genes.

The melanin is also further acted (metabolized) upon by an enzyme controlled by another gene *e* and produces a yellow pigment. The enzyme tyrosinase that catalyzes the production of melanin will be absent in the individual with recessive genotype (cc) and hence the melanin will not be synthesized. Therefore, in the absence of melanin (absence of gene C and presence of genotype cc), the yellow pigment can not be produced, even though the gene *e* is present. The recessive gene c does not allow the production of melanin, so it is epistatic gene. The gene c in homozygous condition (cc) has stopped the action of gene e at another locus and hence enzyme e is not produced. This process of epistasis (gene interaction) has been represented digrammetically as:

(a) **Normal pathway**:

               Gene C   Gene e
             (Enzyme C)   (Enzyme e)
Tyrosine -----------------→Melanin -----------------→Yellow pigment

(b) **Gene interaction (Epistasis between gene c and e)**:

               Genotype cc   Gene e
            (No enzyme C)   (Enzyme e)
Tyrosine ---------------→No melanin ---------------→No yellow pigment

10.1.5 CONTROL OF GENE FUNCTION

The DNA molecule contains the primary genetic information and expresses it by producing the mRNA. Thus, the *gene function is regulated* through the ability to control the mRNA production. According to the theory on gene functioning in microorganisms, the genes may be divided in two major groups as *structural genes* and *control genes*.

The *structural genes* synthesize the various proteins. These genes encode for the synthesis of three types of proteins *viz.* enzymes, non-enzymatic proteins such as hemoglobin and repressor proteins which prevent the activity of other genes.

The *control genes* are of two types, namely *regulator* and *operator genes*. The control genes regulate the activity of structural genes regarding the amount of protein to be produced and called *regulatory genes*. The *operator gene* is considered to be located at the adjacent to the structural genes on the same chromosome and each operator gene control the synthetic activity of adjacent structural genes by acting like a switch to make them *on and off*. An operator with the structural genes under its control is called an *operon*. The adjacent DNA cistron is initiated by operator gene to produce *m*RNA. The regulator gene is considered to be present on the chromosome which is same or different on which are present the structural genes to produce a repressor (protein) which blocks the action of operator gene by not allowing the structural genes to produce proteins because they can not produce *m*RNA. The repressor substance is converted by a depressor substance (substrate or hormone) into an inactive compound which allow the operator gene to function and the structural genes to produce *m*RNA. The repressor substance if present, combine with the operator site and prevent the structural genes to produce mRNA.

10.2 BIOCHEMICAL GENETICS

The study of genetics with biochemistry to explain the nature of metabolic pathways and their control is called as *biochemical genetics.* Thus, biochemical genetics is the study of biochemical pathways of gene function.

The biochemical (metabolic or biosynthetic) pathway constitutes all the steps involved in transforming a precursor to its end product for expression of a trait and each step of the path is catalysed by a specific enzyme which is synthesized under the control of a gene. There is a long path between gene and final product, the primary gene product is the *m*RNA which is complementary to the DNA and controls the assemblage of amino-acids to form enzymes at the surface of ribosomes. Therefore, the phenotypic expression of any character is not a direct effect of a gene (portion of DNA) but it is accomplished by the developmental and biochemical processes through the synthesis of proteins which are under the direct control of genes. Among the proteins synthesized by the genes, some are enzymes and control biosynthetic pathways. Therefore, the genes control the expression of characters by controlling the developmental and biochemical activities of the cells through the enzymes which are synthesized by genes. It was shown by Beadle and Tatum (1941) that the genes express themselves through the synthesis of enzymes. For example, a dominant gene which is the wild type (original gene) synthesizes the enzyme required for the expression of a normal character whereas a recessive gene is unable to synthesize that functional enzyme and so unable to express that phenotype but behaved differently expressing the modified phenotype.

Human Biochemical Genetics: A change in the code sent by DNA causes the production of a different protein in the ribosomes, resulting into a mutation (mutant allele) which causes malfunctioning of gene. Thus, defects are caused by mutant genes as result of blocking a particular step in biochemical reactions. These can be studied when an error in the metabolic pathway due to mutant gene is observed in the form of a defect or disease in an organism. The effect of abnormal gene (mutant) helps to know the normal gene function. The various studis have shown that the function of gene is chemical in nature which means that specific gene controls the specific chemical reaction.

Some inborn errors of metabolism in man have been studied. Bateson (1902) reported the inheritance of a rare human defect (alcaptonuria) as a recessive trait. A book "Inborn Errors of Metabolism" was published in 1909 by Garrod, A.E., English Physician when he noted that some of the hereditary diseases in man are due to the effects of mutant genes on the metabolic system.

As a result of gene mutation, the enzyme responsible to produce normal phenotype is not produced but another enzyme is produced. This alternate enzyme of the biochemical pathway determines the alternate phenotype that may be abnormal. Thus, the normal metabolic pathway is blocked and causes abnormal phenotype (disease). The following are some of the examples-

Examples: The best known example of gene-enzyme relation is the metabolism of the amino-acids, phenylalanine and tyrosine, in man. The phenylalanine is an essential amino-acid of human dietary proteins and it may be incorporated into cellular proteins or converted to phenylpyruvic acid or tyrosine. The following diseases or metabolic disorders are caused by the improper functioning of the enzymes involved in the metabolism of phenylalanine and tyrosine (defective metabolism), as a result of mutant genes which are autosomal recessives in homozygous condition.

(i) **Phenylketonuria (PKU) or Folling' disease:** This disease is hereditary and metabolic, caused by a recessive gene in homozygous condition, and occurs due to accumulation of phenypyruvic acid in the blood, damages the brain tissues causing mental disorders (idiocy). The presence of phenylpyruvic acid in the urine is the symptom of this condition. The enzyme phenylalanine hydroxylase is necessary for the normal metabolism of amino-acid phenylalanine. The phenylpyruvic acid is converted into hydroxy-phenylpyruvic acid in normal persons as a result of catalysis of the enzyme (phenylalnine hydroxylase) controlled by normal gene. But when a defective gene (mutant) is produced, the normal enzyme (phenylalanine hydroxylase) is not produced. There is an accumulation of the amino-acid phenylalanine, derived from the breakdown of proteins, due to inactivation of specific liver enzyme (phenylalanine hydroxylase) which prevents the metabolism of phenylalanine to tyrosine. In the absence of normal enzyme there is no conversion (break down) of the phenylpyrivic acid into hydroxyphenylpyruvic acid. As a result the phenylpyruvic acid accumulates in the blood and causes damage of brain tissues. Such persons fail to break down the phenylpyruvic acid into hydroxyphenylpyruvic acid and suffered with a disease known as *phenylketonuria*. The phenylketonuria (PKU) children are mentally retarded and known as ***phenylpyruvic idiots,*** have recessive genotype *pp* and fails to produce normal enzyme.

(ii) **Alcaptonuria:** This disease is hereditary and metabolic, caused due to the presence of a substance, called alcapton (homogentistic acid), in the urine. The enzyme homogentistic acid oxidase is required to covert alcapton (homogentistic acid) into aceto-acetic acid. In the absence of the catalase enzyme in the person suffering from alcaptonuria, the homogentistic acid is not converted into aceto-acetic acid and passed out in the urine and turns black on exposure to air. The persons carrying the recessive mutant gene can not synthesize the required enzyme and hence the step in the normal degradation of alcapton (homogentistic acid) is blocked. Thus, due to accumulation of alcapton in the patients, the urine turns black after coming in contact with air. This disease is also characterizes by the hardening and blackening of the cartilage of the bones.

(iii) **Albinism:** Absence of melanin pigment causes albinism. The melanine is produced from tyrosine as a result of catalysis of tyrosinase. The person with recessive mutant genotype fails to produce this enzyme required to convert dihydroxyphenylalanine into melanine and hence no pigment is formed. The persons are albino lacking the pigment melanin in their hairs, skin and eyes.

(iv) **Tyrosinosis:** The tyrosine is accumulated in person having recessive gene in homozygous condition *(tt)* due to lack of conversion of hydroxypyruvic acid into dihydroxyphenyl pyruvic acid and person suffer with tyrosinosis.

(v) **Goitrous cretinism:** The conversion of tyrosine into thyroxin (thyroid hormone) is required and the persons lacking the required enzyme for conversion are deficient in thyroid hormone which causes **goitrous cretinism**. This results into the physical and mental retardation and hypertrophy of thyroid gland.

Other examples: There are certain genetic defects, observed in humans and also in animals, which are perhaps the result of the failure to produce a specific protein or due to the production of abnormal protein. The examples are as under-

Sickle cell anemia: This disease occurs in human population of central Africa, Greece and south India. The RBC of patients become sickle shaped instead of normal round shape. The RBC of normal persons contain hemoglobin A (Hb^A) whereas sickle-shape RBC contain an abnormal hemoglobin called hemoglobin C (Hb^C). The persons homozygous for hemoglobin S (Hb^S Hb^S) suffer from anemia characterized by sickle shape of erythrocyte. This disease is fatal in early age. The haemoglobin of homozygote is of the abnormal types. On the other hand, the heterozygote do not suffer from anemia. The selective advantage of heterozygote has been shown to be associated with their resistance to malaria. The sickle cell homozygote (Hb^S Hb^S) is anemic severely, having sickling of RBC but resistant to malaria while the normal homozygote (Hb^A Hb^A) is not anemic, have normal hemoglobin but susceptible to malaria and the heterozygote is mild anemic but resistant to malaria for which the heterozygote has advantage over both homozygote.

It has been found that both the normal as well as abnormal hemoglobin contain same types of amino-acids but different in the amount of one of the polypeptides. The normal hemoglobin contains two glutamic acid units and one valine whereas the abnormal hemoglobin contains one unit of glutamic acid and two units of valine. Thus, one unit of glutamic acid is replaced by the valine unit and this change creates mutation which results in the production of abnormal protein. This gave the evidence that gene carries the code to control the synthesis of a particular protein molecule.

Agammaglobulinemia: This is a genetic defect in human body when the gamma globulin (protein) is not produced in the body. This protein is received by new born baby from its mother and gradually decreases with age reducing to zero at 4 months of age and its production is started at about 3 weeks of age. The absence of this protein increases the susceptibility to bacterial infection.

Lack of sex drive (libido): This is an inherited defect in bulls wherein there is failure of the production of a particular hormone by anterior pituitary gland.

Drosophila-Mutants of Eye colours: The eye transplantation experiments were conducted by Beadle, G.W. with his fellow worker on Drosophila in Europe. They studied the development of pigment in transplanted eyes. On the basis of the results of eye transplantation they suggested the biosynthetic pathway along with the steps in eye pigment formation.

Neurospora-Biochemical mutations: The Neurospora is a unicellular fungus widely used for biochemical mutations affecting amino metabolic pathways, synthesis of molecules of nucleic acids and vitamins. The study of biochemical mutations has suggested the biosynthetic pathways of arginine and tryophan synthesis in Neurospora.

10.3 DEVELOPMENTAL GENETICS

Undoubtedly, the development of embryo into an adult individual is under the control of genes. Thus, the primary function of majority of genes in eukaryotic organisms is the development of adult organism. It is thus important to study the basic principles underlying the role of genes in embryonic development. This branch of genetics is called as *developmental genetics*. In this regard it is important to know that development biology is concerned with which pathways, to know the possible theories of developmental genetics, and to understand the basic principles of induction and autonomous regulation.

The effects of genes are observed from their effect on the phenotype of the juvenile or adult organism. In reality, however, the vast majority of these phenotypes are established during embryonic development, because most genes function to generate the pattern of the developing embryo. Loss of function of a particular gene results in an abnormality in development which manifests itself as an abnormal phenotype in the adult. Therefore, all of eukaryotic genetics is developmental genetics, but it has not been considered in this way.

The developmental biology is concerned with the function of genes in embryogenesis. It is thus important to know the pathway of developing a single cell (fertilized egg, zygote) in to an extremely complex adult organism, which is composed not only of trillions of cells, but of thousands of different types of cells (such as R.B.C., liver cells, muscle cells, nerve cells, etc.).

Development pathways: The cells differentiate, becoming specialized to carry out different functions by following different developmental pathways. The following two theories have been postulated about the development pathways.

(i) Mosaic Theory of Development: The mosaic theory based on qualitative cell division was proposed by Wilhelm Roux and August Weisman in the late 19th century to explain the process of differentiation. According to this theory, there are determinants present in the zygote to specify the various differentiation pathways and these determinants are inherited unequally by the daughter cells at the time of cell division after fertilization. The set of determinants are divided with each cell division until each daughter cell gets the type of determinant required as per the fate of that cell.

The determinants are now considered as the genes which are not divided during cell division. Thus mosaic theory of the division of genes during cell division was disproved. The basis of disproving this theory of gene division rested on the findings of nuclear transplantation studies conducted on amphibians in the early 1960's. In these studies the nuclei from tadpole intestinal cells (which were already differentiated cells) were isolated and injected into eggs after removing the haploid nuclei of the egg. A small percentage of the injected eggs developed into completely normal adult frogs. Therefore, a tadpole intestinal nucleus must have all of the genetic information necessary to allow the differentiation of every cell type in an adult frog otherwise a tadpole intestinal nucleus would have the genetic information necessary only to cause the differentiation of an intestinal cell.

The frogs produced by this procedure would have been genetically identical (clones) to the frog from which the intestinal cell nucleus was obtained. The cloning of Dolly sheep was done on this concept to demonstrate that differentiated mammalian nuclei (from mammary gland in case of Dolly) have all of the genetic information necessary for the development of a normal adult organism.

(ii) Theory of Differential Gene Expression: The differentiation can not occur in case all nuclei in an organism contain the same genetic information. The differentiation occurs as a result of expression of only a subset of the total genes present in the particular cell. For example, if a cell expresses only the set of genes responsible for muscle differentiation, then that cell will differentiate into a muscle cell, but if the cell expresses only the set of genes responsible for spleen cell differentiation, then the cell will differentiate into a spleen cell. This is the theory of differential gene expression.

This is a fairly simple concept based on transcription factors. The activation of a particular set of genes in a cell for directing the differentiation depends on the set of transcription factors present in the cell. A cell activates the specific genes through transcription factors to specifically activate only the genes responsible for development of different specific tissues.

The transcription factors are encoded by genes, and those genes are regulated by other specific transcription factors. Those transcription factors are in turn encoded by other genes, which are regulated by still other transcription factors, etc.

There is a hierarchy of genes within each cell. Genes are expressed that encode transcription factors, which activate other genes that encode transcription factors, which activate other genes that bring about differentiation along a specific pathway. There may be many levels of the hierarchy which seems to go on forever. Each set of transcription factors is encoded by genes that are activated by yet another set of transcription factors. It must end somewhere, but yet not known.

Master Control Genes: A master control gene is the first gene activated in a hierarchy that leads to differentiation along a particular pathway. Master control genes encode the first transcription factor in a hierarchy. The master control gene product activates the next set of genes that encodes the next set of transcription factors, and the cascade of gene expression has been set in motion.

The master control gene for development of particular tissue is actually a family of genes. These genes encode transcription factors. Master control genes have a particular property. They initiate differentiation of particular tissue. If the genes for that tissue are activated in other cell type (say liver or fat cells, etc.), the master control genes cause the liver or fat cells to trans-differentiate into that tissue. For example, liver cells or fat cells that are caused to express the particular genes (using tricks of molecular biology) will change from their normal phenotype into cells of particular tissue.

There are two basic ways that master control genes are regulated. The first is the process of induction which occurs after a cell sends a signal to another cell to differentiate a certain way. The signal is usually a diffusible protein and causes the recipient cell to activate the appropriate master control gene product. This is the way of differentiation of different tissues to occur. Signals from other cells cause the genes responsible for developing a particular tissue to become active in the cells receiving the signal, and differentiation for that particular tissue is initiated.

The inducing cell (which sends the signal) is itself a differentiated phenotype. It might have differentiated because its master control gene products were produced by the second type of regulation by which a cell just knows what to differentiate into. This is known as autonomous regulation. The cell inherits a 'determinant' that causes it to differentiate along a particular pathway. These determinants are generally mRNA molecules that encode transcription factors (*i.e.* master control gene products), and that are produced during oogenesis and stored in localized regions in the egg. Such determinants (*m*RNA) would therefore be unequally inherited by daughter cells during cell division in the embryo. It is thus very much like the mosaic theory, with the difference that it is not the genes which are divided up unequally between cells, but it is the mRNA molecules produced by certain genes.

Conclusively, the developmental biology is concerned with the pathway of the cells to differentiate (become specialized) during embryonic development. The cells differentiate as per the theory of differential gene expression. Master control genes (genes that initiate differentiation pathways) encode transcription factors that activate a cascade of gene expression resulting in differentiation. These master control genes can be activated by *induction* (a signal from another cell) or by *autonomous regulation* (inheritance of a determinant- usually an mRNA that encodes a master control gene product).

10.4 BACTERIAL GENETICS

Earlier it was believed that bacteria and virus reproduce asexually and hence these were not considered as suitable materials for conducting genetic studies. Asexual progeny of a single bacterium are genetically identical except mutants and these progenies (produced asexually) are referred to as a clone or colony when grown on a medium like agar plate.

Later on, the work of American scientists Lederberg, J. and Tatum, E.L. (1946) have demonstrated that sexuality exist in bacteria. This discovery proved very important to include bacteria in genetic studies and hence this opened a new area of research in genetics.

Bacteria- *the suitable material for genetic studies*: The bacteria are unicellular, prokaryotes, rod-like or spherical or spiral shapes, living forms found every where in abiotic (air, water, earth) and biotic (animals and plants) ecological habitats of earth. The prokaryotic bacterial cell has cytoplasm enclosed in an outer covering with three layers. The outer most is the *slimy layer* of polysaccharides, middle strong rigid *cell wall* containing lipids, proteins, carbohydrates phosphorus and some mineral salts, and an inner *plasma membrane* of lipoproteins. The cytoplasm is dense and colloidal having granules of glycogen, proteins and fat but does not have mitochondria, endoplasmic reticulum. The ribosomes occur freely in cytoplasm.

The bacterial cell has no true nucleus (devoid of nuclear envelope). There is nucleus like zone in cytoplasm which contains a single, large, circular chromosome with double-stranded DNA molecule (bacterial chromosome, called nucleoid) but has no protein and RNA around the DNA molecule like eukaryotes. The bacteria are autotrophic, saprotrophic and parasitic to plants and animals, for their nourishment.

The bacteria have a short generation interval, some bacteria divide about once every half an hour and hence about 30 generations are produced within 15 hours period, producing about a billion progeny cells. The sexuality in bacteria, short generation interval, less cost involved in producing large population in a small place with less investment, etc. made the bacteria as the suitable materials for genetic studies. Thus, the experiments which can not be conducted on large animals can be carried out on bacteria easily with lesser cost and man power. This has made advances in the field of basic genetics. The *Escherichia coli* which is a bacillus bacterium and non-pathogenic intestinal parasite of man and other mammals has been very much useful to gain our knowledge of molecular genetics, for its rapid proliferation rate and easy to culture. The *E. coli* has been named after its discoverer's name (T. Escherich) and its habitat (presence in human intestine). The bacteria have proved useful and meritorious organism for conducting genetic experimentation and have been made several landmark discoveries, like identification of DNA and RNA as the genetic material, protein synthesis, genetic code and gene regulation studies, fetching Noble prize.

There are some advantages to use bacteria for the genetic study. These microorganisms are unicellular having only one set of chromosomes and hence like that of the haploid gamete of the diploid organism. Thus there is no phenomenon of dominance among alleles because only one allele is present in individual. This makes the study simple due to the fact that each gene will be expressed in phenotype. The study of mutation on these microorganisms is simple and easy because a large population can be grown in small field (petri dish or tube) and the use of selective media can isolate a mutant easily in a million individuals. The microorganisms like virus provide best opportunity to study the gene structure and function at molecular level.

On the contrary, there are some disadvantages using microbes for genetic study. The first is the relative dearth (scarcity) of regular recombination events. Mostly the mode of reproduction in microorganisms is asexual (vegetative) by amitosis, which is a direct nuclear division, though some sexual reproduction is also known to occur. Secondly, the morphological variations in microorganisms are lacking. The small body size is also a limiting factor to study the gross phenotypic characteristics of an individual. Further, the genetically different clones may respond differently to the nutrients, drugs, pathogens etc. in the culture medium. The wild type and mutant strains grow on different culture medium *viz. mutant strains* (auxotrophic) requires supplementation whereas *wild type* grow on minimal medium (prototrophic) containing the basic growth requirements (glucose, sodium citrate, magnesium sulphate and ammonium sulphate required for growth). Fifth point is that they are though haploid but most of them are multinucleate. The presence of more nuclei makes delay the expression of the traits. Due to their small size, the variation may become clear after several recombination events have occurred and hence the population dynamics may be a source of confusion for genetic analysis. Lastly, they also do not have easily visible chromosomes compared to the organism of higher order.

Genetic material of bacteria: The DNA molecule is the genetic material of all bacteria. The DNA occurs in the ribosome-free cytoplasmic regions of the bacterial cell. The hereditary material (DNA) is a single large circular chromosome (100 μ long) made up of a circular molecule of double strand DNA and called as bacterial chromosome or nucleoid. It has no protein around the DNA molecule as in case of eukaryotes. Some RNA is present. The DNA molecule is super coiled which is due to nucleoid proteins and RNA. The circular chromosome of *Escherichia coli* is of a contour length of about 1.36 μm and 20Å broad has about 50 or more highly twisted or supercoiled loops with about 4 million nucleotide pairs and molecular weight of 2.8×10^9. The DNA is tightly coiled up in nucleiod in such a way that most of the cytoplasmic particles are excluded but without surrounding the nucleoid by a membrane.

Each bacterial cell also contains much smaller circular duplex (1 to 20 in number), DNA molecules, called as plasmids, besides the single large circular chromosome. They are similar to viral DNA in size and replicates in fixed numbers along with the bacterial chromosome. The functions of plasmids vary from being incorporated into the host cell chromosome and called episomes, to being sometimes transferred from one bacterial cell to another during conjugation.

E. coli is most intensively studied bacterium. The *E. coli* is a common colon bacterium. This has a single major chromosome in circular form. The gentle lysis can loosen the tight DNA meshwork of its nucleiods and its enormous circular duplex molecule has been identified among the liberated DNA in many favourable preparations. Its molecular weight is 2.6×10^9 daltons and 1100 *u* length. The *E. coli* cell, besides having a major chromosome often contain one or more minor chromosomes, which are called as plasmids and contain 0.5 to 2 per cent of the DNA of the cell. The plasmids exist distinctly from main chromosome and also replicate independently of the major chromosome. Thus more than one genetic element is transmitted from parent to offspring in *E. coli*. This bacterium though reproduce asexually, but still there are a number of sources by which there is genetic exchange from one bacterial cell to the other *i.e.* DNA from one bacterial cell can undergo genetic exchange with the DNA from another bacterial cell. *E.coli* synthesizes all necessary vitamins and amino-acids required for growth (auxotroph strain). However, there are also some strains of *E. coli* which can not grow on minimal medium and need addition of some particular vitamin or amino-acid which they can not synthesize (prototroph strain).

Asexual reproduction in bacteria: The most bacteria reproduce asexually which is accomplished by *amitosis* and often called as *binary fission*. The mutant cell can be detected by use of selective media in a colony of millions of bacteria grown in liquid media and then distributed over an agar plate *e.g.* mutant strain of bacteria resistant to streptomycin (antibiotic) can be detected in a way that all cells without mutation will be sensitive to antibiotic and hence will not grow but the antibiotic resistant mutant will form a colony. The spontaneous mutations occur in bacteria.

Sexual reproduction in bacteria or **genetic recombination:** In diploid organisms of higher order, there are a number of mechanisms by which the genetic variations are produced *viz.* genes in pairs, multiple alleles, number of gene loci, gene action, mutation and chromosomal aberration, genetic recombination (segregation and crossing over), sex controlled genes and extra chromosomal variation, etc. In case of bacteria, the genetic recombination is important and it has been given here.

The transfer of genetic material from one bacterial cell to another results recombination of genes in bacteria and it was discovered in *E. coli* by Lederberger and Tatum (1946). There are some mechanisms by which recombination and exchange of DNA molecules between two bacterial cells may takes place. These mechanisms are *conjugation, transformation* and *transduction*, etc.

The experiment conducted by Lederberg (1946) contained two mutant strains of E.coli which were biochemically or nutritionally deficient unable to grow on minimal medium because both the strains had one or more mutation in nutritional genes that control synthesis of certain growth substances. One strain (A) was deficient in vitamin biotin (B-) and amino-acid methionine (M-) and hence unable to synthesize them but able to synthesize the amino-acid threonine (T+) and leucine (L+) and vitamin B_1 (B_{1+}). The other strain (B) was able to synthesize vitamin biotin (B+) and amino-acid methionine (M+) but unable to synthesize amino-acids (Thereonine, T-; leucine, L-) and vitamin B_1 (B_1-). The genotypes of the two strains can be written as under:

Strain A: B^- M^- T^+ L^+ B_1^+

Strain B: B^+ M^+ T^- L^- B_1^-

The + and – sign indicates the presence or absence of particular gene responsible for synthesizing or not synthesizing the particular vitamin or amino-acid.

Both the strains were cultured on complete medium having glucose, yeast extract and peptone. The suspensions of the washed cells after one or two generations were placed on minimal medium. It was found that most of the cells failed to grow (multiply) and only a few cells had grown. It was thus postulated that these cells which grew were wild type (auxotroph) with all the genes dominant (B^+ M^+ T^+ L^+ B_1^+) and were formed due to recombination of genes from the two mutant strains.

Recombination mechanisms: There could be the following three mechanisms for transfer of genetic material from one bacterial cell to another cell. These are as:

   (i)   *Conjugation-* This means to unite sexually. This is the unidirectional transfer of DNA from F^+ bacterial cell to F^- bacterial cell.

  (ii)   *Transformation-* This means to make a change. This is the recombination of naked DNA molecule from one bacterial strain to another strain to produce new phenotype.

 (iii)   *Transduction-* This is the gene transfer from one bacterium to another bacterium by intermediate agent (phase).

The workers working on E. coli concluded that bacteria exhibit sexual differentiation behaving like the male and female gametes of higher orders.

Under *conjugation,* the F⁺ means male factor or donor strain possessing sex factor or fertility factor (f factor) containing genes for maleness. The F⁺ strain transfer its genetic material to F⁻ bacterial cell and change them to F⁺. The F⁻ represents the female strain or recipient without f-factor. The sex factor is composed of DNA containing about 100 genes, is autonomous and self replicating, the replication being rapid than the bacterial chromosome. Usually one f-factor is present per male bacterial cell and F⁺ male cells can be changed to F⁻ female cell by losing f-factor. The conjugation is like uniting sexually and it occurs between f+ and F- strains in which the F factor from F⁺ donor is transferred to F⁻ recipient cell producing all the daughter bacterial cells as F⁺.

The *transformation,* is taking up by the living bacterial cells the fragments of DNA released by dead cells and thus the living cells have additional DNA showing additional characteristics (Griffith, 1928). This provides new combination of genes.

The *transduction is the carrying of the* fragments of DNA from one bacterial cell to other through bacterial virus *(bacteriophages).* The bacteriophage attacking the bacteria injects its DNA into bacterial cell. The bacterial DNA breaks down and the viral DNA replicates within the bacterium. The lysis of bacterial cell liberates the phages carrying bacterial DNA. The phage carrying bacterial DNA infects other bacterial cells. The viral DNA which is the donor of genetic information, is incorporated into bacterial DNA (recipient bacterium) and thus provides new combination of genes.

Numerical Exercises

CHAPTER 3: MULTIPLE ALLELES

3.1 If a man belongs to blood group B, his wife to blood group A, can they produce the children of all the four blood groups? Give your answer with full support.

[**Hint:** If both parents are hereozygous, they can produce the children of all blood groups]. This can be proved by making such a cross between AO and BO.

3.2 A young man and a girl, both having AB blood group, are married to each other. Find out the possible blood groups of their children.

3.3 A child having blood group O is claimed by two pairs of couples. The first couple belonged to blood group A whereas the husband of second couple belonged to blood group O and his wife to blood group AB. Give your decision that to which pair of couple the child belonged.

[**Hint:** The child must belong to the first pair of couple having blood group A]

3.4 A child had blood group O, his father had blood group B and his mother belonged to blood group O. Find out whether this child was really of this couple.

[**Hint:** There is 50% probability of the child to belong to this couple provided his father is heterozygous B blood and her mother of O group. Make a cross and check.]

3.5 The crossing of coloured rabbit (full colour, agouti) with albino (pure white) produced the progeny in 1:1 ratio of coloured: albino whereas the crossing of colored rabbit with Chinchilla (silver grey) produced ½ coloured: ¼ Chinchilla: ¼ Himalayan. What were the genotypes of the parents in these crosses?

CHAPTER 4: MONOHYBRID CROSSES: COMPLETE DOMINANCE

4.1. Black wool in sheep is dominant over white controlled by one pair of gene(B). A black sheep was mated with a white ram. Estimate the percentage of black F_2 progeny. What will be the expected mating among F_2 progeny?

4.2. Abnormality of eyelid in human is controlled by a pair of gene with complete dominance. A woman and her father had eyelid abnormality called ptosis (difficulty in opening eye lids completely) but her mother had no problem. Find out the possible genotypes of all the 3 people and the proportion of children to be born if this woman with eye abnormality marries with a man with normal eyes.

4.3. Black colour in guinea pig is dominant over white. Find out the expected genotypes of a black male and black female if their mating resulted into 9 black and 3 chocolate guinea pigs, and the mating of a black male with chocolate female produced 5 black and 5 chocolate offspring.

4.4. Grey colour is dominant over black in fly. The F_2 progenies of a cross resulted into 400 offspring, out of which 300 were grey and rest were black. What were the genotypes of both males and females of parent generation.

4.5. The barking in dogs during training is governed by a singe gene pair (B) expressed in two forms as barking which is dominant and silent expression of the trait. Another trait is erection or dropping of ears which is also governed by single gene pair (E) and erection of ears show dominance over dropping ears. A heterozygous erect eared barker was mated with dropping eared silent nature. Give the genotypes of the mated pair and find out the ratio of progeny with different phenotypes.

Codominance

4.6. The coat colour in Andalusian fowl (black and white) is controlled by one pair of gene with two alleles (B and W) and the heterozygous genotype at this locus (BW) produces blue coat colour which is a mixture of black and white hairs. Work out the genotypic and phenotypic ratio in F_2 progeny and name the type of gene action. .

4.7. The coat colour in Shorthorn cattle (red and white) is governed by single locus two alleles (R and W). The red colour cattle have RR genotype while those of white colour have WW genotype. The crossing of cattle with these two coat colours produce F_1 cattle with roan colour (a mixture of red and white coloured hairs called as roan colour. Find out the results of F_2 generation and interpret the results. Is it possible to develop a breed of cattle that would transmit consistently the roan colour ?

4.8. The ear size in goats is controlled by two alleles at a locus, the long ears and short ears in size are due to homozygous genotypes for the two alleles, respectively while the heterozygote at this locus produces goats for medium size ears. Interpret.

4.9. The yellow guinea pigs breed true, whereas the mating of yellow with cream colour produces yellow and cream coloured progeny in equal ratio. What type of inheritance (gene action) is involved? What is the colour of heterozygote guinea pigs? [Answer Co-dominance and Cream colour].

4.10. A pair of dominant alleles controls the palomino colour. Two homozygote produce *chestnut* (CC) and *white* (cc). Predict the phenotypes from the following crosses, assuming large random mating population: *(i)* White mated with white, and with chestnut *(ii)* Chestnut x chestnut, *(iii)* palomino crossed with palomino, chestnut, and white.

Lethal alleles

4.11. The trait hairless in dogs results from heterozygous genotype (Hh) and the normal condition is due to a recessive gene in homozygous condition (hh) while the dominant homozygous state (HH) is lethal in embryonic stage. Predict the phenotypes from mating of hairless dogs with normal dogs, and inter se mating of hairless dogs.

4.12. In poultry, the creeper condition results from heterozygous genotype (Cc), the homozygous recessives (cc) are normal whereas the homozygous dominants (CC) are died in embryonic stage.

4.13. *Pelger anomaly of rabbit* (abnormal WBC number segmentation) is expressed in heterozygous state of two alleles (Pp) and the normal individuals are homozygous recessive (pp). The homozygous dominant genotype (PP) causes grossly deformed skeleton causing death before or very soon after birth. (a) Whether it is possible to produce a litter with all progeny to be pelgers? (b) Taking the average litter size at weaning as four, what is the expected litter size from a mating of pelgers with pelgers? (c) Predict the average litter size at weaning from mating of pelgers with normals.

Answer: (a) It is not possible because only the pelgers can be produced from a cross of both homozygote (PP x pp) and the dominant homozygote (PP) die before reproducing.

(b) Three. This is because the mating of pelgers with pelgers produces the progeny in a genotypic ratio of of 1 PP: 2 Pp: 1 pp. The genotype PP is lethal before weaning and hence the average litter size at weaning is expected to be three. *(c)* Four. This is because the mating of pelgers (Pp) with normal (pp) will produce 2 pelgers and 2 normal.

Dihybrid crosses: Non- allelic interaction- Epistasis:

4.14. The birds with feathered shanks (Lagshan, Brahama breeds) were crossed with the birds with clean shanks (Cornish, Minorca, Australorp breeds). All the F_1 had feathers on shanks. In F_2 progeny of 325 birds, 19 birds were fond clean shank. What was the type of gene action responsible for expression of this character?.

Solution: Feathered birds in F_2 = 325 − 19 = 306.

Assuming two gene interaction, there will be 16 genotypic classes.

Therefore, 1/16 part of 325 = 325/16 = 20.31.

The ratio of 306 feathered birds based on 16 genotypes will be = 306/20.31 = 15.06

The ratio of feathered to clean shank birds is 15:1.

This ratio of 15:1 indicates duplicate dominant recessive epistasis.

4.15. The genotype AABB produces red phenotype whereas the double recessive genotype produces white phenotype. The recessive homozygous genotype at either gene loci (aa B- and A- bb) produces a third phenotype (sandy). What will be the phenotype of F_1 generation?. Also find out the phenotypic ratios in F_2 generation and name the type of gene interaction.

4.16. The pea comb shape in poultry is dominant (P -) over the single comb shape (pp). Another trait feather colour is due to co-dominance of a pair of alleles producing black (BB), white (bb) and blue (Bb) feathers. In a brood from white hen with pea comb shape, it was observed that 2 birds were blue pea comb, 4 were white pea comb, 2 were white single comb and 1 bird was of the phenotype blue feathers with single comb. (a) Predict the genotype of the hen and the cock. (b) Among the 4 white pea comb progeny, how many are expected to be heterozygous for pea comb gene (c) If each of the four white pea comb progeny are used in a test cross and each produced at least one progeny with single comb, how many of these four progeny are expected to be heterozygous.

Answer: (a) The genotype of the hen is bb Pp and of cock will be Bb pp. This is because the mating with these genotypes will produce the four types of progeny in a ratio of 2: 4: 2: 1. (b) The four white feathers and pea comb progeny from above cross (Bb pp x bb Pp) will be of genotype bb Pp and all the four will be heterozygous for pea comb gene. (c) The test cross will be mating of double recessive (bb pp) with the given genotype (bb Pp). This cross produces at least one progeny with single comb and hence all the original four progeny are expected to be heterozygous.

4.17. A dominant gene (S) produces short hair in rabbit and the recessive allele (s) causes long hairs. The pelger anomaly is caused as detailed in question numbe 4.14. Predict the phenotypic ratios among several litters from mating of heterozygous short-haired pelgers among themselves at weaning.

Answer: The genotype of short-haired pelgers would be Ss Pp. The alleles for length of hais show dominance-recessive relationship and hence will produce 3 short hair: 1 long hair whereas the alleles causing pelger anomaly are dominant lethal (PP) causing death upto soon after birth and hence will produce the 2 pelgers: 1 normal. Therefore, the combined segregation and recombination will produce the progeny at weaning as:

[3 (short): 1 (long hair)] x [2 (pelgers): 1 (normal)] = ? .

CHAPTER 5: LINKAGE AND CROSSINGOVER

Example 5.1: Linkage detection: The following F_2 data on comb shape and leg length in poultry were obtained:

Comb shape (A locus):

	Rose comb (A)	Single comb (a)	Total
Length of legs (B locus):			
(B) Short (Creeper)	62	11	73
(b) Normal	12	11	23
Total	74	22	96

Conduct a test for linkage between two genes affecting comb shape and leg length as well as for the two loci segregating according to expected F_2 ratios.

Solution:

$$\chi^2_{(A\ locus)} = \frac{(C_1 - 3C_2)^2}{3N} = \frac{[(a_1 + a_2) - 3(a_3 + a_4)]^2}{3N}$$

$$= \frac{(74 - 3 \times 22)^2}{3 \times 96}$$

$$= \frac{64}{288} = 0.222^{NS}$$

$$\chi^2_{(B \ locus)} = \frac{(R_1 - 3R_2)^2}{3N} = \frac{[(a_1 + a_3) - 3(a_2 + a_4)]^2}{3N}$$

$$= \frac{(73 - 3 \times 23)^2}{3 \times 96}$$

$$= \frac{16}{288} = 0.055^{\ NS}$$

$$\chi^2_{(Linkage)} = \frac{(a_1 - 3a_2 - 3a_3 + 9a_4)^2}{9N}$$

$$= \frac{(62 - 3 \times 12 - 3 \times 11 + 9 \times 11)^2}{9 \times 96}$$

$$= \frac{(62 - 36 - 33 + 99)^2}{864}$$

$$= \frac{8464}{864} = 9.796^{**}$$

$$\chi^2_{(Total)} = 9.796 + 0.222 + 0.055 = 10.073^*$$

Answer: From the chi-square (χ^2) values calculated for segregation of both the loci, it was observed that neither value is significant tested at 1 degree of freedom, proving that the segregation is in agreement with 3:1 ratio for each loci. The chi-square value due to segregation tested at 1 degree of freedom is highly significant, provides an evidence of linkage of two loci. The total chi-square value (10.073) based on 3 degrees of freedom was significant, showed that the joint segregation is not in agreement with 9:3:3:1 ratio.

Example 5.2 The F_1 dihybrid was produced in coupling phase. Find out the expected F_2 results taking the two loci (A and B) located 20 map units apart.

Solution: The recombinants are 20 % and hence the frequency of Ab and aB gametes will be 0.10 each and the rest 0.80 will be parental types (0.40 each AB and ab). The F_2 results (frequency of different types of progeny) can be obtained from gametic checker board.

Example 5.3 The test cross on three point cross (Trihybrid) gave the following results: Map distance: A – B =10 units and B – C = 20 units, Interference = 40 %.

Find out the frequency of different types of progeny (parental types, SCO at two regions and DCO).

Solution: Coincidence = 1 – interference = 1- 0.40 = 0.60

1. Frequency of expected DCO   = (SCO I) (SCO II)
                                                = 0.1 x 0.2 = 0.02 = 2 %

    Frequency of observed DCO   = Expected DCO x Coincidence
                                                  = 0.02 x 0.6 = 0.012 = 1.2 %

This frequency to be equally divided between two types of double crossovers

2. Frequency of single crossover at region I (SCO I) = 10.0 – 1.2 = 8.8 % and

it will be equally divided between two types of SCO I.

3. Frequency of SCO II = 20.0 − 1.2 = 18.8 % and it will be equally divided between two types of SCO II.

4. Frequency of parental types   = 100 − (SCO I + SCO II + DCO)
                                        = 100 − (8.8 + 18.8 + 1.2) = 71.2 %

and it will be equally divided between two parental types.

Example 5.4 The data given below were recorded on trihybrid crosses on fruit flies in four experiments. The numbers of progenies recorded in these four experiments belonging to different 8 classes have been given below. Find out the frequencies of progenies of different four classes, the crossover percentage, the map distance between loci and the position of 3 genes on the chromosomes.

Classes No.	Class types	No. of progenies (Different Expt.)			
		I	*II*	*III*	*IV*
1	Parental	828	2207	725	580
2		810	2125	719	592
3	S.C.O.(I)	62	273	419	45
4		88	265	383	40
5	S.C.O.(II)	103	217	134	89
6		89	223	109	94
7	D.C.O.	0	3	34	3
8		0	5	32	5
Total progenies recorded:		1980	5318	2555	1448

Find out the frequency of observed DCO types, if the map distance is 10 and 20 units with 40 % interference.

Hint: Find out frequency of expected DCO as 0.10 x 0.20 = 0.02 (= 2 %);

[Frequency of observed DCO types = Coincidence x frequency of expected DCO.]

5.5 The full colour (black) in mice is dominant over Himalayan (white with coloured extremities). Another character of hair texture is expressed as frizzy hair coat by recessive allele and normal smooth hair by dominant allele. A number of dihybrid mice were test crossed and the resultant progeny produced were:

14 mice with Himalayan colour smooth hairs

12 mice with black colour frizzy hairs

4 mice with Himalayan colour frizzy hairs

3 mice with black colour smooth hairs

(a) Conduct a test of linkage between two loci and if linkage is present, calculate the crossover percentage.

(b) Write down the genotypes of the parents of the dihybrid mice.

CHAPTER 6: SEX CONTROLLED INHERITANCE

6.1. The coat colour pattern in Ayrshire breed of cattle is a sex influenced character. A cow with pure mahogany coat colour was mated with a bull with red coat. Find out the genotypes and phenotypes for F_1 and F_2 animals.

6.2. Baldness in human is a sex influenced trait. A normal man is married with a normal girl for this character. They produced one son and one daughter. The baldness appeared in their son at the age of 30. What were the genotypes of the parents?

6.3. A shepherd has a horned ram in the flock of hornless ewes. He is interested to develop a polled flock. He succeeds in his goal. What were the genotypes of male and females of parental generation?

6.4. Cock feathering in poultry is a sex limited character recessive to hen feathering. Give the genotypes and phenotypes of all the progeny of two sexes produced from heterozygous mating.

6.5. What will be the genotypes of the parents of colour blind boy who had his colour blind sister and normal brother? [Hint: XX^{c} and $X^c Y$].

6.6. Hemophilia in man is caused due to recessive sex-linked gene. List out the genotypes and phenotypes of different sons and daughters born from *(i)* the marriage of carrier women with a normal man and *(ii)* if a carrier girl marry with a man suffered with the disease (hemophilia).

6.7. Three sons and two daughters were born to a couple. One of their sons was found hemophilic. What is the chance of their other children to develop this ailment?.

[**Hint:** Since one of the two sons develops the disease, so father should be normal and the mother seems to be carrier].

6.8. A normal girl having a colour blind brother marries with a normal boy. Predict the chance of developing colour blindness in her son.

[**Hint:** The girl may be normal or carrier].

6.9. Barred feather pattern is dominant sex linked character over black. What will be the genotypes and phenotypes of the progeny of two sexes produced by mating of two birds differing in feather pattern?

PART II
POPULATION GENETICS

Chapter 11

Genetic Structutre of Population

The study of genetic structure of population over space and time is known as *Population genetics.* A population is described for its equilibrium genetic structure under random mating assuming the population size to be so large that the sampling variation does not occur. It is important to know the meaning of population before the details of genetic structure of population and then to know the importance of Population Genetics.

11.1 DEFINITION OF POPULATION

A population geneticist considers the population as the *"Mendelian Population"* which is defined as a *group of interbreeding individuals developed over both space and time* sharing a common pool of genes, from which meaningful samples can be drawn and within which the characteristics under study follow the Mendelian rules of inheritance.

Now, it would be better to know about the gene pool and the individual. The *gene pool* is taken as the sum total of genes in a population. In other words, the gene pool includes collectively all the genetic information. The individual is a unit of population which by forming the group constitutes the population.

11.2 GENETIC STRUCTURE OF POPULATION

The genetic structure of a population is given by the population parameters for one or more traits for a given period of time and the changes that occurred over space and time. These population parameters are character specific which means different for qualitative and quantitative characters, for the reason of the number of gene loci affecting these characters. The population parameters for any trait are described for a population in genetic equilibrium state, in general. Such a population is a large random mating population with Hardy-Weinberg equilibrium genotype frequencies in which no special breeding method (selection, mating system) is applied and hence there is no change in the genetic structure of the population over time.

11.2.1 Genetic Structure of Qualitative Traits

The population parameters used to describe the genetic structure of a population for a qualitative trait are the *gene distribution (array of genes)* and the *array of genotypes* for a character. The array of genes (known as genic structure) is given by mentioning the number of loci affecting a trait, the number of alleles present at a locus (two or more *viz.* multiple alleles) affecting the character and the frequencies of different alleles at each locus in a population. The *genotypic array* (known as the genotypic structure of a population) is given by mentioning the numbers of genotypes and their frequencies among all the genotypes for a trait in the population.

Gene frequency and genotype frequency: Any combination of two alleles (in a diploid bisexual individual) in a pair or a series of multiple alleles at one or more loci affecting a character is called the *genotype.*

The *gene frequency* or more specifically and more correctly, the allele frequency is defined as the proportion of a given allele in a pair of alleles or in a series of multiple alleles at a locus to the total number of alleles at that locus in the population. The *genotype frequency* is the percentage or proportion of a particular genotype, formed by a pair or series of alleles at one or more loci affecting a character, among the total individuals in a population.

The gene frequency is represented as p_i whereas the genotype frequency is represented as P_{ii} for homozygous genotypes and as p_{ij} for the heterozygous genotypes, where i indicate the ith allele at a locus.

(i) **Genetic structure - single locus two alleles:** The two alleles at a locus on autosomes may be same (identical) or different in structure and function (A_1 and A_2). These two alleles are represented as A_{ij}. An individual is called as homozygous that carries identical alleles at the locus (A_{ii}). The individual is called heterozygous that carries different alleles at a locus (A_{ij}). The frequencies of two alleles (A_1 and A_2) are represented as p and q, respectively.

In case of diploid population of N individuals, the two alleles in an individual may be present in either combination (genotypes) as A_1A_1, A_1A_2 and A_2A_2. These gene combinations in an individual are called as genotypes represented as D, H and R, and their respective frequencies are denoted as p^2, 2pq and q^2.

Now considering N total individuals in a population, there will be N_{11} individuals with A_1A_1 genotype (gene combination), N_{12} individuals with A_1A_2 genotype and N_{22} individuals with A_2A_2 genotype, so that the genotype frequencies of three genotypes will be as:

Genotypes	No. of individuals	Genotype frequencies
A_1A_1	N_{11}	$N_{11}/N = D = P_{11} = p^2$
A_1A_2	N_{12}	$N_{12}/N = H = 2P_{12} = 2pq$
A_2A_2	N_{22}	$N_{22}/N = R = P_{22} = q^2$
Total	N	. 1.0

(ii) **Genetic structure – single locus multiple alleles:** The multiple alleles are represented as $A_1 A_2,..., A_i, A_j, A_k$ with their respective frequencies as p, q, ..., r, s, t. As only two alleles are present in an individual at a locus, the different alleles of multiple series will be present at that locus in different individuals. The homozygote having identical alleles at a locus is represented to have $A_i A_i$ genotype with its frequency as P_{ii} whereas the heterozygote having different alleles at the locus is represented as $A_i A_j$ with frequency as P_{ij}.

The frequency of an allele (p_i) will be obtained as:

$P_i = P_{i1} + P_{i2} + ... + P_{ik} = \Sigma P_{ij}$, provided all the genotypes are identifiable as in case of no dominance or co-dominance. Therefore, the frequency of an allele is obtained by adding the frequency of the homozygote for that allele to half of the frequency of all the heterozygote having that allele.

The representation of different alleles at a locus along with their frequencies is known as the genic structure of a population.

Considering 3 alleles at a locus in an equilibrium population, the frequencies of 3 alleles and 6 genotypes with their frequencies can be represented as:

Alleles & their frequencies			Genotypes & their frequencies					
A_1	A_2	A_3	$A_1 A_1$	$A_2 A_2$	$A_3 A_3$	$A_1 A_2$	$A_1 A_3$	$A_2 A_3$
p	q	r	p^2	q^2	r^2	2pq	2pr	2qr

The sum of gene frequencies of all the multiple alleles must equal to unity and hence p + q + r = 1.0. The sum of all the genotype frequencies must also be equal to unity and hence $p^2 + q^2 + r^2 + 2pq + 2pr + 2qr = 1.0$. The homozygotes are equal to the sum of the squared allelic frequencies $(p^2 + q^2 + r^2)$ while the heterozygote are equal to the twice of the product of each pair of frequencies *viz.* 2 (pq + pr + qr).

The possible genotypes with multiple alleles at a locus are more in number compared to that of two allelic genetic system. Taking k as the number of multiple alleles at a locus, the number of homozygote $(A_i A_i)$, heterozygote $(A_i A_j)$ and total genotypes produced will be as under:

No. of total genotypes = No. of homozygote + No. of heterozygote

$$\frac{k (k + 1)}{2} = k + \frac{k (k - 1)}{2}$$

The number of total phenotypes depends on the dominance between multiple alleles and they are lesser than the number of genotypes if dominance of alleles exists.

11.2.2 *Genetic structure for Quantitative Traits*

The gene array and genotype array for the gene loci affecting any metric trait can not be estimated. This is because these are polygenic traits. Therefore, some other population parameters are used to describe the genetic structure of a population for metric trait. These are as under-

(i) The mean of the trait: The population mean indicates the mean genotypic value of the population for a trait. The group means are used to estimate the effect of a certain factor *viz.* the effect of genetic and environmental factors. The genetic factors are those which influence the phenotypic value of a character via genes effect *viz.* the effect of sire (family), line, strain and breed whereas the environmental factors are all those other than genes effect, which affect the phenotypic value.

Each genetic and environmental factor has a number of levels of the effect *viz.* number of breeds, number of strains, number of sires within a breed, number of seasons in a year, number of years for entire period of study, number of feeding levels, number of age groups, etc. The means are estimated for different levels of a factor, known as group means. The magnitude of an effect of a certain factor is reflected in the form of the different means of the different levels of an effect (group means). These group means are compared to know the effects of these genetic and environmental factors. Now, the important point to consider is to know whether the means of different levels of an effect (group means) differ in real sense due to the causing factor or the differences in group means have arisen just by chance. This is tested by partitioning the variance in a character arisen due to the different levels of the causing factor.

(ii) The variance in the trait: The variability in a character is caused by different levels of certain causing factor and it is measured by the variance. The variance, caused in a character due to causing factor, is used to test the reality of an effect of a certain factor. This is known as significance of difference, in statistical term. This is done by partitioning the variance into different causal factors causing the differences in a character. The partitioning means the analysis and thus the analysis of variance means the partitioning of variance. The analysis of variance is done to partition the total variance into causing factors *viz.* genetic variance and environmental variance (variation due to breed, sire, season, year, age, etc.) and the significance of the proportionate contribution of a particular factor is tested by F-test developed by Fisher. The genetic variance is further partitioned into additive genetic variance, dominance variance and interaction variance to describe the genetic structure of a population.

(iii) The covariance among relatives and among traits as well: The covariance among relatives is estimated as a measure of resemblance among relatives and to estimate the different components of genetic variance (additive, dominance and interaction variance). The covariance among traits is estimated to know the association between them and to know the extent of genetic cause of correlation among characters.

The mean, variance and covariance for a metric trait are the population parameters whose genetic basis explains the properties of genes controlling a trait so as to specify the genetic architecture (structure) of a population for any metric trait. The main *properties of genes* are the manner in which the genes produce their effects for expression of phenotypic value of a trait, like additive effect, the degree of dominance at a locus, combined effect of genes at different loci (epistasis), pleiotropic gene action, the linkage relationship among genes involved, effect of genes on fitness (fitness of genotype), and finally their interaction with the environment.

Change in population: The change in population is the second aspect to describe a population. The change may be evolutionary or historical. The history of breeds/ plants varieties/races in humans cover the historical description for a short period of time while evolutionary change includes the change extending over long period of time. When the genetic structure does not change over time from generation to generation but remain constant over generations, it is said to be at equilibrium (or genetic equilibrium) or equilibrium gene frequencies and equilibrium genotypic frequencies.

A number of biological factors determine the *maintenance* or *change* in the genetic equilibrium. These forces may be grouped as: the *breeding behaviour* (breeding systems *viz.* inbreeding and out breeding), the *genetic* factors (mutation), the *environmental factors* (selection and migration) which act on individuals of a population affecting their survival and fertility, and the *population size*. The study of the effect of these factors on change in genetic structure of a population is the subject matter of population genetics.

11.3 POPULATION GENETICS

The *population genetics* is the integral part of the Mendelian Genetics in the sense that the concepts of Mendelian genetics are extended from the individual mating to the mating in the population. Thus, the Mendelian genetics is extended to the population level. The science of genetics studied at the population level to study the genes in a population covering the genetic structure of a population over time and space is thus known as 'population genetics'. Therefore, *population genetics is the study of genetic structure (genetic variation) of a population and changes that occur from generation to generation.*

In a population, the genes interact with size of population, systems of mating and evolutionary forces (selection, migration and mutation) to lead fixation of desirable genes as well as to eliminate undesirable ones. Therefore, *population genetics is fundamentally concerned with the study of the amount of genetic variation within a population along with the effect of forces that change and shape the genetic variation of the population over space and time.*

Population genetics versus quantitative genetics: The population genetics is trait specific. When the traits are governed by major genes showing discontinuous variation (qualitative characters) it is of interest to know the frequency of individual genes at a locus (allele frequencies) as well as of different genotypes (genotype frequencies) in a population as a whole together with the changes in the gene and genotype frequencies over time (from generation to generation) under the influence of certain forces. However, the main aim of animal breeder is to make improvement in useful (economic) traits which are governed by many pairs of genes (polygenes) showing continuous variation. It is not possible to study the individual polygenes affecting the economic traits because these genes have their minor, small and unappreciable effects and further their effects are modified by the variation in environment to express the character. The presence of such minor genes (polygenes), the magnitude of their effect and the potentiality for change in genetic structure for these genes can be demonstrated by using special statistical methods. Thus, statistical methods are used to study Mendelism at population level to know the genetic structure of a population for quantitative traits. This covers the *quantitative analysis of population's genetic potential*. The population genetics thus also study the hereditary phenomenon of quantitative characters based on Mendelism at a population level using statistical methods. The population genetics dealing with inheritance pattern of quantitative characters is called as the *"Quantitative genetics"* or *"Biometrical genetics"*. The quantitative genetics is thus more statistical as it involves the application of statistical methods to study the inheritance pattern of quantitative traits in a population. Thus, the quantitative genetics is a part of population genetics limiting to the study of inheritance of quantitative characters.

Chapter 12
Hardy - Weinberg Law

The Hardy-Weinberg law states about the genetic equilibrium of a population. The equilibrium is a state of no change and hence a constant state. The genetic equilibrium is thus concerned with the constant state of the genetic structure of population. Yule (1902), Castle (1903) and Pearson (1904) reported the genetic equilibrium for the special case of equal gene frequencies ($p = q = 0.5$) at a locus with two alleles. W.E. Castle (1903) of USA was actually the founder of genetic equilibrium principle. He worked on the genetic equilibrium for a case of equal gene frequencies at a locus based on the inheritance of coat colour in mice. He supported Mendelism and rejected Galtonian basis of blending inheritance by illustrating the law of purity of gametes. Karls Pearson (1904) though rejected Mendelism but generalized the principle of segregation and showed that the F_2 ratio of ¼ AA: ½ Aa: ¼ aa is maintained with random mating in a large population.

12.1 GENETIC EQUILIBRIUM

The *genetic equilibrium* means no change in genetic structure of population (gene and genotype frequencies) from one generation to the next. In a genetic equilibrium population, there is a relationship between the gene frequencies and the genotype frequencies. According to the relationship between the two, the genotype frequencies are equal to the square of the sum of gene frequencies and hence this is also called as *square law*. This is as under:

$$(p + q)^2 = p^2 + 2pq + q^2 \dots\dots\dots\dots\dots\dots\dots\dots\dots\dots\dots 12.1$$

in such a way that $p + q = 1.0$ and $p^2 + 2pq + q^2 = 1.0$

A population with above relationship between gene frequencies and genotype frequencies is said to be in genetic equilibrium. This relationship is the *test of genetic equilibrium*. The genetic equilibrium is observed under many possible conditions. The first equilibrium condition is the random mating for different genetic systems *viz.* single locus for two alleles and multiple alleles (autosomes), single locus for sex linked genes, and two or more loci case.

12.2 HARDY- WEINBERG LAW

Hardy, G.H. (1908) of England (Mathematician of Cambridge University) and W. Weinberg (1908) of Germany (Physician) reported independently that the principle of genetic equilibrium in a large random mating population can be applied for any value of gene frequencies. They also described the genetic equilibrium under certain conditions. Therefore, this law of genetic equilibrium under random mating is known as the Hardy-Weinberg law or Hardy-Weinberg principle.

The *HW law states* that in a large random mating population the genetic structure (gene and genotype frequencies) remains constant from generation to generation in the absence of evolutionary forces (mutation, migration and selection).

Random mating is the simplest form of mating behavior in which no principle for mating is followed and hence any individual of one sex has equal chance to mate with any individual of the opposite sex. Thus, no restriction is imposed on mating. The main genetic consequence of random mating in a large population is that the genetic structure of population remains constant from generation to generation. No change in genetic structure (*genetic equilibrium*) is the consequence of random mating in a large population and hence it is also called as *"random mating principle"*.

The Hardy-Weinberg Equilibrium is an extension of Mendel's laws of inheritance, describes the consequence of random mating in a large population, and gives the expected relationship between the gene frequencies and the genotype frequencies in the population. Hardy-Weinberg Law is really a corollary of Mendel's law of segregation. The significance of Hardy-Weinberg equilibrium was appreciated and popularized through the "Genetics and the origin of species" (Dobzhansky, 1937).

The necessary conditions, to hold true the Hardy-Weinberg law, are the large population, random mating among parents and absence of evolutionary forces (migration, mutation and selection). The Hardy-Weinberg Equilibrium is disturbed leading to the change in the genetic properties of a population if the population is of small size, mating is non-random (inbreeding and out breeding) and any of the evolutionary forces (migration, mutation and selection) is working. Therefore, in a population which is not in Hardy-Weinberg Equilibrium any of the above conditions may be operating for not holding true the Hardy-Weinberg Equilibrium.

12.3 APPLICATION OF HARDY-WEINBERG LAW

This law is useful in following ways-

1. Maintenance or change in genetic structure: This law suggests that the genetic structure of a population can be conserved under random mating. Therefore, random mating is advocated if the population has optimum fitness. However, the animal breeder remains interested to change the genetic structure in a desired direction and level. The knowledge of this law is useful for bringing a change in genetic structure by violating one or more conditions under which the Hardy-Weinberg law holds true. The most important and practical force is the selection for bringing genetic change.

2. *Estimation of genotypic frequencies:* Provided the population is larg, the mating is random and no evolutionary force is operating, the genotypic frequencies in progeny generation can be estimated from gene frequencies in parent generation by using the square law as:

$(p + q)^2 = p^2 + 2\ pq + q^2$

3. *Estimation of gene frequencies:* When there is dominance relation between alleles, the estimation of gene frequencies is not possible either by gene counting or from genotypic frequencies. Under this situation, the frequency of recessive allele may be estimated as square root of frequency of recessive homozygote. $q =$

$\sqrt{q^2}$

4. *Test of genetic equilibrium:* The genotypic frequencies observed in a population should be in agreement to that expected genotypic frequencies based on the square law of the relation of gene and genotypic frequencies. This is the test of genetic equilibrium. A significant deviation between observed and expected genotypic frequencies indicates that the population is not in genetic equilibrium and hence one or more of the assumptions of holding true the Hardy-Weinberg law is not satisfied. The chi-square test is applied to test the significance of deviation as-

$$\chi^2 = \frac{(A - p^2N)^2}{p^2N} + \frac{(B - pqN)^2}{2pqN} + \frac{(C - q^2\ N)^2}{q^2N}$$

$$= \frac{A^2/}{p^2N} + \frac{B^2}{2pqN} + \frac{C^2}{p^2N} - N$$

where, A, B and C are the observed numbers of three genotypes

N = A + B + C It is tested at one degree of freedom.

5. *To establish Mendelian hypothesis i.e. mode of inheritance:* The Hardy-Weinberg law can be used to establish a genetic theory about the mode of inheritance of a trait. Some of the examples are as under -

(i) Wright (1917) established the evidence of genetic basis of coat colour in Short horn breed of cattle being controlled by a single locus with two co-dominant alleles rather than two loci hypothesis proposed earlier. He compared the observed frequencies of progeny genotypes with the expected frequencies based on square law for 2 loci and single locus.

(ii) The genetic basis of A, B, O blood antigens as three alleles of a single locus was established by Bernstein (1925) rather than two loci hypothesis proposed in 1911. He found that the observed frequency data was not in agreement with that expected on two loci with dominance.

(iii) The monofactorial basis of thalassemia major and minor was established by Neel (1950) which was earlier thought to be the result of interaction of two dominant factors.

(iv) Snyder (1932) collected family data on the ability to taste the synthetic substance *phenylthiocarbamide* (PTC) and studied the inheritance of human ability

to taste PTC. He found that the ability to taste PTC is dominant (TT, Tt) over lack of ability to taste the substance by people (tt).

Establishment of hypothesis for the genetic determination of a character controlled by major genes requires the analysis of family data for two generations or for a single generation (sib pairs), particularly when there is dominance so as the heterozygotes are distinguished from dominant homozygotes by conducting genetic analy s or by biochemical and immunological techniques.

He worked out the ratio of recessive progeny expected from D x R (dominant x recessive) mating as $S_1 = \dfrac{q}{(1 + q)}$ and the recessive progeny expected from D

x D (dominant x dominant) mating as $S_2 = \dfrac{q^2}{(1 + q)^2}$. The S_1 and S_2 are the expected ratios of recessive progenies from two type of mating. These were called as Snyder's ratios or population ratios which are expected based on square law, for single locus two allelic system with dominance. The value of q is estimated from observed sample data as;

$$q^2 = \frac{\text{No. of recessive parents and progeny}}{\text{No. of total parents and progeny}}$$

$$q = \sqrt{q^2}$$

The observed ratios of recessive progenies out of total progenies from D x D mating (S_2) and from D x R mating (S_1) are worked out from sample data. The observed and expected ratios are compared.

6. Square law and *mating system:* A close fit between the observed and expected frequency data indicates the random mating whereas the deficiency of heterozygotic frequency is an indication of inbreeding, in general.

7. Estimate the micro-evolutionary changes: The micro evolutionary changes can be known by comparing the gene and genotype frequencies of different samples from two localities to know the change over space or from one population at different times (generations) to know the changes over time. Significant deviation between populations over space and over time for a population indicate the micro-evolutionary changes in space and time.

The **Hardy-Weinberg law for single locus two allele and multiple allele** (equilibrium proportions, proof of law, approach to equilibrium and estimation of gene frequencies) have been described here as under-

12.4 HARDY-WEINBERG LAW FOR SINGLE LOCUS TWO ALLELES

In case of single locus with two alleles (A_1 and A_2) with allele frequencies as p and q, three genotypes are produced (A_1A_1, A_1A_2 and A_2A_2) with their respective frequencies denoted as p^2, $2pq$ and q^2.

12.4.1 Equilibrium proportions

Single locus two alleles: When Hardy-Weinberg law holds true in a population, the genotype frequencies among progeny are obtained by the square of the sum of gene frequencies among parents as-

Square of sum of gene frequencies	= Genotype frequencies
(among parents)	(among progeny)

$$(p + q)^2 = p^2 + 2pq + q^2$$

such that $p + q = 1.0$ and $p^2 + 2pq + q^2 = 1.0$

The *homozygote* are equal to the sum of the squared gene frequencies (p^2+q^2) whereas the *heterozygote* are equal to twice of the product of frequencies of two allele *viz.* 2pq. The number of phenotypes depends on the dominance effect of alleles.

12.4.2 Proof of Hardy-Weinberg Law

Single locus two alleles: The random mating in a large population leads to Hardy-Weinberg genotype frequencies in the absence of evolutionary forces. The random mating can be considered in either of two ways. The first is the random union of gametes and second is the random mating among individuals (genotypes). However, the genetic consequences of both the ways are equal in the sense that both ways lead to Hardy-Weinberg genotype frequencies. This can be shown by two ways as under

(i) Random union of gametes: A gamete produced by an individual for one locus is equivalent to one gene. The frequencies of two gametes produced by an individual are thus equal to the allele frequencies. The A_1 gamete will carry A_1 allele while A_2 gametes will carry A_2 allele. The frequency of A_1 gametes is taken as p and frequency of A_2 gametes as *q* among parents. The genotypes and their frequencies, the gametes produced and their frequencies are considered as:

Genotypes (parents)	Frequencies of genotypes	Gametes produced by parents	Frequencies of gametes (genes)
A_1A_1	$p^2 = D$	A_1	p
A_1A_2	$2pq = H$		
A_2A_2	$q^2 = R$	A_2	q

The arithmetic of probability of uniting two gametes of opposite sex is used to predict the zygotic (genotype) frequencies in the progeny. The probability that a zygote of a particular genotype will be formed is the joint probability of random union of two gametes. The probability that a male gamete carrying either A_1 or A_2 gene will unite with the female gamete carrying either A_1 or A_2 gene to from a zygote are independent of each other and hence these are two independent events. Thus, the zygote (genotype) frequencies are the product of the frequencies of the gametes. For example, if A_1 male gamete with its frequency p unites with the female gamete carrying A_1 gene with frequency p, then the zygotic frequency of A_1A_1 zygote will be p x p = p^2 and when A_1 gamete with its frequency p unites with

A_2 gamete with its frequency q, then the zygotic frequency will be p x q = p q. Therefore, the genotype frequencies among the progeny produced by random union of gametes can be determined by multiplying the frequencies of the two uniting gametes. This has been illustrated with equal gene frequencies in two sexes:

Table 12.1: Random union of gametes

		Male parent		
	Genotypes	A_1A_1	A_1A_2	A_2A_2
	Gametes		$A_1(p)$	$A_2(q)$
	A_1A_1			
Female	A_1A_2	A_1 (p)	A_1A_1 (p^2)	A_1A_2 (pq)
parent		A_2 (q)	A_1A_2 (pq)	A_2A_2 (q^2)
	A_2A_2			

The total results in the progeny (the genotypes produced among progeny generation and their frequencies) as a consequence of random union of gametes among parents will be as under:

p^2 (A_1A_1) : 2pq (A_1A_2) : q^2 (A_2A_2)

The third step assumes equal survivability of all the zygotes till identified as genotypes. This results the same genotype frequencies among adults as were among zygotes. Lastly, these genotype frequencies among the progeny generation will give the following gene frequencies:

Freq of A_1 allele $= p^2 + \frac{1}{2}$ (2pq) $= p^2 + p\,q = p\,(p + q) = p$

Freq. of A_2 allele $= q^2 + \frac{1}{2}$ (2pq) $= q^2 + p\,q = q\,(q + p) = q$

This has proved the Hardy-Weinberg law because the gene frequencies in the progeny generation are the same as were in the parent generation and that the genotype frequencies in the progeny depend on the gene frequencies among the parents irrespective of the genotype frequencies. Thus when the mating is random the gene frequencies do not change from one generation to the next and the genotype frequencies among progeny are the square of the sum of gene frequencies among parents as per equation 12.1.

(ii) Random mating of genotypes: This procedure has two steps.

(a) *Mating types and their frequencies*: With complete random mating among the three genotypes taking their equal frequencies in two sexes, the mating frequencies of different genotypes will be obtained by multiplying together the frequencies of the 3 genotypes as under:

Table 12.2: Mating types and their frequencies with random mating

	Genotypes & their frequency in female parent				
	Genotypes	*Frequencies*	*D*	*H*	*R*
Genotypes,	A_1A_1	D	D^2	DH	DR
their freq. in	A_1A_2	H	HD	H^2	RH
male parent	A_2A_2	R	RD	RH	R^2

There are nine types of mating with some mating types being equivalent by ignoring the sex of the parent and thus these nine types are reduced to six types for the reason that DH = HD, DR = RD and HR = RH. This is because the mating of A_1A_1 male with A_1A_2 female (DH) will give equivalent result to the mating of A_1A_1 female with A_1A_2 male (HD). Thus DH = HD and similarly DR = RD and HR = RH.

(b) *Genotypes and their frequencies among progeny*: The above six mating types will produce the genotypes of offspring. Among the total progeny produced the frequency of different genotypes in the progeny can be obtained. The A_1A_1 will produce only A_1 gametes and hence the A_1A_1 x A_1A_1 mating will produce only A_1A_1 progeny. The frequency of A_1A_1 genotype among parent is D and hence in the total progeny the A_1A_1 genotype will be in the frequency of D^2 from A_1A_1 x A_1A_1, mating. Likewise, the mating of homozygous parents with A_2A_2 genotype (A_2A_2 x A_2A_2) will produce only A_2A_2 progeny with a frequency of R^2 among total progeny. Similarly, the mating of parents homozygous for different alleles (*e.g.* A_1A_1 x A_2A_2) will produce only the heterozygous progeny of A_1A_2 genotype with a frequency of 2 DR in the total progeny.

The mating of heterozygous with heterozygous parent (heterozygous mating of A_1A_2 x A_1A_2) will produce three genotypes in a ratio of 1:2:1 according to the Mendelian ratios. Thus ¼ of the total progeny from this mating will be of A_1A_1 genotype, ½ of A_1A_2 genotype and ¼ of A_2A_2 genotype.

The mating of heterozygous parent (A_1A_2) with homozygous parent for A_1 allele (A_1A_1) will produce progeny of two genotypes, A_1A_1 and A_1A_2 in equal frequency (DH). Similarly, the mating of heterozygous parent (A_1A_2) with another homozygous parent for A_2 allele (A_2A_2) will produce progeny of two genotypes (A_2A_2 and A_1A_2) in equal frequency (HR). The total results explained above have been shown in a tabular form below:

Table 12.3 Genotypes and their frequencies among progeny

| Mating among parents | | Genotypes & their frequency in progeny | | |
Types	Freq.	A_1A_1	A_1A_2	A_2A_2
A_1A_1 x A_1A_1	D^2	D^2	--	--
A_2A_2 x A_2A_2	R^2	--	--	R^2
A_1A_1 x A_2A_2	2DR	--	2DR	--
A_1A_2 x A_1A_2	H^2	¼ H^2	½ H^2	¼ H^2
A_1A_2 x A_1A_1	2DH	DH	DH	--
A_1A_2 x A_2A_2	2HR	--	HR	HR

The frequency of each genotype in the total progeny can be obtained by adding the frequencies of progeny of different genotypes produced by each type of mating as under-

$$\text{Frequency of } A_1A_1 \text{ genotype: } = D^2 + DH + ¼ H^2 = (D + ½ H)^2$$
$$= p^2 \text{ Since } D + ½ H = p$$
$$\text{Frequency of } A_1A_2 \text{ genotype: } = DH + 2DR + ½ H^2 + HR$$
$$= 2(½ DH + DR + ¼ H^2 + ½ HR)$$

$$= 2[D (½ H + R) + ½ H (½ H + R)]$$
$$= 2(½ H + R) (D + ½ H)$$
$$= 2pq \text{ since } ½ H + R = q$$
$$\text{and } ½ H + D = p$$
$$= q$$

Frequency of A_2A_2 genotype: $= R^2 + HR + ¼ H^2 = (R + ½ H)^2$
$$= q^2$$

The frequencies of the three genotypes so obtained are the Hardy-Weinberg equilibrium frequencies. This proves that Hardy-Weinberg E. frequencies are approached by one generation of random mating, irrespective of the genotype frequencies in the parent generation.

The important point about the two ways of random mating (random union of gametes and random mating of individuals) is that they have equivalent genetic consequences. The total result of random mating between individuals and the subsequent random union of gametes produced by the mates is equivalent to complete random union of all the gametes produced by population. The use of the principle of random union of gametes is simple particularly when dealing with multiple alleles, sex linked genes and two or more loci case because the complete enumeration of all possible mating frequencies become tedious. The genetic composition of any generation can be determined by the principle of random union of gametes very easily by finding the total gametic output of the parent generation and uniting the gametes at random.

12.4.3 Approach to equilibrium – Single locus two alleles system

The rate of approach to equilibrium depends upon whether the gene frequencies in the population are equal or different in the two sexes.

(1) Equal gene frequencies in two sexes: If a population is not in equilibrium, it will attain equilibrium in next generation if the random mating is followed and the gene frequencies are equal in two sexes. For example, take the gene frequencies as p=0.2 and q=0.8 in both the sexes of a population. With these gene frequencies, the three genotypes may be in any proportion (frequencies) in the population as given under-

(0.10, 0.20, 0.70); (0.18, 0.04, 0.78); (0.0, 0.4, 0.6);

(0.20, 0.00, 0.80); (0.05, 0.30, 0.65); etc.

Now, if each of these populations are allowed to mate randomly, they will all reach the genetic equilibrium in next generation having the genotype frequencies as, 0.04, 0.32, 0.64 following the square law: $(p + q)^2 = p^2 + 2pq + q^2$. This has been shown in solved example 12.1. The genetic equilibrium will be maintained in successive generations till random mating will be allowed. Thus, the equilibrium condition is immediately established under random mating of one generation.

(2) Unequal gene frequencies in two sexes: When the gene frequencies are not equal in the two sexes, it requires two generations of random mating to approach genetic equilibrium. The initial gene frequencies may be different in the two sexes particularly when most of the migrants are of one sex (like crossbreeding zebu

cows with bulls of European breeds). In such cases, the Hardy-Weinberg frequencies for an autosomal locus are attained in two generations of random mating. The gene frequencies in the first generation of random mating become equal in both sexes which are the average of the frequencies in the parents of two sexes. The genotype frequencies in the second generation of random mating attain equilibrium. This has been verified by an example 12.2.

It may be concluded that if genotype frequencies confirm to the HW law or the square law, then mating among parents was probably at random and other conditions of HW equilibrium were probably met. However, such a conclusion should not be based on a single generation.

12.4.4 Estimation of gene frequencies - Two allelic systems: This depends on the type of gene action *i.e.* dominance-recessive relation between alleles.

1. Co-dominance system: There are two methods to estimate gene frequencies:

(i) Gene counting method: Each individual has two alleles at each locus and so there are a total of 2N genes at a locus in the population of N diploid individuals.

Genotypes	No. of individuals	No. of genes
A_1A_1	N_{11}	$2 N_{11}$
A_1A_2	N_{12}	$2 N_{12}$
A_2A_2	N_{22}	$2 N_{22}$
Total	N	2 N

$$\text{Frequency of } A_1 \text{ gene (p)} = \frac{(2 N_{11} + N_{12})}{2N} = \frac{\text{No. of } A_1 \text{ alleles}}{\text{Total alleles}}$$

$$\text{Frequency of } A_2 \text{ gene (q)} = \frac{(2 N_{22} + N_{12})}{2N} = \frac{\text{No. of } A_2 \text{ alleles}}{\text{Total alleles}}$$

(ii) Genotype frequency method: The A_1A_1 genotype contains all the A_1 alleles, A_2A_2 has all the A_2 alleles while A_1A_2 genotype has one half of the A_1 alleles and another half the A_2 alleles. The Mendelian segregation partitions the two alleles of heterozygote into the two gametic pools. Thus, the frequency of an allele equals to the sum of the frequency of homozygous genotype plus half of the heterozygous genotype. Therefore, the frequency of A_1 allele (p) and A_2 allele (q) are estimated as:

$$P_{(A1)} = D + \tfrac{1}{2} H \text{ and}$$

$$q_{(A2)} = R + \tfrac{1}{2} H$$

As an example to illustrate the numerical estimation of gene frequencies from genotype frequencies, the case of MN blood group of 1000 British people may be cited here from example 12.3.

2. Dominance System: The gene frequencies for the two alleles with dominance can only be estimated in a population with H.W.E. genotype frequencies. The dominance at a locus does not change the equilibrium genotype frequencies because equilibrium is a function of the gene frequencies and not of the type of gene action.

Assuming that the population is in H.W.E., the gene frequency of recessive allele is estimated as the square root of the genotype frequency of the recessive homozygote. This is because the frequency of recessive homozygote (R) is taken as q^2 and the frequency of recessive allele is q.

Therefore, $q = \sqrt{R} = \sqrt{q^2}$ and p = 1-q

The *coat colour* in Aberdeen-Angus cattle has only two phenotypes *viz.* black and red. The red colour is recessive and hence the red cattle are homozygous recessive. Another example is *dropsy in calf* of Ayrshire breed of cattle caused by a recessive autosomal gene. Other examples (characters) showing dominance of alleles have been cited in chapter 4.

12.5 HARDY-WEINBERG LAW FOR MULTIPLE ALLELES

Weinberg (1909) showed that HW law is also applicable to the case of multiple alleles. The zygotes are formed proportionately by random mating of two sexes having equal allelic frequencies according to the square of sum of gene frequencies.

12.5.1 Equilibrium proportions for multiple alleles: The genotype frequencies for multiple alleles in a large random mating population are obtained from expansion of square of a multinomial. For example, consider 3 alleles (A_1, A_2 and A_3) at locus A with their frequencies as p, q, r, respectively such that p +q +r =1.0. In this case there will be 6 genotypes with their proportions as under:

Alleles and their frequencies			Genotypes and their frequencies					
A1	A_2	A_3	A_1A_1	A_2A_2	A_3A_3	A_1A_2	A_1A_3	A_2A_3
$(P + q + r)^2$		=	$p^2 +$	q^2	$+ r^2$	$+ 2pq$	$+ 2pr$	$+ 2 qr$

The sum of allelic frequencies must be equal to unity and hence p + q + r =1.0. Likewise, the sum of genotype frequencies must also be equal to unity and hence $p^2 + q^2 + r^2 + 2pq + 2pr + 2 qr = 1.0$. The homozygote are equal to the sum of the squared frequencies ($p^2 + q^2 + r^2$) while the heterozygote are equal to the twice of the product of each pair of frequencies *viz.* 2 (pq + pr + qr). In general, the genotypes in an equilibrium population can be represented as:

$$(\Sigma p_{i\ Ai})^2 = \Sigma p_i (A_i A_j)^2 + 2 \Sigma (A_i A_j)$$
$$= \text{Homozygote + Heterozygote}$$

Taking k as the number of multiple alleles at a locus, the number of homozygote, heterozygote and total genotypes produced will be as under:

No. of total genotypes = No. of homozygote + No. of heterozygote

$$\frac{K (K+1)}{2} = K + \frac{K (K-1)}{2}$$

The number of total phenotypes depends on the dominance between multiple alleles and they are lesser than the number of genotypes if dominance of alleles exists.

12.5.2 Proof of Hardy-Weinberg Law-Single locus multiple alleles

The HW equilibrium frequencies of genotypes can be proved by the same two methods as done in case of two allelic system *i.e.* random union of gametes and random mating of genotypes. However, the random mating procedure becomes cumbersome due to more number of combinations of 6 genotypes as there will be 36 combinations of mating out of which 15 combinations are similar and therefore there will be only 21 different types of mating among 6 genotypes. Therefore, random union of gamete method is used to test the equilibrium condition as was done for two allelic systems.

12.5.3 Approach to equilibrium for multiple alleles

(i) Equal gene frequencies in two sexes

A population which is not in genetic equilibrium state will approach equilibrium after one generation of random mating provided there are equal gene frequencies in two sexes.

(ii) Unequal gene frequencies in two sexes

When the gene frequencies are not equal in two sexes, it requires 2 generations of random mating to reach the equilibrium state. In the first generation of random mating, the gene frequencies become equal in the two sexes which will be equal to the average of the gene frequencies in parents. In the second generation (produced by random mating) the genotypic frequencies attain equilibrium state. The approach to equilibrium state for multiple alleles having their unequal frequencies can be illustrated by taking 3 alleles at a locus as:

Gene frequencies in male sex $\quad$ $P_{m(A1)} = 0.3$, $q_{m(A2)} = 0.5$ and $r_{m(A3)} = 0.2$

Gene frequencies in female sex $\quad$ $P_{f(A1)} = 0.4$, $q_{f(A2)} = 0.3$ and $r_{f(A3)} = 0.3$

Gene frequencies in G_1 generation after one generation of random mating will be:

$$P_{1(A1)} = 0.35, \quad q_{1(A2)} = 0.4 \text{ and } r_{1(A3)} = 0.25.$$

These new gene frequencies in G_1 generation under random mating are equal to the average of the respective gene frequencies among the parents but these are not in accordance to the square law and hence G_1 generation is not in HWE. But in next generation (G_2), produced after random mating, the genotype frequencies can be determined from the gene frequencies in G_1 generation. The genotype frequencies in G_2 generation will be as under:

$$(p_1 + q_1 + r_1)^2 = p_2^2 + q_2^2 + r_2^2 + 2 p^2 q^2 + 2 p^2 r^2 + 2 q^2 r^2$$
$$(0.35 + 0.40 + 0.25)^2 = 0.122 + 0.16 + 0.062 + 2 (0.14 + 0.087 + 0.10)$$

The above values can be verified from random mating among two sexes of G_1 generation. The genotypic frequencies in G_2 generation are in HWE and will remain in equilibrium in future generations under the conditions of HW law. Thus, if the gene frequencies in a population are not equal in the two sexes for multiple alleles, it will require two generations of random mating to attain H.W.E.

12.5.4 Estimation of gene frequencies for multiple alleles

The gene frequencies are estimated considering the type of allelic interaction among multiple alleles *viz.* absence or presence of dominance and mixture of dominance and co-dominance.

1. Absence of dominance (co-dominance): The number of different alleles are counted and divided by the total number of genes in the population at that locus. Considering 3 alleles at a locus (A_1, A_2 and A_3), there will be 6 genotypes (A_1A_1, A_2A_2, A_3A_3, A_1A_2, A_1A_3 and A_2A_3) and 6 corresponding phenotypes. The frequency of A_1 allele can be estimated as a proportion of total number of A_1 alleles to that of total number of alleles (2N) in the population. Thus, the frequencies of three alleles will be estimated as-

$$P_{(A1)} = \frac{2A_1A_1 + A_1A_2 + A_1A_3}{2N}$$

$$q_{(A2)} = \frac{2A_2A_2 + A_1A_2 + A_2A_3}{2N}$$

$$r_{(A3)} = \frac{2A_3A_3 + A_1A_3 + A_2A_3}{2N}$$

The allele frequencies can also be estimated from the genotypic frequencies. For example, with 3 alleles at a locus, the frequency of A_1 allele will be estimated as-

$P_{(A1)} = f\ A_1A_1 + \frac{1}{2}\ (f\ A_1A_2 + f\ A_1A_3)$,

where, f stands for the frequency of the genotype.

Example: There exists hemoglobin polymorphism in human controlled by 3 alleles at a locus represented by A (normal hemoglobin), S (sickle cell hemoglobin), and C (mild anemia). The hemoglobin S is produced when the amino-acid, glutamic acid present in normal hemoglobin at 6th place is replaced by amino-acid valine, and causes RBC to be sickle under reduced oxygen tension resulting in hemolytic anemia which is fatal before the age of 20. The hemoglobin C (Hb [C]) produced at the same position (residue 6 of â chain) causes mild anemia with hemolysis. The 6 possible genotypes produced for hemoglobin are AA, SS, CC, AS, AC, SC and all these 6 genotypes produce 6 phenotypes due the co-dominance among the 3 alleles.

2. Dominance System: The number of phenotypes equals the number of alleles. The genotypes of some heterozygote show dominance which results in some of the heterozygote and homozygote to become indistinguishable. This makes difficult the estimation of gene frequencies. In such case, random mating is assumed. In case of complete dominance between 3 or more alleles in succession, the gene frequency can be estimated with the most recessive allele by taking its frequency as square root of the recessive genotype frequency.

Let the 3 alleles at a locus be designated as A, a' and a with dominance hierarchy as A>a'>a and representing their corresponding frequencies as p, q, r. This condition of dominance hierarchy produces only 3 phenotypes from 6 genotypes. The genotypes AA, Aa' and Aa with their respective genotypic frequencies of p^2, 2pq, 2pr are grouped in first phenotypic class, denoting the observed number of individuals of this phenotypic class by the letter a or A. and its proportion as $\dfrac{a}{N} = \dfrac{A.}{N}$. The genotypes a'a' and a'a with their respective genotypic frequencies of q^2 and 2pr are grouped in second phenotypic class, denoting the observed number of individuals of this class by the letter b or a'. and its proportion as $\dfrac{b}{N} = \dfrac{a'.}{N}$. The remainder sixth genotype (aa) with its genotypic frequency of r^2 is the third phenotypic class, denoting the observed number of individuals of this third class by the letter c or a. and its proportion as $\dfrac{c}{N} = \dfrac{a.}{N}$ which is the frequency of most recessive genotype (r^2). The proportion of three phenotypic classes will be as under:

$$\text{Proportion of phenotype I} = \frac{AA + Aa' + Aa}{N} = \frac{A.}{N} = \frac{a}{N}$$
$$= p^2 + 2pq + 2pr$$

$$\text{Proportion of phenotype II} = \frac{a'a' + a'a}{N} = \frac{a'.}{N} = \frac{b}{N}$$
$$= q^2 + 2qr$$

$$\text{Proportion of phenotype III} = \frac{aa}{N} = \frac{a.}{N} = \frac{c.}{N}$$
$$= r^2$$

(i) The frequency of most recessive allele (a) represented by r is estimated as-

$$r = \sqrt{r^2} \text{ or } r = \sqrt{\frac{c}{N}}$$

(ii) The frequency of second recessive allele (a') in hierarchy represented by q is estimated as-

$$q = \sqrt{\frac{b}{N} + \frac{c}{N}} - \sqrt{\frac{c}{N}} \quad \text{(Wiener and others)}$$

$$= 1 - \sqrt{\frac{b}{N} + \frac{c}{N}} \quad \text{(Bernstein, 1925)}$$

(iii) The frequency of most dominant allele (A) is estimated by difference as:
p = 1- (q + r).

Examples: Coat colour of rabbit is controlled by three alleles at a single autosomal locus with dominance hierarchy as: $C > c^h > c$ occurring with the frequencies as p. q and r, respectively. The allele c produces albino rabbits, c^h produces Himalayan colour rabbit (white body with pigmented ears, nose, tip of feet and tail) and C produces agouti (full colour).

Land snail (Copaca memoralis in Europe) is the most common polymorphic for shell color and banding patterns. The colour pattern is governed by 3 alleles at a locus showing dominance hierarchy as Brown > pink > yellow and thus yellow colour is the most recessive. The 3 alleles can be represented as B, P, Y, respectively for the 3 colours with frequencies as p, q and r, respectively. The phenotypes of 3 colours in land snail are represented as-

Phenotypes	Brown	Pink	Yellow
Frequencies	$p^2 + 2pq + 2pr$	$q^2 + 2qr$	r^2

3. *Mixture of dominant and co-dominant allele*: The alleles A and B are co-dominant to each other but both are dominant over the third allele O. The procedure to estimate the allelic frequencies is the same as that for the alleles showing dominance. The best example is the *ABO blood group system in man* controlled by three alleles (A, B and O).

Bernstein (1930) explained that the allele A produces antigen A, B allele produces antigen B while the allele O does not produce any antigen. The three alleles follow the ordinary Mendelian inheritance that the alleles A and B are co-dominant to each other but both are dominant over the allele O. The frequencies of three alleles A, B and O are taken, respectively, as p. q and r. This produces six genotypes but only four phenotypes known as blood groups. The phenotypes, genotypes and their frequencies are as under:

Phenotypes (Blood Groups)	Genotypes	Genotype frequencies	Serum agglutinin
A	AA, AO	$p^2 + 2 pr$	Anti- B
B	BB, BO	$q^2 + 2 qr$	Anti- A
AB	AB	$2 pq$	None
O	OO	r^2	Anti- A and Anti-B.

The genotype frequencies are obtained from the expansion of gene frequencies as:

$$(p + q + r)^2 = p^2 + q^2 + r^2 + 2pq + 2pr + 2qr$$

This can be obtained from random union of gametes.

Estimation of allelic frequencies for multiple alleles

(i) Freq. of most recessive allele,

$$r = \sqrt{r^2}$$

$$= \sqrt{\text{genotypic freq. of blood group O}}$$

(ii) Frequencies of co-dominant alleles are estimated from combined frequencies of either of the co-dominant allele (p or q) with most recessive allele (r) as:

(a) Freq. of one co-domonant allele (p) is estimated from combined frequency of blood group A and O as:

$$(p + r)^2 = p^2 + 2pr + r^2$$

Thus, $(p + r) = \sqrt{(A + O)}$ and hence

$$p = (p + r) - r = \sqrt{(A + O)} - \sqrt{O}$$

= frequency of A allele.

(b) Similarly, freq. of other codominant allele (q) is estimated from combined frequency of blood group B and O as:

$$(q + r)^2 = q^2 + 2pq + r^2$$

Thus, $(q + r) = \sqrt{(B + O)}$ and hence

$$q = (q + r) - r = \sqrt{(B + O)} - \sqrt{O}$$

= frequency of allele B

Alternately, the frequency of codominant allele can be estimated by difference as:

$$p = 1 - (q + r) = 1 - \sqrt{(B + O)}$$

and $q = 1 - (p + r) = 1 - \sqrt{(A + O)}$

where, A, B, and O are the frequencies of blood groups.

Chapter 13

Factors Affecting Gene Frequency

The random mating in a large population results in a genotype distribution according to the frequencies of genes (square law) and hence maintains the genetic equilibrium in a population. The Hardy-Weinberg law states about the conditions necessary to maintain the genetic equilibrium. The population in which these conditions are met is called the ideal or idealized population. The *ideal population* is one which produces progeny with all possible relationships between uniting gametes including self fertilization. Therefore, in idealized population, the opposite sex gametes have equal chance of uniting in all possible combinations and the genetic equilibrium is maintained because no disturbing force comes in action to upset the genetic equilibrium. The specified conditions for ideal population are large population size, random mating (mating with all possible combinations including selfing), absence of evolutionary forces of selection, migration and mutation, and the distinct generations without overlapping.

The populations of farm animals are not ideal and hence the genetic equilibrium is disturbed. First of all, the farm animals are bisexual and hence selfing can not take place. Secondly, the *real population* of farm animals also differs from idealized population because the population size is small, the migration takes place, selection is occurring, mating is restricted, mutation may takes place though very rare and the generations are not distinct but overlap.

It is thus obvious that most of the conditions for ideal population to maintain the genetic equilibrium are not fulfilled. Therefore, the genetic equilibrium is not observed in population of farm animals and the change in genetic structure of population is likely to occur in all practical situations. The change can be brought to favourable direction and magnitude after having the knowledge of the genetic effects of the breeding policy which mainly covers the selective breeding and migration. Therefore, it is most essential for a breeder to know the effects of different forces leading to the change in genetic structure of population under.

EFFECT OF FORCES CHANGING GENE FREQUENCIES

The forces changing genetic structure of population are divided into two groups. The first group is the deterministic or *systematic forces* which are also called as the vectorial process. The *systematic forces* tend to change the gene frequencies predictable both in amount and direction, and consequently the genotypic frequencies are changed. These are *mutation, migration and selection.*

The second category is the stochastic process or random or *dispersive process.* This process arises in *small population* from the effect of sampling. The change brought by dispersive process is predictable only in amount but not in direction.

13.1 MUTATION AND CHANGE OF GENE FREQUENCY

The mutant allele has an effect different to that of the normal allele. Thus, mutation provides new genetic material on which the different forces can operate. The effect of mutation to cause genetic change depends upon the recurrence of mutation (single mutation or recurrent mutation) and the balancing effect of reverse mutation.

1. Single mutation (Non-recurrent) and its fate: A newly mutant gene arisen in an individual may increase or decrease in its frequency or may be lost from the population by chance elimination in the progeny generation. The frequency of the mutated gene (q) when it occurs just once as a single mutational event is extremely low and it is equal to $\dfrac{1}{2N}$. However, it has an equal chance of survival or being lost. The probability of extinction of a single mutant gene is 0.3679 in the first generation. The probability that it should be lost in second generation after its occurrence is 0.5315 and the probability of being loss of new mutant gene goes on increasing in subsequent generations if the gene is neutral (neither beneficial non harmful) to the individual carrying it. The great majority of single mutant alleles are lost within a few generations after they arise and only a few ones become established in the population. This is because the loss of a new single mutant gene is an irreversible process. Thus, a unique mutation has no effect on the change of population without its selective advantage. The fate of a new single mutation depends on population size, selective advantage, and the distribution of progenies in families in which the mutant gene is arisen.

2. Recurrent mutation: When each mutational event recurred regularly it is called as recurred mutation. This result to an increase in the frequency of the mutant allele in a large population and thus the mutant allele can not be lost by sampling but they established in the population. The frequency of this mutant allele depends on the rate of mutation per generation which varies for different loci but seems to be reasonably constant for a particular locus from generation to generation under constant environmental conditions.

Considering, the original gene A which mutates to its alternate form a (A mutates to a) with a mutation rate denoted by u per generation (u is the proportion of all A genes that mutate to *a* between one generation and next) and the frequency of A in one generation as *p*, then the frequency of a gene (mutant gene) in the next generation will be equal to u p. This makes the frequency of A gene in the next generation as p – *u p* because the change of gene frequency is – *u p*. The frequency of allele *a* in the next generation is increased by the amount *u p*. At any generation (t) the frequency of mutant gene (*q*) may be obtained as:

$$q_t = q_{t-1} + u \; p_{t-1}$$
$$q_t = q_{t-1} + u \; (1-q_{t-1}) \text{ By putting } p_{t-1} = 1- q_{t-1}$$
$$= q_{t-1} + u - u \; q_{t-1}$$
$$= u + (1-u) \; q_{t-1}$$

and $q_{t+1} = u + (1-u) \; q_t$

where, up_{t-1} is the proportion of A allele mutated to *a* in the population of t generation,

q_{t-1} is the frequency of allele a in previous generation (t-1).

This can be taken as the sequence equation of the q_i series. The repeated substitution of the values of q_{t-1} and q_{t-2} expresses the general terms as:

$$q_t = u + (1 - u) \; [u + (1- u) \; q_{t-2}]$$
$$= u + (1-u) \; u + (1- u)^2 \; q_{t-2}$$

This produces a geometric series in which the last term with q_0 refers to the initial generation:

$$q_t = u + (1- u) \; u + (1-u)^2 \; u + (1-u)^3 u + \ldots + (1-u)^t \; q_0$$

The sum of t terms in this geometric series is $(1-u)^t$. Therefore,

$$q_t = 1 - (1-u)^t + (1-u)^t \; q_0$$
$$= 1 - (1-u)^t \; (1 - q_0)$$

Secondly, if the recurrent mutation from A to a in one generation is not opposed, all the A genes in the population will be a after a number of generations. The amount of change in q per generation (Δq) is:

$$\Delta q = q_{t+1} - q_t = u + (1 - u) \; q_t - q_t$$
$$= u \; (1 - q_t)$$

The amount of increase in allele a per generation is larger in the beginning when the frequency of q is small than when allele a in the population is abundant.

3. Reverse mutation: The gene mutates in both directions which means that all the alleles mutate. The mutation of A gene into a allele is termed as *forward mutation* and when *a* mutates into A, it is called as *backward* or *reverse mutation*. It has been indicated that when forward mutation (A to a) is slow, reverse mutation (a to A) is usually slow. Therefore, when an allele is rare the detection of mutation to other alleles is difficult because of low rate of mutation. The forward and backward mutations are important when the frequency of an allele is abundant.

Taking Δq as the increment in a allele, *u* = mutation rate per generation from A to a, and v = reverse mutation rate from a to A, then the gain in a = up and loss in a = *v q*. This means that after one generation there is a gain of a genes equal to *u p* as a result of mutation (A $\rightarrow$ a) and loss in a genes equals to *v q*. The net amount of change in the frequency of a gene per generation (Δq) is:

$$\Delta q = \text{gain} - \text{loss}$$
$$= up - v\, q$$

Thus the relative magnitude of gain or loss per generation will decide the increase or decrease of q. If the forward and backward mutation rates are equal then q will become zero. This will lead no change in q in next generations and thus an equilibrium condition will be reached.

Thus, $\Delta q = 0$ which makes $u\, p = v\, q$ when population is at equilibrium.

and $v\, q = u\, (1\text{-}q)$
$$= u\text{-}u\, q$$
$$u = v\, q + u\, q$$
$$= q\, (v + u)$$

Thus, at equilibrium, $q = \dfrac{u}{u+v}$ and $p = \dfrac{v}{u+v}$

It has been observed that the reverse mutation (a to A) is less frequent than forward mutation (A to a). Thus *the mutation rate from wild type mutant is more* than mutation from mutant to the wild type. This resulted in to mutants as the more common form whereas the wild type as rare form. However, this condition is not observed in natural populations due to the effects of selection.

Regarding equilibrium state of q, it is important to note that:

- The rate of forward and backward mutation (u and v) are constant which makes the q stable
- The value of q at equilibrium is independent of initial gene frequencies in the population but determined entirely by the relative amount of u and v.
- The value of q is decreased in subsequent generation if q is higher than the value of q at equilibrium and the decrease continues till the value of q at equilibrium is reached and the vice versa is also true. As a result if the value of q deviates from q value at equilibrium $\left(\hat{q}\right)$ due to some reason, it comes back to equilibrium slowly as soon as the causal factor ceases to operate.
- The rate of approaching the q value to equilibrium is low and depends on the extent of deviation of the actual q from its equilibrium value. Thus if the actual q deviates from q at equilibrium, the change per generation (Δq) will be:

$$q = u\, p - v\, q = u\, (1 - q) - v\, q = u - u\, q - v\, q$$

$$= (u + v)\, \hat{q} - (u + v)\, q \text{ since } u = (u + v)\, q$$

$$= - (u + v)\, (q - \hat{q})$$

13.2. MIGRATION AND CHANGE OF GENE FREQUENCY

The migration means the transfer or movement of animals from one herd (sub population) to another. The animals are migrated through purchase or under exchange system in trade breeding by transferring the animals to another breeder's herd. The sires of improved breeds are imported from other countries for genetic improvement of other breeds. The technique of AI and frozen semen has made possible the transfer of semen from one herd to another inside or outside the country and has made faster the process of gene flow or gene migration without migration of the animals. Therefore, the genes of improved breed (s) migrate by transferring or importing the sire as such or their semen to be used in an inferior herd under the crossbreeding programme or upgrading the local, the non- descript animals. Thus, there is gene flow or gene migration through migration of animals or their semen.

The individuals leaving the population are called as *emigrants* but after joining other population are called *immigrants* or *migrants*. Thus the individuals which enter a population are called immigrants or migrants. The individuals of the population to which the migrants join are called the native animals or the native population. The migrants take part in reproduction of the native population. Thus the migrants genetically unit the native one and make a link to cause all members of a species to share a common pool of genes. The migration is closely related to mating systems because migrants take part into the breeding structure of a population.

Patterns of migration and their effects: The migration may occur in several ways *viz.* one way migration, mutual or reverse migration, fusion of sub populations etc. The effects of all the three types of migration on genetic structure of population have been discussed separately.

1. **One-way Migration:** When migration occurs from one population to another without any amount of migration in the reverse direction, it is called as the one-way migration.

Single Versus Recurrent Migration: The single and recurrent migration has their different effects. As a result of single migration of animals, genetic structure of population is changed but a new genetic equilibrium is attained after random mating on the basis of new genetic structure. In case of recurrent migration (when local sires are completely replaced by extraneous ones) the mixed population eventually turned like that of migrants resulting the ousting of the existing genes of native population. The recurrent (continuous) migration is like that of upgrading through continuous back crossing of crossbred progeny with sires of improved breed. This leads the genes of the improved breeds to be fixed in the progeny generations and the frequency of the alleles introduced become $(2^t-1)/2^t$ after *t* generation. The genes from migrants become numerous gradually and thereby reducing the frequency of the existing genes of native population. Thus migration brings about change in gene frequencies and so creates differences in genotypic relationship. This results in a marked reduction in the frequency of occurrence of original genotypes and appearance of new genotypes due to combination of the original and new genes if the migrants belong to other breed.

Consider two populations before migration. Some individuals migrate from one population (population 2, migrants) to join the other population (population

1, which is native population) and form a mixed population:

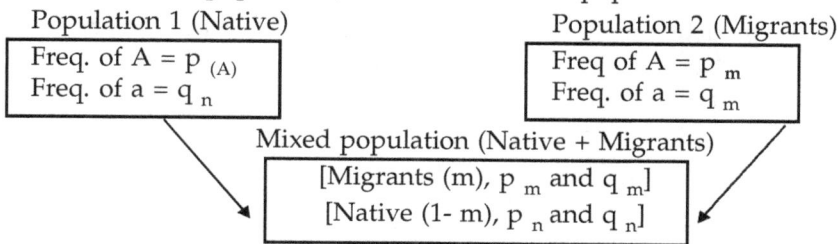

Population 1 (Native)   Population 2 (Migrants)

Freq. of A = $p_{(A)}$ Freq of A = p_m
Freq. of a = q_n Freq. of a = q_m

Mixed population (Native + Migrants)
[Migrants (m), p_m and q_m]
[Native (1- m), p_n and q_n]

The migration is similar to mating system *viz.* crossbreeding, out crossing and grading up. The genetic effect of migration can be described considering the rate of migration (m) and the gene frequencies (q) in the two populations as:

$$m = \text{Migration rate} = \text{proportion of migrants} = n_2/n_1 + n_2$$

where, n_1 are the number of individuals in the native population,

n_2 are the number of migrants, respectively.

1 - m = Proportion of native individuals in the mixed population,

q_m = gene frequency among migrants

q_n = gene frequency among natives.

q_1 = gene frequency in mixed population

q_t = gene frequency in t[th] generation of recurrent migration

Among humans, the migration rate depends upon a number of factors *viz.* sex, marital status, socio-economic status, population density, education, business etc.

The gene in the mixed population after migration will come from native (host) population with a probability of (1-m) q_n and from the migrants with a probability of m q_m. Therefore, the *gene frequency after migration* (q_1) will be:

$$q_1 = m\, q_m + (1\text{-}m)\, q_n = m\, q_m + q_n - m\, q_n$$
$$= q_n + m\, (q_m - q_n),$$
$$= q_n - m\, (q_n - q_m) \qquad\qquad 13.1$$

This can be expressed in terms of the proportion of natives (1- m). This is done by adding and substracting q_m in the above equation of q_1. On solving, the value of q_1 will be as

$$q_1 = q_m + (1\text{-}m)\,(q_n - q_m) \qquad\qquad 13.2$$

The gene frequency with recurrent migration for a number of *t* generations (q_t) will be as:

$$q_t = q_m + (1\text{-}m)^t\,(q_n - q_m) \qquad\qquad 13.3$$

The *rate of change in gene frequency* in one generation (" q) can be estimated from the difference in gene frequency in natives before and after migration as

$$\Delta q = q_1 - q_n = q_n + m\,(q_m - q_n) - q_n$$
$$= m\,(q_m - q_n) \qquad\qquad 13.4$$

It can be noted that the change in gene frequency (" q) depends on the migration rate (m) and the difference in gene frequency of all natives and migrants ($q_m - q_n$). The change is proportional to both these factors.

The *difference in gene frequency* between two populations is reduced after migration. The reduction in difference of gene frequency can be obtained as:

$$q_1 - q_m = [q_n + m (q_m - q_n)] - q_m$$
$$= q_n - q_m - m (q_n - q_m)$$
$$= 1 (q_n - q_m) - m (q_n - q_m)$$
$$= (1-m) (q_n - q_m) \qquad 13.5$$

Thus, the difference in gene frequency between two populations after migration depends upon the initial difference in gene frequency between two populations and the proportion of natives. The difference is reduced with increase in migration rate. Therefore, further migration will reduce the difference in gene frequency between two populations. It means that the two populations tend to become alike genetically or homogeneous in gene frequencies with continuous migration. The frequency of gene present in the migrants is continuously increased with continuous migration. Therefore, in a crossbreeding programme with continuous use of sires of improved breed on the crossbred female progeny or in an upgrading programme of mating non descript female with sires of improved breed will increase the frequency of genes of improved breed leading to fixation of the allele after a number of generations. The frequency of alleles introduced becomes $(2^t - 1)/2^t$ after t generations if the gene frequency is zero in the native or non-descript population. In this situation of crossbreeding and upgrading, the migration rate is 0.5 because the sires of improved breed contribute half of the total genes of the next progeny generation.

The amount of genetic migration (m) can be estimated from the allele frequency data as:

$$m = \frac{\Delta q}{q_m - q_n} \text{ from equation 13.4}$$

$$m = \frac{q_1 - q_n}{q_m - q_n}$$

The migration rate (m) per generation after a number of t generations of continuous migration can be estimated using the equation of q_t from equation 13.3 as:

$$q_t = q_m + (1- m)^t (q_n - q_m)$$

$$(1-m)^t = \left[\frac{(q_t - q_m)}{(q_n - q_m)} \right]$$

$$(1-m) = \left[\frac{(q_t - q_m)}{(q_n - q_m)} \right]^{\frac{1}{t}}$$

and $\quad m = 1-(1-m)$

2. **Reverse or Mutual migration:** The migration is also practiced for another purpose when the individuals of two or more populations are reciprocally exchanged under associated herd improvement programme. The interchange of individuals between sub populations is called as reverse migration.

To estimate the effect of reverse or mutual migration the migration rate (m) from and to both the populations is taken equal as well as both these populations are assumed to be large enough to ignore the effect of random drift. In this case, the frequency of gene among the migrants is taken equal to the average gene frequency in the two populations (q). Therefore, substituting the value of q in place of q_m in the estimation equation of (q_1) obtained in case of one-way migration (eq. 13.1). The gene frequency after migration can be estimated as:

$$q_1 = m \bar{q} + (1-m) q_0 = q_0 - m (q_0 - \bar{q})$$

$$\Delta q = q_1 - q_0 = m (\bar{q} - q_0)$$

where, q_0 is the initial gene frequency in a particular group.
The gene frequency after t generation will be

$$q_t = m \bar{q} + q_{t-1} (1-m)$$

$$= \bar{q} + (1-m)^t (q_0 - \bar{q}) \text{ equivalent to eq. 13.3}$$

The example of reverse or mutual migration is the associated herd progeny testing programme under which the animals are exchanged between two or more herds. The migration rate is fixed for all generations to estimate the change in gene frequency over time (generations). The gene frequency is increased in herd (s) having initial gene frequency lower than average gene frequency of all herds ($\bar{q}$) while the gene frequency is reduced towards q in those herds having initial gene frequency higher than $\bar{q}$. The magnitude of convergence in t generation (q_t-$\bar{q}$, measured as Δq_t) towards equilibrium gene frequency can be estimated as:

$$q_t - \bar{q} = (1-m)^t (q_0 - \bar{q}) \text{ equivalent to eq.} \qquad 13.5$$

The average gene frequency in two or more subpopulation (q) remains same in all the generations ignoring the random genetic drift, mutation and selection.

The possibility of reverse migration is more among the neighbouring herds or subpopulations. The change in gene frequency is less because the neighbouring sub populations have lesser differences in their gene frequency due to frequent interchange of animals. Therefore, with frequent exchange of animals between any sub populations, the gene frequencies become nearly equal in such populations.

3. Fusion of subpopulations- Isolate breaking: As a result of migration, the isolated subpopulations are fused (mixed) together which is known as isolate breaking. The subdivision of a population into small lines produces a deficiency of heterozygote and a corresponding increase of the frequency of homozygote. The fusion of two subpopulations into one has an effect, reverse to that of the subdivision, on the genotype frequencies. Fusion of populations reduces the frequency of homozygote and a consequent increase in the frequencies of heterozygote. This phenomenon is called as *Wahlund's principle*.

In human populations, there are a number of harmful recessive genes at high frequency in certain population *viz.* sickle-cell anemia in Negros (q= 0.05 to 0.1), Tay-Sachs disease (q=0.013) which is degenerative disorder of brain among Ashkenazi Jews, albinism in Southwest American Indians (q= 0.077), antitrypsin deficiency (q = 0.024) and cystic fibrosis (q= 0.022) in Caucasians. The frequency of children born with genetic defects resulting from these recessive genes in homozygous condition can be reduced by migration in the population in which they are relatively high.

Illustration of Wahlund's Principle: The reduction in the frequency of homozygote in fused population after random mating (Wahlunds principle or isolate breaking) can be expressed in terms of variance of gene frequency. The variance measures the degree of dispersion of a set of numbers from the mean or how closely the individual numbers cluster around the mean. Mathematically, the variance is calculated as the mean of the squared values minus the square of mean.

Thus, $\sigma^2_q = \overline{q^2} - (\overline{q})^2$. This can be shown as:

	Sub population 1	Sub population 2	Mixed population
Freq. of allele a	q_1	q_2	q
Freq. of genotype aa	$(q_1)^2$	$(q_2)^2$	q^2

Av. freq. of allele a $(\overline{q})^2 \quad = \frac{1}{2}(q_1 + q_2)$

Av. freq. of genotype aa $(\overline{q^2}) \quad = \frac{1}{2}[(q_1)^2 + (q_2)^2] = $ Mean of squares

The frequency of recessive homozygote (aa) before fusion is the mean of squares of gene frequency $(\overline{q^2})$ whereas after fusion is the square of the mean gene frequency $(\overline{q})^2$. Therefore, the difference in the frequency of recessive homozygote before and after fusion is the variance in allele frequency among sub populations which indicate the reduction in the frequency of recessive homozygote due to fusion:

Variance of gene freq.

= Freq. of aa before fusion – Freq. of aa after fusion

= [Mean square, q^2] - [Square of the mean, $(\overline{q})^2$]

Example 13.1: Isolate breaking and reduction of frequency of homozygous genotypes:

Sub population	Allele frequency		Genotypic frequencies		
	p	q	p^2	2pq	q^2
1	0.9	0.1	0.81	0.18	0.01
2	0.3	0.7	0.09	0.42	0.49
Average of two sub populations	0.6	0.4	0.45	0.30	0.25
After fusion and random mating	(0.6+0.4)2		0.36	0.48	0.16
Difference in genotype frequencies			0.09	-0.18	0.09

The average frequency of aa homozygote in two sub populations ($\overline{q^2}$) is higher (0.25) than the frequency of homozygote (0.16) after fusion of two sub-populations and one generation of random mating ($\overline{q}$)^2.

In the above example of fusion of two sub populations in which the frequency of recessive allele was 0.1 and 0.7, the average of the variance of gene frequency (mean of squares) before fusion was 0.25 (average frequency of homozygote, $\overline{q^2}$) and the variance of the average gene frequency [square of the mean gene frequency, ($\overline{q}$)^2] was 0.16 which is the frequency of recessive homozygote after fusion. Thus the variance equals 0.25 - 0.16 = 0.09. Therefore, the variance between sub populations is 0.25 - 0.16 = 0.09. This indicated that the frequency of recessive homozygote has been reduced by 9 percent as a result of fusion of two sub populations..

The variance is large when the individual values deviate more from the mean. Similar is the case for the variance of gene frequency among sub populations. When sub populations have higher variation in allele frequencies, the variance in allele frequency among sub populations will be more and if the allele frequencies are equal in all sub-populations the variance in allele frequency among subpopulations will be zero.

Significance of migration: The genetic change in the population is quite obvious in case of migration. Consequent to migration, new alleles are introduced which *create new genetic variation* like mutation, equalize the genetic differences between sub population and continuous migration reduces the chance of fixation of alleles. The migration holds the subpopulations close together genetically reducing the genetic divergence and hence it is a sort of *genetic glue (gelatin).* It has its homogenizing effect on the genetic structure of subpopulations. On the other hand, the species become differentiated genetically in the absence of migration, as in case of self fertilizing plants, asexual reproducing organisms and closed populations. The migration between two populations located distantly is practiced to introduce, from other population, the genes that are rare or absent in one population (native or non-descript). This purpose is achieved by one- way migration. The main effect of isolate breaking is to decrease the overall frequency of homozygote. This information can be utilized to reduce the frequency of genetic defects arise from homozygous recessive genes that are harmful.

13.3 SELECTION AND CHANGE OF GENE FREQUENCY

It is common practice to speak of selection in the sense of favouring the objects. When the objects are animals, the selection implies for giving preference to certain animals in the population to reproduce than others and hence selection is the choice of individuals to produce the next generation. In genetic term, the selection is the differential propagation of genetic material to the next generation. This is the result of differential survival and reproductive ability of different individuals of a population. Some are capable to survive and to produce their progeny while others are not, and some others produce more numbers of their progeny than others. Therefore, there is differential survival and reproductive success among the individuals of a population. The *process of differential survival and reproduction of individuals is known as "selection"*.

13.3.1. FORCES OF SELECTION

The survival and reproductive success are two ingredients of selection and these are influenced by a number of life-cycle components at gametic and zygotic stage which in turn are under the control of two forces *viz.* natural and artificial.

(i) **Natural Selection:** The first force of selection is the natural force whose principle is the *"survival of fittest"* in a given environment. The survival of the fittest is determined / measured by the survivability and the reproductive success of the individual. Therefore, survivability and the reproductive status are known as the fitness or adaptive characters. The individuals that are not fit to survive and to produce their progeny died of their genetic death without leaving any offspring. This is known as natural selection.

(ii) **Artificial selection:-** The second force of selection is man made which depends upon the choice of the breeder to allow the animals to produce the next generation. This is called as artificial selection. It is under the control of breeder who decides that to which animals he wants to retain and allow becoming parents of next generation. The choice of the breeder is objective specific and hence the artificial selection has certain purposes. Some genotypes are either more attractive, productive or more efficient functionally and hence preferred by the breeder.

13.3.2. FITNESS AND SELECTION COEFFICIENT

A genotype is better fit and adapted if it produces and leaves more number of offspring than others in the same environment. Therefore, the *fitness (adaptive value* or *selective value)* of an individual indicates its contribution of offspring to the next generation and hence is a function of the survival and reproductive success of the individual in a given environment compared to its competitors in the population The simplest form of measuring the fitness is counting the number of offspring produced and left by an individual (genotype) compared to that produced by another optimum genotype (individual) having maximum number of offspring. The *fitness of an individual is thus defined as the overall relative survival and reproductive success*. This is also known as *Darwinian fitness* of individual.

The fitness of different genotypes of a population is known as the *absolute fitness* which is converted into relative (Darwinian) fitness by assigning the value 1.0 to the absolute fitness of the genotype with largest fitness in the population For example, the females of three genotypes at a locus (AA, Aa and aa) produces on an average 80, 80 and 64 eggs, respectively. These values are the absolute fitness. Now assigning the females with highest absolute fitness (AA and Aa) as 1.0, the relative fitness of three genotypes will be $\frac{80}{80}, \frac{80}{80}, \frac{64}{80}$ or 1.0, 1.0 and 0.8, respectively. Thus, the *relative fitness (Darwinian fitness) is* assigned to the individual of each genotype within a population. These are called as the adaptive values or selective values denoted by W_1. Thus, W_1 indicates the relative fitness (selective value) of ith genotype of a population.

In the example above, the aa genotype had less contribution to next generation and hence the selection is operating against *aa* genotype. The *proportionate reduction in contribution of a genotype selected against compared to the contribution of the favoured genotype with highest fitness* (W = 1.0) is called as the selection coefficient denoted by S. The selection coefficient indicates the strength of selection. The contribution of a favoured genotype (fitness) is taken as 1.0 and that of the genotype with reduced contribution (for the genotype selected against) is taken as 1 – S. Thus, 1-S is *fitness of the genotype* selected against relative to the other favoured genotype. Therefore, the fitness of a genotype is conversely related to the selection coefficient as: Fitness (W) = 1-S

The S is the selection coefficient and indicates the fraction of genotypes which do not reproduce and hence do not contribute to the next generation. Thus S expresses the amount by which the adaptive value is reduced and is used to measure the relative fitness of different genotypes as $1\text{-}S_1 = W_1$, $1\text{-}S_2 = W_2$, ..., etc. taking the relative fitness of the most favoured genotype as 1.0.

However, the term adaptability is used for population rather than for the individual because an individual is powerless to respond to changed environment in changing its genotype in a favourable direction. The population or species, through ages, evolves new gene pools which become adapted under new environmental conditions. It is, therefore, estimated the average fitness of a population denoted by $\overline{W}$. The *average fitness of a population (W) is* taken as the mean of the selective value of different genotypes weighted by their relative frequencies. Thus

$$\overline{W} = p_i\, p_j\, w_{ij} = \Sigma f_i\, w_i$$

Marginal fitness of genotypes: The marginal fitness of the alleles A and a equals to the average fitness of all the genotypes containing that allele (A or a) weighted by their relative frequency and the number of A or a alleles they contain. For example, allele A occurs in AA and Aa genotype with relative proportions p and q. Therefore, the marginal fitness (W_1) of genotype containing A allele equals:

$\overline{W}_1 = p\, w_{11} + q\, w_{12}$. Similarly, marginal fitness of genotype containing a allele ($\overline{W}_2$) is: $\overline{W}_2 = p w_{12} + q\, w_{22}$.

The s is the selection coefficient against the aa genotype and h indicates the degree of dominance of allele a. The value of h may be either 0, or ½ or 1. When h=0, the fitness of AA, Aa, aa will be 1, 1, 1 - s, respectively and a is recessive. When h =1, the fitness will be 1, 1 - s and 1 - s, respectively for the three genotypes and a is dominant to A or say that A is favoured recessive. When h= ½, the fitness will be respectively as 1, 1 - ½ s, 1 - s, the allele A is favoured and the alleles are additive in effects on fitness.

13.3.3. CHANGE OF GENE FREQUENCY BY SELECTION

General case: The effect of selection depends on the intensity of selection (selection coefficient) and initial frequency of allele. The change in gene frequency per generation (Δq) is taken as the difference in gene frequency between two generations. Thus, $\Delta q = q_t - q_{t-1}$.

Steps to estimate the effect of selection: General expression

When all the genotypes are equally fit, the population remains in HWE. Therefore, the effect of selection is described by the relative fitness of each genotype. The steps involved to determine the effect of selection on gene frequencies are as:

1. Initial gene frequencies, genotype frequencies (p_{ii}, p_{ij}) and relative fitness (W_{ij}) of different genotypes are determined.

2. Gamete contribution (frequencies) of genotypes after selection ($p_i p_j W_{ij}$):- These are obtained as the products of genotypic frequencies and corresponding fitness values as $p_i p_j W_{ij}$. These are the proportion of each genotype after selection. Therefore, the ratio of AA: Aa: aa among adults is $p^2 w_{11}$: $2pq\ w_{12}$: $q^2 w_{22}$

The ratio of A : a among the gametes of the next generations is therefore, as under:

$$p^2 w_{11} + \frac{1}{2}\ (2pq\ w_{12}) : q^2 w_{22} + \frac{1}{2}\ (2pq\ w_{12})$$
$$= (p^2 w_{11} + p\ q\ w_{12}): q^2 w_{22} + p\ q\ w_{12}$$
$$= p\ (p\ w_{11} + q\ w_{12}): q\ (q\ w_{22} + p\ w_{12})$$

3. Average Fitness ($\overline{W}$): This is the total gamete contribution of the population of different genotypes after selection and is equal to the sum of the products of genotype frequencies with their fitness as

$$\overline{W} = \Sigma p_i p_j\ w_{ij} = p^2 w_{11} + 2pq\ w_{12} + q^2 w_{22}$$

4. New genotypic frequencies after selection: These are the relative genotypic frequencies obtained for each genotype as the fitness of each genotype multiplied by its frequency (gamete contribution after selection, $p_i p_j W_{ij}$) divided by the average fitness of the population ($\overline{W} = \Sigma p_i p_j\ w_{ij}$). Thus, new genotypic frequency

$$= \frac{p_i p_j W_{ij}}{\overline{W}}$$

5. Gene frequency after selection (q_1): It is obtained by dividing the gamete ratio with average fitness as: $q_1 = \dfrac{q^2 w_{22} + pqw_{12}}{\overline{W}}$

6. Change in gene frequency (Δq):
$$\Delta q = q_1 - q_0 = pq\ [p\ (w_{12} - w_{11}) + q\ (w_{22} - w_{12})]$$
$$\Delta p = 1 - \Delta q$$

13.3.4. Patterns of selection and their effect

There are three modes of selection depending on the choice of individuals to be selected on the basis of their phenotypes. These are directional selection, disruptive selection (Bidirectional selection) and stabilizing selection. The directional selection is when the genotype selected against is either homozygous recessive or the genotype may contain dominant gene (homozygous dominant or heterozygote). Thus, the selection is in one direction. The bidirectional selection is when selection is against the heterozygote and in favour of both types of recessives. Thus, the selection is in both the directions. The stabilizing selection is when the selection is in favour of heterozygote and against both the homozygote.

I. Directional selection: This is also called as *one way upward selection* or linear or dynamic selection in which the individuals with extreme phenotypic values (genetically superior) are selected. This type of selection is of great importance to the animal breeder.

(i) Selection against the recessive genotype: The gene A is favoured dominant: This can be taken under two situations – first is when there is partial selection against recessive and second is when there is complete elimination of recessive individuals due to lethal gene when all the recessive homozygote die prior to breeding age and also due to artificial selection against unwanted recessive in a breeding programme.

(a) Partial selection against recessive genotype (Recessive with low fitness):- The dominant are favoured because of their higher fertility or lower mortality or better performance. The A allele is favoured dominant and hence $h = 0$. As a consequence the recessives produce only $(1-s)$ offspring where s is the coefficient of selection. The effect on population can be obtained from the table given below.

		AA	Aa	aa
1.	Genotypes	AA	Aa	aa
	Initial frequencies	p^2	$2pq$	q^2
	Relative fitness	(w_{11})	(w_{12})	(w_{22})
	(Darwinian)	1	1	$1-s$
2.	Gamete contribution	p^2w_{11}	$2pqw_{12}$	q^2w_{22}
	after selection	$= p^2$	$2p$	$q^2(1-s)$

3. Average Fitness (W) $= p^2 + 2pq + q^2(1-s) = 1-s\,q^2$

4. New genotypic frequencies $\dfrac{p^2}{1-sq^2}$ $\dfrac{pq}{1-sq^2}$ $\dfrac{q^2(1-s)}{1-sq^2}$

(Zygotes after selection) $= \dfrac{P_iP_jW_{ij}}{W}$

5. Gene frequency after selection (q_1 or p_1)

$$P_{1\,(A)} = \frac{p^2}{1-sq^2} + \frac{pq}{1-sq^2} = \frac{p}{1-sq^2}$$

$$q_{1\,(a)} = \frac{pq}{1-sq^2} + \frac{q^2(1-s)}{1-sq^2} = \frac{q-sq^2}{1-sq^2}$$

6. Change in gene frequency:

$$\Delta q = q_1 - q$$

$$= \frac{q - sq^2}{1 - sq^2} - q$$

$$= \frac{-sq^2(1-q)}{1 - sq^2}$$

$$= \frac{-spq^2}{1 - sq^2}$$

The frequency of recessive allele decreases by the amount per generation equal

to $\frac{-spq^2}{1 - sq^2}$. When the recessive allele is at low frequency the amount of change in

its frequency per generation (" q) will be low and equal to $-sq^2$ but appreciable when the q is at intermediate. The change will also be low if q is higher than intermediate. Therefore, selection is most effective for common traits in a population but less effective for rare traits.

(b) Complete elimination of recessive genotype (Lethal recessive):- This would apply to natural selection against a recessive lethal. In this case the fitness (w) of recessive genotype is zero (w = 1- s = 1-1= 0). This is because the selection coefficient is equal to one (S = 1). The frequency of recessive allele in the progeny generation (q_1) under partial selection against recessive phenotype is obtained as under.

$$q_1 = \frac{q - sq^2}{1 - sq^2}$$

Now taking, *s* = 1 (complete elimination of recessives)

$$q_1 = \frac{q - q^2}{1 - q^2} = \frac{q(1-q)}{(1+q)(1-q)}]$$

$$= \frac{q}{1+q}$$

Likewise, $p_1 = \frac{p}{1 - q^2} = \frac{1}{1+q}$

The change in gene frequency $(\Delta q) = q_1 - q = \frac{q}{1+q} - q$

$$= \frac{-q_2}{1+q}$$

There is only the heterozygote which produces the gametes having *a* gene, when the aa individuals either die or are completely infertile. In each generation the aa progeny die or fail to reproduce and therefore there is a consequent decrease in the frequency of recessive allele. The gene frequency in t generation can be predicted taking S=1 as:

$$q_1 = \frac{q}{1+q} = \frac{q_0}{1+q_0}$$

$$q_2 = \frac{q_1}{1+q_1} = \frac{q_0}{1+2q_0}$$

Likewise, $qt = \dfrac{q_0}{1+tq_0}$ where, t indicates the number of generations.

(ii) Selection against recessive gene, a: No dominance. The allele A is favoured and will be fixed by this type of selection. When there is no dominance of fitness, the heterozygote is distinguishable and has it fitness (phenotypic value) exactly intermediate between the fitness (phenotypic values) of two homozygotes. It this case of additive effects, the *h* = ½. This is partial selection against incomplete recessive.

		AA	Aa	aa
1.	Genotypes			
	Initial frequencies	p^2	2pq	q^2
	Relative fitness	1	1- ½ s	1- s
2.	Gamete contribution	p^2	2pq (1- ½ s)	q^2 (1- s)
3.	Average Fitness, W =	1- sq		
4.	Genotypic proportion	$\dfrac{p^2 (1-s)}{1-s\,(1-q^2)}$	$\dfrac{2pq (1-s\frac{1}{2})}{1-s\,(1-q^2)}$	$\dfrac{q^2(1-s)}{1-s\,(1-q^2)}$

5. Gene freq. after selection

$$P_{1\,(A)} = \frac{p^2 + pq\left(1-\dfrac{1}{2}s\right)}{1-sq}$$

$$= \frac{p - \dfrac{1}{2}spq}{1- sq}$$

$$q_1(a) = \frac{q - \frac{1}{2}pqs - sq^2}{1 - sq}$$

6. Change in gene frequency $(\Delta q) = q_1 - q$

$$= \frac{q - \frac{1}{2}spq - sq}{1 - sq} - q$$

$$= \frac{spq^2}{1-s\,(1-q^2)}$$

(iii) Selection against dominant phenotype: There may be complete dominance or co-dominance in fitness:

(a) Complete dominance of Fitness: This has equal meaning that selection is in favour of recessive phenotype and against A gene. When the selection is against the dominant phenotype, the fitness is equal for AA as well as for Aa genotype which will be (*1-s*) and the fitness for aa genotype will be equal to 1.0. Consequently the frequency of recessive allele will be increased in each generation till it becomes fixed. The change is demonstrated as follows:

1. Genotypes

	AA	Aa	aa
Initial frequencies	p^2	$2pq$	q^2
Relative fitness	$1-s$	$1-s$	1

2. Gamete contribution after selection (final frequencies)

	AA	Aa	aa
	$p^2(1-s)$	$2pq(1-s)$	q^2

3. Average Fitness = $W = 1 - s(1-q^2)$

4. Genotypic proportion

$$\frac{p^2(1-q)}{1-s(1-q^2)} \qquad \frac{2pq(1-s)}{1-s(1-q^2)} \qquad \frac{q^2}{1-s(1-q^2)}$$

5. Gene freq. after selection $p_{1\,(A)} = \dfrac{p^2(1-s) + pq\,(1-s)}{1-s\,(1-q^2)} = \dfrac{p\,(1-s)}{1-s\,(1-q^2)}$

$$q_{1\,(a)} = \frac{q - spq}{1-s\,(1-q^2)}$$

6. Change in gene frequency $\Delta q = q_1 - q$

$$= \frac{q - sq + sq^2}{1-s(1-q^2)} - q$$

$$= \frac{spq^2}{1 - s(1-q^2)}$$

When the dominant phenotypes will be completely eliminated, there is no source for dominant gene to be transmitted in the progeny. Therefore, after one generation of complete elimination of dominant phenotype the frequency of recessive allele will become one. *It is thus more effective.* If s is small, the change in gene frequency become nearly equal but in opposite direction in two cases when the selection is against the recessive phenotype. The ultimate fate of A allele is that it will lost.

(b) No *dominance of Fitness:* In this case, the value of Δq = ½ sq (1-q).

II. Bidirectional selection: *Selection against heterozygote*: This is also called as *two way selection* or diversifying or centrifugal or *disruptive selection*. The individuals having extreme phenotypes on both the sides are selected and thus, the heterozygote is at selective disadvantage. This selection favours two diverse types at a time and results in two populations with better and poor performance. This selection results in little change in the phenotypic values in the next generation. This type of selection does not change the gene frequency but change the genotypic frequencies, the heterozygous genotypes are reduced. Under some circumstance the fitness of heterozygote may be less than the homozygote. In a random mating population the effect of selection against heterozygote on the change of gene frequency can be estimated as:

	AA	Aa	aa
1. Genotypes			
Initial frequencies	p^2	$2pq$	q^2
Relative fitness	1	1-s	1
2. Gamete contribution after Selection (Final frequencies)	p^2	$2pq(1-s)$	q^2

3. Average Fitness = W = 1− 2pqs

4. Genotypic proportion $\dfrac{p^2}{1-2pqs}$ $\dfrac{2pq(1-s)}{1-2pqs}$ $\dfrac{q^2}{1-2pqs}$

5. Gene freq. after selection $p_{1(A)} = \dfrac{p^2 + pq(1-s)}{1-2pqs} = \dfrac{p-pqs}{1-2pqs}$

$$q_{1(a)} = \frac{q^2 + pq(1-s)}{1-2pqs} = \frac{q-pqs}{1-2pqs}$$

6. Change in gene freq.   Δq = q₁ − q

$$= \frac{pqs(q-p)}{1-2pqs}$$

The interesting point of this type of selection is that when q = ½ the proportion of the two homozygote are equal (p^2 = q^2= ¼) in a random mating population. Consequently there will be no change in gene frequency either the heterozygote is eliminated completely or partially and the population will be in its original genotypic frequencies in the next generation on random mating. Secondly if p# q but p>q the heterozygote will have higher proportion of the rare alleles than in the recessive homozygote. Thus selection against heterozygote will affect more to the rare alleles than the common allele and consequently the frequency of rare allele will be further decreased.

The *well known example of selection against the heterozygote* is the *Rh factor in men*. The heterozygous baby (Rh rh) from Rh negative women (*rh rh*) will have a hemolytic disease known as *"Erylhroblastosis fatalis"* which may cause death of the child. Such a case will only happen if such heterozygous baby is born to a recessive mother (rh rh, said to be rh-negative). Thus, in this case the selection is against the heterozygote born to recessive mothers rather than born to heterozygous or Rh positive mothers which will bore normal children.

III. Stabilizing selection: *Selection favours the heterozygote (over dominance)* and is against both the homozygote. This is also called as *balanced selection* or centripetal or unifying selection. This is the case of over-dominance (heterosis). This selection is based on the superiority of heterozygote over both the homozygote. This selection does not change the mean of progeny generation and to some extent reduces the variance. This selection preserves both the alleles, so it is called as *balanced polymorphism or balanced lethal* when both the homozygotes are lethal and the only surviving genotype is the heterozygote.

The change in gene frequency due to this selection will be:

1. Genotypes

	AA	Aa	aa
	AA	Aa	aa
Initial frequencies	p^2	2pq	q^2
Relative fitness	1- s$_1$	1	1- s$_2$

2. Gamete contribution $\qquad$ p^2 (1- s$_1$) $\qquad$ 2pq $\qquad$ q^2 (1- s$_2$)
   after selection (final frequencies)

3. Average Fitness = $\overline{W}$ =1- s$_1$p^2 – s$_2$q^2

4. Genotypic proportion

$$\frac{p^2 (1-s_1)}{[1-s_1p^2 - s_2q^2]}, \frac{2pq}{1-s_1p^2 - s_2q^2}, \frac{q^2 (1-s_2)}{1- s_1p^2 - s_2q^2}$$

5. Gene freq. after selection $p_{1(A)} = \dfrac{p^2 (1\text{-} s_1) + pq}{1\text{-} s_1p^2 - s_2q^2}$

$$= \frac{p-s_1p^2}{1-s_1p^2 - s_2q^2}$$

$$q_{1\,(a)} = \frac{pq + q^2\,(1-s_2)}{1 - s_1 p^2 - s_2 q^2}$$

$$= \frac{q - s_2 q^2}{1 - s_1 p^2 - s_2 q^2}$$

6. Change in gene frequency

$$\Delta q = q_1 - q$$

$$= \frac{pq\,(s_1 p - s_2 q)}{1 - s_1 p^2 - s_2 q^2}$$

Equilibrium Condition: The consequence of this type of selection leads to an increase in the frequency of heterozygote and a consequent decrease in frequency of both homozygote than their initial frequencies. The gene frequencies will eventually reach a stable equilibrium value instead of being lost or increased to unity as a limit. The increase or decrease in gene frequency depends on the proportion of $s_1 p$ and $s_2 q$. When $s_1 p = s_2 q$, then there will be no change in gene frequency and an equilibrium state will reach and thus q = 0. This will happen when $s_1 p = s_2 q$ or $p\,(w_{11} - w_{12}) = q\,(w_{12} - w_{22})$. Therefore, at equilibrium

$$\frac{p}{q} = \frac{s_2}{s_1}, \quad \frac{1-q}{q} = \frac{s_2}{s_1} \text{ putting } p = 1 - q$$

$$s_1 - s_1 q = s_2 q \text{ and hence } s_1 = s_2 q + s_1 q = q\,(s_1 + s_2)$$

$$\hat{q} = \frac{s_1}{s_1 + s_2} \text{ and } \hat{p} = \frac{s_2}{s_1 + s_2}$$

Therefore, the values of p and q at equilibrium are independent of their initial values in the population but entirely determined by the intensity of selection (selection coefficient) of two homozygotes. Thus, it is not only the degree of superiority which decides the gene frequencies at equilibrium but it is the relative disadvantage of one homozygote (1- s_1) compared with that of others (1- s_2). When the heterozygote is superior irrespective of its degree, the gene frequencies at equilibrium are at more or less intermediate gene frequencies. The intermediate gene frequencies are expected when selection favours the heterozygote.

The effect of birth weight on infant mortality is an example of stabilizing selection, because the new born having their extreme birth weight to either side are likely to die. The best example of *superiority of heterozygote is the sickle cell anemia*. The homozygote suffers from anemia characterized by sickle shape of erythrocyte. This disease is fatal. The hemoglobin of homozygote is of the abnormal types. On the other hand, the heterozygotes do not suffer from anemia. The selective advantage of heterozygote has been shown to be associated with their resistance to malaria. The sickle cell homozygote ($Hb^S\,Hb^S$) is anemic severely having sickle of RBC but resistant to malaria while the normal homozygote ($Hb^+\,Hb^+$) is not anemic have normal hemoglobin but susceptible to malaria and the heterozygote is mild anemic but resistant to malaria for which the heterozygote has advantage over both homozygote.

13.4 SMALL POPLATION AND CHANGE OF GENE FREQUENCY (GENETIC DRIFT)

The gene frequencies remain constant from generation to generation in large population, excluding the effects of migration, mutation and selection. The effects of systematic processes were studied considering large population. On the contrary, if the population size is small, the gene frequencies are subject to change, even the systematic processes are not operating.

The small population is a dispersive process. It differs from the systematic processes discussed so far in this chapter. The effect of small population is predictable in amount but it has random effect in direction which implies that the direction of change in gene frequency is not predictable. The gene frequencies are under random fluctuations as a result of the sampling of gametes in small population. The gametes carry the genes and transmit to the next generation. These gametes formed in parent generation in small population are a sample of the genes of the parent generation and do not represent the whole population. It is well known that if the sample is small, the gene frequencies between two generations are subject to change and this change is random. Thus, the small population is the dispersive process because it disperse the gene frequencies in either direction which is unpredictable

The whole population of any species is divided into sub-populations for certain reasons *viz.* geographical or ecological reason under natural conditions or as a result of breeding plans in domestic and laboratory animals. These sub-populations are the various herds/flocks of domestic animals which are small in numbers. Thus, all the populations of domestic animals are constituted of small number of animals.

13.4.1 SAMPLING

The herds or flocks of farm animals having less number of animals thus represent only a sample of individuals of the whole population. In a population of small size (sample), the total gene pool is thus divided into samples (herds/flocks) of small size. This results in random sampling from the whole population and hence subjected to sampling error because small sample deviates randomly from expectations/reality as per theory of the probability of events. The theory states that there is an increasing consistency to the actual trend (reality) with the increase in number of trails. For example, the occurrence of heads and tails of a coin by tossing come close to the expectation of 50:50 when the numbers of tossing become large but deviates from expectation when the numbers of tossing (trials) are small. Therefore, the consistency of the actual with the expectation depends on the numbers of events (trails) under observation. In case of population of domestic animals, the frequency of two or more alleles of a gene pair are like head or tail of a coin and the number of animals of population are like the numbers of tossing of a coin.

The numbers of breeding animals constitute the size of a population. In small size population, the numbers of breeding animals (parents) are small. Therefore, the allelic frequencies among parents do not represent the actual frequencies of alleles of the whole population but they deviate from the actual frequencies. These allelic frequencies among parents determine the genotypic frequencies of the progeny generation which in turn decides the gene frequency among progeny generation. Therefore, the allele frequencies in different samples of small size are expected to deviate from that of the whole population.

13.4.2 SAMPLING EFFECT

The effect of sampling can be seen by considering a single locus with two alleles (A and a) with their equal frequencies as p = q = ½. In a large random mating population, the frequency of A allele is expected to be ½ in progeny generation. But when the population is small, a few numbers of offspring are raised from the large pool of gametes produced from parental population. The gametes are sampled in a small population and the sample is not expected to have equal number of two alleles at a locus but either of the two alleles may be transmitted in greater numbers than the other. This will lead to change the gene frequency from ½ to any other value, *viz.* if the offspring raised are 20 which are formed by the union of 40 gametes drawn at random from the large pool of gametes produced by the parent generation. Exactly equal number of *A* and *a* gametes may not be sampled out of 40 gametes but it may happen that 30 gametes may carry A allele and the rest 10 gametes may carry *a* allele. Thus, the gene frequency of A allele

(p) in progeny generation will be $\dfrac{30}{40}$ = 0.75 instead of 0.5. Therefore, the frequency

of A allele has changed from 0.5 to 0.75 in one generation. This change in gene frequency is the sampling error whose extent depends upon the size of the population. Thus, when population size is small, the sampling error occurs leading to the gene frequencies to change or fluctuate in any direction and hence the change is random.

The contribution of all the breeding individuals to the gamete pool is equal due to random mating among the individuals of a line. A large number of gametes are produced in each line and unite together at random to form the zygote. However, all the zygotes formed do not survive to become parents due to the restriction of constant population size in all generations. The survival of zygote is random and only some of them become the breeding individuals of the next generation. As a result of random survival of the zygote, the parents do not have their uniform contribution to the progeny generation but contribute according to the probability of survival of their progeny. The number of progeny is one per parent or two per mated pair to maintain the constant population size in all the generations. The random survival of the zygotes is the stage at which the sampling error occurs and has its consequence to influence the gene frequencies. This process results in the reduction of a large number of gametes produced by parental generation to a smaller number of progeny reaching the breeding age. This reduction may occur at several stages but the final number of the breeding progeny determines the consequences of the sampling process if the sampling is random at each stage.

13.4.3. CONCEPT AND DEFINITION OF GENETIC DRIFT

In a population of small size, the numbers of parents (breeding animals) are small and hence the gene frequencies among them do not represent the actual gene frequencies of the whole population but deviate from the of the whole population. The gametes carry a sample of genes of the parental generation of small size and hence the gene frequencies are liable (expected) to change between one generation and the next.

The change in gene frequency occurs due to sampling and hence it is a sampling error. Any change has its magnitude and direction. The magnitude of the change in gene frequency due to sampling process is predictable but its direction cannot be predicted. The direction of change in gene frequency as a result of sampling error is random which means that the change may take place in any direction. Thus, there is no trend in change of gene frequency (increase or decrease) but it may take any value in either direction in the next generation. Therefore, the gene frequency may deviate (move or drift) randomly to any direction and thus the change of gene frequency from one generation to the next is random. The random change of gene frequency in small population, resulting from the sampling process, is called the *random drift* or the *"genetic drift"*, the term coined by Wright (1931).This is also called as *random genetic drift.* The random change of gene frequency due to genetic drift is the *dispersive process as* it disperses the gene frequencies.

13.4.4. CONSEQUENCE OF GENETIC DRIFT

The consequences of dispersive process (random genetic drift) to change the genetic structure of population are as under-

(i) **Direction of change is random:** The direction of change in gene frequency is random due to random genetic drift, because the gene frequency in successive generation seems to drift about in either direction. The change in gene frequency is continuous till it lies between $\dfrac{1}{2N}$ to $\dfrac{2N-1}{2N}$. The gene frequency may increase or decrease in next generation and hence there is no trend in change of gene frequency but the change is in a erratic manner from generation to generation. The gene frequency is drifted back and forth and does not reach equilibrium.

Amount of change and its measurement: *It is measured as variance of gene frequency.* The mean gene frequency of all the lines (q) does not change and remains equal to the gene frequency in the base population (q_0). The gene frequencies in different lines (q_i) deviate from the gene frequency in base population (q_0) due to sampling error and distributed around the mean gene frequency of the base population with a variance $\dfrac{pq}{2N}$ which indicates the deviation (change) in gene frequency among lines after one generation. The variance is the variance of gene frequency in different lines after one generation. All the lines have the same gene frequency initially represented as q_0 and hence the variance $\dfrac{pq}{2N}$ is the variance of the change of gene frequency $(q_1 - q_0 = \Delta q)$ which is due to sampling in one generation. Therefore, the change of gene frequency due to dispersive process can be stated in terms of the variance as:

$$\sigma^2 \Delta q = \frac{p_0 q_0}{2N}$$

The change of gene frequency in one generation is denoted by $\Delta q = (q_1 - q_0)$ and measured by $\sigma^2 \Delta q = \dfrac{pq}{2N}$.

***(ii)* Decay of genetic variability within line:** A population of small size ultimately reaches complete homozygosis which is known as decay of variability or genetic decay because the line losses its capacity to change genetically. The rates of loss and fixation of an allele are equal because the distribution of gene frequency is uniform. Thus, the total rate of decay is $\dfrac{1}{2N}$ per generation which means that the average heterozygosity decreases at this rate per generation out of which 50% $\left(\dfrac{1}{4N}\right)$ is the rate of fixation and the remaining 50% is the rate of loss of an allele per generation. Thus, the average homozygosity is increased equal to the loss in average heterozygosity and this increase in homozygosity is measured in terms of F. Thus F for one generation is $\dfrac{1}{2N}$ and it is the F of the progeny. This loss of heterozygosity on random mating in small population is also called as the rate of "disintegration".

***(iii)* Genetic differences between lines:** The large population is divided in to small groups for a number of reasons. The mating occurs more often among the animals of the same sub-population residing in the same locality. This subdivision and restricted mating within line results the differences in gene frequencies among sub-populations, for the number of individuals in the line is small. Thus, the random drift leads to genetic differences between lines (sub-populations).

***(iv)* Increased Homozygosity: Change in genotypic frequencies:** (*Wahlund formula of breeding structure*): The mating is assumed to be random within each sub population (line) and hence the genotypic frequencies in any one line are in Hardy-Weinberg proportions determined by the gene frequencies in the previous generation of that line. Therefore, the change in gene frequencies leads to a change in genotypic frequencies. The direction of change in genotypic frequencies is towards an increase of homozygous genotype and decrease of heterozygous genotype. The increase in frequency of homozygote in the population as a whole occurs in spite of the fact that random mating is followed in all the sub populations into which a large population is subdivided. The reason is very simple. The gene frequencies drift apart from intermediate value towards the extreme due to random drift. The frequency of heterozygote in a random mating population (2pq) is maximum at intermediate gene frequencies (p = q = 0.5) and decreases with movement of gene frequencies towards the extremes. Thus the random drift or dispersion of gene frequencies towards the extreme as a result of subdivision of population into lines leads to a decrease in the frequency of heterozygote and an excess of homozygote compared with a large single population (in which all lines are considered together as a random mating entirely). Thus the subdivision of a large population into small groups increases the frequency of homozygote and a deficiency of heterozygote. This effect of subdivision of population is like the effect of inbreeding. This is known as *stratification principle or Wahlund's principle.*

Gene fixation–final consequence: All the lines become fixed for one or the other allele. The proportion of lines that become fixed for A allele is p_0 and those for a allele is q_0 because the average gene frequency of A remains p_0 even when all lines become fixed. Thus the probability of fixation of an allele in any ideal line is equal to the frequency of that allele in the initial population. In small population the process of random sampling is continuous till the population eventually becoming fixed for one of the existing allele at a locus. Thus the genetic drift leads to each allele ultimately becoming either fixed or lost. This is the genetic drift theory which states that with or without natural selection, a gene locus with alternate alleles in a small population will become fixed for one of the existing alleles sooner or later. The fixation of one or the other allele is the final consequence of genetic drift. Once of gene is lost (q=0) or fixed (q=1.0), there is no variability in gene frequency and the points of q=0.0 and 1.0 are said to be the dead points or dead ends or points of no return which indicates that the change in gene frequency has been stopped and the process of genetic drift is irreversible at dead ends with out mutation or introduction of an allele. When the lines become genetically uniform (fixed) the variance of gene frequency among lines is equal to pq.

Numerical Solved Examples

CHAPTER 12: HARDY-WEINBERG LAW

Example 12.1 The genotypic frequencies for 3 genotypes in a population were recorded to be as 0.0, 0.4 and 0.6. Estimate the frequencies of two alleles, apply test of genetic equilibrium and show the approach to genetic equilibrium under random mating.

Solution: *(i)* The frequencies of two alleles can be estimated from genotypic frequencies as:

Frequency of A_1 allele (p) = ½ (0.4) = 0.2

Frequency of A_2 allele (q) = ½ (0.4) + 0.6 = 0.8

(ii) Test of genetic equilibrium:

The genotypic frequencies as per square law will be as:

$$(p + q)^2 = p^2 + 2pq + q^2$$
$$(0.2 + 0.8)^2 = 0.04 + 0.32 + 0.64$$

Thus, the genotype frequencies in parental generation were not in genetic equilibrium because they were not in accordance with the gene frequencies.

(iii) Approach to equilibrium: The genotype frequencies in the progeny generation, produced by random union of gametes of parental population, which was not in equilibrium, become as:

$p^2_{(A1A1)} = 0.04$; $2pq$ $(A_1A_2) = 0.32$; q^2 $(A_2A_2) = 0.64$.

These genotype frequencies are in genetic equilibrium because these can be determined by the square of gene frequencies in the parent generation as shown above. Thus, the population (which was not in equilibrium but had equal gene frequencies in the two sexes) attains an equilibrium structure of the genotype proportion in the first generation of random mating.

Example 12.2: The gene frequencies in two sexes were found unequal as under - Gene frequencies in males $p_{m (A1)} = 0.4$, $q_{m (A2)} = 0.6$;

Gene frequencies in females $p_{f(A1)} = 0.8$, $q_{f(A2)} = 0.2$

Where, the subscripts *m* and f indicate the male and female sex.

Show the approach to genetic equilibrium of this population under random mating.

Solution: The population will attain the genetic equilibrium in two generations of random mating. This can be verified by random union of gametes. The new gene frequencies in the first generation will be the average of the respective gene frequencies among the parents and hence

p_1 = ½ (0.4 + 0.8) = 0.6 and $\qquad$ q_1= ½ (0.6 + 0.2) = 0.4

It is interesting to note that the genotype frequencies in the progeny of first generation do not follow the square law because these are not in accordance of the gene frequencies in the parents. Thus the progeny of first generation are not in HW Equilibrium.

Second generation of random mating:- In the next generation (G_2) the genotype frequencies can be determined from the gene frequencies in G_1. Thus the genotype frequencies in second generation (G_2) will be as:

$(p_1 + q_1)^2 = p_1^2 + 2p_1q_1 + q_1^2$ and hence,
$(0.6 + 0.4)^2 = 0.36 + 0.48 + 0.16$

Now this second generation (G_2) is in HW equilibrium and will remain in equilibrium under the assumption of HW law.

Example 12.3: The *MN blood group* in man is controlled by two alleles at one locus showing no dominance. These two alleles are designated as M and N; having co-dominance relation between them, producing molecules (substance) on the surface of RBC of the individuals. The genetic structure for this locus of MN blood group in a sample of 1000 Britishers (Race and Sanger, 1975) is given as under:

Phenotypes (Blood Groups)	M	MN	N	Total
Genotypes				
No. of individuals	298	489	213	1000
% individuals	29.8	48.9	21.3	100.0
Proportion of individuals	0.298	0.489	0.213	1.0
(genotype frequencies)	P_{11} = D	$(2P_{12})$ = H	(P_{22}) = R	

Estimate the frequencies of two alleles from above data.

Solution:

Total numbers of genes (2N) = 2x 1000 = 2000.

Number of M alleles = (2 x 298) + 489 = 1085

Frequency of M allele (p) = $\dfrac{1085}{2000}$ = 0.5425.

Number of N alleles out of total 2000 alleles = (2 x 213) + 489 = 915

Frequency of N allele (q) = $\dfrac{915}{2000}$ = 0.4575.

Example 12.4. The frequency of dropsy calves is about 1 in 300 births. Find out the frequencies of two alleles showing dominance relation and proportion of heterozygous.

Solution: $R = q^2 = \dfrac{1}{300} = 0.0033$ and so

$$q = \sqrt{\dfrac{1}{300}} = \sqrt{0.0033} = 0.057$$

The frequency of dominant allele (p) will be 1-q = 1-0.057 = 0.943

Example 12.5. The following data on hemoglobin polymorphism contolled by multiple alleles showing co-dominance in American Negros was given by Levingstone (1967) as:

Hb Types	AA	AS	AC	SS	SC	CC	Total
Numbers	2501	213	64	14	4	4	2800
Frequencies	0.8932	0.0776	0.0229	0.0059	0.0014	0.0014	100.0

Find out the frequencies of three alleles.

Solution: The gene frequencies can be estimated by gene counting method as well as from the frequencies of genotypes containing that particular allele. The allelic frequencies so estimated were found as:

Hb A = 0.9427; Hb S = 0.0437; Hb C = 0.0136

Example 12.6. The following data on shell colour polymorphism of land snail were recorded. Estimate the frequencies of 3 alleles, showing dominance relationship.

Phenotypes   Brown   Pink   Yellow   Total
Observed No.   173   443   115   731

Solution: The allelic frequencies of 3 alleles producing shell colour can be estimated as under:

Frequency of most recessive allele $(r_Y) = \sqrt{r^2}$

$$= \sqrt{\left(\dfrac{115}{731}\right)} = 0.3966$$

$$\text{Frequency of pink gene (q)} = \sqrt{\left(\dfrac{443+115}{731}\right)} - \sqrt{\left(\dfrac{115}{731}\right)}$$

$$= 0.8737 - 0.3966 = 0.4771$$

Frequency of brown gene (p) = 1- (q + r)

$$= 1-(0.4771 + 0.3966) = 0.1263$$

Example 12.7. ABO blood group in human follow the dominance relation as A=B > O. Find out the allelic frequencies in a population having 29 AB, 371 B, 54 A and 436 O blood groups. Find out the frequencies of 3 alleles.

Solution: Frequency of most recessive allele O (r)

$= \sqrt{\text{genotype frequency of blood group O}} = \sqrt{r^2}$

$= \sqrt{(436/\ 890)} = \sqrt{0.4898} = 0.699$

$$\text{Frequency of A allele (p)} = \sqrt{\left(\frac{54+436}{890}\right)} - \sqrt{\left(\frac{436}{890}\right)}$$

$$= \sqrt{\left(\frac{490}{890}\right)} - 0.4898$$

$$= 0.742 - 0.699 = 0.043$$

$$\text{Frequency of B allele (q)} = 1\text{-}\ p - r = 1\text{-}\ \sqrt{A+O}$$

$$= 1 - 0.043 - 0.699 = 0.258$$

EXERCISES

12.8. Find the proportions of 3 genotypes in next generation produced by random mating in case of the following populations- 0.25, 0.0, 0.75; 0.20, 0.40, 0.40 ; 0.15, 0.25, 0.60

12.9. From the composition of the following population, check whether these are in genetic equilibrium. Estimate the equilibrium proportions for those which are not in equilibrium:

0.20, 0.32, 0.48; 0.30, 0.25, 0.35; 35, 15, 5; 0.50, 0.50, 0;

0.9, 0.10, 0.81; 27, 36, 12

12.10. Verify the following relations, taking p + q =1:

(i) $p^2q + pq^2 = pq = q\text{-}q^2$

(ii) $p^2\text{-}q^2 = p\text{-}q$

(iii) $p+2q = 1+2q = 1+q = 2\text{-}p$

(iv) $(1\text{-}2q)^2 = (1\text{-}2p)^2$

(v) $1\text{-}2q = p\text{-}q = 2p\text{-}1$

(vi) $2q\text{-}p = 1\text{-}2p = q\text{-}p$

(vii) $p (1+q) = 1\text{-}q^2 = p (2\text{-}p)$

12.11. A character is controlled by single locus with no dominance among alleles. The genotypic frequencies in the population observed were 0.24 AA, 0.52 Aa and 0.24 aa in one population while these were 0.27 AA, 0.20 Aa and 0.53 in second population. Examine whether these population are in HWE state and find out the genotypic frequencies in next generation assuming random mating.

12.12. Calculate gene frequency of A and a gene in a population having 500 AA, 600 Aa and 950 aa individuals. What is the expected number of each genotype if the population is in HWE state? If the frequency of allele **a** is 0.45, estimate the expected number of individuals of three genotypes in a random mating population of 500 size.

12.13. What is the frequency of heterozygote (Aa) in a random mating population *(i)* if the frequency of recessive phenotype aa is 0.09 and *(ii)* when the frequency of all dominants is 0.19?

12.14. Consider three genotypes (PP, Pp and pp) in a random mating population of cattle in which the individuals with genotypes PP and Pp were polled but with genotype pp were non-polled. It was found that 36 out of 100 cattle were non-polled. Calculate the relative frequencies of *(i)* two alleles, *(ii)* 3 genotypes and *(iii)* two phenotypes.

12.15. In a random mating population, it was found that one-fourth of the normal individuals were carrier for a defect controlled by single locus two alleles. Find out the frequency of recessive allele.

(**Hint:** The frequency of recessive, normal and carrier individuals among normal individuals is q^2, $1-q^2$ and $\dfrac{2pq}{1-q^2}$, respectively. Thus equate $\dfrac{2pq}{1-q^2} = \dfrac{1}{4}$ and find the answer).

12.16. Consider that black coat color in cattle is dominant over red colour. Find out the gene frequency of black and red alleles and the number of heterozygous individuals if there were 72 red animals in a sample of 450 animals.

12.17. The followings are the allelic frequencies at a locus with 3 alleles: A_1 = 0.3, A_2 = 0.5 and A_3 = ? Calculate the frequency of A_3 allele and write down all the possible genotypes at this locus with their frequencies.

12.18. The *RBC acid phosphatase enzyme* controlled by three alleles (A, B, and C) at a locus with co-dominance among all of them have been recorded in a sample of 500 people with the following results

Genotype	AA	BB	CC	AB	AC	BC
Frequency	0.09	0.35	0.0	0.48	0.03	0.05

Estimate the frequency of three alleles. Explain why there was no CC individual in the sample.

12.19. The three alleles at a locus (A, S and C) control the *Hb variation in humans* showing co-dominance among all the three alleles. Calculate the frequencies of three alleles from the data. Find out the percentage of heterozygote and predict the genetic structure of next generation assuming random mating.

Genotype	AA	AS	AC	SS	SC	CC
Numbers	719	199	114	2	5	3

12.20. Taking the order of dominance as agouti (full colour) > Himalayan >albino colour allele in rabbit, estimate the frequencies of 3 alleles in a sample having 570 agouti, 140 himalayan and 20 albino rabbits and in another sample of 500 rabbits having 25 % agouti, 50% Himalayan and rest albino in a random mating population.

12.21. A survey of 600 people conducted on ABO blood group system revealed that there were 37 people having blood group A and rest had the group O. Find out the allelic frequencies assuming random mating.

CHAPTER 13. FACTORS AFFECTING GENE FREQUENCY

Example 13.1: The initial gene frequencies of the two alleles (A and **a**) are 0.4 and 0.6. The A allele mutates to allele **a** with the mutation rate of 6 per thousand with reverse mutation rate from **a** to A as 2 per thousand. Estimate the change in gene frequency in one generation due to mutation. What will be the equilibrium gene frequency ?

Solution:

$$\Delta q = up - vq$$

where, $u = \dfrac{6}{1000} = 0.006$

and $v = \dfrac{2}{1000} = 0.002$

$p = 0.4$ and $q = 0.6$

Therefore, $\Delta q = up - vq = 0.006 \times 0.4 - 0.002 \times 0.6$
$$= 0.0024 - 0.0012 = 0.0012$$

Gene frequency at equilibrium (p)

$$= \frac{v}{u+v} = \frac{0.002}{[0.006+0.002]}$$

$$= \frac{2}{8} = 0.25$$

and $q = 1 - p = 1 - 0.25 = 0.75$

Example 13.2: Consider initial gene frequency of an allele among the native population as 0.3 and among immigrants as 0.6. The migration rate was 10 %. What will be the frequency of gene in the mixed population?. Estimate the change in gene frequency.

Solution: Given migration rate = 10% = 0.1; $q_n = 0.3$; $q_m = 0.6$

Gene frequency after migration (mixed population, q_1) = $q_n - m (q_n - q_m)$
$= 0.3 - [0.1(0.3-0.6)] = 0.3 + 0.03 = 0.33$

Change in gene frequency **(Δq)** = m $(q_m - q_n)$ = 0.1 (0.6-0.3) = 0.03

or $\Delta q = q_1 - q_n = 0.33 - 0.30 = 0.03$

Example 13.3:   The gene frequency of an allele in native population was observed as 0.4 whereas among migrants it was 0.7 and after migration the gene frequency in mixed population was 0.5. Find out the migration rate.

Solution: Migration rate (m) = $\dfrac{(q_1 - q_n)}{(q_m - q_n)}$

$$= \frac{(0.5-0.4)}{(0.7-0.4)} = \frac{0.1}{0.3} = 0.33$$

Thus, the migration rate was 33%.

Example 13.4: The cy allele producing curly wings in D. melanogester is lethal with fitness (w_{12}) as 0.5 relative to a value of $w_{22} = 1.0$ for ++ genotype. Estimate (i) the frequency of cy allele (p) after one generation, (ii) the change in gene frequency ("p).

Solution:

Genotypes:	cy cy	cy+	++
Frequency:	0	0.67	0.33
Fitness:	0	0.5	1.0

Frequency of cy allele (p_0) = ½ (0.67) = 0.335

Therefore, $q_0 = 1 - p_0 = 0.665$

$$W = p^2w_{11} + 2pqw_{12} + q^2w_{22}$$
$$= (0.335)^2 (0) + 2(0.335) (0.665) (0.5) + (0.665)^2 (1.0)$$
$$= 0 + 0.223 + 0.442 = 0.665$$

(i) Gene freq. after one generation (p_1) = $\dfrac{[p_0(p_0w_{11} + q_0w_{12})]}{W}$

$$= \frac{\{0.335\ [(0.335)\ (0) + (0.665)\ (0.5)]\}}{0.665}$$

$$= \frac{[0.335\ (0 + 0.3325)]}{0.665} = \frac{0.1114}{0.665} = 0.167$$

(ii) Change in gene freq. (Δp) = $p_1 - p_0$ = 0.167 - 0.335 = - 0.168

Example 13.5. The gene frequency of recessive allele in a population was 0.3. This population was under selection with selection coefficient against homozygous recessive (aa) as 0.4. Estimate the change in gene frequency per generation of selection.

Solution: Change in gene frequency (Δq) = $\dfrac{-spq^2}{1 - sq^2}$

Here, q = 0.3, hence p = 0.7, and s = 0.4

$$\Delta q = \frac{-spq^2}{1 - spq^2}$$

$$= \frac{[-0.4 \times 0.7 \times 0.09]}{1 - 0.4 \times 0.09}$$

$$= \frac{-0.0252}{1 - 0.036} = \frac{-0.0253}{0.964} = -0.0262$$

Therefore, the change in q per generation will be equal to – 0.0262.

EXERCISES

13.5. If the frequency of recessive gene in a population was 0.2, compute its frequency after 5 generation of selection against homozygous recessive.

13.6. In a cattle farm of 500 animals, the frequency of red gene is 0.6, what will be the frequency of red and white genes after the death of 60 white and 40 roan animals due to sudden outbreak of a contagious disease?

13.7. In every 1000 calves born in a large cattle herd, 10 suffer from muscular hypertrophy and if they are totally eliminated from breeding, what will be the frequency of the gene concerned with this ailment in the fifth generation taking the generation in which selection is given effect as generation zero.

13.8. The frequency of white animals in a Shorthorn cattle population is 4%. What will be its frequency after two generations of complete selection against white animals?

Chapter 14

Quantitative Inheritance

The genetics of quantitative characters in a population is called as Quantitative Genetics. This deals with the study of the inheritance of quantitative characters by applying the statistical methods to estimate the genetic parameters *viz.* the mean, variance and covariance. These parameters characterize the genetic variation in any quantitative character in a population and the changes that take place from time to time. It is thus important to know about the term character, types of characters and relation between gene and character before discussing the mode of inheritance of quantitative inheritance. It is also relevant to know the sources of genetic variation in a character and its significance.

14.1 RELATION BETWEEN GENE AND CHARACTER

The existence of genes to play their role in controlling the development and expression of character is inferred from the effects of genes to cause differences in the expression of the character by changing the phenotype or phenotypic value. However, the genes do not produce the character directly but via some specific biochemical reactions. The primary function of a gene is to produce a specific protein which often acts as an enzyme or more precisely a polypeptide. The enzyme has an effect to catalyze a specific biochemical reaction that leads to the development and expression of the character. Thus, enzyme production is the primary function of a gene while the development and expression of a character is the ultimate function of a gene through intermediary biochemical reactions. However, such relation between these functions of a gene is known only for a very few characters.

The following principles have emerged to establish the relation between gene and character.

First is the *Like begets like*. This means that every newly formed or developed living individual has the characteristics similar and common to its species, breed, race or line. Thus, in plant kingdom, the plants of a species give rise to the plants of its own species *viz*. a mango give rise to mango tree, wheat to wheat plant and so on whereas the different animal species produce the animals having the similar and common characteristics of their own species, such that the cats beget cat, dogs produce dogs, human to human.

Secondly, there exist *genetic differences* between different breeds, strains / race / lines. The different breeds of a species show genetic differences *viz*. the animals of two breeds have distinctive genetic differences in color, appearance and other physical characteristics like horn pattern, body size and body weight. Likewise, the mean performance for a quantitative trait (body weight, milk yield) is different for different breeds, lines and strains within a species. These are the indications of the genetic differences among breeds, strains and lines.

Thirdly, there also exist *genetic variations* between the animals of a breed / strains / lines as well as between any two closely related individuals belonging to the same sire–dam families, even if they are raised together under the same environment, especially for quantitative traits.

Fourth, there is some *resemblance (similarity)* between genetically related individuals compared to unrelated individuals within a breed / race. Not only this, but the progeny resemble to their parents, though the resemblance among relatives in not exact. This indicates that genetically related individuals within a breed / strain / line share common genes responsible for producing the similarities among relatives.

Fifth, the quantitative characters show *response to selection* which indicates the genetic differences among different individuals (selected and culled).

Finally, the quantitative characters also show *inbreeding depression* on crossing the closely related individuals and its converse *hybrid vigour* on crossing the individuals with different genetic background.

14.2 MODE OF INHERITANCE OF QUANTITATIVE CHARACTERS

The quantitative characters show the following mode of inheritance which are taken as the essential features of multiple factors hypothesis or polygenic inheritance or quantitative inheritance.

1. Controlled by genes: The quantitative characters are *controlled by genes and hence inherited*. The relation between gene and character has been described under paragraph 13.3 above.

2. Controlling genes are poly genes: The quantitative characters are controlled by many gene pairs (*poly genes* or minor genes). This was made evident from chromosome assays in Drosophila for abdominal chaetae number. The chromosomal assays had shown (using of major genes as marker) that the genes for continuous variation in abdominal chaetae number are nuclear born and that all the 3 major chromosomes carry poly genes (at least 11 in number) affecting this quantitative character, abdominal chaetae number.

3. Controlling genes are nuclear born: The quantitative characters show Mendelian inheritance pattern which means that the genes controlling these characters are *nuclear born* (genes are present on the chromosomes). The nuclear born genes show two properties *viz. segregation and linkage.*

The variability shown by quantitative traits in F_2 generation is more than parental (P_1 and P_2) and F_1 generation. This comparatively higher genetic variability in F_2 generation is an indication of segregation and recombination of genes. The evidence of linkage of poly genes with major genes is also evident from the significant differences among the means of quantitative traits corresponding to the phenotypic classes based on qualitative trait. Sax (1923) gave first the evidence of linkage of major genes (controlling the colour of seed bean) with poly genes (controlling seed size showing continuous variation). Similarly, Rasmusson (1935) in a study of flower colour (coloured and white phenotype controlled by major gene) and flowering time (controlled by poly genes showing continuous variation) also observed linkage between pigmentation gene and the poly genes responsible for the flowering time.

4. *Environmental effects:* The quantitative characters are affected by *environmental factors.* The evidence of the environmental effect on the phenotypic expression of the genotype come from the fact there are significant differences in the phenotypic value of a quantitative trait of the same individual recorded at different ages or under different environments (feeding regimes). The genotype of an individual is fixed at zygotic stage but an individual receives different environment at its different ages and hence the differences in the performance of the individual indicate the effect of environment *viz.* milk yield in different lactations, body weight recorded at different time intervals.

5. *Continuous variation:* The quantitative characters show *continuous variation.* There is a continuously graded series of phenotypic values of quantitative traits from one extreme to the other within the range. This continuous variation in phenotypic values for quantitative trait (like milk yield, wool yield etc.) is caused by the effect of poly genes and by the effect of environmental factors. The simultaneous segregation of poly genes controlling a character creates many genotypes and hence many expressions (phenotypes) of the character and the environmental factors further modify the expression of poly genes (genotypes). The number of phenotypic classes (phenotypic values) depend on the number of genes affecting a trait, number of alleles at a locus and the interaction between genes.

14.3 SOURCES OF GENETIC VARIATION

The different sources of genetic variation can be grouped as-

(1) Primary level and ultimate sources at *individual level:* There are some genetic factors which shape and change the genetic structure of the individual. These include the chromosomal variation and extra chromosomal variation.

(a) The *chromosomal variation* includes the existing genetic variation, variation arising from segregation and recombination of genes during meiosis, variation within individual arising fron gene action and the future variation caused due to change in genetic materials. The genetic variation thus exists and occurs due to the followings:

- existence of genes in pairs at all loci in every diploid individual,
- existence of multiple alleles at a locus,
- number of gene loci (polygenes) affecting a character,
- gene action or expression *viz.* additive and non additive gene action,
- genetic recombination (crossing over, segregation and recombination),
- gene mutation and chromosomal aberration, though very rare.

The genetic variation due to gene action constitutes most of the part of genetic variation for quantitative characters. The existing genetic variation and that arising from gene recombination are considered basic causes of genetic variation although all types of genetic variations are basic.

(b) The *extra chromosomal variation* involves the plasmatic factors called as plasma genes or cytogenes or plasmids. The inheritance due to cytoplasmic factors is called as *cytoplasmic inheritance.* The cytoplasmic factors do not contribute to genetic variation in metric traits. Therefore, the variation in metric traits due to cytoplasm is not important.

(2) *Secondary level of processes* creating genetic variation at *population level*: These processes change the genetic structure of population rather than of the individual. These are population size, mating systems, selection, migration, mutation and geographical isolation. These processes are capable of molding the genetic make up of a population into new species in conformity with the environment and ecology of species.

The genetic variations produced by either level of process are ultimately fixed. The isolating mechanism either geographical or reproductive helps in the fixation of genes.

14.4 EXTRA-CHROMOSOMAL INHERITANCE

This is also known as *cytoplasmic inheritance.* The DNA present in nucleus is the genetic material and is the store house of genetic information for most of the characteristics. The chromosome complement in the male and female individual is the same and the genes from male and female parents contribute equally to the genetic constitution of the progeny. Based on equal contribution of two parents to progeny generation, the reciprocal crosses between homozygous parents should give the similar results in terms of producing the progeny of identical phenotype except the sex linked inheritance.

However, the studies have shown that some *extra-nuclear genes* or DNA molecules are also present outside the nucleus in the cytoplasm. Thus, the genetic information exists in the cytoplasm also. The female parent contributes more cytoplasm to the zygote, it is expected that there should be differences in the results of reciprocal crosses for the characters having cytoplasmic control. If the reciprocal crosses differ, it would be due to the maternal effect as a result of contribution of cytoplasm. This is the *criteria for testing the role of cytoplasmic inheritance.* The possible role of cytoplasm in the inheritance of certain characters was suggested by Sanger and his colleagues in 1950's. The inheritance of certain characters in *Chlamydomous* is controlled by non-chromosomal genes and hence the inheritance is independent of chromosomal genes. It was found that in these cases the cytoplasm has self perpetuating hereditary factors which are formed of DNA. The factors of cytoplasm are called plasmon or plasmagenes.

The *cytoplasmic DNA* has been indentified by autoradiography, electron microscopy, Feulgen staining and density gradient centrifugation. The DNA is present in plastids of plants and algae, and mitochondria of both plant and animal cells. These cell organelles in plants contain their own DNA and protein-synthesizing apparatus. The electron microscopy have shown that cytoplasmic DNA is thinner than the protein covered DNA rnolecule of eukaryotic nuclear chromosome and have thin fibrous appearance believed to be naked nonhistone DNA like that found in prokaryotes. The DNA present in chloroplasts and mitochondria is double stranded which replicate semi-conservatively. The chloroplasts of certain algae contain circular DNA.

The characters showing cytoplasmic inheritance have *maternal influence*. This is because most of the cytoplasm of the zygote is contributed by the female gamete and moreover if there are some hereditary units present in the cytoplasm, these are expected to be transmitted to the progeny.

The cytoplasmic inheritance and or maternal effects are indicated from the followings-

* The characters controlled by cytoplasmic inheritance show differences between the results of reciprocal crosses. The reciprocal crosses produce the progeny of different phenotypes.
* The characteristics of the mother are exhibited by the progeny for such characters. Thus, the genotype of the mother determines the phenotype of the progeny rather than genotype of the progeny.
* Such characters do not show segregation like nuclear born genes.

The expression of certain characteristics of the progeny is not due to their own genotype but is influenced by their maternal parent. Such an effect is produced by egg cytoplasm and called as the *maternal effect* which may be for a short period of life or throughout lifespan of the organism. In mammals, the mother may affect the development of their offspring through egg cytoplasm as well as through the intra-uterine environment *viz.* maternal effect on the birth weight via maternal nutrition, embryonic defects caused by maternal diabetes. The maternal effects are taken when the mother contributes to the phenotypic expression of her progeny besides that being produced from nuclear borne genes.

Causes of maternal effects: The maternal effects may arise in one or other of the following manners:

1. Maternal environment: (i) The **maternal nutrition** in mammals produces the maternal effects. The example is the birth weight. The heavier dams produce heavier progeny because of the better and adequate nutrients supply to the foetus. The reciprocal crosses have shown that birth weight is a maternal trait, because the birth weight of the progeny deviates from mid parent value towards the birth weight of maternal breed (Tomar, 1979).

(ii) *Seedling characters* in plants show maternal effects which are diminished with increasing age, *e.g.* radio sensitivity in the tomato (first true leaves and second leaves), length of first seedling leaf in the rye grass, juvenile traits (characters expressed in early age) of sexual offspring.

The cytoplasmic inheritance differs from the maternal effects in the sense that the effects of plasmagenes do not disappear after one generation but persist as long as such plasmogenes are capable to self perpetuation.

2. **Cytoplasmic Factors:** The following are some of the examples of maternal effects with *cytoplasmic factors:*

(i) *Maternal effect for shell coiling in snail (Limnaea pegera- water snail):* In snails, there are two types of cleavages depending on the direction of coiling and these are genetically controlled. The first type is known as dextral coiling in which the coiling is clockwise (shell coiled towards the right), found in most of the cases and controlled by dominant gene (*D*) so that dextral is *DD*. The second type is known as sinistral cleavage (coiling) in which coiling is counter clockwise (shell coiled towards left), found in some cases and controlled by recessive allele (*d*) so that sinistral is dd. Thus normal dextral are normal while sinistrals are mutants. The two alleles of a gene pair are inherited as per Mendelian laws. However, the action of any genetic combination is expressed only in the next generation after the one in which a given genotype is found. It is important that the type of coiling produced from reciprocal crosses is determined by the genotype and not by the phenotype of the female parent.

In the reciprocal crosses, the heterozygous genotype *Dd* may be dextral as well as sinistral and this depends on the genotype of the female parent. In a similar way, if the genotype of the female parent is dominant (*Dd*), the dd genotype will turn in dextral coiling. Thus the phenotype of the female parent does not have any effect on the phenotype of the progeny but it is the genotype of the female parent which decides the coiling pattern of the progeny. Therefore, the coiling character is determined by the gene of the mother and not by the individual's own gene. The results of reciprocal crosses can be summarized as under-

(a) **Dextral female (DD) x Sinistral male (dd):** All the F_1 snails (*Dd*) have dextral coiling because of dextral mother. Also all the F_2 progeny have dextral coiling irrespective of their homozygous recessive genotype (*dd*) that are supposed to develop sinistral coiling with genotype dd. But the dd genotype of F_2 snail develops dextral coiling because the F_1 mother was dextral (*Dd*).

(b) **Sinistral female (dd) x Dextral male (DD):** All the F_1 progeny (Dd) developed sinistral coiling, though they were heterozygous and should have develop dextral. This was because the genotype of mother was sinistral (*dd*). All the F_2 developed dextral coiling irrespective of their recessive genotype (*dd*) supposed to develop sinistral. This was because the F_2 progeny were produced from F_1 mother having genotype as Dd (dominant gene for dextral).

(ii) *Maternal effect of pigment production in eyes of flour mouth and in water flea:* The normal eye colour is dark due to dominant gene, in both these invertebrates. The pigment production in eyes depends on a pair of nuclear born gene showing dominance relation between two alleles. The dominant gene (*A*) produces a substance known as kynurenine which produces pigment (causing production of dark eyes) whereas the recessive allele fails to develop the pigment. In this case, the gene A produces a pigment precursor, kynurenine which is involved in pigment synthesis and restored in the cytoplasm of the eggs (ova). In flour moth, the gene A is dominant to produce darkening of skin producing dark eyes. The recessive gene **a** fails to develop pigment and the eyes are red. A cross of recessive female (*aa*) with heterozygous male (*Aa*) produce only half of the larvae which show dark pigment in the eye while the reciprocal cross (Aa female with aa male) produces all larvae with dark eyes.

(iii) Respiratory deficiency in yeast (**Sacchoromyces cerevisiae**): In the presence of sugar glucose, the growth of yeast is slow (producing very small size colonies, called petites) because they are unable to utilize oxygen in the metabolism of carbohydrates. This respiratory deficiency in yeast prevents growth on carbon source and prevents them (petite) from producing spores. The petites mated with normal yeast cells produce three petite varieties. (a) *nuclear or segregational petites* which follow Mendelian segregation; (b) *neutral petities*, caused by an extra chromosomal factor because they follow non-Mendelian behaviour, produces only wild type ascospores and colonies and the petite trait never appears, in further generation; (c) *suppressive petities* which suppress normal respiratory behaviour in crosses with normal strains, and segregate in a manner different from chromosomal genes.

(iv) Slow and normal spore germination trait in fungus (**Neurospora**): The reciprocal crosses of two stains of fungus with slow and normal germination produce different results. The slow germination trait persists when each generation is backcrossed with normal germination strain. Such type of inheritance is possible if the cytoplasm carries some hereditary units for the expression of a character.

(v) Male sterility in plants: There is pollen failure in many flowering plants which results in male sterility, like wheat, maize, onion etc. In these plants, the fertility is controlled to some extent by cytoplasmic factors. The male sterility in corn which is caused by the failure of pollen to develop properly (the sterility factor is not present on chromosome) is the example of cytoplasmic inheritance. The male sterility in maize is controlled by cytoplasm. If the female parent is male sterile, the F_1 progeny will also be male sterile. This is due the reason that cytoplasm was of the female parent male sterile.

(vi) Plastid inheritance in four o'clock and corn: There are three types of leaves and branches in four O'clock plant depending on the presence of plastids *viz. full green leaves* having chloroplast, *white* (pale leaves and branches) with no chloroplast and the *variegated branches* (mixed green white) with leucoplast in white area and chloroplast in green area. The colour of the progeny depends on the branch colour of female parent and is due to the production of chlorophyll by the plastids (chloroplast) present in cytoplasm. The chlorophyll pigment of chloroplast is related to photosynthetically prepared food. The leucoplast can not perform photosynthesis and the pale parts take food from green parts. Similarly, is the inheritance of the character (Iojap) with contrasting strips of green and white colours on leaves of maize plants.

3. Transmission of pathogens and antibodies via prenatal blood supply or by post natal feeding *(Infective particles):* The followings are the examples of cytoplasmic inheritance-

(i) Inheritance of kappa particle in Paramecium (Protozoa): Some cytoplasmic factors are not self perpetuating and disappear unless replaced by the effect of a nuclear gene. Thus, there is a relationship between a nuclear gene and a cytoplasmic factor. The Kappa particle in paramecium is reproduced in the presence of a nuclear gene (K) in some strains of protozoa (paramecium). Kappa seems to be an infective particle with a hereditary continuity of its own. The kappa particle produces a toxin substance (paramecin) which is toxic to the strains not having kappa (sensitive strain). The kappa particle is produced by a dominant allele.

(ii) Inheritance of sigma particle in Drosophila: There are certain characters for which the cytoplasmic characters are transmitted by infection *e.g.* sensitivity of CO_2 in a strain of Drosophila. The sensitivity factor is carried by the cytoplasm of the egg. Injection of body fluid from a sensitive fly into a non-sensitive fly can induce sensitivity. This sensitivity to CO_2 is inherited on a maternal basis called infective heredity.

(iii) Breast tumor in mice: The maternal effect by way of transmission of pathogen through parental blood supply or post-natal feeding is evident in case of development of *mammary cancer or breast tumor in mouse.* The cross fostering experiment showed that the development of mammary cancer is responsible to carry the factor. It is found to be maternally transmitted or having cytoplasmic inheritance. Recent studies have shown that this character is developed through mother's milk, known as milk fever, as a result of the presence of some virus in milk.

Chapter 15

Determinants of Phenotypic Values

The phenotypic expression (phenotype) of a metric trait is called as the phenotypic value and it is developed by the joint action of the genotype of the individual and the environmental circumstances under which the individual lived during the course of development and expression of the trait. The principle that the phenotypic value of a quantitative character is a joint product of genotype (gene effects) and environment was first given by Johannsen.

The expression of a genotype is called the phenotype. The genotype determines the genotypic value of an individual for the trait and the environment modifies this genotypes value causing a deviation in the genotypic value in either direction depending upon the type of environment (favourable or unfavourable). This deviation results continuous variation in metric traits caused partly by heritable and partly by non-heritable (environmental) factors. The heredity and environment are therefore considered two main factors and the only determinants of the phenotypic value for metric traits. Therefore, any change either in genotype or in environment would result a change in the phenotypic value of individual for the quantitative trait.

The observable or measurable differences among individuals of a population for a particular character (phenotypic values) are called as *phenotypic variation*. The variation in genotype and environment gives rise to the variation in phenotypic values.

15.1 COMPONENTS OF PHENOTYPIC VALUE (P)

The variability in phenotypic values due to gene's effects and environmental effects can not be separated by simple observations but it requires the statistical analysis of data. The division of phenotypic value into its component parts attributable to different factors (genotype and environment) is essentially required to know how the population parameters are influenced by the properties of genes for metric trait and how these parameters can be utilized for genetic improvement of livestock.

1. Genotypic Value (G): The genes controlling a character have their effect on the development of the character of an individual and confer a certain value to the character. This value of the character conferred by the genotype attributed to the effect of genes is called as the genotypic value, denoted by the letter G. The effect of genotype is studied in terms of the genetic differences between different genetic groups of individuals. These genetic groups may be different breeds of a species (breed differences), different herd of a breed (herd differences or strain or line differences) and sire differences based on different families (half-sib families) of a sire or likewise dam families in a herd.

2. Environmental Deviation (E): The term environment includes all the non-genetic factors other than genes effect that influence the phenotypic value of the character. The *gene expression is environment specific*. The environmental factors play an important role in the expression of genotype in the form of a character (phenotypic value) and cause variation in gene expression among individuals. The differences caused by environmental factors are known as environmental variations which are not heritable but limited up to the individual.

The *evidences of the effects of environmental factors* to influence the performance (phenotypic value) of individuals are observed daily in our life *viz.* variation in body weight, milk production, work capacity, etc. The environmental conditions are not similar but changing every time in the life of an individual and between different individuals. The environmental conditions are thus different for different individuals of a population, even in the same herd, in spite of the best effort of the owner. This may be due to the differences in age, climatic (cold and heat stress, humidity, etc.) and nutritional factors, physical condition, thermo-regulatory mechanism and genotypes, etc. Temperature affects the morphological and physiological expression of animals and plants. The climatic and nutritional factors are different to the animals born in different years and seasons.

The environment modifies the genotypic value of a character during the course of its development but before it is expressed and measured in the form of phenotypic value on an individual. For example, the genotype of the cow is fixed and hence the genotypic value for milk yield is also fixed but the amount of milk produced by a cow on different days is different. The milk yield produced by a cow is increased or decreased depending on the amount and quality of ration fed to the cow. Thus, the environment (amount and quality of ration) changes the milk yield of the cow by modifying the genotypic value of cow for milk yield. Therefore, there are two determinants (components) of the phenotypic value (P) corresponding to the value assigned by the genotype called as genotypic value (G) and the deviation caused in genotypic value by the environment called as environmental deviation (E).The phenotypic value has thus its component parts as-

$$P = G + E$$

where,   P = Phenotypic value

G = Genotypic value (a part of the phenotypic value due to the gene's effect).

E = Environmental deviation in genotypic value.

Mean environmental deviation is zero: The individuals of a population are exposed to different environmental conditions normal for a population, like different feeding regimes, different seasons of the year, etc. These different environments have their different effect on gene expression. The better environment favours the full expression of the genotype while the poor environment does not. As a result, the phenotypic value is expected to be higher under favourable environment (say, balanced diet) but to be lower under poor environment (say, deficient diet). Therefore, the environmental effects are cancelled in taking the average of the phenotypic values of all the individuals exposed to different environmental effects, and thus the mean environmental deviation in the population as a whole is zero ($\Sigma E = 0$). This is because the environmental effects are measured as deviation and the sum of mean deviation is always zero in a normal population. *Therefore, the environmental deviations (effects) do not contribute to the population mean.* This results the mean phenotypic value (P) equal to the mean genotypic value (G) and represents the population mean (M). In other words, the population mean refers equally to the mean phenotypic value as well as to the mean genotypic value.

3. Genotype - Environment Interaction (I_{GE}): The partitioning of phenotypic value into genotypic value and environmental deviation assumes that these two effects are additive and independent in their effect to produce the phenotypic value. The additive and independent effect means that the environmental effect is associated irrespective of the genotype on which it acts and that there is no correlation between these two effects of genotype and environment. Under the assumptions of additivity and independency of genetic and environmental effects in their action, the relative performances of two genotypes should remain the same in a changed environment. However, different genotypes (breeds) may perform differently in different environments. This means that a specific difference in environment produces different effects on different genotypes. Thus, the genotype and environment are not independent to develop a phenotype (phenotypic value) and hence these two assumptions do not hold true in practical situation.

The G-E interaction is based on the logic that different sets of genes express under different environments and this is also the basis of adaptability of different breeds (genotypes) of a species under different environments. In population genetics, the G-E interaction is taken in terms of the performance of two genotypes under two environments. All the genotypes are affected by change in environment but to a different degree. A specific difference of environment (say, change in any environmental effect like temperature) has its different effect on different genotypes (native and exotic breeds reared under native environment) for the reason that different genes (genotypes) express themselves differently in different environments. Thus, a genotype fails to give the same response in different environments. *The differential response of different genotypes under different environments is called the G-E interaction by a biologist.*

A naturalist considers the response in terms of adaptability. Thus, the *differential adaptability of different genotypes (breeds) in different environmental conditions* is called as the *G-E interaction* by a *naturalist*.

As a result of genotype - environment interaction the *genetic and environmental factors do not combine their effects linearly (additively)* to produce the phenotypic value. This means that the phenotypic value is not simply the sum of the effects of genotype and environment but an addional component is associated to phenotypic value. This is the meaning of G – E interaction to a *statistician and animal geneticist* (Breeder). The additional part of phenotypic value is represented as I_{GE}. This makes the mathematical model of phenotypic value as:

$$P = G + E + I_{GE}$$

15.2 ESTIMATION OF THE COMPONENTS OF PHENOTYPIC VALUE

The different components of the phenotypic value are estimated based on a number of genotypes, say *n* genotypes (n = 1, 2, ...i...,n) measured in a number of environments, say *m* environments (m = 1, 2, ..j,...m).

Table 15.1: Phenotypic values (p_{ij}) of n genotypes measured under m environments

Genotypes	Environments					Mean genotypic value ($G_i=P_i$)
	1	2	3	j	m	
1	P_{11}	P_{12}	P_{13}	P_{1j}	P_{1m}	$P_{1.}$
2	P_{21}	P_{22}	P_{23}	P_{2j}	P_{2m}	$P_{2.}$
3	P_{31}	P_{32}	P_{33}	P_{3j}	P_{3m}	$P_{3.}$
i	P_{i1}	P_{i2}	P_{i3}	P_{ij}	P_{im}	$P_{i.}$
n	P_{n1}	P_{n2}	P_{n3}	P_{nj}	P_{nm}	$P_{n.}$
Envro. Deviation $E_j = P_{.j} - P_{..}$	$P_{.1}$	$P_{.2}$	$P_{.3}$	$P_{.j}$	$P_{.m}$	$P_{..} = u$

$P_{ij} =$ $G_i + E_j + I_{Gi\,Ej}$. It is the phenotypic value of ith genotype under j^{th} environment.

$G_i =$ Genotypic value of ith genotype
$=$ $P_{i.}$. It is the average of phenotypic values of ith genotype over all environments normal to this genotype.

$E_j =$ Environmental deviation for j^{th} environment
$=$ $P_{.j} - P_{..}$. It is the environmental deviation, due to j_{th} environment for all the genotypes, taken as deviation of the mean of all genotypes in j^{th} environment ($P_{.j}$) from population mean ($P_{..}$)

$I_{Gi\,Ej}$ $=$ $P_{ij} - (G_i + E_j)$
$=$ $P_{ij} - (P_{i.} + P_j)$. It is the interaction component between i^{th} genotype with j^{th} environment. This is estimated as the difference of the phenotypic value of ith genotype measured under j^{th} environment (P_{ij}) from sum of the genotypic value of ith genotype and j^{th} environmental deviation.

When a particular genotype is measured under different normal environmental conditions, the mean over all environments is the mean genotypic value of that genotype and the deviation of mean phenotypic value under one environment from the mean genotypic value is the environmental deviation (e). The mean of all environmental deviations would be zero. ($\Sigma e_j = 0$). This has been made clear in solved example 15.1

15.3 GENOTYPIC VALUE (G) AND ITS COMPONENTS

The genes constituting the genotype of a character may have their independent and individual effect (additive gene effect) or they may produce their effect by interacting with each other (interaction effect).

Models of genotypic value: The genotypic value is partitioned to the corresponding additive effect and non additive effects of genes (interaction effect). The absence of interaction effect of genes means that the genes concern act additively within locus or between loci. The additive gene action with reference to genes at one locus means the absence of dominance whereas with reference to genes at different loci means the absence of epistasis.

(i) *Additive-dominance model of genotypic value*: In case of no epistasis, there are only two kinds of gene effects *viz.* additive and dominance. Thus, the genotypic value is defined in terms of additive genetic effects (A) and dominance deviation (D) as:

$$G = A + D$$

Estimation of additive and dominance effect of genes: This model of genotypic value can be illustrated taking a trait governed by single locus with two allele *viz.* A_1 and A_2. The population will have three genotypes (A_1A_1, A_1A_2, A_2A_2). It is assumed that the A_1 allele is favourable allele (having positive effect on the character) whereas A_2 allele is undesirable (with negative effect on the phenotypic value). The effect of gene difference on the phenotype is specified by two parameters (**a** and **d**) which describe the additive and dominance effects. The genotypic value of a genotype, carrying i^{th} and j^{th} allele and denoted by (G_{ij}), is measured in units of the character as the deviation of the phenotypic value of G_{ij} genotype from the mid point of the phenotypic values of two homozygotes. The average of the two homozygotes is called as the *mid point (m)* or as the point of origin (o).

The following *assumptions* are taken to estimate the genotypic value and its components as well as the population mean:

- The population is in Hardy-Weinberg equilibrium,
- There is no effect of environment,
- There is no measurement error, and
- The character is affected by single locus with two alleles (A_1 and A_2).
- The A_1 allele increases the value while A_2 allele decreases the value of the character.

Genotypes	Frequency	Phenotypic value (P_{ij})	Genotypic value $(G_{ij} = P_{ij} - m)$
A_1A_1	p^2	P_{11}	$G_{11} = P_{11} - m = a$
A_1A_2	$2pq$	P_{12}	$G_{12} = P_{12} - m = d$
A_2A_2	q^2	P_{22}	$G_{22} = P_{22} - m = -a$

$$m = \text{Mid Parent Value} = \tfrac{1}{2}\,(P_{11} + P_{22}) = \tfrac{1}{2}\,(P_{A1A1} + P_{A2A2})$$
$$= \text{Average of two homozygotes}$$
$$a = \text{Additive gene effect} = \tfrac{1}{2}\,(P_{11} - P_{22}) = \tfrac{1}{2}\,(P_{A1A1} - P_{A2A2})$$
$$= \text{Average of the difference of two homozygotes}$$
$$d = \text{Dominance effect} = P_{12} - m = \text{Deviation of heterozygote}$$
$$(P_{ij})\ \text{from the mid parent value (m).}$$

The *m* is independent on the distribution of genes between the genotypes and thus m is the natural zero point from which measurements can be expressed as deviations. The value of m is constant depending on the action of genes not under consideration and taken as the basis of measurement of two parameters (a and d). However, both *a* and *d* are independent of m.

The a indicates the gene's contribution to the additive genetic effect among the individuals of a population whereas d indicates the dominance effect of the gene. Therefore, the value of d depends upon the degree of dominance between two alleles. The degree of dominance is taken as d/a. The relationship of d to a *i.e.* degree of dominance (d/a) defines the kind of dominance *viz.* complete dominance of A_1 or complete dominance of A_2, no dominance or co-dominance between two alleles, over-dominance of A_1 or over-dominance of A_2, incomplete dominance of A_1 or incomplete dominance of A_2.

The parameters a, d and –a are also taken as the arbitrarily assigned values of genotypes, used to describe the effect of gene difference on the phenotype. The arbitrary assigned genotypic values (a, d and -a) are helpful to obtain general formulae of values, mean, and variance of a population for a trait. The a represents an increment in a constant direction whereas d represents a change in either direction depending upon the dominance of either increasing or decreasing allele. The genotypic values a and d are estimated as deviation of phenotypic value of a particular genotype from mid point between to homozygote.

(ii) *Interaction model of genotypic value*: When the different loci interact together to produce the genotypic value of a quantitative character, it means the effects of genes at different loci are dependent on each other and this is called as epistasis. In such a case, an additional component, known as interaction effect, also influence the genotypic value. This represents the epistatic effect or epistatic deviation or the interaction effect of genes between loci, denoted by I. Thus, the genotypic value becomes as: G = A + D + I

15.4 POPULATION MEAN

The population mean *(M)* for a quantitative trait is the mean of phenotypic values of all the individuals of different genotypes in a population. However, the population mean in quantitative genetics is expressed as a deviation from mid parent value or mid point (m) as $M = P - m$. Now considering a trait being controlled by a single locus with two alleles, there will be three genotypes in the population assuming their frequencies in HW proportions.

The *population mean (M) in absolute term* will be:

$$M = P^2 P_{11} + 2pq P_{12} + q^2 P_{22}$$
$$= p^2 (m + a) + 2pq (m + d) + q^2 (m - a) \text{ taking } P_{ij} = m + g_{ij}$$
$$= p^2 m + p^2 a + 2pq m + 2pq d + q^2 m - q^2 a$$
$$= m (p^2 + 2pq + q^2) + a (p^2 - q^2) + 2pq d$$
$$= m + a (p - q) + 2pq d$$

The *population mean (M)* as a *deviation from mid point (m)* will thus be:

$$M = P - m = a (p - q) + 2pq d$$

The *m* is the mid point between two homozygotes and it is a fixed part of the population mean. The rest quantity *[a (p - q) + 2pq d]* is the population mean (M) taken as a deviation from the mid value of two homozygote (*m*). The *population mean (M) as a deviation* from mid point is the weighted average of the genotypic values and obtained as the sum of product of the assigned value of each genotype with its frequency as shown below:

Genotypes	Frequency	Value	Freq. x value
A_1A_1	p^2	a	$p^2 a$
A_1A_2	$2pq$	d	$2pq\ d$
A_2A_2	q^2	−a	$−q^2 a$
Sum		= Mean =	$a (p - q) + 2\ pq\ d$

The population mean (M) is the mean genotypic value (G) and the mean phenotypic value of the population (P) for the character concerned. Thus, the population mean (M) = P = G. The genotypic values a and d are deviations in phenotypic values from the mean *value of two* homozygote (m). However, the population mean is expressed as a deviation from mid point of two homozygote (m) and it is the weighted average of genotypic values as: $[a (p - q) + 2\ pq\ d]$.

The *population mean has two parts viz. a (p - q)* attributable to homozygote and *2pqd* attributable to heterozygote. Therefore, the gene effect (a and d) and the gene frequency (p and q) are the two factors affecting the population mean. The gene effects influence the population mean in a way that:

$M = a (p - q) + 2pq d$ when there is incomplete dominance,

$M = a (p - q) = a (1 - 2q)$ when there is no dominance (d=0) and

$M = a (1 - 2q^2)$ when there is complete dominance (d= a).

The effect of gene frequency on the population mean can be understood taking one of the two alleles as fixed in the population (*i.e.* either p = 1.0 or q = 1.0). The population mean will then be as under-

$M = a$ when the allele A_1 is fixed (p= 1) and

$M = -a$ when the allele A_2 is fixed (q= 1).

Therefore, the total range of values attributable to the locus is 2a in the absence of over-dominance but the mean will be found beyond this range in case of over-dominance in an unfixed population (no allele is fixed in the population).

15.5 ESTIMATION OF GENOTYPIC VALUE

The genotypic values of three genotypes are measured as a deviation of the assigned genotypic values (a, d, -a) from population mean The arbitrarily assigned genotypic values of three genotypes are a, d and –a, whereas the population mean (M) is [a (p - q) + 2pq d]. The genotypic values of three genotypes as a deviation from population mean will be as under:

Genotypes	Assigned value	Genotypic values *
A_1A_1	a	a - M = a-[a (p - q)+2pqd] = 2q (a - pd)
A_1A_2	d	d - M = d- [a(p-q)+2pqd] = a (q - p) + d (1 - 2pq)
A_2A_2	-a	- a - M = -a- [a (p - q) + 2pq d] = -2p (a + q d)

*Taken as deviation from population mean, M = a (p - q) + 2 pq d

Mean genotypic value is zero: Since the genotypic values are taken as deviations from population mean, the mean genotypic value of the population would be zero. This can be verified by multiplying the genotypic value of each genotype with its frequency and summing the cross product. This is because the sum of the cross products of the value with its frequency is the mean.

15.6 ESTIMATION OF THE COMPONENTS OF GENOTYPIC VALUE

There are two models to describe the genotypic value. These are additive-dominance model used for singe locus two alleles system and interaction model used for two or more loci.

Method of estimation: The components of genotypic value (A, D, and I) are estimated by the gene effect and gene frequency method:

Single locus with two alleles:

1. Additive component of genotypic value (Breeding value): This is the first and most important part of genotypic value and it depends on the *average effect of genes*. The sum of the average effect of the genes affecting a character and carried by an individual is called by various names *viz. additive genetic value* or a*dditive genetic merit* or *breeding value (B.V.) of genotype (individual)*. This is because the B.V. is caused by additive gene effect (average effect of genes) irrespective of the presence or absence of another gene at one locus or many loci. The average effect of gene is a fixed part of genotypic value, represented by "A" and it is transmitted to the progeny as such.

The breeding value is taken in two ways *viz.* in terms of the average effect of genes carried by the individual and second in terms of measured value of the progeny of the individual. However, both these ways are related to each other because the mean value of the progeny of the individual depends on the sum of the average effect of genes carried by the parent.

(i) Progeny performance and breeding value: The most accurate and practical method of estimating the breeding value of an individual is the mean performance of its large number of progeny. The mean value (performance) of the progeny is due to the average effect of genes carried by the parent. The progeny performance of any parent deviates from the population mean and this deviation is called as the transmitting ability of the parent. *Twice the transmitting ability is the breeding value* for the reason that the progeny contains only a sample half of the genes of the parent, due to the halving nature of inheritance.

The breeding value of different individuals though may be taken as absolute value but is generally taken as the deviation from population mean in order to know the relative superiority of different individuals under testing. The breeding value is the property of the individual and the population to which the individual belongs.

(ii) Average effect of gene and breeding value: The breeding value represents the *sum of the average effect of genes of all loci affecting a character.* The magnitude of the average effect of gene has two characteristic features. The first is that it depends on the gene frequency. The second is that it is related to the genotypic value, a and d, used to express the population mean. Thus, it is the property of genes and the population.

Estimation: The average effect of genes, the transmitting ability and the breeding value (B.V.) of the genotypes (individuals) can be estimated by taking a male parent either with genotype A_1A_1 or A_2A_2 at a locus mated to the females of all the 3 genotypes. The mean value of progeny of the male parent of genotype A_1A_1 taken as deviation from population mean is the average effect of the gene A_1. This is because it will introduce the A_1 gene and will replace the A_2 gene.

(a) Average effect of A_1 gene: This can be estimated by mating A_1A_1 individuals with females of 3 genotypes as shown below-

	Genotypes of		Genotype frequency among progeny	Genotypic value
Male parent	Female parent	Progeny		
A_1A_1	A_1A_1	A_1A_1	p	a
A_1A_1	A_1A_2	A_1A_2	q	d
A_1A_1	A_2A_2	--	0	0

This mating has introduced the A_1 allele in place of A_2 allele. The expected mean of the progeny produced from A_1A_1 male (P_{11}) is the sum of cross product of genotypic value with its frequency as: $P_{11} = p\,a + q\,d$

Transmitting ability of A_1A_1 male (T_{11}) is taken as deviation of expected mean of progeny (P_{11}) from population mean and it will be:

$$T_{11} = P_{11} - M = p\,a + q\,d - [a\,(p - q) + 2pq\,d]$$
$$= q\,a + q\,d - 2pq\,d = q\,a + q\,d\,(1 - 2p)$$
$$= q\,[a + d\,(q - p)]$$
$$= q = \alpha_1$$

The quantity $\{q\ [a + d\ (q - p)]\}$ is the *average effect of A_1 gene denoted by $á_1$*. The $á_1$ is also called as the transmitting ability of A_1A_1 genotype which equals to half of the B.V. Therefore, the *B.V. of A_1A_1 male* is twice of its transmitting ability as-

B.V.$_{11}$ = 2 T_{11}
$$= 2q\ \ \alpha = 2q\ \{a + q\ (q - p)] = 2\ \alpha_1.$$

(b) Average effect of A_2 gene: This can also be estimated by the similar method used to estimate the average effect of A_1 gene. The A_2A_2 male is mated to the females of 3 genotypes of the population:

	Genotypes of		Genotype frequency among progeny	Genotypic value
Male parent	Female parent	Progeny		
A_2A_2	A_1A_1	---	0	0
A_2A_2	A_1A_2	A_1A_2	p	d
A_2A_2	A_2A_2	A_2A_2	q	-a

This mating has introduced the A_2 gene in place of A_1 gene. The expected mean of progeny (P_{22}) produced from A_2A_2 male is the sum of cross product of genotypic value with its frequency as: $P_{22} = p\ d - q\ a$

The *transmitting ability of A_2A_2 male (genotype)* denoted as T_{22} taken as deviation from population mean will be-

$T_{22} = P_{22} - M = pd - qa - [a\ (p - q) + 2pqd]$
$= -ap + pd - 2pqd = -p\ [a + d\ (-1 + 2\ q)]$
$= -p\ [a + d\ (q - p)]$
$= -p\alpha = \alpha_2$

The quantity $\{- p\ [a + d\ (q - p)]\}$ is the *average effect of A_2 gene denoted by $á_2$* and is also the transmitting ability of A_2A_2 genotype. The breeding value of A_2A_2 genotype

B.V.$_{22}$ = 2 T_{22}
$$= - 2p\ \alpha = - 2p\ [a + d\ (q - p)] = 2\alpha_2$$

(c) B.V. of A_1A_2 genotype: The B.V. of heterozygous genotype is estimated by adding the average effect of both the gene as:

B.V.$_{12}$ = $\alpha_1 + \alpha_2$ = $q\ \alpha + (-p\ \alpha)$
$= (q - p)\ \alpha = (q - p)\ [a + d\ (q - p)]$

It is obvious from the above that the B.V. of an individual is the sum of the average effect of genes carried by the individual. The B.V. is the function of gene frequencies and genotypic values, and hence population specific.

Summary: The different values *viz.* average effect of two genes, transmitting ability and breeding values of 3 genotypes can be summarized as:

Average effect of A_1 gene = Transmitting ability of A_1A_1 genotype = α_1 = $q\ \alpha$

Average effect of A_2 gene = Transmitting ability of A_2A_2 genotype
$$= \alpha_2 = -p\ \alpha$$

Transmitting ability of A_1A_2 genotype = ½ $[\alpha_1 + \alpha_2]$
$$= ½\ (q - p)\ \alpha$$

B.V. of A_1A_1 genotype $= 2 \alpha_1 = 2q \alpha$

B.V. of A_1A_2 genotype $= \alpha_1 + \alpha_2 = (q-p) \alpha$

B.V. of A_2A_2 genotype $= 2 \alpha_2 = -2p \alpha$

Average effect of gene substitution: The change in population mean is described by the average effects of two genes at a locus when one allele (A_1) replaces the other allele (A_2). This change in population mean is indicated by the difference in the average effect of two alleles at a locus $(A_1$ and $A_2)$ *viz.* $\alpha_1 - \alpha_2$. Therefore, the *difference in the average effect of two alleles is called as the average effect of gene substitution*, denoted by the letter α. The average effect of gene substitution (α) can be estimated in either of the following ways:

(i) The average effect of gene substitution (α) is equal to the difference in the average effect of two alleles.

$= \alpha_1 - \alpha_2 = \alpha =$ Average effect of gene substitution

(ii) The average effect of gene substitution (α) is equal to the difference in B.V. of heterozygote from the B.V. of either homozygote:

$= \text{B.V.}_{11} - \text{B.V}_{12}$ or $BV_{12} - BV_{22}$

$= 2q \alpha - (q - p) \alpha = 2q \alpha - q \alpha + p \alpha$

$= \alpha (q + p) = \alpha$

$=$ Average effect of gene substitution

Mean B.V. equals zero: The mean breeding value can be obtained by multiplying the BV with the frequency of each genotype and summing them all. This will equal zero.

II. Non additive component of genotypic value (*Gene combination value***):** The non additive gene effect is due to the effect of genes when combined to form the genotype and show interaction. Thus, let it be taken as the *gene combination value* **(G.C.V.)** which is a value produced by gene combination.

(i) **Dominance deviation (D):** The dominance effect is indicated when the phenotypic value of heterozygote deviates from the average phenotypic value of two homozygote. The dominance effect arises because the effect of gene substitution is not additive for all genotypes in the population.

This can be understood in another way, that when two alleles unite to form the genotype, they interact with each other and produce an additional effect which is not accounted for by the effect of two alleles taken singly. This is the value of gene combination in the genotype produced due to gene interaction. Thus, the dominance deviation can be defined as the gene combination value of the genotype at a locus.

This dominance deviation is added to the BV of genotype to obtain the genotypic value. Thus, $G = A + D$. Therefore, in the absence of dominance, the BV and genotypic value are equal and hence the difference in BV and genotypic value of a particular genotype indicates the dominance deviation (D). In other words, *the deviation of genotypic value from BV for a locus is called dominance deviation*. This is the meaning of dominance deviation or dominance effect. According to additive-dominance model of genotypic value, the genotypic value (G) is taken as:

$G = A + D$ and so, $D = G - A.$

Thus, dominance deviation can be estimated by difference as:
$[D_{ij} = G_{ij} - (\alpha_i + \alpha_j)]$. These have been given as under:

Genotypes	Genotypic value (G_{ij})	Breeding value (A_{ij})	Dominance deviation $[D_{ij} = (G_{ij} - A_{ij})]$
A_1A_1	2q (a - pd)	2q α*	$-2q^2d$
A_1A_2	a (q -p) + d (1-2pq)	(q - p) α	2pqd
A_2A_2	-2p (a + qd)	-2p α	$-2p^2d$

* $\alpha = a + d (q - p)$

It is clear that all the dominance deviations are functions of gene frequency and *d*. This is because the average effect of genes, BV and genotypic values depends on the gene frequency. The dominance deviations are zero when the alleles do not show dominance (d = 0) and in this case the genotypic values and BV will be equal as:

G = A + D = A + 0 = A

Mean dominance deviation is zero: This is based on the fact that mean BV and mean G.V. are equal. It can also be proved by multiplying the dominance deviation by the frequency of each genotype and summing the cross products.

BV and dominance deviation are not correlated: This can be verified by multiplying the two values (A and D) with frequency of each genotype and summing the cross products.

(ii) **Interaction deviation (I): Two loci case:** In addition to the additive (A) and dominance (D) effects of genes, the epistasis also affect the quantitative traits and is generally referred as the interaction deviation, in case a trait is affected by polygenes. The interaction effect denoted by I is caused by interaction of loci and indicated when the sum of the genotypic effect of all loci deviate from the phenotypic or genotypic values. The aggregate genotypic value for two loci taken together (G_{AB}) can be defined as:

$$G_{AB} = G_A + G_B + I_{AB}$$

Where, G_A and G_B are genotypic values of two loci (A and B) combining the A and D effects together in G_A and G_B.

I_{AB} is the interaction effect of two loci indicating the deviation from additive combination of genes.

The interaction effect (I) are indicated when the additive-dominance effects are not sufficient to account completely the genotypic value. The genotype can be defined as:

G = M + A + D + I

The interaction component (I_{AB}) can be obtained as a difference:

I = G - A - D = G - (A + D)

The kind of interaction depends on the mode of gene action of loci which mean whether the interaction is involved between additive or dominance deviation. This can be seen in two loci case as:

$$I_{AB} = I_{AA} + I_{AD} + I_{DD}.$$

Where, I_{AB} is the interaction effect of genes between two loci, I_{AA} is the interaction between additive values of two loci, I_{AD} is the interaction between additive value of one locus and dominance deviation of other locus or vice versa, I_{DD} is the interaction between dominance deviations of two loci.

Indication of interaction effect of two loci (I_{AB}): The magnitude of epistasis is indicated by the difference between two homozygotes for one locus with respect to genotype for the other locus. When these differences (between two homozygotes) at one locus are independent of the genotype of the other locus, it then indicates that the two loci act additively (independently) and they do not interact. In this case, the difference between two homozygotes for one locus will be equal for each genotype of other locus. For example, the difference in phenotypic value between two homozygotes (A_1A_1 and A_2A_2) or between heterozygote and one homozygote (A_1A_2 and A_2A_2 or A_1A_2 and A_1A_1) will be of equal size for B_1B_1, B_1B_2, and B_2B_2 genotypes. Similarly, the difference between any two genotypes for B locus will be similar for any genotype at A locus (A_1A_1, A_1A_2 and A_2A_2). On the contrary, if the differences are not equal but vary for different genotypes of other locus, it then indicates that the effects of the two loci are dependent on each other and hence shows that they do not act additively but interact with each other showing epistasis. Therefore, *the epistasis is indicated when the phenotypic expression of any genotype at one locus is dependent upon the genotype of the other locus.*

15.7 ENVIRONMENTAL EFFECTS

Some environmental factors cause variations in gene expression. The effects of environmental factors to modify the genotypic value along with the evidences of environmental effects have been discussed in the beginning of this chapter. Now, it is also important to classify the different environmental factors and the estimation of their effects.

15.7.1 CLASSIFICATION OF ENVIRONMENTAL FACTORS

The environmental factors have been classified on different basis *viz.* known or unknown causes, duration of their effect, and internal or external to the individual.

(i) Known (tangible) and unknown (intangible) causes: The know causes of environmental factors are the feeding levels, age, climatic factors, farm effect, management factors *viz.* open versus close housing, weaning practice, frequency of milking, floor versus battery system. These environmental factors causes major effect and their effects can be partly reduced by designing proper experiment. The effect of some environmental factors are beyond control, like maternal effect arises in mammals due to sharing common environment by relatives during prenatal and postnatal period, mainly due to nutrition of the mother. The common intra-uterine environment of litter mates, milk yield and mothering ability of sows also provide a common environment to the piglets. Moreover, some variation is caused due to error in recording the observations either due to carelessness involved in recording as human error or machine error (balance for recording body weight and milk yield) or due to error in recording the score characters (judging of animals, carcass quality characters, organolaptic quality, etc.). The environment variance caused by tangible factors can be eliminated statistically by adjusting the data.

Unknown factors (intangible causes) also cause variation in phenotypic values of quantitative traits, though the variation due to them is small (micro-environmental effects), *viz.* variation between monozygotic twins.

(ii) Permanent versus temporary environmental factors: *The* examples of these environmental factors have been given under repeatability. The effect of these two types of factors can be separated off by partitioning the total phenotypic variance into variance within individual and between individuals by analysis of variance

(iii) Random and fix environmental factors: The environmental effects which are *internal to the individual* and vary between individuals of a population influencing the individual records of animals (for which called as *variable effects*) are categorized as *random effects*. These factors affect the individual alone and not the whole population. The examples of random or internal environmental factors are hormone level, sex hormone, diet of the animal, sex, lameness and minor infection, chance accident, social dominance, age, reproductive status (pregnancy or lactation effect), maternal effect (age and weight of dam, type of birth), sire of animal, inbreeding level of individual. The random environmental factors cause a large share of the total variation.

The other category of environmental factors is the *fix effects* which includes those factors which *affect the whole population*, and are constant for a group of animals. These factors include climatic factors (temperature, humidity, season, sun light), years, regions, feeding level, husbandry practices and other factors which affect all the animals.

The occurrence of disease affects both individual as well as population as a whole and hence taken accordingly.

15.7.2 Estimation of environmental effects

The environmental effects are estimated by the same methods used to estimate the G – E interaction effects. See section 15.8.3 below.

15.8. GENOTYPE-ENVIRONMENT INTERACTION

The concept of this interaction component of phenotypic value has already been given earlier in this chapter (section 14.1).

15.8.1 *Types of G – E interaction*

To be more illustrative about the G - E interaction, it is better to describe this component with a simple case of 2 genotypes (G_1 and G_2) reared in two environments (E_1 and E_2) as given by Haldane (1946) and to know the type of G – E interaction based on the relative magnitude of differences between genotypes and between environments as given by Mc .Bridge (1958) and Dunlop (1962). With 2 genotypes and 2 environments, the 4 phenotypes will be produced having 6 different relationships among them which can be grouped into 4 types of relation and interaction as under-

1. Linear or additive relationship: This is grouped as *type A interaction*. When there is no change in the phenotypic values between 2 genotypes in either environment and it indicates the linear relation between genotype and environment. The phenotypic values of 2 genotypes may be either increased or decreased equally, with similar change, in the same direction under changed environment (E_2). Thus, there was *no change in the rank* order of two genotypes under changed environment. This occurs when the animals of a single herd / flock with small genetic differences (being intra population genotypes, of the same herd or flock) are kept and tested in two environments with small differences (micro-environment), like antibiotic feeding, individual versus group housing, birds housed on floor versus cages. This type of relation indicates that some genes (genotypes) are favourable and others are unfavourable in all or most of the environments.

2. *Non-linear or non-additive relationship:* This type of relation can be grouped in 3 different types.

(i) Type B interaction: The change in phenotypic values of genotypes may be in the same direction (increase or decrease in phenotypic values) but unequal in magnitude under changed environment (E_2) with a *change in rank order* of two genotypes. This occurs when the animals of a single herd with small differences are tested under macro-environment in order to test the genetic variation in the ability to respond to different environments with large differences *e.g.* interaction between families and date of hatching affecting sexual maturity in poultry, families reared in batteries or on floor, etc., sire - herd interaction, sire - ration (half sib families reared on 2 plane of nutrition), sire - year - season, sire – region interaction, etc. This type of interaction is more important when selection is within a breed under different environments. This interaction indicates that some genes (genotypes) may have effects that differ under different environment.

(ii) Type C interaction: When the change in phenotypic values of two genotypes under changed environment (E_2) is in the opposite direction but *does not change the rank order* of two genotypes. This involves when animals with large differences (breeds, strains, lines, crosses) are tested under different environments having very less differences (environmental variability within herd). The micro environmental variation cause some genotypes (pure breds) to vary but without producing any effect on hybrids (heterozygotes), as the hybrids are well buffered to micro-environmental changes. This type of interaction is not important practically.

(iii) Type D interaction: This occurs when the change in phenotypic values of 2 genotypes under changed environment is in the opposite direction leading to a *change in rank order*. This includes when breed differences are tested under macro-environmental changes. These are important in animal breeding. If this type of interaction exists, it will then decide the relative merits of breeds or lines in different environments and will be required to evaluate the adaptability of different genotypes under different environments. This interaction indicates that some genes may have effects that differ from environment to environment.

Genetic slippage: The G – E interaction may cause decrease in performance of quantitative characters when a population is shifted to a new environment which indicates the regression in performance. Dickerson (1962) called the regression in performance under changed environment as the *"genetic slippage"* .which indicates the reduction in breeding value.

15.8.2 *Examples of G – E interaction*

The G – E interaction are considered when there is a change in the rank order of genotype under changed environment. The followings are some of the examples of G – E interaction in terms of differential response of different genotypes. The change in normal environmental conditions for living to unhealthy and unhappy environment changes the phenotype of the character *viz.* the occurrence of disease, the healthy body (resistance, normal health) become sick (susceptible) in man, animals and plants, and untoward happenings, life turns to end of life (death) through heart attack and suicide in humans, etc.

The resistance and susceptibility to *various diseases* in man, animals and plants are the examples of G - E interaction. This can be understood that a healthy individual has a normal phenotype (normal health, resistance) for the character health. Normal health (resistance) and sickness (susceptibility) are two forms (phenotypes) of health, as a character. The environmental stress due to any type of unhealthy environments (cold and heat stock, hunger, excessive smoking, excessive alcoholic use, food poisoning or a toxin produced by disease causing organism) causes an individual to become sick and finally may lead to death.

Likewise, some people (genotype) remain least affected mentally while others are very much sensitive, responding very much erratically to unhappy environment and *untoward happening* (heart felt incidents) like, demotion or termination of their service, harassment / atrocity (physical, mental and sexual), death of their very near and dear, being cheated and looted for their capital, defamation in the society for their honesty and sexual character, etc. The sensitive people become either victim of heart attack or they took very much drastic step like suicide and thus there is end of their life.

There are some more extreme examples of G – E interaction. These are the production of phenocopies, discussed already in chapter 4.

15.8.3 Estimation of G – E interaction

The presence of G – E interaction to affect a trait can be estimated by applying any of the following methods, depending on the type of population available:

(i) Biometrical approach: The means of non-segregating generations *viz.* two pure parental lines measured in two environments are used to estimate additive value, environment deviation and additive x environment components; whereas the means of F_1 generation measured in two environments are used to estimate dominance deviation, environment deviation and dominance x environment components. The means of segregating generations (F_2 and first generation back crosses *viz.* B_1 and B_2) are used to estimate additive value, dominance deviation, environment deviation and additive x environment as well as dominance x environment components. The non-segregating and segregating generations can also be measured in more than two environments and all the components of phenotypic value can be estimated.

(ii) Mixed population: The procedure to estimate genotypic value, environmental deviation and G – E interaction components of phenotypic value for a character from the means of a number of genotypes (mixed population) kept under a number of environments has been given above under section 15.2 and illustrated with a solved example no. 15.1.

15.8.4 Implication of G – E interaction: The significance of this component of phenotypic value depends on the magnitude of the change under changed environment. The consequence that it changes the rank of different genotypes under changed environment, reduces the correlation between genotype and phenotype and also leads to inaccurate or biased estimate of the variance components. Each of these consequences has its implication.

(i) It is well known that all the genotypes (breeds / species) are not equally adapted to all the environments. Each genotype (Breed) has its certain environmental requirements and a change in that environment has adverse effect on the performance of the genotype. The G-E interaction is responsible for this adverse effect on performance. This creates the problem of adaptation of genotype in new environment and therefore it emphasizes on the correct choice of a breed in the breeding policy for genetic improvement. In the presence of G – E interaction, a breed can not be shifted to a new environment after improving the breed in another environment because of genetic slippage in performance under new environment. The cause of genetic slippage is that the physiological mechanism for the development and expression of the character in adverse environment is different to some extent and it is likely that separate set of genes express themselves under different environments. In the presence of G-E interaction, *the change in environment thus becomes a cause of insufficient response to selection.* The G-E interaction is a potential cause of disagreement, leading to disappointment and financial loss to the breeder.

(ii) Many traits deteriorate as a consequence of inbreeding and show heterosis as a consequence of line crossing. However, the effect of mating system (inbreeding depression and heterosis) depends on the environmental conditions. The adverse effect of inbreeding and the beneficial effect of cross breeding are more under substandard (poor) environmental conditions. Cunningham (1978) had explained that the difference between the F_1 and the local breed in poor environment is largely due to heterosis whereas in better environment is due to additive genetic effects. Thus, interaction of mating system with environment leads to disagreement between actual and expected response.

(iii) The presence of G-E interaction reduces the correlation between genotype and phenotype and hence the breeding worth (B.V.) of a breed can not be accurately estimated whose ranking is changed under changed environment. The performance of a breed can be predicted in the absence of G – E interaction because it will also have similar performance under changed environment. A breed with high adaptability in changed environment is generally a low producer or vice versa. There is a negative relationship between adaptability and productivity.

(iv) The estimation of G-E interaction requires at least two environments. In case of testing a breed in one environment, the G-E interaction component remains included in the genotypic variance as: $\sigma^2_G = \sigma^2_G + \sigma^2_{GE}$. Therefore, the σ^2_G is biased upward and hence, it will lead to disagreement between observed and expected response to selection. Thus, in this situation, it will be the wastage of time, resources and efforts, because the selection programme will not be effective if there exist G-E interaction.

Chapter 16

Components of Phenotypic Variance

The phenotypic value and its causal components along with mean have been discussed in the last chapter. The population mean is used to compare the two or more populations for certain purpose. However, the mean value of a character is not sufficient to describe a population for a quantitative character because the population mean does not give any idea about the variability in phenotypic values of different individuals. The differences in phenotypic values of different individuals of a population are called as variation. The variation may be classified as group variation recorded on individuals of different groups, individual variation recorded on different individuals within a group and within individual variation recorded at different times on the same individuals. The degree or amount of variation is measured by various statistical measures *viz.* range, mean deviation, variance including the standard deviation, standard error and coefficient of variation. The variance is more useful population parameter because it has the properties of additivity and sudivisibility.

The properties of variance are used to partition the total variance in to its causal components analogous to the partitioning of phenotypic value. These various components of phenotypic variance are very useful for their exploitation in bringing genetic improvement of a character.

16.1 ESTIMATION OF VARIANCE

The variance is taken as the mean of squares of the deviated values from their mean and referred as the mean of squared values (Mean square, M.S.). Thus, in estimating the variance, the phenotypic value of each individual is taken as its deviation from population mean, the deviated values are then squared and added, and the sum of squared values is divided by the number of individuals (on which the phenotypic values were taken). Thus, it will give the mean of the squared values and hence called as mean square. This mean square is the variance. The variance is calculated by denoting the phenotypic values as X rather than P:

$$V_X = \frac{\Sigma(X - \overline{X})^2}{N} \qquad \dots\dots\dots (i)$$

$$= \frac{\Sigma X^2 - \frac{(\Sigma X)^2}{N}}{N} \qquad \dots\dots\dots(ii)$$

$$= \overline{X^2} - (\overline{X})^2 \qquad \dots\dots\dots(iii)$$

= Mean of square values – Square of Mean Values

The numerator in *(i)* and *(ii)* is the sum of squares. This is divided by N – 1 instead of N when the sample variance is estimated.

16.2 MAJOR COMPONENTS OF PHENOTYPIC VARIANCE

The variance has its components corresponding to the components of phenotypic values. The reason is that the phenotypic values are changed as a result of change in either genotype or in environment. Thus the genotype and environment are two major component of variation and form the basis to partition the variance in to corresponding components as the variance has the property of its sudivisibility. Here it is important to note that variation in genotypes causes variation in genotypic values which give rise to the genotypic (genetic) variance and the variation in environment causes the environmental variance. Therefore, the phenotypic variance which is the variance in phenotypic values is partitioned analogous to the partitioning of the phenotypic values as:

P = G + E

$(P)^2 = (G + E)^2 = G^2 + E^2 + 2GE$

$V_P = V_G + V_E + 2Cov_{GE}$

where, V_P is the phenotypic variance,

V_G is the genotypic variance

V_E is the environmental variance

Cov_{GE} is the interaction variance (genotypic values and environmental deviation).

However, it is assumed that the genotype and environment are independent and hence the G - E interaction is taken as absent or treated as part of environmental variance. For practical purpose, independence is assumed. However, the genotypic and environmental variance components are further partitioned analogous to the partitioning of the genotypic value and environmental deviations, respectively.

16.3 COMPONENTS OF GENOTYPIC VARIANCE

The genotypic variance is composed of the components of genotypic value *viz.* additive gene effects, dominance deviations and interaction deviations. These components of genotypic value create variation in genotypic value of different individuals. Therefore, the variance of genotypic values called as the genotypic or genetic variance is partitioned into variance components associated to these components of genotypic values according to Fisher (1918) as:

$$V_G = V_A + V_D + V_I$$

Where, V_A is the additive genetic variance caused by differences in breeding values of different individuals,

V_D is the dominance variance caused by differences in dominance deviations of different individuals or due to deviation of genotypic value from additive gene action.

V_I is the interaction variance caused by epistasis of genes for two or more loci case.

The sum of $V_D + V_I$ is called the non-additive genetic variance which is due to the gene combination value of the genotypic value. This part is neither fixed nor inherited whereas the additive part of variance (V_A) is fixed and heritable.

Additive - dominance variance (*Single locus variance*): The genotypic value and its variance for a single locus two alleles system ($A_i A_j$ genotype) can be represented as:

$$G_{ij} = a_{ij} + d_{ij}$$
$$(G_{ij})^2 = (a_{ij} + d_{ij})^2 = a_{ij}^2 + d_{ij}^2 + 2 a_{ij} d_{ij}$$
$$= a_{ij}^2 + d_{ij}^2 \text{ since } a_{ij}^2 \text{ and } d_{ij}^2 \text{ are uncorrelated}$$

Thus $V_G = V_A + V_D$

This is for single locus and further assuming that there is no epistasis.

Interaction variance (*Two or more loci*): The genotypic value for polygenes taking their interaction effect is: $G = A + D + I$

Therefore, the genotypic variance will be accordingly partitioned as:

$$V_G = V_A + V_D + V_I$$

since A, D and I parts are not correlated

The magnitude of V_I is very small and negligible and hence ignored. However, the interaction variance can be divided on two bases, according to the number of loci *viz.* two loci interaction, three loci interaction and so on and according to the interaction involved between breeding values and dominant deviation. Considering two loci interaction, the interaction may be between breeding values of two loci causing additive x additive variance (V_{AA}), interaction between BV of one locus and dominance deviation of other locus producing additive x dominance variance (V_{AD}) and the interaction between dominance deviations of two loci causing dominance x dominance variance (V_{DD}). Hayman and Mather (1955) partitioned the epistatic (interaction) variance as under:

$$I = I_{AA} + I_{AD} + I_{DD}$$
$$V_I = I_{AA} + I_{AD} + I_{DD}$$

Now, the total phenotypic variance (V_P) can be partitioned on the assumption of independence of genotype and environment as:

$$V_P = V_G + V_E = V_A + V_D + V_E \text{ for single locus}$$
$$= V_A + V_D + V_I + V_E \text{ for two or more loci.}$$
$$= V_A + V_D + (V_{AA} + V_{AD} + V_{DD}) + V_E$$

16.4 COMPONENTS OF ENVIRONMENTAL VARIANCE

The environment surrounding the individual during the course of development and expression of a character form a major part of the phenotypic variance. The environmental factors in this respect can be considered as permanent environmental effects (E_P) and temporary environmental effects (E_T). The corresponding variance caused by these two types of environmental effects is termed as the permanent environmental variance (V_{EP}) and the temporary environmental effects (V_{ET}). The permanent environmental factors affect the phenotypic values in general and hence the variation caused by them is also termed as the general environmental variance (V_{EG}). On the other hand, the temporary environmental factors affect the individual for a specific or temporary period and hence the variation caused by them is also termed as special environmental variance (V_{ES}). Thus, the environmental variance is partitioned as:

$$V_E = V_{EP} + V_{ET} = V_{EG} + V_{ES}$$

With this the partitioning of phenotypic variance is complete and can be shown as:

$$V_P = V_G + V_E + V_{GE}$$
$$= (V_A + V_D + V_I) + (V_{EP} + V_{ET}) + V_{GE}$$

Thus, it is clear that the phenotypic variance which is the variance of phenotypic values, can be partitioned corresponding to the components of the phenotypic value. This indicates the differences in phenotypic values attributed to the genes or genotypes of the individuals and the environment received by the individual. The phenotypic variance is partitioned by the statistical technique known as the analysis of variance, more popularly known as Fisher's F-test.

16.5 IMPORTANCE AND USE OF VARIANCE COMPONENTS

The genetics of a quantitative character is the study of the component parts of variance attributed to different causes. The magnitude of the different parts of the variance provide very useful information as given below-

Total genetic variance: It is necessary to know the effect of different factors causing variation in a character and to estimate the percent contribution of different causing factors to the total variance. This helps to provide the information about the relative role of heredity and environment to affect a trait and to know the degree of genetic determination expressed in terms of heritability in broad sense which is the ratio of genetic variance to the phenotypic variance as: $H^2 = \dfrac{VG}{VP}$. It is important to know the relative role of heredity (genetic variation) and environment to cause the variation as it is specific to the character, the population and the individual.

Additive genetic variance: This part of variance is expressed in terms of heritability in narrow sense (h^2) as: $h^2 = \dfrac{V_A}{V_P}$. The use and importance of additive genetic variance in terms of heritability in narrow sense (h^2) have been discussed in chapter 17 under section 17.2.

Non-additive genetic variance: The non-additive genetic variance is not fixed and not inherited. Thus, higher amount of this variance component is not desirable and selection based on this part of variance is not effective. It is thus not desirable to plan for genetic improvement based on the amount of non-additive genetic variance. However, this part of variance can be exploited by special breeding methods.

G-E Interaction variance: When the expression of genotype (phenotype) becomes dependent on the environment, then different genotypes respond differently to the specific differences of environments and this causes an additional component of variance known as genotype-environmental interaction variance (V_{GE}) which is included with the V_E. The knowledge of the existence of G - E interaction is very useful in a number of ways:

- When the V_{GE} exist, a genotype proved to be better in one environment can not be expected to be better in another environment. Therefore, selection should be carried out in an environment where improved genotype has to live. This information is thus helpful to decide the choice of environment for selective breeding and testing of sires. The existence of G-E interaction modifies the gene effects in a different environment while the absence of G-E interaction indicates the same gene effects in all the environments.
- The herd comparison is not advisable in the presence of G-E interaction.
- The data correction for environmental effects is advisable in the presence of G-E interaction.

Environmental Variance: The environmental variations are also very important for the following reasons:

- The environmentally determined superiority (environmental variation) is not transmitted to the progeny and hence the parents showing environmental superiority do not produce superior progeny. Thus, the selection based on environmental variance in a trait is not effective.
- *The environmental variations overshadow the genetic variation and hence* reduce the correspondence between genotype and phenotype. Thus, the phenotype does not reveal the genotype when environmental effects are more important. The degree of correspondence between genotype and phenotype is increased with the decrease in V_E. The phenotype is produced and completely determined by the genotype when V_E is zero. Thus, the V_E shadows the V_G.
- The proper environment is essentially required to an individual for expression of full genetic potential.
- However, the uniform and better environment given to a breeding stock reduces environmental variability and is helpful for improvement for livestock production efficiency.
- The presence of environmental variance suggests the data adjustment for environmental effects.

16.6 ESTIMATION OF COMPONENTS OF PHENOTYPIC VARIANCE

The methods used to estimate the variance and its major components are the elimination method and analysis of variance method.

1. Elimination method: The two major component of variance *viz.* genetic and environmental can be estimated by eliminating the either component in planning the experiment. It is clear that environment is beyond control and hence can not be eliminated. The genotypic variance can be eliminated experimentally by taking the individuals with identical genotype like highly inbred lines or F_1 of a cross between two such lines or clones from a single individual or identical twins (monozygotic). The individuals having identical genotypes will have no genetic variance and hence their phenotypic variance will be entirely the environmental variance. Such a population having no genetic variance is called as the genetically uniform population. The genetically mixed population (random bred population) will have both parts of phenotypic variance (genetic and environmental). The two major components will thus be estimated as:

Mixed population (random bred) $V_{P(M)} = V_G + V_E$

Uniform population $V_{P(U)} = V_E$

Difference $V_{P(M)} - V_{P(U)} = V_G$

This method is of very limited use and scope due to non availability of highly inbred lines in large farm animals except identical twins which are very rare. This method is thus limited to plants and small lab animals.

2. Analysis of variance method: The data on at least two genotypes (breeds, lines, strains) are collected. The phenotypic value is described by a mathematical model to conduct the statistical analysis of variance. The phenotype is expressed as: $X = \mu + G + E + I_{GE}$

Where, μ is the population mean (all the genotypes),

      G is the effect of a particular genotype,

      E is the effect of environment,

      I_{GE} is the effect of genotype-environment interaction,

Assuming that all the effects except U are random and independent (uncorrelated), the phenotypic variance will be:

$$\sigma^2_P = \sigma^2_G + \sigma^2_E + \sigma^2_{IGE}$$

The F test is applied by conducting the analysis of variance as follows-

Sources of variation	DF	SS	MS	EMS
Genotypes	G-1	SS_G	MS_G	$\sigma^2 e + K \sigma^2_G$
Error	N-G	SS_E	MSe	$\sigma^2 e$
Total	N-1			

where, K is the number of observations per genotype,

   G is number of genotypes,

   N is total number of records,

   $\sigma^2_G = (MS_G - MSe) / K$, is the genotypic variance due to genetic difference among genotypes,

$\sigma^2 e = \sigma^2_E$ = MSe, is the environmental variance including I_{GE}
component and the error in sampling as well as in recording
the data.

3. Generation analysis method: The phenotypic variance into its components,
in case of large livestock and plants, can be partitioned and estimated from the
phenotypic variance of parental lines or breeds (say P_1 an P_2), their F_1 and F_2
generations. The phenotypic variance of P_1, P_2, and F_1 (non-segregating generations)
is entirely environmental in nature (V_E) because all the individuals have the same
genotype. The F_2 is genetically mixed population whose individuals have different
genotypes and hence contains both genetic and environmental parts of phenotypic
variance. These information are used to estimate the two major component of
phenotypic components as under:

$$\frac{V_{p1} + V_{P2} + V_{F1}}{3} = V_E$$

$$V_{F2} = V_G + V_E$$

$$V_{F2} - \frac{V_{p1} + V_{P2} + V_{F1}}{3} = V_G$$

16.7 ESTIMATION OF COMPONENTS OF GENETIC VARIANCE

There are four methods for estimating the various components of genetic
variance. These are gene frequency and gene effect method, genetic covariance
method successive generation data, and multiple mating methods.

16.7.1. Gene effects and gene frequency method

The additive genetic values due to additive effect of genes (BV) and the
dominance deviation of three genotypes are expressed as deviation from the
population mean. The mean BV as well as dominance deviation over the 3 genotypes
are both zero, because these are estimated from deviated values. Therefore, no
correction for assumed mean is required and the variance is simply the mean of
squared values. The variance is thus obtained from sum of cross product of the
value of genotypes with their frequency.

Additive genetic variance (V_A): The variance of breeding values is obtained
as:

Genotypes	Frequency	B.V.	Freq x $(B.V.)^2$
AA	p^2	$2q\alpha$	$4p^2q^2\alpha^2$
Aa	$2pq$	$(q-p)\alpha$	$2pq(q-p)^2\alpha^2$
Aa	q^2	$-2p\alpha$	$4p^2q^2\alpha^2$
		Sum =	$2pq\alpha^2 = V_A$
		=	$2pq[a + d(q-p)]^2$

Dominance variance (VD): The variance of dominance deviation is obtained as:

Genotypes	Frequency	Dom. Dev.	Freq x (Dom. Dev.)^2
AA	p^2	$2q^2d$	$4\,p^2q^4\,d^2$
Aa	$2pq$	$2pqd$	$8\,p^3q^3\,d^2$
Aa	q^2	$-2p^2d$	$4\,p^4q^2\,d^2$
		Sum =	$(2pqd)^2$ $= V_D$

Total genetic variance: The genetic variance can be obtained from genotypic vales but it being lengthy and hence V_G is estimated as sum of V_A and V_D.

Therefore,

$$G^2 = (A + D)^2.$$
$$V_G = V_A + V_D + 2Cov_{AD}$$
$$= V_A + V_D \text{ since } Cov_{AD} = 0$$
$$= 2pq\,\alpha^2 + (2pqd)^2.$$

The breeding value and dominance deviation are not correlated and their covariance is zero.

16.4.2. Genetic covariance method: (Resemblance among relatives)

The genetic covariance among relatives computed from phenotypic values recorded on relatives is translated into the genetic variance. This procedure is based on resemblance among relatives. The resemblance among relatives for metric traits, measured from observations recorded on individuals of a population, is basic to the study of the genetics of quantitative characters, is the property of the character, and used to estimate the amount of additive genetic variance which forms the basis of estimating heritability. Thus, the degree of resemblance provides the method to estimate the amount of additive genetic variance and can be determined from simple measurements (observations) recorded on the population.

Causes of resemblance: A group of related individuals have similar performance for any metric character because they have common genes and also share common environment.

The similarity in performance (resemblance) is increased with increase in number of common genes. The more close relatives have more number of common genes and hence show more resemblance (similarity). Therefore, the relatives will have similar performance in metric traits, if they share more common genes and also they will inherit more number of common genes to their progeny. On the contrary, the relatives will have less similarity in performance if they share less common genes and hence they will inherit less common genes to their progeny.

The common environment shared by relatives has also the similar effect to cause similarity among relatives. The resemblance among relatives is increased with increase of common genes and common environment or vice versa. Therefore, the common genes and the common environment shared by the relatives are the determinants of resemblance among relatives.

Measurement of resemblance: The resemblance (similarity) in performance of relatives shows that they (relatives) are correlated genetically as well as environmentally to some degree with each other. The correlation is measured by the covariance. The covariance among relatives is thus used to measure the resemblance among them (relatives). Therefore, it requires the estimation of the amount of covariance (co-variation) that exists among the individuals of a group. *This covariance is taken as a proportion of the total variance because the covariance is a part of the total phenotypic variance.* The covariance between relatives is equivalent to the corresponding variance components. That is to say, the covariance of members of the groups expresses the amount of variation common to the members of a group. The degree of resemblance among relatives, in terms of the covariance, is thus measured by partitioning the phenotypic variance into its components corresponding to the grouping of individuals. Therefore, a population is divided in a number of groups depending on the genetic relationship between relative *viz.* parent-offspring groups, half sib groups and full sib groups.

Covariance among relatives equals to the variance component corresponding to group of relatives: The amount of variation common to members of a group is taken as the covariance of members of the group. The covariance measures the resemblance and equals the variance (which measures the differences among the groups of relatives). Thus, the co-variance between certain groups of relatives is equivalent to the variance component between that group of relatives attributed to the cause and the extent of genetic relationship. For example, the covariance between half sib is equal to the sire variance because the half sibs are related by sharing common genes of the same sire and hence common sire is the cause of relation. Further, the genetic relationship between half sibs is 25% which is due to their common breeding values. Thus, $\text{Cov}_{.(HS)}$ = sire variance = $\frac{1}{4} V_A$. Likewise, the covariance between full sibs is equivalent to sire plus dam variance because they are the progenies of same sire and same dam and they have 50 % genetic relationship due to their common breeding values. Therefore, $\text{Cov}_{(FS)} = \sigma^2_S + \sigma^2_d = \frac{1}{2} V_A + \frac{1}{4} V_D$. Mathematically, this has been shown in estimating the genetic covariance among relatives. It is therefore that the covariance measures the degree of resemblance between relatives based on the grouping of individuals into families *viz.* sire families or dam families (half-sibs), sire-dam families (full sibs), parent-offspring (sire-daughter groups) etc. and used to partition the phenotypic variance into variance components corresponding to the group of relatives.

The analysis of variance is conducted to partition the total observed variance of a character into the following two components:

Between group variance (σ^2_B), is variance of means of groups about the population mean,

Within group variance $(\sigma^2_W = \sigma^2 e)$, is variance of individuals of a group about their group mean.

The between group component of variance (sire variance, etc.) as a proportion of the total variance expresses the amount of variation common to members of the same group and called as the intra – class correlation. This amount of variance is equally referred to as the covariance of members of the same group.

The variance components estimated directly from observations (phenotypic values) based on the degree of resemblance between relatives are called as observational components denoted as σ^2. On the other hand, the components of phenotypic variance attributable to different genetic and environmental causes (estimated in preceding chapter) are called as causal components denoted by the symbol V.

The resemblance (similarity) between relatives can be taken either as similarity of individuals of a group or as the differences between individuals of different groups. It is proper to think that more is the similarity within a group of relatives more will be the differences between groups. Therefore, the amount of similarity (degree of resemblance) can be expressed as a proportion of between group components to the total variance, expressed by intra class correlation (t) as:

$$t = \frac{\sigma^2_B}{\sigma^2_B + \sigma^2_W}$$

The variance among groups is a measure of the similarity within group. This can be understood by two situations of large and small variation between groups. When there will be large variance between groups, the variance within group will be small. This indicates that individuals belonging to a group (*e.g.* half sibs or full sibs) are more similar (resemble) in their performance and hence they will have their phenotypic values closer to the mean phenotypic value of their group and so variance within group will be small but the variance among group means will be large which will result into large intra class correlation. On the contrary, when variation between groups is small but within group variation is large, this will indicate that individuals of a group are not similar in their performance (phenotypic values) but have their phenotypic values widely spread about their group mean. This will result small variance among means of the groups compared to the variance among individuals within group and thus intra class correlation will be small. Therefore, the magnitude of intra class correlation is an indication of the amount to which the means of the different groups differ from each other.

I. Phenotypic Resemblance: The significance of estimating the degree of resemblance between relatives is to estimate the proportionate amount of additive genetic variance to the total phenotypic variance, V_A / V_P. The degree of resemblance is observed from measurements of phenotypic values of relatives and expressed as a regression or correlation by dividing the covariance among relatives with the appropriate variance. The degree of resemblance is taken as a correlation between sibs. The correlation of half sibs is: $r_{HS} = \dfrac{\frac{1}{4}\sigma^2_A}{\sigma^2_\rho}$ and the correlation of full sibs is:

$r_{FS} = \dfrac{\frac{1}{2}\sigma^2_A + \frac{1}{4}\sigma^2_D + \sigma^2_E}{\sigma^2_\rho}$. The degree of resemblance between offspring and parent

is expressed as the regression of offspring on parent. The regression is $b_{OP} = \dfrac{\frac{1}{2}\sigma^2_A}{\sigma^2_\rho}$

for single parent for which the $\sigma^2{}_p$ is the phenotypic variance of parent. In case of mid parent value, the covariance is to be divided by the variance of mid parent value which is half the phenotypic variance giving the regression as:

$$b_{OP} = \frac{\sigma_A^2}{\sigma_p^2}.$$

II. Genetic Resemblance (covariance): The common genes shared by the relatives cause the genetic covariance between relatives and hence it is estimated depending on the type of relatives. The genetic covariance can be estimated directly from observations (phenotypic values) recorded on relatives and this comes equal to the genetic covariance obtained using the concept of the coefficient of parentage based on the probability of genes, identical by descent, possessed by two related individuals.

(i) **Covariance between offspring and one parent**: This is equal to the cross product of the genotypic value of parent and its offspring. Expressing the value as deviation from population mean, the mean genotypic value of offspring is half of the breeding value of the parent. Thus the covariance of parent offspring (Cov $_{OP}$) is:

$$\text{Cov}_{OP} = \text{Parent value x Mean of offspring}$$
$$= G\ (\tfrac{1}{2}\ A) = [(A + D)\ \tfrac{1}{2}\ A]$$
$$= \tfrac{1}{2}\ A^2 + \tfrac{1}{2}\ AD$$
$$= \tfrac{1}{2}\ V_A \text{ since, A and D are uncorrelated}$$

where, G is the genotypic value of parent and equals to A + D

A^2 is the sum of squares of breeding values

The regression coefficient as an expression of degree of resemblance will be obtained by dividing the Cov $_{OP}$ with the total phenotypic variance of parent (V $_p$). Thus

$$b_{OP} = \frac{Co\ v_{op}}{V_p} = \frac{\tfrac{1}{2}V_A}{V_p}$$

$$r_{OP} = \frac{Cov_{op}}{V_p V_o} = \frac{Cov_{po}}{V_p} \text{ Since } V_p = V_o$$

$$= \frac{\tfrac{1}{2}V_A}{V_p}$$

(ii) **Genetic Covariance between offspring and mid parent:** This covariance (Cov $_{OP}$) will be obtained as:

$$\text{Cov}_{OP} = \text{Offspring with } \tfrac{1}{2}\ (P_1 + P_2)$$
$$= (\text{Cov}_{OP1} + \text{Cov}_{OP2})$$

Now taking that both parents have equal variance

$$\text{Cov}_{OP1} = \text{Cov}_{OP2} \text{ and Cov}_{OP} = \text{Cov}_{OP}$$

Since,
$$Cov_{OP} = \tfrac{1}{2} V_A \text{ so } Cov_{OP} = \tfrac{1}{2} V_A$$

The regression of offspring on mid parent as a measure of degree of resemblance between them will be:

$$b_{\overline{op}} = \frac{Co\, v_{\overline{op}}}{V_p} = \frac{\tfrac{1}{2} V_A}{\tfrac{1}{2} V_p} = \frac{V_A}{V_p}$$

The variance of mid parent value (V_p) is half the phenotypic variance ($\tfrac{1}{2} V_P$).

(iii) **Genetic Covariance between half sibs**: Suppose there are **m** numbers of sire families and each sire family have **n** individuals. The measurement recorded on j th progeny of i th sire (X_{ij}) can be written as:

$$X_{ij} = \mu + S_i + e_{ij}$$

where, μ = Overall mean of the population

$\quad\quad$ Si = effect of ith sire common to all members of ith family *i.e.* true mean of ith sire about the population mean

e_{ij} = random error with the jth progeny of ith sire family

The covariance between half sibs, Cov ($X_{ij} X_{ik}$), will be the covariance between the measurements of two members of the family:

$$\begin{aligned}
Cov_{HS} &= Cov\,(X_{ij} X_{ik}) = E\,[(S_i + e_{ij})\,(S_i + e_{ik})] \\
&= E\,S_i^2 + E(S_i\, e_{ij}) + E\,(S_i\, e_{ik}) + E\,(e_{ij}\, e_{ik}) \\
&= \sigma^2_S \text{ Since other products are not correlated} \\
&= \tfrac{1}{4} V_A \text{ Since half sibs are 25 \% genetically correlated.}
\end{aligned}$$

Thus, the covariance between half sibs is equal to the variance between half sib family means (σ^2_S) which is equal to $\tfrac{1}{4} V_A$. The covariance is thus the variance of the means of the half sib groups as:

$$(\tfrac{1}{2} A)\,(\tfrac{1}{2} A) = \tfrac{1}{4}\, \sigma^2_A.$$

Therefore, the variance of half the B.V. of parent is equal to the covariance between half sibs which is a quarter of the additive genetic variance.

The variance of phenotypic values (σ^2_X) which is the phenotypic variance (σ^2_P) is:

$$\sigma^2_P = \sigma^2_S + \sigma^2_W = Cov_{HS} + \sigma^2_W$$

Therefore, $\sigma^2_W = \sigma^2_P - Cov_{HS} = \sigma^2_P - \sigma^2_S$

The two variance components of phenotypic values estimate the followings:

$$\sigma^2_S = Cov_{HS} = \tfrac{1}{4} V_A + 1/16\, V_{AA} + 1/64\, V_{AAA}$$

$$\begin{aligned}
\sigma^2 e &= \sigma^2_P - Cov_{HS} \\
&= \tfrac{3}{4} V_A + 15/16\, V_{AA} + 63/64\, V_{AAA} + V_D + V_{AD} + V_{DD} + \ldots + V_E
\end{aligned}$$

Thus, the σ^2_W or $\sigma^2 e$ estimates the remainder of the genetic variance plus all the environmental variance. The correlation between two half sib is:

$$r_{(X\,ij,\, X\,ik)} = \frac{Cov_{(Xij, Xik)}}{\sigma^2_s + \sigma^2_w} = \frac{\sigma^2_s}{\sigma^2_s + \sigma^2_w}$$

Thus, $r_{(HS)} = \sigma^2{}_s / \sigma^2{}_p = \frac{1}{4} \sigma^2{}_A / \sigma^2{}_p$. This correlation measures the degree of relationship (resemblance) expressed as the between group component of variance as a proportion of total variance and known as intra class correlation. Here, $\sigma^2{}_s$ is the between group component of variance and $\sigma^2{}_w$ is the within group component of variance. It is thus the between group component of variance (covariance between group) which expresses the amount of variation common to members of the same group. The components of variance (between groups, $\sigma^2{}_s$ and within group, $\sigma^2{}_w$) are estimated by analysis of variance.

(iv) **Genetic Covariance between full sibs**: A number of sires are each mated to many dams and a number of progeny from each dam are born and measured for metric trait. The measurement recorded on k th progeny of j th dam mated to i th sire (X_{ijk}) is described by the following mathematical model:

$$X_{ijk} = \mu + s_i + d_{ij} + e_{ijk}$$

where, $\quad \mu \;=\;$ Overall mean of the population

$\qquad s_i \;=\;$ Sire effect (deviation) common to members of ith sire family

$\qquad d_{ij} \;=\;$ Dam effect (deviation) common to members of jth dam family mated to i^{th} sire

$\qquad e^{ijk} \;=\;$ error component associated with k^{th} progeny of jth dam family of ith sire.

So that the phenotypic variance ($\sigma^2{}_X = \sigma^2{}_p$) is

$$\sigma^2{}_X = \sigma^2{}_p = \sigma^2 s + \sigma^2 d + \sigma^2 w$$

Now, the covariance between two members (full sibs) for their measurements (X_{ijk1} and X_{ijk2})

$$\begin{aligned}
Cov_{FS} &= Cov\,(X_{ijk1}, X_{ijk2}) \\
&= E(s_i + d_{ij} + e_{ijk\,1})\,(s_i + d_{ij} + e_{ijk2}) \\
&= E\,(\sigma^2 s_i) + E\,(\sigma^2 d_{ij}) \text{ since other products are zero being} \\
&\quad \text{uncorrelated.}
\end{aligned}$$

The sire and dam progenies having both parents in common also contains the half sibs, so the covariance between half sibs for their measurement X_{ijk1} and X_{ijk2} is:

$$Cov_{HS} = \sigma^2 s = \tfrac{1}{4}\,V_A \text{ as obtained earlier}$$
$$\sigma^2 d = Cov_{FS} - Cov_{HS} = (\sigma^2 s + \sigma^2 d) - \sigma^2 s$$
$$\sigma^2 w = \sigma^2 P - Cov_{FS}$$

The phenotypic variance of a trait ($\sigma^2{}_X$) is partitioned in to observational components attributable to the followings-

(i) $\sigma^2 s = Cov_{HS} =$ variance between the means of half sib families (sire families) which represents the difference between the progeny of different sires (known as sire component of variance) and estimates the covariance of half sibs (Cov_{HS}). Thus, $\sigma^2{}_s = cov_{HS} = \tfrac{1}{4}\,V_A$.

(ii) $\sigma^2 d = Cov_{HS} =$ variance between the means of half sib families (dam families) which represents the differences between the progenies of different dams mated to same sire (known as between dam within sire component), obtained as:

$$\sigma^2 d = Cov_{FS} - Cov_{HS} = \sigma^2 T - \sigma^2 w - \sigma^2 s = \tfrac{1}{4}\,V_A + \tfrac{1}{4}\,V_D.$$

(iii) $\sigma^2 w$ = error variance = variance within progeny component which represents the differences between different progenies of the same dam (known as within progeny component) and obtained as: $V_P - \text{Cov}_{FS} = \frac{1}{2} V_A + \frac{3}{4} V_D$. This is because the progenies of dams are full sib families and the within group component is obtained by substracting the covariance of members of the groups from total variance. $\sigma^2_W = \frac{1}{2} V_A + \frac{3}{4} V_D$.

(iv) $\sigma^2 s + \sigma^2 d$ = Cov FS = variance of the means of full sib families and equals the sum of between sire and between dam components of variance.

Thus, $\text{Cov}_{FS} = \sigma^2 s + \sigma^2 d = \frac{1}{2} V_A + \frac{1}{4} V_D$.

The correlation (two full sibs) is: $r (X_{jki1}, X_{ijk2}) = \dfrac{\text{Cov}(X_{ijk1}, X_{ijk2})}{\sigma^2 P}$

$$= \dfrac{\sigma^2 s + \sigma^2 d}{\sigma^2 s + \sigma^2 d + \sigma^2 w}$$

Interaction component of covariance: The genetic covariance among different sort of relatives also contains the interaction component of variance besides additive and dominance components. The interaction is not considered an important part of genetic variance but neglected for two reasons. The first is that it has very small contribution to the covariance and hence it contributes very less to the resemblance between relatives. Secondly, the separation of interaction variance from other components is not easy but requires special experimental techniques, like use of inbred lines which partition the V_P into V_A and the rest $(V_D + V_I + V_E)$.

III. Environmental Resemblance (Covariance): In addition to the genetic causes, there are environmental circumstances which make the relatives to resemble each other. The members of a family born in the same gestation in litter bearing animals (mice and pigs) or reared together (half sibs and full sibs, particularly in human and farm animals) share a common environment. The other sources of common environment are the cultural influences in humans, nutrition and climatic factors etc. The common environment shared by relatives tends to make the relatives more resemble with each other and hence contribute to the resemblance (covariance) between relatives but causes differences between groups of unrelated individuals and hence contributes to the variance between means of different families. All sort of relatives are subject to environmental source of resemblance due to sharing common environment.

Considering the resemblance (covariance) between relatives due to sharing common environment, the environmental variance (V_E) can be partitioned into the following two components:

$$V_E = V_{EC} + V_{EW}$$

where, V_{EC} represents the environmental causes of similarity between relatives arise due to sharing common environment. It is the part of environmental variance between the means of groups and is taken in terms of environmental cause of similarity between members of a group (relatives) rather than a cause of differences between members of different groups (un-relatives). This contributes to the variance between the members of the families and hence contributes to the covariance of the related individuals. This is the part of variance between the means of groups due to common environment.

V_{EW} is the remainder of environmental variance known as within group component of variance, arises from causes of differences not connected with the relationship among individuals and it does not contribute to the variance of the true means of the group (between group component).

The V_{EC} component of environmental variance contributes more to the covariance of full sibs than any other types of relatives because they have a common maternal environment which is most difficult to remove by experimental design. The V_{EC} component can be reduced or eliminated by proper designing the experiment to distribute the relatives over a range of environment but the component due to maternal effect (a source of common environment) is not possible to eliminate. The V_{EC} can be measured by replication.

The environmental effects that are same (common) to all members of the family, are called as the common environmental effects (C – effect, Lerner, 1950). There are two types of the common environment. The first is the common maternal environment called as maternal effect and second is the contemporary environment effect.

(i) The *common maternal environment* (prenatal and post natal, shared by litter mates) causes *maternal effect*. The litter mates share common maternal environment mainly through the nutritional effect of the mother on the foetus and the young, common intra uterine environment, milk yield and mothering ability in mammals (pigs and sheep). This maternal environment results the offspring of the same dam to show resemblance to each other more than others which differentiate the members of different dams or families. Thus, maternal effects influence the resemblance of maternally related individuals like full sibs, maternal half sibs and dam-daughter groups and thereby contribute to their covariance.

The maternal effects cause a resemblance between parents – offspring as well as among the offspring themselves (full sibs and maternal half sibs) particularly for body size in mammals. The phenotypic value of the mother may have its effect on the phenotypic value of the offspring *e.g.* large mothers produce heavier progeny with faster growth rate in early ages because the larger mothers provide better nutrition to their foetus and young ones during early ages compared to small mothers. This makes the mother and its offspring to resemble each other in body size. Secondly, the maternal effect causes the resemblance among the offspring of the same dam but not between the mother and offspring. The maternal effect thus increases the covariance (resemblance) between relatives The maternal effects have more impact on juvenile (young hood) characters and are reduced in later ages.

(ii) The *contemporary environmental conditions* to a group of individuals provide similar environment to contemporary individuals, thereby increasing the similarity (resemblance) within members of the same group born and grow together. The *contemporary environment* includes the same period of birth and rearing (members of a family like half sibs in animals are more likely to be born and reared during the same period), the same conditions of culture media in Drosophila, same cultural and soil conditions for plants. The c - effects create differences between families

not receiving contemporary environment and are difficult to separate. The increase in size of families also does not decrease the c- effects. The c- effects among full sib families are larger than half sib families.

The within family selection eliminates the non genetic familial effects, because within family selection is made among the individuals having C-effects. Therefore, the familial effects being held in common to all members of the family, individuals deviating from family mean is supposed to be genetically superior and deserve to be selected. The within family selection is used when the environmental effects are large (low h^2) but common for all family members.

It is important to remove the environmental likeness before estimating genetic covariance. This can be done by comparing the animals within the same environment or to adjust the data for the environmental effects.

Chapter 17
Genetic Parameters: Heritability

Selective breeding primarily aims the genetic improvement of progeny generation. The genetic progress in performance is estimated by genetic parameters which are heritability, repeatability and genetic correlation among characters. The heritability estimates the progress in terms of response to selection in a trait whereas the heritability along with genetic correlation estimates the progress in terms of correlated response in a correlated character. The repeatability is used to estimate the increase in mean performance at subsequent age whereas the repeatability along with phenotypic correlation estimates the improvement in correlated trait at subsequent age of the animal. The use and importance of all these three genetic parameters have been discussed in subsequent chapters along with their definition and methods of estimation.

17.1 DEFINITION OF HERITABILITY

The term heritability was coined by Lush (1948). The heritability can be defined in a number of ways *viz.* mathematically, statistically and genetically. However, it expresses the genetic part of the phenotypic variance. It is taken in two sense *viz.* broad sense and narrow sense heritability according to that which genetic part (genotypic or additive) is to be expressed as a ratio to the total phenotypic variance. When only additive genetic part is taken, it is then said as the heritability in narrow sense and denoted as h^2. Thus, $h^2 = \dfrac{\sigma^2 A}{\sigma^2 P}$. Whereas when the complete genetic part is taken, it is then said as the heritability in broad sense denoted H^2 = $\dfrac{\sigma^2 G}{\sigma^2 P}$. The heritability is taken in narrow sense unless other wise mentioned.

Mathematically, the heritability is the portion of differences in phenotypic values of a character attributable to the differences in breeding values for that character. In other words, the heritability is the ratio of additive genetic variance to the phenotypic variance as:

$$h^2 = \frac{\sigma^2 A}{\sigma^2 P}.$$

Statistically, the heritability is defined in terms of the relationship between breeding value and phenotypic value of a character in a population taken as a measure of the strength (reliability or correspondence or consistency) of relationship between the two values. Heritability in term of the correlation between the two values is taken as square of the correlation between phenotypic value and breeding value. Therefore, $h^2 = (r_{AP})^2$.

The heritability, also in statistical term, is the regression of breeding value (A) on the phenotypic value (P) and hence, $h^2 = b_{AP}$.

Genetically, the heritability measures the degree to which the offspring resemble their parents in performance traits. Thus, it indicates the resemblance between parent and offspring. It is also taken as the portion of genetic superiority of parents inherited and expressed by the offspring.

17.2 IMPORTANCE AND USE OF HERITABILITY

The heritability is the most useful genetic parameter in providing a number of useful information, predicting the possibility and amount of change in the genetic improvement of population as result of selection, and deciding the proper selection and mating systems.

1. The heritability of a character provides the following *useful information* -

(i) The heritability is a *measure of the genetic variability* in phenotypic values of a character among the individuals of the population. Heritability is expressed in two ways *viz.* the amount of total variability attributed to differences in genotypic values (H^2) or attributed to differences in the breeding values (h^2) of the individuals of a population.

(ii) The heritability is a *measure of the portion of phenotypic superiority of parents passed on to the progeny.* This is because the heritability expresses the amount of phenotypic superiority of parents (the phenotypic differences between the selected and rejected parents), attributed to additive effect of genes which is transmitted (inherited) by the parents and hence passed on to their progeny,.

(iii) The heritability is a *measure of the degree of correspondence (agreement) between the phenotypic value and breeding value.* This is because the square root of heritability (h) is equal to the correlation between breeding value and phenotypic value (h = r_{AP}) which indicates the association between the two values.

(iv) The heritability is a *measure of accuracy of selection* based on phenotype. This is because if the heritability is higher, the r_{AP} will also be higher, the higher proportion of superiority will be transmitted and hence the selection will be more accurate.

(v) The heritability is an *indication of type of gene action*. This is because the heritability indicates about the amount of additive genetic variance arising due to additive gene action. Thus, the magnitude of heritability indicates the type of gene action and hence the heritability is an indicator of the extent to which the trait is affected by additive gene action. This is because the heritability estimate is a measure of the proportion of the total variation in a trait due to additive genetic variance arises due to additive gene action.

(vi) The heritability indicates about the *basic genetic phenomenon* of metric traits which are resemblance between relatives and the inbreeding depression with its converse hybrid vigour (heterosis). The resemblance between relatives is due to the effect of genes commonly shared by them. The heritability is a measure of the additive genetic variance which arises due to additive gene effect that are measured in terms of the breeding value. The relatives sharing common genes will have similar additive gene effects and hence similar breeding values. This similarity in breeding values causes resemblance between relatives which forms the basis to estimate the heritability (additive genetic variance arising due to additive gene effects). Secondly, the inbreeding depression or its converse heterosis depends on the type of gene action. Both these phenomenon (I.D. and heterosis) depend on the occurrence of dominance and epistasis. The loci with additive gene action cause neither inbreeding depression nor heterosis. Therefore, the magnitude of heritability which is a measure of gene action gives an indication of the inbreeding depression or heterosis likely to occur or not.

2. The heritability has its *predicting role* to predict the breeding value of an individual and the genetic gain in a population through selection.

(i) The heritability is used to predict the *breeding value* of an individual. The heritability is regression of breeding value on phenotypic vale ($h^2 = b_{AP}$). The regression is used to predict the unknown variable based on the value of a known variable as: $Y = b_{YX} (X_i - X)$. The animal breeder is interested to know the breeding value (A) which is unknown for an individual based on its phenotypic value which is known. Thus, the estimate of an individual's breeding value for a trait is obtained by multiplying its phenotypic value (P) with heritability (h^2) of the trait as: $A = h^2 P$ by taking both the values as deviation from population mean Thus, the best estimate of the absolute B.V. from phenotypic value for each individual in a population can be obtained as: B.V. $= \mu + h^2 (P_i - \mu)$. Where, μ is population mean and P_i is the phenotypic value of ith individual.

(ii) The heritability is used to predict the *genetic gain* (ΔG). The genetic gain or genetic improvement or response to selection is the average superiority of the progeny over their parental generation mean before selection ($X_O - X_P$). The mean of progeny generation is estimated as:

$$X_O = \mu + h^2 (X_S - \mu)$$
$$X_O - \mu = h^2 (Xs - \mu)$$
$$\Delta G = h^2 (Xs - \mu)$$

where,　μ = Population mean before selection;
　　　　X_S = Mean of selected parents;
　　　　X_O = Mean of offspring.
　　$X_S - \mu$ = Phenotypic superiority of selected parents known as selection differential

$X_O - \mu$ = Response to selection or genetic gain per generation

ΔG = Av. superiority of progeny.

3. The heritability plays very important role in *decision on the criteria of selection and the mating system*. The uses of heritability in designing the breeding programmes are as under-

(i) Selection of traits: In a breeding programme, multiple trait selection is preferred rather than single trait selection. The heritability is one of the most important aspects to decide that which traits should be included in construction of selection index for multi trait selection. The traits with high heritability are included.

(ii) Optimum family size: The sib selection is applied for selection of sex limited traits *viz.* milk yield etc. The efficiency of sib selection depends on family size (no. of sibs to be recorded) because larger is the family size larger will be the response. However, it is not so in practice because of the involvement of intensity of selection as a factor affecting response. There is a limitation of breeding space or facilities to evaluate so many families with larger number of individuals. This impose restriction either to maintain smaller families (having lesser number of individuals per family) or to maintain larger families (having more number of individuals per family). The larger families (having more number of individuals per family but with few numbers of families) result in low intensity of selection. Thus, family size and family number has a conflict and a compromise has to be made. Taking into consideration the intensity of selection, there should be optimal family size to give maximum expected response. The optimal family size (n) depends on the magnitude of heritability as:

$$n = 0.56 \sqrt{\frac{T}{Nh^2}}$$

where, N = no. of families to be selected;

T = Family size (total number of individuals to be measured)

(iii) Selection criterion: The selection criterion may be the phenotypic value or the combining ability of the trait. The selection based on phenotypic value rests on the breeding value of the individuals estimated either from individuals' own record (individual selection) or from the records of its relatives (sib selection, within family selection). The selection criterion based on phenotypic value are often used to improve the performance of a single purebred population whereas the selection criterion based on individuals' combining ability for the trait is used in commercial application involving crossbreeding.

(a) *Pure breeding programme*- The choice of selection criterion based on phenotypic value depends on the magnitude of heritability of the trait. The individual selection based on individuals' own performance is advocated for traits with high heritability whereas family selection is used to improve the traits with low heritability. In family selection, either whole family is selected or few individuals with phenotypic values above family mean are selected.

When the selection criterion is the family mean and whole family is selected and it is called as family selection. This family selection is used when environmental effects are large (low h^2) but different for different individuals so that the environmental effects are cancelled by taking the mean. On the contrary, when one or few individuals of a family are selected it is known as within family selection. The few individuals within a family are selected on the basis of the deviation of their phenotypic values from the family mean. This within family selection is also used when the environmental effects are large (low h^2) but these environmental effects should be same (common) to all the family members (C-effects). The C-effects are eliminated in within family selection because the selection is made among the individuals having C-effects. Therefore, individuals deviating from family mean are supposed to be genetically superior and should be selected.

(b) Crossbreeding programme: The individuals combining ability for a trait is measured as the average phenotypic value of the progeny produced by crossbreeding the individuals of two populations. The combining ability is of two types *viz.* general (G.C.A.) and special (S.C.A.). The selection is done for utilizing both types of combining ability using two selection procedures *viz.* recurrent selection (RS) and reciprocal recurrent selection (RSS). Both these selection procedures are useful for improvement in traits affected by non additive gene action like over dominance *i.e.* traits with low heritability.

17.3 METHODS OF ESTIMATING HERITABILITY

The resemblance among relatives is measured as the covariance among them and this covariance is equated to the additive genetic variance depending upon the types of relatives on whom the observations are recorded. Thus, the resemblance is used as a basis to estimate the additive genetic variance in a trait and hence the heritability of the character. This has been shown below:

Table 17.1: Relation of resemblance (covariance) among relatives with variance components

Relatives	Covariance = Variance components	Resemblance as correlation (r) or regression (b)
Offspring-one parent	$Cov_{Op} = \frac{1}{2} V_A + \frac{1}{4} V_I$	$b = \dfrac{\frac{1}{2} V_A}{V_P}$
Offspring-mid parent	$Cov_{o\hat{p}} = \frac{1}{2} V_A$	$b = \dfrac{V_A}{V_P}$
Half-sibs	$Cov_{HS} = \frac{1}{4} V_A + \frac{1}{16} V_I$	$r = \dfrac{\frac{1}{4} V_A + \frac{1}{4} V_I}{V_P}$
Full sibs	$Cov_{FS} = \frac{1}{2} V_A + \frac{1}{4} V_D + V_{Ec}$	$r = \dfrac{\frac{1}{2} V_A + \frac{1}{4} V_D}{V_P}$

where, $V_{\hat{p}} = \frac{1}{2} V_P$ but $V_P = \sigma^2_{P;} V_P = V_O$ and hence $\sqrt{V_P V_O} = \sigma^2_P$

$h^2 = 2\, b_{OS} = 2b_{OD} = b_{O\hat{p}} = 2\, r_{OP}$

Thus, the covariance of relative is identical to variance components relating to group means (intra class correlation) as shown in table above.

The method of estimation of heritability used depends on the type of data available *i.e.* whether the data is available on parent offspring performance or sib performance (half sibs or full sibs). The heritability can also be estimated from analysis of twin data as well as from the data collected on selection experiment by comparing the performance of parental and progeny generation.

1. Regression of offspring on dam: This method is used when the mating are random, all groups are raised together and each dam (X) has one offspring (Y).

The parent offspring covariance (Cov$_{OP}$) contains $\frac{1}{2} V_A + \frac{1}{4} V_{AA}$. The observation of progeny (Y_i) is taken as: $Y_i = b\, X_i + e_i$. The heritability is estimated as: $h^2 = 2\, b_{OP}$

where,

$$b_{OP} = \frac{Cov_{XY}}{V_X} = \frac{\Sigma xy}{\sigma^2 x}$$

$$\sigma^2 x = \Sigma X^2 - \frac{(\Sigma X)^2}{N}$$

$$\Sigma xy = \Sigma XY - \frac{(\Sigma X\, \Sigma Y)^2}{N}$$

N = No. of dam – daughter pairs

Merits of the method: The estimate does not contain dominance variation but only a little amount of epistatic variation. It is free from environmental covariance because parent and offspring are measured at different times. Selection of parents does not affect the estimate because selection has same effect to reduce the V_p and Cov$_{OP}$. The estimate is least affected by mating system. It provides a more precise estimate od heritability.

Demerits and limitations: It requires the records on two generations. The information on two generations are rarely available for certain traits like slaughter traits. The method of regression of offspring on mid parent value can not be used for sex limited traits. The regression of offspring on dam (b_{OD})and on mid parent value for traits particularly measured early in life (birth weight, litter weight etc) are affected(by maternal effects, though regression of offspring on sire(b_{OS}) does not contain maternal effect .but its application is limited due to lesser degree of freedom and for sex-limited traits.

2. Intra – sire regression of offspring on dam ISRD): This method is used when a number of sires are mated to several dams and each mating produces one offspring. It is estimated as the average regression of daughter on dam within sire. The regression of daughter on dam is calculated separately for each dam group mated to one sire and then the regression from each sire is pooled in a weighted average. This removes the sire effect.

The parent offspring covariance (Cov_{op}) estimated by this method contains $\frac{1}{2}V_A + \frac{1}{4}V_{AA} + \frac{1}{2}V_M$.

The observation of progeny (Y_{ij}) born to j^{th} dam mated to i^{th} sire is taken as: $Y_{ij} = \mu + S_i + b(X_{ij} - X..) + e_{ij}$. Thus, the sire effect (S_i) is removed. The heritability is estimated by conducting the analysis of variance and covariance as:

Table 17.2: ANOVA and Covariance.

S.V.	D.F.	Dam (X)	Daughter (Y)	Cross product (XY)
Sires	S-1	SS $_{S(X)}$	SS $_{S(Y)}$	SCP $_{S(XY)}$
Dams / sire	D-S	SS $_{DS(X)}$	SS $_{DS(Y)}$	SCP $_{DS(XY)}$
Total	D-1	SS $_{T(X)}$	SS $_{T(Y)}$	SCP $_{T(XY)}$

The procedure to calculate sum of squares for sires (SS_S), dams within sire (SS_{DS}) and total sum of squares (SS_T) on both dams (X) and daughter (Y) observations, and the sum of cross products (SCP_{XY}) is given as under-

Sum of Squares for dams:

$$\text{Correction Factor (C.F.)} = \frac{(\Sigma X_{ij})^2}{D} \; ;$$

$$S.S._{T(X)} = \Sigma X^2_{ij} - C.F.;$$

$$S.S._{S(X)} = \frac{\Sigma X^2_{i.}}{n_i} - C.F.$$

$$S.S._{DS(X)} = S.S._{T(X)} - S.S._{S(X)} = \Sigma X^2_{ij} - \frac{(\Sigma X_{i.})^2}{n_i}$$

Sum of Cross Product (C.S.P.$_{XY}$)

$$\text{Correction Factor} = \frac{(\Sigma X_{ij})(\Sigma Y_{ij})}{D}$$

$$S.C.P._{T(XY)} = \Sigma X_{ij} Y_{ij} - C.F.;$$

$$S.C.P._{S(XY)} = \Sigma \frac{(X_{i.}Y_{i.})}{n_i} - C.F.$$

$$S.C.P._{DS(XY)} = S.C.P._{T(XY)} - S.C.P._{S(XY)} = \Sigma X_{ij} Y_{ij} - \frac{\Sigma X_{i.} Y_{i.}}{n_i}$$

where, D = number of total dams = Σn_i
n$_{i.}$ = number of dams mated to i^{th} sire
X$_{i.}$ = Sum of phenotypic values of all the dams mated to i^{th} sire
Y$_{i.}$ = Sum of phenotypic values of all the daughters born to i^{th} sire
X$_{ij}$ = phenotypic value of jth dam mated to i^{th} sire
Y$_{ij}$ = phenotypic value of jth daughter born to i^{th} sire

$$b_{YX} = \text{Intra-sire regression of daughter on dam} = \frac{S.C.P._{DS\,(XY)}}{S.S._{DS\,(X)}}$$

$$h^2 = 2\,b_{YX}$$

Merit of method: No correction is required for mating system because progeny within group are produced from the same sire. Environmental component is much reduced by similar environment within group. This provides useful estimate of h^2 when there is no selection in progeny generation and when the maternal effects are absent. This method is most suited for sex – limited traits. This method is more appropriate and suitable than other regression estimate. This method has the limitation that it does not eliminate the environmental differences among dams within sires. The characters measured in early life may have some maternal effect which may increase the estimate.

3. Paternal half sib correlation method: The half sibs are of two types *viz.* paternal and maternal. The intra class correlation between paternal half sibs is more commonly used to estimate heritability for the reason that these are available more in numbers than maternal half sibs. The correlation compares the resemblance among HS to the resemblance among individuals related by an average amount for the population. This is the ratio of variance component among the means of HS group to the total phenotypic variance. The variance among groups is a measure of similarity within group.

The half sib covariance (Cov_{HS}) equals to sire variance (σ^2_S). This is the variance of sire group means about the population mean whereas the rest of the variance ($\sigma^2 e$) indicates the difference between the progeny of the same sire group (within sire component) and is the variance of individual observations about their group mean. The $\sigma^2 e$ estimates the remainder of the genetic variance plus all the environmental variance.

This method is used when a number of females (dams) are mated to a number of sires and each dam produces one offspring. The observation (phenotypic value) on a progeny (X_{ij}) born to j^{th} dam mated to i^{th} sire is taken as: $X_{ij} = \mu + S_i + e_{ij}$. The analysis of variance is conducted to partition the phenotypic variance (σ^2_p) into $\sigma^2 S$ and $\sigma^2 e$ after estimating the sum of squares.

Table 17.3: Analysis of variance for half sib groups

S.V.	D.F.	Sum of square	Mean Square	Expected Mean square
Sires	S – 1	Sire S.S.	M.S.$_S$ = Sire S.S. / S-1	$\sigma^2_e + K\sigma^2_s$
Progeny /Sire	N – S	Error S.S.	M.S.$_e$= Error S.S. / N	$S\sigma^2_e$
Or Error				

The procedure to calculate the sum of squares and different component of variance is as under-

$$\text{Correction Factor (C.F.)} = \frac{(\sum X)^2}{N} = \frac{X^2..}{n..}$$

Total sum of squares (T.S.S.) $= \Sigma X^2_{ij} - C.F.$

Sire Sum of square $= \Sigma X^2_{i.} - C.F.$

Progeny within sire S.S. (Error S.S.) = T.S.S. – Sire S.S.

$$\sigma^2_S = \frac{(M.S._S - M.S._e)}{K} \; ; \; \sigma^2 e = M.S. \; e; \; \sigma^2_S + \sigma^2_e = \sigma^2_P$$

$$t = \frac{\sigma^2 S}{\sigma^2 P} \text{ and } h^2 = 4 \; t.$$

where, K = Average number of daughters per sire

$$= \frac{1}{S-1}\left(n. - \frac{\Sigma n_{i.}^2}{n.} \right)$$

n_i = number of daughters of ith sire and n. $= \Sigma n_i$

Merits of Half sib method: Half sib correlation does not contain variance due to dominance or maternal effect or common environment and hence free from bias. The records of all daughters are recorded at the same time. The method can be used for sex limited traits and for the traits expressed after death of animals. It does not require records of parents. The heritability estimated from maternal HS correlation contains epistatic effect, maternal effect and some common environmental effect. The paternal Hs method is more commonly used than maternal HS because the maternal HS are not available in dairy animals, measured in different years and contain maternal effect as well as common environmental effect. This inflates the heritability. The maternal component of variance can be obtained as a difference in the intra class correlation between maternal Hs and paternal HS.

Limitations of half sib method: Half sib correlation contains some interaction component. The heritability is obtained by multiplying the intra class correlation with 4, the environmental correlation (if exist) are also quadrupled. The environmental correlation exists in the presence of genotype–environment correlation. It is valid in random mating population because the estimate is affected by selection. The heritability estimated based on small numbers of sire increases the chance of large error.

3. Full sib method: This method is used in species with larger full sib families like pigs, poultry, mice, goat when each male is mated to several females and each mating produces several offspring (multiple births) or when repeated mating are common. The progenies are arranged as full sib families according to sires taking that i^{th} sire is mated to a set of d_i dams producing n_{ij} progeny. The observation (X_{ijk}) on the kth progeny from j^{th} dam mated to i^{th} sire can be described as:

$$X_{ijk} = \mu + S_i + d_{ij} + e_{ijk}$$

The analysis of variance is conducted to partition the phenotypic variance into observational components.

Sources	D.F.	S. S.	M.S.	E. M. S.
Sire	S -1	Sire S.S.	$M.S._S$	$\sigma^2_e + K_2 \sigma^2_D + K_3 \sigma^2_S$
Dam / sire	D – s	Dam S.S.	$M.S._D$	$\sigma^2_e + K_1 \sigma^2_D$
Progeny / dam (full sibs within dam)	N- D	Error S.S.	$M.S._e$	σ^2_e

where,

$$K_1 = \frac{1}{D-S}\left(n.. - \frac{\sum n^2_{ij}}{n_{i.}}\right)$$

$$K_2 = \frac{1}{S-1}\left(\frac{\sum n^2_{ij}}{n_i} - \frac{\sum n^2_i}{n..}\right)$$

$$K_3 = \frac{1}{S-1}\left(n.. - \frac{\sum n_{i.}^2}{n..}\right)$$

$N = n.. = $ Total no. of progeny; $n_{ij} = $ No. of progeny per dam;
$n_{i.} = $ No. of progeny per sire.
$S = $ No. of sires; $D = $ No. of dams

The different components of variance are computed as:

$$\sigma^2_D = \frac{MS_D - MS_e}{K_1}$$

$$\sigma^2_e = MS_e$$

$$\sigma^2_s = \frac{MS_s - MS_e - K_2^2\sigma^2_D}{K_3}$$

$$= \frac{MS_S - MS_D}{K_3}$$

$$\sigma^2_P = \sigma^2_S + \sigma^2_D + \sigma^2_e$$

$$h^2_s = 4\frac{\sigma^2_s}{\sigma^2_P}; h^2_D = 4\frac{\sigma^2_D}{\sigma^2_P};$$

$$h^2_{(S+D)} = 2\frac{\sigma^2_s + \sigma^2_D}{\sigma^2_P}$$

The heritability estimated from full sib component [$h^2_{(S+D)}$] contains twice the maternal effects and ½ V_D whereas h^2_D contains four times the maternal effects plus all the dominance effect because the σ^2_D contains variance due to the maternal effect (V_M) during early life of an individual.

Merits and Limitations of full sib method: The h^2_{FS} contains dominance deviations, epistatic variance, maternal effects and common environment in addition to the additive gene effects. Maternal effects strongly bias the covariance in full sibs. Thus the estimate is subject to bias and is least reliable to other methods. Therefore, this method is not valid for traits affected by maternal effects. This method requires larger full sib families which are, in general, not available particularly for large farm animals. The difference between dam and sire component of variance from FS analysis indicates the importance of dominance, interaction and maternal effects. The h^2_s is relatively unbiased, whereas h^2_D is biased in comparison to h^2_s and $h^2_{(S+D)}$. Combined estimate of non-additive and maternal effects can be known from three estimates of heritability.

Chapter 18

Repeatability

The characters recorded at different times (at different ages of the animals) take different phenotypic values in spite of the fact that genotype of an individual is fixed at fertilization and do not change throughout the life. In view of no change in genotype, the only source of differences in phenotypic values of the same individual at different age is the variation in environment to which the individual is exposed.

These environmental sources of variation are of two types *viz.* permanent and temporary. Accordingly, the environmental variance can be partitioned into its two components. The permanent environment effects cause permanent environment variance denoted as V_{EP} and these are general environmental effects. The permanent environmental effects have persistent effect over repeated records of the same animal like udder damage has persistent effect on milk production in subsequent lactations, nutrition during calf hood age, training of horses, etc. The temporary environment effects cause temporary environment variance denoted as V_{ET} and these effects are for a specific time and not always, for which it is called as special environmental variance. These factors vary from time to time. The V_{ET} indicates the within individual variation. Theses temporary factors are climatic factors (temperature, humidity, season of a year), feeding regime, age, management practices which vary from year to year and their effects are likely to have positive and negative effects on the phenotypic values and hence their effects tend to average zero over several periods.

In spite of the different phenotypic values of repeated character at different ages, there exist a correlation between any two records of the individual and the magnitude of correlation between two records depends on the adjacency of the two records as well as the extent of genetic and permanent environment effects affecting the repeated records. More adjacent are the two records and more is the variability due to permanent environment effects between two records, higher will be the correlation between the two records. The magnitude of correlation between two records will indicate the reliability of the previous record in predicting the future record. When the correlation is higher, the reliability of previous record will be higher in predicting the future record or vice versa. The correlation between two repeated records will be higher if these records are less affected by temporary environmental effects existing for a specific time. Therefore, it is required to measure the extent to which the observed variations among the repeated records are affected by permanent (both genetic and environmental) and temporary environmental effects. It is measured by the term called repeatability coined by Lush (1937) and denoted by either t or r.

18.1 DEFINITION OF REPEATABILITY

The repeatability is the correlation between two records of individuals of a population. Thus repeatability indicates the strength of relationship (consistency, reliability) between repeated records of the same trait on the individuals of a population. Now take X_1 and X_2 as the two records which are the two observations of the same trait on the same individual recorded at different ages.

Thus, $t = r_{X1X2,}$

The repeatability has its meaning and use depending on the magnitude of repeatability value in a way that if the repeatability is high, it then indicates that the first record on an individual (X_0) is a good indicator of the second record of the same individual (X_2).

Secondly, the repeatability can be taken as the regression of second record on the first record and this will indicate the change in future record per unit change in first record. Thus, $t = b_{X2\ X1}$.

In this way the repeatability is useful in predicting the future performance (record) from early record. Thus the repeatability is the fraction of differences between single records of individuals that are likely to occur in future record.

Thirdly, the repeatability can also be expressed as a ratio of variances *viz.* proportion of differences in first record that are attributable to the differences in future record which is influenced by the permanent effects (genetic and environmental).

Thus, $t = (\sigma^2_G + \sigma^2_{EP}) / \sigma^2_P$.

Thus the repeatability indicates the fraction of the observed differences among individuals due to permanent effects and the remaining portion of the differences (1-t) caused by temporary environmental effects. Therefore, the repeatability provides information on the relative importance of permanent effects and temporary environmental effects.

Like heritability, the repeatability is also a population parameter. It is the property of a trait, population and the environmental effects. It is also affected by same factors which affect the heritability and thus the repeatability can be increased by the same ways that are used to increase the heritability.

18.2 IMPORTANCE AND USE OF REPEATABILITY

1. The main property of repeatability is that its *determination is much easier* than heritability in either sense. The repeatability can be estimated without information on groups of relatives (parentage information) and hence the repeatability can be used as the preliminary genetic information.

2. The repeatability sets *upper limit of heritability* in broad sense because the repeatability includes all the genetic (additive and non additive components) plus permanent environmental influences which contribute to the real differences among animals whereas the heritability in broad sense includes only the genetic differences.

3 *(i)* The repeatability is used in *prediction of future performance based on single record* on the individuals. The future record (X_2) may be estimated as;

$$X_2 = \mu + r\,(X_S - \mu)$$
$$X_2 - \mu = r\,(Xs - \mu) = \Delta S$$

where, $\quad \mu$ = Population mean before selection;

$\quad\quad Xs$ = Mean of selected individuals;

$\quad\quad X_2$ = Mean of 2$^\text{nd}$ record.

$\quad Xs - \mu$ = Phenotypic superiority of selected individuals (selection differential)

$\quad X_2 - \mu$ = ΔS Gain in future record by selection

$\quad\quad\quad$ = Av. superiority of future record (X_2).

This indicates the change in population mean for a future record due to selection based on previous record and hence it is the portion of superiority or inferiority of a given record on an individual as compared to the herd average.. Thus it is used to estimate the performance potential or real producing ability of an animal and to compare different individuals in the same herd / flock.

(ii) The repeatability can also be used in *prediction of future performance based on more records* known as multiple records. The repeatability is useful to know the relative efficiency of using single or more records. The repeatability estimate gives an indication that how many records on an individual should be taken before it is culled. The *repeatability for multiple records (R) is estimated* as;

$$R = \frac{nr}{1 + (n-1)\,r}$$

where, $\quad n$ = number of records;

$\quad\quad r$ = repeatability based on one record.

The estimate R increases with increase in number of repeated records. When repeatability is high, single record is enough to predict the real producing ability (future performance) than considering several records of a trait on an individual with low repeatability will do. This means that when temporary environmental effects are less important (high repeatability) the use of repeated records is of no use, hence the repeated records do not add to the accuracy and the culling on the basis of single record will be effective. On the other hand, the repeated records add to the accuracy when repeatability is low. This is because increase in multiple records cancelled out the temporary environmental effects. The change in later record based on repeatability of multiple records (R) can be predicted by replacing the value of **r** with R in the above equation of predicting of future record (performance) based on single record. The gain due to selection (mean change) at subsequent age based on **n** records is estimated as:

$$\Delta S = \frac{nr}{1 + (n-1)\, r} \,(X_S - \mu)$$

Lush (1945) used the above relationship to predict the *most probable producing ability (MPPA)* for dairy cows with varying number of records per cow to adjust the records of cows to the same basis for comparing the different cows in the same herd. The MPPA is estimated as:

$$MPPA = \mu + \frac{nr}{1 + (n-1)\, r}\,(C - \mu)$$

where, C is the cow's own average = X_i

μ is the herd average i.e. average before selection.

The temporary environmental effects vary from time to time and are likely to have positive and negative effects at different times and thus cancelled. The permanent environmental effects together with genotypic effects decide an individual's performance potential for its whole life. This performance potential is called as the 'real producing ability".

4. The *relative efficiency of using different records* taken at different ages can be tested for predicting life time production by estimating the correlation between total and individual record (r_{Ti}) as:

$$r_{Ti} = \sqrt{\frac{1}{n}\,[\, + (n-1)\, r]}$$

where, n is the number of records used to calculate total of all the records (T). The relative efficiency of ith record to predict the total production is estimated as:

$Q = \dfrac{r_{Ti}}{r_{Tj}}$. This relationship (Q) is used to determine that which record is more

useful in selection.

5. The repeatability is used in taking *culling decision*. The animals with low performance based on first record can be culled when repeatability of the trait is high but when repeatability is low the poor performing animals based on first record should not be culled because the first record is not reliable of the producing ability of the animal in case of low repeatability. Thus when repeatability is low the additional future record has its importance. On the other hand, the additional future record is not important when repeatability is high to take the culling decision.

6. The repeatability estimate can be used in *feed* lot *experiment*. In case of high repeatability, it requires to distribute equally the different progeny of a sire or dam in different lots. This is because in case of high repeatability the lot differences will largely be due to the genetic differences rather than actually to the lot differences ascribed to treatments.

18.3 METHODS OF ESTIMATION OF REPEATABILITY

The estimation of repeatability requires multiple/ repeated records of the same character on the same animal. The number of records per animal may vary but at least two records per animal are required in estimation of repeatability.

(i) **Inter class correlation:** This method is used when only two records per animal are available. Considering X and Y as the two records taken at different ages, the repeatability (r) is estimated simply as the correlation coefficient between X and Y as:

$$r_{XY} = \frac{\Sigma xy}{\sqrt{\Sigma x^2 - \Sigma y^2}}$$

Actually the inter class correlation between repeated records on an animal is the ratio of variance attributable to permanent differences (genetic and environmental) among animals to the phenotypic variance. This ratio is repeatability. Thus the correlation between repeated records on animals is the repeatability. The inter class correlation in terms of variance components can be understood from the measurement P_{ij} taken on i th animal at jth time and the measurement (P_{ij}) is defined as:

$P_{ij} = \mu + S_i + e_{ij}$.

The measurements at two different times (j and j') on the same individual will differ because e_{ij} and $e_{ij'}$ are different representing temporary environment effects but theses two measurements will be similar to some extent because two measurements have S_i in common representing permanent effects (genetic and environmental) as:

$P_{ij} = \mu + S_i + e_{ij}$ and $P_{ij'} = \mu + S_i + e_{ij'}$

Therefore, the covariance of two measurements taken at two times will be as:

Cov $(P_{ij} \, P_{ij'})$ = Cov $(S_i + e_{ij})$ $(S_i + e_{ij'})$ = $\sigma^2 \, _S$.

The correlation (r) between two records: r = $\dfrac{Cov P_{ij} P_{ij'}}{\sigma P_{ij} \sigma P_{ij'}}$ = $\dfrac{\sigma^2 \, _S}{\sigma^2 \, _P}$

(ii) **Regression Method:** In case of two repeated records (X and Y), the correlation between them is equal to the regression of one record on the other. This is based on the assumption that the variances of the two records (V_X or V_Y) do not differ ($\sigma^2 \, _X = \sigma^2 \, _Y$), so $\sigma_X = \sigma_Y$. Therefore, $r_{XY} = b_{YX}$. Thus the regression of future record (second record) on the previous record (b_{YX}) = correlation between these two records (r_{XY}). This regression of future record on the previous record (b_{YX}) estimates the repeatability of the character. The repeatability in terms of regression can be

used to predict gain in future performance of the herd by selection based on earlier record. The repeatability is the predicted deviation in future records per unit deviation in early records. This means that repeatability is the proportion of individual's deviation from population mean for one record expected to be retained in future record of the same trait. The predicted superiority in future record will be equal to b_{YX} time superiority of the individual for first record over the population mean. Thus, the repeatability is the proportion of an individual superiority for a trait expected to be retained in future record.

(iii) **Intra class correlation:** Most often more than two records per animal are available. The repeatability, based on n animals with *k* records each, is estimated considering the k records of each animal as the records of a family and conducting the analysis of variance similar to the estimation of intra class correlation of paternal half sibs, as done for estimation of heritability. Here, in this case of estimation of repeatability by intra class correlation based on the different records of the same animal, the expected composition of variance between animals will be different from the expected composition of variance between half sibs because in case of repeatability estimation the different records are of the same animal whereas in case of heritability estimation the different records are of the different progeny of the same sire or dam. Therefore, in this case of estimation of repeatability, the expected composition of variance between animals denoted by $\sigma^2 a$ will be equal t   o
$\sigma^2_G + \sigma^2_{EP} = V_A + V_D + V_I + V_{EP}$. Thus, the between animal variance ($\sigma^2 a$) will estimate all the genetic variance (V_G) plus the permanent environmental variance peculiar to the animal (V_{EP}). The $\sigma^2 e$ is the within animal variance due to temporary environmental effects and so, $\sigma^2 e = \sigma^2_{ET}$. The ANOVA is conducted as:

.SV.	D.F.	S.S.	M.S.	E. M.S.
Animals	n-1	S.S.a	M.S. a	$\sigma^2 e + K \sigma^2 a$
Within animal	n (k-1)	S.S.e	M.S. e	$\sigma^2 e = \sigma^2_{ET}$

$$\text{where, } \sigma^2 a = \frac{MS_a - MS_e}{k}$$

$$K = \text{Av. no. of records per animal} = \frac{1}{N-1} n. - \frac{\Sigma n^2_i}{n.}$$

N   =   No. of animal;

n.   = Σn_i = Total no. of records;

n_i =   no. of records of j [th] animal

The **repeatability is the intra class correlation** among the repeated measurements of a trait. This intra class correlation is the proportion of the variation containing genotypic variation (V_G) plus variation due to permanent environment effects (V_{EP}) to the total phenotypic variation (V_P). Therefore, repeatability (t) is:

$$t = r = \frac{\sigma^2_G + \sigma^2_{EP}}{\sigma^2_P} = \frac{\sigma^2_a}{\sigma^2_a + \sigma^2_e}$$

Thus, the repeatability indicates the proportion of total phenotypic variance attributed to the variance among individuals due to permanent differences (both genetic and environmental). This is because the repeated records are the records of the same animal and hence the same sets of genes are responsible for the expression of the character at different times.

This procedure of analysis of variance of repeated records for estimating repeatability is useful in estimating the variance due to temporary environmental effects (σ^2_{ET}).

Chapter 19

Correlation Among Characters

It is of great interest and importance to know the variation in two character that how they vary together. This helps to understand the relationship between them. When two characters have a relationship between them, it means there is a simultaneous change (variation) in both the characters. Two characters may be related to each other in such a way that a change (variation) in one character correspond with a particular directional change in the other. If there is an increase in the value of one character, it may lead to either an increase or decrease in the value of other character. The simultaneous change (variation) in two characters is measured in terms of *covariation* which indicates that the two characters vary together and have dependency on each other. The measurement of the covariation is the *covariance* which is the mean product between the deviations of the two characters measured on the same individuals. The covariance indicates the direction and strength of relationship between two characters and used to estimate the degree of relationship (dependency) between two characters in terms of correlation and regression.

The degree of mutual relationship between two characters is measured by *correlation coefficient* which is a measure of simultaneous change in two characters *i.e.* how the two characters change together. It is denoted by the letter *r*. The coefficient of correlation is defined as the ratio of the variance common to the two characters (Covariance) to the geometric mean of the variances of the two characters. This is the *simple correlation*. Thus, the correlation is obtained by dividing the covariance of two characters (Cov_{XY}) by the square root of the product of the variances of two characters as:

$$r_{XY} = \frac{Cov_{XY}}{\sigma_X \, \sigma_X}$$

Thus, the correlation is the covariance between two characters when compared to their average variability and it is the covariance per unit of average change.

In general, most of the economic characters have correlation with each other to some extent. The existence of correlation among characters requires certain common causes which affect the different characters. The correlation between two characters assumes the cause and effect relationship between them and depends upon their biological relationship, but the correlation coefficient does not give the idea that which one is the cause and which is the effect.

19.1 PHENOTYPIC CORRELATION, CAUSES AND TYPES

The correlation directly estimated from phenotypic values of two characters is known as the *phenotypic correlation.* This measures the linear association between two traits and indicates the deviation of an individual's phenotypic value of one trait from the population mean in relation to the deviation of other trait from its population mean. Therefore, the phenotypic correlation (r_p) gives the idea of the extent to which the individuals above average in one trait are above, below or near average in the other trait.

The two characters are correlated due to certain causes. It is known that the different causes which affect the different characters are the genetic and environmental in origin. Thus, the characters showing correlation among them selves may be affected by some common genes and by some common environment. Therefore, the genetic and environmental factors are two possible causes which are responsible to create a correlation between characters. The correlations are of two types *viz.* genetic and environmental, depending upon the nature of the factors which cause correlation.

The genetic factors causing correlation are the pleiotropic effect and linkage of genes. Some genes affect more than one character. Such gene effect is called pleiotropy and it is common property of genes based on their biochemical characteristics biological functions. The linkage is the presence two genes located adjacently on the same chromosome with their tendency to inherit together.

The environmental factors causing correlation include the common environment *viz.* the sheep raised under poor nutrition are likely to have low body weight and produce low fleece weight, better nutrition regime results more milk yield and more butter fat yield, and likewise poor health of animal influence appetite, growth, reproduction and production traits simultaneously.

It is important to know the relative importance of two types of causes responsible for causing correlation between characters. The portion of phenotypic correlation caused by genetic causes (pleiotropic action of genes and linkage of genes) is termed as *genetic correlation,* whereas the part caused by common environment shared by two characters is termed as *environmental correlation.*

Phenotypic correlation (r_p) is the correlation between the observed phenotypic values of two characters, arising from the combined effect of genotypes and environment affecting two characters. Thus, it has both genetic and environmental components.

19.2 CAUSES OF GENETIC CORRELATION

The possible genetic factors causing genetic correlation among characters may thus be that the same genes or same set of genes either affect the two or more characters or they may be closely linked on the same chromosome at adjacent loci.

Pleiotropy: It has been observed that some genes affect two or more characters and such type of gene effect is known as pleiotropy. This pleiotropy is the property of gene and said to be the most important cause of genetic correlation. The effect of some genes may be synergetic or antagonistic. The synergetic effect means that the gene affects two or more characters in the same direction and cause positive correlation whereas the antagonistic effect means that the genes affect the two or more characters in opposite direction causing negative correlation. Therefore, the pleiotropy may not be necessarily causes an appreciable amount of correlation and the magnitude of correlation due to pleiotropy depends on the direction of their effect. The genetic correlation caused by pleiotropy is similar in magnitude and direction in any population and does not change in random mating population over generations. It is difficult to identify and measure the pleiotropy as a cause of correlation but it can be imagined *viz.* the genes responsible for growth also increase the stature or size and weight of the animal and it is interpreted that such genes cause a correlation between these traits. Likewise, the genes affecting the digestion or absorption of nutrients may affect the growth and other traits like production of milk, wool, eggs, meat or work efficiency. A mutant gene for coat colour also causes sleepiness. Male sterility and weak pulmonary function are caused by a gene in human. The sickle cell gene in homozygous condition causes abnormal shape of RBC (sickle shape) also reduces the O_2 carrying capacity of blood, causing anemia and finally become fatal to the animal.

Linkage of genes: It is another genetic cause of genetic correlation and it causes a transient correlation which is decreased in every generation due to crossing over. The rate of decrease in correlation depends on the crossing over distance between two genes affecting different characters. The genetic correlation caused by linkage of genes affecting two characters is expected to differ both in amount and direction in different populations which are of different genetic background. Moreover, the genetic correlation due to linkage is also decreased in the same population over generations of random mating due to crossing over.

Types of genetic correlation: The genetic correlation is of two types. The first is the *genotypic correlation* which is estimated based on genotypic values of two characters. It is analogous to heritability in broad sense and hence can be estimated from specialized populations which are not possible in farm animals. It is not useful to measure genotypic correlation because its causation includes the interaction effect of genes which are broken in next generation. The second type of genetic correlation is the *additive genetic correlation* (r_A) which is the correlation between breeding values of two characters and hence it is analogous to heritability in narrow sense. It is used to predict

19.3 USE AND APPLICATIONS OF GENETIC CORRELATION

The knowledge of genetic correlation among traits is of great interest and use to the breeders for the following purposes:

1. The knowledge of genetic correlation can be used for *counter selection measures* to prevent any harmful correlated change. The magnitude of genetic correlation indicates that how the characters are likely to be changed in the next generation as a result of selection. For example, selection for higher growth rate by selecting heavier bulls will increase the size and birth weight of calves which may result in calving difficulty.

2. The genetic correlation can be used to apply *indirect selection based on correlated response.* This is useful for traits which are difficult to improve through direct selection which may be either due to low heritability of a trait, or when it is difficult to measure the trait for want of information or the information are made available in old age or after death of the animal or it is costly to measure the trait. The correlated response will be helpful in these cases. Moreover, the indirect selection based on correlated response is helpful to reduce generation interval *e.g.* selection based on part lactation (milk production) or part year production (egg production).

3. The genetic correlation is used *to predict the breeding value* for net economic merit based on two or more traits by constructing an index using the genetic correlation. The genetic correlation can also be used to estimate the breeding value of a sex limited trait which can not be measured *viz.* ovulation rate. The ovulation rate has high genetic correlation with an easily measured trait in opposite sex *viz.* weight or diameter of male genetalia in sheep and mice. The sex limited trait can be improved by selection for a genetically correlated trait of other sex.

$$\text{B.V. for sex limited trait} = r_A \, h_M \, S_M \, h_F \frac{\sigma_{P(M)}}{\sigma_{P(F)}}$$

Where, M and F indicates males and females, respectively.

S indicates the selection differential.

4. The genetic correlation of the performance of a genotype in two environments can be used to *measure the G-E interaction,* considering the performance in two environments as two traits. If the genetic correlation is low, it indicates that the performance in two environments indeed are two different traits governed by different sets of genes (G-E interaction exists) and hence in case of low genetic correlation the selection should be done under the environment in which the improved population has to live.

19.4 ESTIMATION OF GENETIC CORRELATION

The genetic correlation is estimated in a manner analogous to the estimation of h^2 based on resemblance between relatives. The estimation of genetic correlation requires the estimation of observational components of covariance of two traits in addition to variances of both the traits. The components of covariance of two characters are computed from analysis of covariance similar to ANOVA. The covariance of two characters is partitioned according to the sources of variation.

1. Sib analysis: The composition of variance of both the traits is similar to that for estimating heritability. The composition of the covariance, in terms of expected covariance components, corresponds to that of the variance components. The analysis of covariance on FS and HS is similar to that shown in estimating heritability except that components of covariance also occur in addition of variance components. There is similarity in the variance components and covariance components.

The covariance component between sires $[Cov_{S(X,Y)}]$ is caused mainly by association of additive effects of genes and composed (in terms of expected genetic composition) as:

$$Cov_{S(X,Y)} = \tfrac{1}{4} Cov\ A_{(XY)} + \tfrac{1}{16} Cov\ AA_{(XY)} + ...$$

Where, A_{XY} indicates additive effects of traits (X and Y)

AA_{XY} indicates the additive by additive interaction

The covariance component between dams $[Cov_{D(X,Y)}]$ is expected to contain additive effects, dominance effects, interaction effects and maternal effects and hence composed of genetic components as:

$$Cov_{D(X,Y)} = \tfrac{1}{4} Cov\ A_{(X,Y)} + \tfrac{1}{4} Cov\ D_{(X,Y)} + \tfrac{1}{16} Cov\ AA_{(XY)} +$$
$$\tfrac{1}{16} Cov\ AD_{(XY)} + \tfrac{1}{16} Cov\ DD_{(XY)} + ...$$

The expected composition of mean cross product [E (MCP)] for all the sources of variation is also the same as that of the expected composition of the variance components for both half sibs and full sib analysis.

(i) Half sib correlation: The ANOVA for estimating genetic correlation for two traits is similar to that given for estimating heritability. The analysis of covariance is also similar and given below-

Table 19.1: Analysis of covariance for half sibs

Sources of variation	D.F.	MCP	E (MCP)
Between sires	S – 1	MCP_S	$Cov_W + K\ Cov_S$
Within sires	N – S	MCP_W	Cov_W

$Cov_W = MCP_W = Cov_{W(XY)}$

The components of variance for $\sigma^2_{S(X)}$, $\sigma^2_{S(Y)}$, $\sigma^2_{W(X)}$, $\sigma^2_{W(Y)}$ are estimated from analysis of variance as given earlier.

The genetic correlation from sire component of variance and covariance as:

$$r_A = \frac{Cov_{S(XY)}}{\sigma_{S(X)}\ \sigma_{S(Y)}}$$

The environmental correlation from error variance and covariance as:

$$r_E = \frac{Cov_{W(XY)}}{\sigma_{w(X)}\ \sigma_{w(Y)}}$$

The phenotypic correlation from phenotypic variance and covariance as:

$$r_P = \frac{Cov_{S(XY)} + Cov_{W(XY)}}{\sqrt{[\sigma^2_{S(X)} + \sigma^2_{W(X)}][\sigma^2_{S(Y)} + \sigma^2_{W(Y)}]}}$$

(ii) Full sib correlation: The analysis of full sib data give the information on two traits as covariance of two traits between sires, between dams within sires, and

between progenies within sires, similar to the ANOVA for single trait. The analysis of covariance is given below-

Table 19.2: Analysis of covariance for full sibs

Sources of variation	D.F.	MCP	E (MCP)
Bet. Sires	S – 1	MCP_S	$Cov_W + K_2\ Cov_D + K_3\ Cov_S$
Bet. Dams/sire	D – S	MCP_D	$Cov_W + K_1\ Cov_D$
Bet Sibs	N – D	MCP_W	Cov_W

$$Cov_W = MCP_W = Cov_{W\ (XY)}$$

$$Cov_D = \frac{MCP_D - MCP_W}{K_1} = Cov_{D\ (XY)}$$

$$Cov_S = \frac{MCP_S - MCP_D}{K_3} = Cov_{S\ (XY)}$$

The components of variance for $\sigma^2{}_{S(X)}$, $\sigma^2{}_{S(Y)}$, $\sigma^2{}_{D(X)}$, $\sigma^2{}_{D(Y)}$, $\sigma^2{}_{W(X)}$, $\sigma^2{}_{W(Y)}$ are estimated from analysis of variance as given earlier.

The *genetic correlation* (r_A) is estimated from sire component, dam component, Sire + dam component of covariance and variance.

$$r_A\ (\text{sire component}) = \frac{Cov_{S(XY)}}{\sqrt{\sigma^2{}_{S\,(X)}\sigma^2{}_{S\,(X)}}}$$

$$r_A\ (\text{dam component}) = \frac{Cov_{D(XY)}}{\sqrt{\sigma^2{}_{D\,(X)}\sigma^2{}_{D\,(X)}}}$$

$$r_A\ (\text{sire + dam component}) = \frac{Cov_{S(XY)} + Cov_{D(XY)}}{\sqrt{\left[\sigma^2{}_{S\,(X)}\sigma^2{}_{D\,(X)}\right]\left[\sigma^2{}_{S\,(Y)}\sigma^2{}_{D\,(Y)}\right]}}$$

2. Parent offspring correlation: Hazel (1943) had shown the method of estimating the genetic correlation between two traits (X and Y) from cross covariance of one trait in progeny (O) and other trait in the parent (P). The following four parent- offspring covariances are estimated: Cov $P_X O_{Y;}$ Cov $P_Y O_X$; Cov $P_X O_X$; Cov $P_Y O_Y$

The genetic correlation (r_A) is estimated as:

$$r_A = \frac{Cov\ P_X O_Y}{\sqrt{(Cov\ P_X O_X)(Cov\ P_Y O_Y)}}$$

$$\text{or}\quad r_A = \frac{Cov\ P_Y O_X}{\sqrt{(Cov\ P_X O_X)(Cov\ P_Y O_Y)}}$$

$$\text{or}\quad r_A = \frac{\tfrac{1}{2}(Cov\ P_Y O_X + Cov\ P_X O_X)}{\sqrt{(Cov\ P_X O_X)(Cov\ P_Y O_Y)}}$$

Numerical Examples

CHAPTER 15: DETERMINANTS OF PHENOTYPIC VALUE

Example 15.1: Birth weights are given below for 4 genetic groups of calves born during 4 seasons. Find out the major components of phenotypic value and test their values for the second group calves born in first season.

Genetic / seasons groups	S_1	S_2	S_3	S_4	Total	$G_{i.}$ = Mean
G_1	18	20	20	22	80	20.00
G_2	17	19	18	20	74	18.50
G_3	19	18	19	19	75	18.75
G_4	17	19	20	19	75	18.75
Total	71	76	77	80	304	
Mean $(E_{.j})$	17.75	19.0	19.25	20.0		19.0
$= e_j$	-1.25	0.0	0.25	1.0		

Solution: Population Mean = 19.0

(1) Genotypic values: These are the mean birth weight of the corresponding genetic group over all the environments $(G_{i.})$: G_1= 20.0; G_2= 18.5; G_3= 18.75 and G_4= 18.75 kg.

(2) Environmental Deviations: These are the deviations of the mean of all genotypes under one environment from the population mean *viz.* $e_j = M - E_{.j}$,

$E_1 = 17.75 - 19.0 = -1.25$; $E_2 = 19.0 - 19.0 = 0.0$;
$E_3 = 19.25 - 19.00.25$ and $E_4 = 20.0 - 19.0 = 1.0$ kg.
(Note that $\Sigma e_j = 0$)

(3) G-E interaction: These will be 4 x 4= 16 in numbers corresponding to the P_{ij}.

$I_{G2 EI} = P_{21} - (G_{i.} + E_j)$
$I_{G2 EI} = 17 - (18.5 - 1.25) = 17 - 18.5 + 1.25$
$= 18.25 - 18.5 = - 0.25$ Kg.

(4) Expected Phenotypic value (P_{21})
$= G_{2.} + E_{.1} + I_{G2EI}$
$= 18.5 - 1.25 - 0.25 = 18.5 - 1.5 = 17.0$

The genotypic values for the different genotypes can also be expressed as deviation from the population mean (19.0 kg in this example).

The genotypic worth of the second genetic group was to produce the calves weighing 18.5 kg (G_2=18.5 kg) but when this genotype (G_2) was measured in different environments the birth weight of valves born in different seasons varied from 17 to 20 kg. The environment in different seasons experienced by the animals of this genotype modified the genotypic value in both the sides of the genotypic value of 18.5 kg.

In this example, it can be observed that mean environmental deviation equal to zero.

E_j = -1.25 + 0 + 0.25 + 1.00

The same is true for the environmental deviations for each genotype. For example, the mean phenotypic value for the second genetic group (G_2) is 18.5 which is the population mean or the mean genotypic value for this genotype (G_2). The environmental deviations for the first through fourth environments (seasons) are -1.5, 0.5, -0.5 and 1.5, respectively whose sum is zero. Therefore, the environment has no contribution to the population mean.

Now considering a breed as a single genotype, for example take 2nd breed as a single cow which produced calves sired by same bull in different lactations (example 10.1) with birth weight of 4 calves as 17, 19, 18 and 20 kg, respectively during different lactations. The average birth weight of 4 calves produced by this cow is 18.5 kg which is the population mean or mean phenotypic value or the mean genotypic value of this cow which was modified by the effect of lactation number (environment). Thus 18.5 kg birth weight was the genotype value for birth weight of 2nd cow in this example.

Example 15.2. The following is a random mating population with phenotypic values of 3 genotypes and the numbers of individuals of each genotype.

Genotypes	No. of observations	Phenotypic values (P_{ij})
A_1A	27	50
A_1A_2	36	36
A_2A_2	12	18

From the information about the above population estimate the following population parameters:

Gene frequencies of two allele, population mean, average effect of gene substitution, average effect of two alleles, genotypic values, breeding values and dominance deviation of 3 genotypes.

Solution:

(i) Gene frequencies: $p = \dfrac{27(2+36)}{75(2)} = \dfrac{90}{150}$

$= 0.60$ and hence q = 1- 0.6 = 0.4

(ii) Mid point (m = o) = Mean of two homozygote

$= \frac{1}{2}(50+18) = 34$

(iii) Gene effects and Genotypic values of 3 genotypes (a, d and –a):

Genotypic value of G_{11} = Av. effect of dom. allele (a)

= Av. of the diff of 2 homozygote

= ½ (50 - 18) = 16

Genotypic value of G_{12} = dominance effect(d)

= Heterozygote - mid point

= P_{12} - m = 36 - 34 = 2.

Genotypic value of G_{22} = Av. effect of recessive allele (-a)

= Equal to a but with –ive sign = - 16

These genotypic values indicate the additive effect of A and a allele, and the dominance effect (d).

Or, Genotypic values of 3 genotypes (a, d and –a) estimated as a deviation of the respective phenotypic value (p_{ij}) from mid point (m) value as:

$G_{ij} = P_{ij}$ - m

a = 50 – 34 = 16;

d = 36 – 34 = 2

and – a = 18 – 34= -16.

(iv) Population mean (Absolute mean) = Arithmatic average

$$= \frac{\text{Sum of cross product of phenotypic values and numbers of individuals}}{\text{total number of individuals}}$$

$$= \frac{(27 \times 50) + (36 \times 36) + (12 \times 18)}{75} = 38.16$$

Population mean (M) = Arithmatic mean as deviation from mid point (m)

= Arithmetic Av. – mid point = 38.16 – 34.0 = 4.16

Population mean (M) = Weighted average of all genotypic values

= Sum of product of genotypic freq. x assigned genotypic values   (fv)

= (0.36 x 16) + (0.48 x 2) + (0.16 x -16)

= 5.76 + 0.96 – 2.56 = 4.16

Population mean in absolute terms = 34.0 + 4.16 = 38.16

Average effect of gene substitution (α)

= a+d (q-p) = 16+2 (0.4–0.6)=15.6

Average effect of A_1 allele (α_1) = q [a + d (-p)] = qα

= 0.4 x 15.6 = 6.2

Average effect of A_2 allele (α_2) = - p [a + d (q - p)] = - pα

= - 0.6 x 15.6 = - 9.36

Genotypic value of A_1A_1 genotype = 2q (a - pd)

= 2 x 0.4(16 – 0.6 x 2) = 11.84

Genotypic value of A_1A_2 genotype = a (q- p) + d (1- 2pq)

= 16 (0.4–0.6)+2 (1 – 0.48)= -2.16

Genotypic value of A_2A_2 genotype = - 2p (a + qd)

= -2 x 0.6 (16 + 0.4 x 2) = -1.2 x 16.8 = - 20.16.

The genotypic values can also be obtained as deviation of assigned genotypic values from mean (4.16) *viz.* g_{11} = a - m ; g_{12} = d – m; and g_{22} = - a – m.

Similarly, the breeding values and dominance deviation of 3 genotypes can also be estimated by putting the numerical values in their respective formulae.

Heritability

Example 17.1. Six sires were mated to 21 dams which produced different number of progenies. Estimate the heritability of the character (birth weight) from half sib (paternal and maternal) and full sib components.

Sires	Dams	Progeny records (X_{ij})					Totals	
		1	2	3	4	n_{ij}	$X_{ij.}$	$X_{i..}$
1	1	21	24	-	-	2	45	
	2	16	18	20	-	3	54	
	3	16	17	19	20	4	72	171
2	4	18	20	19	-	3	57	
	5	17	21	23	-	3	61	
	6	23	21	-	-	2	44	
	7	19	19	-	-	2	38	200
3	8	17	20	20	-	3	57	
	9	18	19	22	23	4	82	139
4	10	16	19	20	21	4	76	
	11	20	20	24	-	3	64	
	12	22	22	-	-	2	44	
	13	19	20	21	-	3	60	
	14	19	20	22	23	4	84	328
5	15	19	20	19	-	3	58	
	16	22	23	23	-	3	68	
	17	20	22	-	-	2	42	168
6	18	22	23	21	20	4	86	
	19	20	22	23	-	3	65	
	20	21	22	24	-	3	67	
	21	20	21	22	22	4	85	303
						64		1309
Total								

Solution:

K_1 = Av. no. dams per sire

$= 64 - [(2^2 + 3^2 + 4^2)/ 9 + (3^2 + 3^2 + 2^2 + 2^2)/10 + .. + (4^2 + 3^2 + 3^2 + 4^2)/ 14] / (21 - 6)$

$= (64 - 19.09) / 15 = 2.99$

K_2 = Av. no. of progeny per dam $= 19.09 - [2^2 + 3^2 + 4^2 + 3^2 + 3^2 + 2^2 + .. + 3^2 + 3^2 + 4^2] /64 - 6 - 1$

$= (19.09 - 3.22) / 5 = 3.17$

K_3 = Av. no. of progeny per sire $= \{64 - [9^2 + 10^2 + 7^2 + 16^2 + 8^2 + 14^2)] / 64 \} / 6 - 1$

$= (64 - 11.66) / 5 = 10.47$

Correction Factor $= (1308)^2 / 65 = 26773.14$

T. S. S. $= \Sigma X^2_{ijk} - C.F. = 27029 - 26773.14 = 255.86$

Sire S.S. $= (171)^2/9 + ... + (303)^2/14 - C.F.$

$= 26818.93 - C.F. = 45.78$

Dam /Sire S.S $= [45^2 /2 + 54^2/ 3 + ... + 67^2/ 3 + 85^2 / 4] -$ Sire S.S. (crude)

$= 26903.75 - 26818.93 = 84.82$

Progeny / within dam S.S. = T.S.S. (crude) – Dams / Sire S.S. (crude)

$$= 27029 - 26903.75 = 125.25$$

ANOVA table for HS and FS for computing different components of varianc:

Sources of variation	D.F.	S.S.	M.S.
Sires	5	45.787	9.157
Dams / sire	15	84.821	5.654
Progenies / dams	43	125.25	2.912

$$\sigma^2_S = \frac{9.157 - 2.912}{10.47} = 0.596$$

$$\sigma^2_D = \frac{5.654 - 2.9120}{2.99} = 0.917$$

$$\sigma^2 e = 2.912$$

$$\sigma^2 = \sigma^2_S + \sigma^2_D + \sigma^2 e = 4.425$$

$$h^2_S = \frac{4\sigma^2_S}{\sigma^2_P} = \frac{4 \times 0.596}{4.425} = 0.539 \ ;$$

$$h^2_D = \frac{4\sigma^2_D}{\sigma^2_P} = \frac{4 \times 0.917}{4.425} = 0.828$$

$$h^2_{(S+D)} = \frac{2(\sigma^2_S + \sigma^2_D)}{\sigma^2_P} = \frac{2(0.596 + 0.917)}{4.425} = 0.684$$

Example 17.2: Find out the heritability of the character based on paternal half sib correlation from the data given above in example 17.1 considering that the information on dam number's of progenies (Dam's identity) were not available, for which full sib pairs can not be sorted out.

Solution : In this case, the C.F., T.S.S. and sire S.S. will remain the same but the error sum of squares and error variance will be higher and it will be calculated as:

Error S.S. = T.S.S. – Sire S.S. (crude) = 27029- 26818.93 = 210.07

ANOVA table for paternal HS

Sources of variation	D.F.	S.S.	M.S.
Sire	5	45.78	9.157
Progeny / sire	58	210.07	3.62

$$\sigma^2_S = \frac{9.157 - 3.63}{10.66} = \frac{5.527}{10.66} = 0.58$$

$$\sigma^2 e = 3.62$$

$$h^2_S = 4\left(\frac{\sigma^2_S}{\sigma^2_P}\right) = \frac{4 \times 0.518}{3.62} = \frac{2.072}{3.62} = 0.572$$

Example 17.3: Lactation length of 22 dams and their daughters sired by 6 sires were recorded Find out the heritability of the character based on regression of daughter on dam mrthod and also from intra sire regression of daughter on dam method.

Sire no.	Pairs	Dams L.L.		Daughter's L.L.	
		(X_{ij})	$X_{I.}$	(Y_{ij})	Y_i
1	1	308		300	
	2	290		295	
	3	288		302	
	4	307	1193	285	1182
2	5	290		297	
	6	305		295	
	7	310	905	315	907
3	8	277		291	
	9	280		295	
	10	397		305	
	11	310	1264	295	1206
4	12	282		295	
	13	280		285	
	14	300	862	295	875
5	15	270		260	
	16	282		298	
	17	310		304	
	18	292	1154	297	1159
6	19	267		270	
	20	280		285	
	21	301		310	
	22	290	1138	300	1165
Total		6516		6494	

Solution: S.S. (Dams):

$$C.F. = \frac{(6516)^2}{22} = 1929920.7$$

$$T.S.S. = (308)^2 + \ldots + (290)^2 - C.F. = 1944322 - C.F. = 14401.27$$

$$Sire\ S.S. = \frac{(1193)^2}{4} + \ldots + \frac{(1138)^2}{4} = 1932615.9 - C.F. = 2695.2$$

Dams/sire S.S. = T.S.S. − Sire S.S. = 14401.27 − 2695.2 = 11706.07

Sum of Crossproducts:

$$C.F. = \frac{6516 \times 6494}{22} = 1923404.7$$

Total S.C.P. = (308 × 300) + ….. + (290 × 300) − C.F.
$$= 1927008 - 1923404.7 = 3603.3$$

$$\text{Sire S.C.P.} = \frac{1193 \times 1182}{4} + \dots + \frac{1138 \times 1165}{4} - \text{C.F.}$$

$$= 1924469.9 - 1923404.7 = 1065.2$$

$$\text{Dams/sire S.S.} = \text{Total S.C.P.} - \text{Sire S.C.P.} = 3603.3 - 1065.2 = 2538.1$$

ANOVA table:

S.V.	D.F.	S.S.	dams S.C.P.
Sires	5	2695.2	1965.2
Dams/sire	16	11706.07	2538.1
Total	21	14401.27	3603.3

Answer: Intra-sire regression of daughter on dam method:

$$b_{OD} = \frac{2538.1}{11706.07} = 0.2168$$

$$h^2 = 2\, b_{OD} = 2 \times 0.2168 = 0.4336$$

Regression of daughters on dam method:

$$b_{OD} = \frac{3603.3}{14401.27} = 0.25$$

$$h^2 = 2\, b_{OD} = 2 \times 0.25 = 0.50$$

Repeatabilty

Example 18.1: The data given below are the records of birth weights of 70 calves born to 10 Hariana cows. Estimate the *repeatability* of the character by different methods *viz.* regression of second record on the first, correlation between first two records and based on all the records using intra class correlation method.

Cows	Records								Total
	1	2	3	4	5	6	7	8	$\sum X_i$
1	22	22	22	23	24	22	--	--	137
2	21	22	21	22	22	--	--	--	108
3	24	25	25	24	26	25	23	--	172
4	23	22	23	24	24	24	--	--	140
5	21	21	21	22	22	21	22	23	173
6	21	22	22	23	23	23	23	--	157
7	24	25	26	25	26	27	26	28	207
8	22	23	22	24	23	23	24	--	161
9	21	22	22	23	24	22	21	23	178
10	22	22	23	24	25	22	23	24	185
Total	221	226							1616

Solution:

First record (X_1)

$$\sum X^2 = 4897; \quad \text{C.F.} = \frac{(221)^2}{10} = 4884.1; \quad \text{Corrected S.S.} = 12.9$$

Second record (X_2):

$$\Sigma X^2 = 5124;\ C.F. = \frac{(226)^2}{10} = 5107.6;\ \text{Corrected S.S.} = 16.4$$

Covariance ($X_1\ X_2$)

$$\Sigma X_1\ X_2 = 5007;\quad C.F. = \frac{221 \times 226}{10} = 4994.6;$$

Corrected SCP = 5007 − 4994.6 = 12.4

Repeatability as correlation of first two records ($r_{X1\ X2}$)

$$= \frac{\text{Cov. } X_1\ X_2}{\Sigma \text{Var } X_1\ \text{Var} X_2}$$

$$= \frac{12.4}{\sqrt{(12.9)\ (16.4)}}$$

$$= \frac{12.4}{14.54} = 0.85$$

Repeatability as regression of second on first record ($b_{X2\ X1}$)

$$= \frac{12.4}{12.9}$$

$$= 0.96$$

Repeatability as intra-lass correlation: The calculations are based on all the 70 records of all the 10 cows a under:

$$N = 70\ \Sigma X = 1616;\ C.F. = \frac{(1616)^2}{70} = 37306.51$$

$$\Sigma X^2 = 37436\ \text{T.S.S.} = 37436 - 37306.51 = 129.49$$

$$\text{Cow S.S.} = \frac{(137)^2}{6} + \ldots\ldots + \frac{(185)^2}{8} - C.F.$$

$$= 37423.41 - 37306.51 = 116.9$$

$$\text{Error S.S.} = \text{T.S.S.} - \text{Cow S.S.} = 129.49 - 116.9 = 12.59$$

ANOVA Table

S.V.	D.F.	S.S.	M.S.
Bet. Cows	9	116.9	12.9
Bet records within cows	60	12.59	0.21

$$K = \frac{1}{(C-1)}\ \frac{\Sigma n. - \Sigma n_i^2}{\Sigma n.} = \frac{1}{9}\left[70 - \left(\frac{500}{70}\right)\right] = 6.98$$

$$\text{Cow variance } (\sigma^2_C) = \frac{(12.99 - 0.21)}{6.98} = 1.83$$

$$\text{Total Variance } (\sigma^2_P) = 1.83 + 0.21 = 2.04$$

$$\text{Repeatability} = \frac{\sigma^2_C}{\sigma^2_P} = \frac{1.83}{2.04} = 0.897$$

Genetic Correlation

Example 19.1: Estimate the genetic correlation from the following data recorded on two characters on 12 progenies of 4 sires, (3 progenies of each sire):

Sire	Progeny No.	X	$\Sigma X_{i.}$	Y	$\Sigma Y_{i.}$
1	1	3		3	
	2	4		2	
	3	4	11	4	9
2	1	4		4	
	2	3		4	
	3	3	10	3	11
3	1	3		3	
	2	3		3	
	3	2	8	2	8
4	1	4		3	
	2	3		3	
	3	2	9	2	8

SOLUTION:

$$\Sigma X = 38; \text{ T.S.S.} = 126; \text{ C.F.} = \frac{(38)^2}{12} = 120.33$$

Corrected Total S.S. = 126.0 − 120.33 = 5.67

$$\text{Sire S.S.} = \frac{[112 + \ldots\ldots + 92]}{3} = 122.9 - \text{C.F.} = 1.67$$

$$\text{Error S.S.} = \text{T.S.S.} - \text{Sire S.S.} = 5.67 - 1.67 = 4.0$$

$$\Sigma Y = 36; \text{ T.S.S.} = 114; \text{ C.F.} = \frac{(36)^2}{12} = 108$$

Corrected Total S.S. = 114.0 − 108 = 6.0

$$\text{Sire S.S.} = \frac{[92 + \ldots\ldots + 82]}{3} = 110 - 108 = 2.0$$

$$\text{Error S.S.} = \text{T.S.S.} - \text{Sire S.S.} = 6.0 - 2.0 = 4.0$$

$$\text{Cross Product (XY): T.C.P.} = 117.0 \text{ C.F.} = \frac{38 \times 36}{12} = 114.0$$

Corrected Total C.P. $= 117.0 - 114.0 = 3.0$

$$\text{Sire C.P.} = \frac{[11 \times 9 + \ldots\ldots + 9 \times 8]}{3} = 115.0 - 114.0 = 1.0$$

Error C.P. $=$ T.S.C.P. $-$ Sire C.P. $= 3.0 - 1.0 = 2.0$

ANOVA and ANCOVA Table

S.V.	D.F.	X		Y		Y Cov	
		S.S	M.S.	S.S	M.S.	C.P.	M.C.P.
Sire	3	1.67	0.556	2.0	0.667	1.0	0.333
Error	8	4.0	0.500	4.0	0.500	2.0	0.25

$$\sigma^2_S = (\text{M.S.sire} - \text{M.S.error})/K$$

$$\sigma^2_{S(X)} = \frac{0.556 - 0.50}{3} = 0.0187 \quad \sigma^2_{S(Y)} = \frac{0.667 - 0.50}{3} = 0.0556$$

$$\text{Cov}_{S(XY)} = \frac{0.333 - 0.25}{3} = 0.0276$$

$$r_A = \frac{\text{Cov}_{S(XY)}}{\sqrt{\text{var.}_{S(X)} \text{var.}_{S(Y)}}} = \frac{0.0276}{\sqrt{0.0187 \times 0.0556}}$$

$$= \frac{0.0276}{0.0322} = 0.85$$

Example 19.2. Estimate the genetic correlation from parent-offspring correlation method between X and Y characters from the following data on 10 dams and their daughters.

D – D pairs	Dams records		Daughters records	
	X	Y	X	Y
1	10.0	5.4	12.0	4.7
2	10.5	5.2	10.2	5.8
3	9.0	6.0	10.3	6.4
4	8.7	6.5	11.0	5.9
5	10.2	5.6	11.0	5.4
6	11.5	4.8	12.2	5.0
7	10.5	5.4	12.2	5.0
8	10.2	5.6	11.5	4.6
9	11.8	5.0	10.4	5.8
10	9.1	6.2	10.1	5.6
Totals	101.5	55.7	110.9	54.2

Solution:

Cov $P_X O_X$ = 1128.03   C.F. = 1125.63 = 2.4
Cov $P_Y O_Y$ = 303.01   = 301.89 = 1.12
Cov $P_X O_Y$ = 548.23   = 550.13 = - 1.9
Cov $P_Y O_X$ = 616.20   = 617.71 = - 1.51

$$r_A = \frac{\frac{1}{2}(Cov\ P_Y\ O_X + Cov\ P_X\ O_Y)}{\sqrt{(Cov\ P_X\ O_X + Cov\ P_Y\ O_Y)}}$$

$$= \frac{\frac{1}{2}[(-1.9) + (-51)]}{\sqrt{(2.4 \times 1.12)}}$$

$$= \frac{\frac{1}{2}[(-341)}{\sqrt{(2.68 \times 1.12)}} - \frac{1705}{1637} = -1.041$$

PART III
ANIMAL BREEDING

Chapter 20

Economic Characters and Farm Records

There is a long list of characters of farm animals, man, laboratory animals and plants. The characters are species specific and hence the different species are known for their specific characters. It is essential to know about the term character, measurement, types and mode of inheritance of characters.

20.1 CONCEPT AND DEFINITION OF CHARACTER

In general, the term character is any of the observed or measured property of an organism. However, in genetics and population genetics, all the characters of living being are not of interest for the reason that the aim of a population geneticist is to bring a change in a character in a desirable direction for the benefit of mankind. The change in a character is only possible if there exist differences in that character among the individuals of a population and further these differences must be genetically controlled, known as genetic differences. There are some characters which do not show differences among individuals of a population e.g. number of legs in farm animals and humans; number of fingers, eyes, ears and hands in man, etc, their number being fixed. These characters are undoubtedly determined by genes and hence inherited (transmitted from parents to progeny) but do not have differences and hence no change can be made in these characters. Therefore, the word difference is important to define the term character and further the differences in a character should be genetically controlled. The genetically determined character is called the *inherited character* and the character having genetic differences among individuals is called the *heritable character*. A character may be inherited but it may not be heritable which means that all the inherited characters do not show genetic differences. Thus, the characters having genetic differences are called the heritable characters. In quantitative genetics, the term character is thus defined as:

Any measurable or observable property of living individual having genetic differences is called the character. The *synonym* of the term character is the *trait*.

Phenotype verses phenotypic value: The expression of genotype in terms of a character is called the phenotype. A character is observed or measured either in terms of the *phenotype* or in terms of *phenotypic value,* depending on the type of character. Each character has at least two phenotypes (qualitative characters) or many phenotypic values (quantitative characters) measured on different individuals of a population. But one individual has only one phenotype or one phenotypic value for a character.

20.2 MEASUREMENT OF CHARACTERS

The different characters are measured in different ways. Some characters are measured by simply making observations or examining the individuals for the character with naked eyes or by chemical test or other test. The observation on an individual of such character is called the phenotype *viz.* red or white coat colour, tall or dwarf plants, round or wrinkled shape of seed, horned or polled condition, hemoglobin type A or B etc. Thus, the pea plants studied by Mendel were either tall or dwarf in their height. The tall and dwarf forms of plants are two phenotypes of the character while the plant height is the character. Thus, the phenotype is one of the two or more forms of a character.

On the other hand, some characters are measured / recorded in metric units, like gm., kg, cm, days. The measured value in metric unit for a character on an individual is a numerical value. This numerical value is called as the phenotypic value *viz.* amount of milk produced by a cow at one time or in one day or in one lactation or during life time; body weight or length or height of a cow (or any individual); number of days a cow remained in milk in a lactation or during life time, etc. Thus, phenotypic value is the amount or quantity of milk, in kg., produced at one time or in one day or during the lactation by milking animal while the milk production is the character.

Animal Breeding Data: The phenotype which is an observation and the phenotypic value which is a measurement of a character on an individual is the record of the individual for that character. The records (observations or measurements) for a character on all the individuals of a population are called as the *data.* This data on all individuals of a population for a character have genetic differences and used for the genetic improvement of animals through breeding plan. Therefore, this data is called as the *Animal breeding data.*

20.3 ECONOMIC CHARACTERS OF DIFFERENT SPECIES

There are different species of farm animals kept for various purposes and hence the traits of economic importance are different in different species.

(1) Economic characters in dairy animals (Cattle and Buffaloes): The main dairy animals are cattle and buffaloes which contribute more than 97% of the milk produced in India, the share of goat milk had been nearly 2.5% only. The various

traits of economic importance in dairy animals are the traits influencing reproduction and production. These are as under-

1. Reproductive traits: These are those traits which are directly or indirectly related with reproduction of animals of two sexes.

The reproductive traits in females are age at first heat, age at first service, age at first conception, number of services per conception, conception rate, age at first calving, post-partum breeding interval, service period, gestation period, repeat breeding, calving interval, number of calving in the herd, reproductive health and related ailments.

The reproductive characters in males are age at first service, libido (sex desire), semen volume, semen quality traits (motility, sperm counts, live and dead sperm count), semen freezability.

2. Productive traits: These are the traits which are related to milk production. These include milk yield (in a day, during one lactation, per day of time), peak yield, fat percent, lactation length, dry period, persistency of lactation.

3. Growth traits: Birth weight, body weight at subsequent ages, growth rate.

4. Lifetime traits: Lifetime calf production, lifetime milk produced, life span, herd life, breeding efficiency, days dry, days open, total days in milk.

5. Fitness traits: Mortality at different ages, abnormal calving, selective value, disease incidence.

6. Physical traits: These describe the external physical traits *e.g.* characters describing length, diameter and width of teats and udder, body measurements, measurement of different body parts *viz.* legs, neck, hump, eyes, conformation, visual appraisal, type, etc. These traits are also called as *"eye-ball"* traits.

7. Physiological and biochemical traits: Body temperature, pulse rate, respiration rate are the physiological traits whereas blood groups, hemoglobin, transferin, milk proteins, blood potassium types, antigen production are the biochemical traits.

(2) Economic characters in goats: The goat is called as poor man's cow in India. The traits of economic importance are age at first kidding, kidding interval, multiple births, gestation period, milk yield, lactation length, birth weight, growth rate, body weight at different ages, death losses and survivability at different ages, abnormal kidding, mohair/pashmina production, biochemical traits.

The goat also provides meat (chevion). The, economic traits in goat for meat are also the growth rate and body weight at different ages, meat production and quality characters.

(3) Economic characters in sheep: Sheep are reared for wool production (carpet and fine wool) and meat (mutton) production. The economic characters are age at first lambing, tupping percentage, twinning, number of lambs weaned, lambing percentage, death rate, birth weight, weaning weight, yearling weight, adult body weight, growth rate, fleece production and quality traits (fiber length, number of crimps, fiber diameter, medullation percentage), meat production and meat quality traits, feed conversion efficiency, biochemical traits.

(4) Economic characters in pigs: Meat production is the main purpose in pig farming. Pig rearing is practiced mainly by poor people in India. The economic characters of pigs are litter size and litter weight at birth and at weaning, number of litters per year, feed efficiency, growth rate, weight at slaughter, carcass length, carcass weight and quality traits (dressing percentage, lean meat/muscle meat, backfat thickness).

(5) Economic characters in poultry: The poultry rearing is done for egg and meat production. The most important economic characters are age at first egg, egg size, egg number and egg weight, hen housed average, persistency in egg production for a long time, egg quality traits (shell colour, texture, thickness, yolk colour, thickness of the white, absence of blood and meat spots). The incidence of various diseases and death losses at various ages in poultry is also very important which influence the economic viability of poultry farming.

The broilers are meat birds. The traits of economic importance of broiler are the fast growth rate, feed conversion efficiency, large breast with fleshy thigh and short legs, weight of meat produced and meat quality traits (dressing percentage, tenderness, flavor).

20.4 CLASSIFICATION OF ECONOMIC CHARACTERS

The characters can be classified in to different types in a number of ways. The different classifications have been made on the basis of the description of the characters (subjective and objective characters), development and expression of the character (simple and complex characters), function of the character (physical traits, fitness traits, behabiour traits, biochemical traits, physiological traits, reproductive and productive traits) and mode of inheritance of characters (qualitative, quantitative and threshold characters).

20.4.1 Descriptive Classification

The traits may be classified according to the ways they need description. Based on description all the characters can be classified as subjective and objective.

(i) Subjective Traits: These characters need some description and opinion of the person recording the character greatly affects the assessment of the character. These characters are measured by scores, grades, proportions.

The examples are breeds specification, temperament and behavioral traits (social, psychological and sexual behavior) which are measured by assigning scores or grades, resistance to disease and toxic agents, ability to infect a host, and act as a vector of disease, conformation traits *viz.* udder, teat and structural soundness, carcass conformation – shape and proportion, comb shape in poultry, noise making (barking) and eye characters in dogs, horn shape and pattern, etc.

(ii) Objective Traits: These traits are measured in quantitative terms like, days, cm, gm, kg, etc. These characters are little affected by error of measurement.

The examples are milk yield per day or in a lactation or in lifetime, fat percent in milk, body weight and measurements, calving interval and its components (service period and gestation length, lactation length and dry period), age at which reproduction and production starts, fertility measures, wool yield and quality traits (staple length, fiber diameter, medullation percentage, follicle parameters *viz.* secondary to primary ratio, teat number, egg weight, size and number, egg quality characters (yolk weight, yolk colour), egg shell texture and colour, feed conversion, feather colour, body colour, litter size, number of litters per year , carcass weight and quality characters.

20.4.2 Developmental classification

The characters are classified as simple and complex traits depending on the development of characters.

(*i*) *Simple traits:* The development and expression of such traits is simple and independent of other traits.

The examples of simple traits are eye colour, colour of hair and wool, presence or absence of horns.

(*ii*) *Complex traits:* The development and expression of such traits depends / associated with other traits.

The examples of complex traits are growth, survival, maternal ability, lactation, etc.

20.4.3 Functional classification

The traits can be classified depending on the function to which a trait is associated like, physical traits, fitness traits, behavioural traits, biochemical traits, physiological traits, reproductive traits and productive traits.

(*i*) *Physical traits:* These traits are related to the external morphology of the animal *viz.* characters describing the length, diameter and width of teats, udder, ears, horns; body colour, various body measurements and measurement of other external body parts (legs, hump, neck, mouth, eye), external soundness, conformation, visual appraisal, type, etc. These visual traits are also called *"eye-ball traits"*.

(*ii*) *Fitness traits:* These traits affect the adaptability of individuals in a given environment *e.g.* susceptibility or resistance to various disease, survival, type of birth (normal vs. abnormal), etc. The productive and reproductive traits are also sometimes regarded as fitness traits because an organism can not said to be adapted to a new environment if it does reproduce and produce well.

(*iii*) *Behavioural traits:* The traits indicate the behavior of the organism *e.g.* temperament, mothering ability, learning ability.

Temperament: This indicates the cooperation of the individuals. A good temperament is non-aggressive and required in farm animals for completing every farm operation like treating them from ailments, milking, riding, moving them around the farm, yoking them for pulling, handling them for dehorning and castration, etc. On the other hand, an aggressive behavior of an animal is dangerous to the animal itself, to other fellow animals and to the people handling them. Fighting of mother to protect its young ones from predators is an important quality. A quit temperament of draft animals (bullocks, horses, ass) is essentially required.

There is usually a social hierarchy or peak order in groups of animals in which the aggressive animals push to docile animals to the order. The docile or less aggressive animals eat less, less active, less productive particularly in draft capacity (speed) but produce more milk more prone to diseases. The temperament is controlled by hormonal condition of the animal.

Learning ability: This is also sometimes considered as the intelligence. This quality is important for farm dogs particularly used for gathering, driving and penning the sheep; training of horses and bullocks. The ability of farm animals to learn the simple routines *viz.* operating, watering and feeding devices, drinking of milk is important.

Maternal ability: A good maternal ability (mothering) is required in animals which suckle their own offspring. This trait is subjective and difficult to describe objectively. Mothering is measured indirectly like weight of the offspring at weaning. It is a complex trait associated with survival and milk production. Young ones survival is affected by its dam's ability to feed, shelter and protect it from predators, etc. The dams with poor milk production have poor mothering instinct. The feeding level of the dam during later stages of pregnancy and during lactation is important and measured by her weight and body condition.

(iv) Biochemical traits: The are related to blood groups or types, blood hemoglobin, transferrins, milk proteins types, blood potassium types and concentration, milk fat, antigen production, etc.

(v) Physiological traits: These include the body temperature, pulse rate, respiration rate indicating the physiological functions.

(v) Reproductive traits: The traits indicate the ability of the animal to produce offspring. These traits include the age of puberty, age of producing first offspring (age at first calving in cattle and buffalo, age at first lambing in sheep, age at first kidding in goat, age at first farrowing in pigs, age at first egg in poultry), number of services required to conceive, post partum breeding interval (number of days from calving to first heat after calving), service period (number of days from calving to conception), gestation period (number of days from conception to calving), calving/lambing/kidding interval (number of days between two successive calving/lambing/kidding). Mating capacity, semen production and quality, libido in males, number of eggs shed from ovary (ovulation rate), number of fertilized ova implanted in uterus, conception or pregnancy rate (number of females conceived per 100 inseminations), number of offspring born per animal, number of offspring weaned in pigs and sheep.

(vi) Productive traits: These characters are related to the production of animals and these are different in different species. The examples are-

Lactation: This involves whole complex reproductive system, mammary tissue development during pregnancy, milk flow rate, milk yield, fat percent, fat yield, lifetime yield, persistency, lactation length, dry period, milk yield per day of time.

Wool production: This includes greasy fleece weight, clean fleece weight, fiber length, fiber diameter, medullation percentage (% of hairy fibers that have medulla/hollow space).

Egg production: These are the egg size, egg weight, egg number in a given period of time (week, month, year), egg quality traits (shell colour, shell texture, shell thickness, yolk height, yolk color, thickness of egg white, absence blood and meat spots), good hatchability, age at first egg without pausing (stop laying) during production and broody, persistency of egg production for a long time. The body weight and feed conversion are also important.

Meat production: The meat production trait of primary importance is the growth rate, body weight and measurements. These are also important in dairy animals and sheep for wool also. The traits of importance in meat are the growth rate, feed conversion efficiency (feed consumed per unit of carcass weight or dressed weight), dressed weight, dressing percentage, muscle part of the carcass, more lean and less fat, tenderness, flavor.

In broilers, the growth rate is of great concern. The fast growing bird converts its feed efficiently. The bird besides fast growth should have large breast with fleshy thigh and short legs. There should be no fat deposit inside the body cavity so that the feed consumed is converted directly to edible meat. The traits of interest in broilers are the body weight, carcass weight, proportion of breast meat to total carcass, shank length, feed consumed per unit of dressed carcass and disease resistance.

In sheep for meat (mutton), the important traits are birth weight, weaning weight, yearling weight, weight of lambs weaned per ewe.

In pigs, the important traits are litter size, number of litters/ year, birth weight of piglets, number of piglets weaned/litter, litter weight at weaning, weaning weight per piglet, rate of gain from weaning to market, feed efficiency, weight at slaughter, cold carcass weight, carcass length, dressing percentage, lean meat (muscle meat), back far thickness, meat to bone ratio .

Draft and speed: The draft or pulling power can be measured objectively by a dynamometer and the speed can be measured by distance travelled in a given time period.

20.4.4 Classification based on measurement of traits

(i) Directly observed traits: These are as under-

The traits which are measured in metric units directly are birth weight, body weight at different ages, type of calf born (normal and abnormal), sex of calf, mortality, body colour, body measurements (length, height etc.), measurement of various body parts, physiological traits (body temperature, pulse rate, respiration rate), greasy fleece weight, litter size, carcass weight, egg weight, incidence of diseases, presence or absence of any part on the body, physical defect, etc.

The traits which are measured by some tests are milk composition, biochemical/ molecular polymorphic traits *viz.* blood groups, protein polymorphism, hemoglobin and potassium types in blood, transferrin types, molecular markers, etc.

(ii) Traits which are generated: The economic traits in farm animals are also generated from the information available in history cum pedigree sheets or other registers. Some of the traits are generated for each individual while others are for

the whole population. The procedures to generate the important traits are given as under:-

Age at sexual maturity: Date of birth to date of first heat.

Age at first calving: Date of birth to date of first calving.

Post partum breeding interval: Date of calving to date of first heat.

Service period: Date of calving to date of conception.

Gestation period: Date of calving to date of calving.

Calving interval: Interval between two calving.

$$\text{Fertility Index (bulls)} = \frac{\text{No. of conceptions}}{\text{No. of inseminations}} \times 100$$

This is the fertility index for i^{th} bull if estimated for the entire herd; it is then called as the conception rate or pregnancy rate or fertility rate.

Breeding efficiency: This is estimated for the lifetime of a cow/buffalo based on the number of calving in lifetime as under:

(1) Tomar's formula (1965)

$$\text{Zebu cows} = \frac{n\,365 + 1020}{AFC + C1} \times 100$$

$$\text{Buffaloes} = \frac{n\,365 + 1040}{AFC + C1} \times 100$$

where, n = No. of calving intervals

CI = sum of all calving intervals in days

AFC = age at first calving in days

(2) Wilcox formula (1957)

$$\text{B.E.} = \frac{365\,(n-1)}{D} \times 100$$

where, n = No. of total calving

D = No. of days from first to last calving

Lactation length: Date of calving to date of dry.

Dry period: Date of dry to date of next calving.

Sex ratio: Percentage of male births to the total normal births.

Abnormal calving: Percentage of abnormal calves to the total calves born.

$$\text{Mortality rate} = \frac{\text{No. of animals died in particular age group}}{\text{Total no. of animals present in the beginning of that age group}} \times 100$$

$$\text{Incidence of disease} = \frac{\text{No. of animals affected}}{\text{Total no. of animals}} \times 100$$

Replacement rate: Percentage of animals calves reached to AFC out of total female calves born or total calves born (pregnancies) after Tomar and Verma (1988).

Replacement index: Ratio of heifer calving to the total adult females left the herd in a year or a period of many years after Ram and Tomar (1993).

Selective value: It is estimated as the number of females' adult born reaching milking herd from each cows/buffalo after Ram and Tomar (1995).

Lactation milk yield: Total milk yield produced by a cow/buffalo in complete lactation.

305 days milk yield: Milk yield produced up to 305 or less days of lactation.

Milk yield per day of time: It may be calculated for per day of lactation length or per day of calving interval or per day of lactation during life time or per day of herd life.

Total days in milk: This is the sum of lactation length for all the lactations.

Days open (%): It is the percentage of total service period (days) in all lactation to the herd life.

Days dry (%): It is the percentage of total days dry in all lactations to the herd life.

Life time milk production: It is estimated as the sum of milk produced in all lactations.

Herd life: Difference in days between date of first calving to the date of disposal (death or culling).

Longevity or life span: Date of birth to the date of disposal.

Peak yield: Maximum amount of milk produced on any day in a lactation.

Days to attain peak yield: Date of calving to date of attaining peak yield.

Persistency of lactation: The rate of milk secretion throughout the lactation period follows certain patterns which are called as the lactation **curves**. These are the graphical representations of the daily or weekly milk production. The rate of milk secretion is maintained by some cows for longer period in lactation and known as persistent producers while others go dry or drop the yield within a few weeks and hence called as non persistent. The persistent cows are more economical as it influences the lactation milk production. The degree with which the rate of milk secretion is maintained during decline phase of the lactation with the advancement of lactation is called the persistency of lactation. Mathematically, it is expressed as the average percentage or degree of decrease in milk production each month to that of the pervious month.

It is more convenient to study the lactation curves from weekly milk yields. Different mathematical models describing the shape of lactation curve have been given by different workers. The different functions employed are exponential function, parabolic exponential, gamma type and polynomial. The last two functions gave higher R^2 values.

The lactations milk yield is the function of initial milk yield (first week milk yield), ascending phase milk yield which includes initial milk yield plus the yield to the day of attaining peak yield and it is about 25% of the 305 day yield, and the descending phase milk yield starting from the day of peak yield to the date of drying or 305 days of lactation which accounts nearly 75% of 305 days milk yield. The descending phase milk yield decides the persistency of lactation. Different methods are available for estimating the persistency of lactation *viz.* lactation curves, regression coefficients phase milk yield on the descending phase milk yield, the ratio of production of different months of lactation and the analysis of variance method by estimating the intra-class correlation. The last two methods are mostly used for calculating the persistency of lactation (P).

Johansson and Hansson (1940) estimated the persistency as a ratio of production during 101 to 200 days to the production during first 100 days (P2:1) and the ratio of production during 201-300 days to the production of first 100 days (P 3:1).

Ludwick and Peterson (1943) defined the persistency of the consecutive ratios obtained by comparing production of each individual sub division of lactation period with the preceding one as:

$$P = W_1 \frac{P_2}{P_1} + W_2 \frac{P_3}{P_2} + W_3 \frac{P_4}{P_3}$$

where, P_1 = milk yield during 2nd + 3rd month

P_2 = Milk yield during 4th + 5th month

P_3 = milk yield during 6th + 7th month

P_2 = milk yield during 8th + 9th month

$W_1 = R_1 / R_1 + R_2 + R_3$

$W_2 = R_2 / R_1 + R_2 + R_3$

$W_3 = R_3 / R_1 + R_2 + R_3$

$$R_1 = \frac{\overline{P}_2}{\overline{P}_1}; \; R_2 = \frac{\overline{P}_3}{\overline{P}_2}; \; R_3 = \frac{\overline{P}_4}{\overline{P}_3}$$

Mahadeven (1951) used the initial milk yield of ascending phase (up to peak yield) and 180 days milk to estimate persistency as:

$$P = \frac{A - B}{A}$$

where, A = first 180 days milk yield

B = first 10 weeks or ascending phase (60 days) milk yield up to peak yield]

EXTENDING AND PREDICTING MILK YIELD

It is sometime of interest to know 305 days milk yield of dairy cow/buffalo before she has completed a lactation. This is possible to know the ratio factors which are developed on large number of data for different herds/breeds. The ratio factors are developed each 10 days intervals of lactation period and for different lactations. The factors are obtained as:

$$\text{Factor} = \frac{\text{Av. 305 days milk yield}}{\text{Milk yield in days}}$$

It is not possible to record the daily milk yield particularly in field conditions. However, one day milk yield in each month of lactation (known as test day milk yield) can be recorded easily.

Test day milk yield is multiplied by the number of days in that month (cow days in a month). The first test day should be after one week of calving at least. The product of test day yield and number of days the cow being milked in that month yield the total yield for that month

Average egg production per bird is calculated on three bases *viz.*

(*i*) *Monthly average*: This is estimated as the total number of eggs laid during the month divided by the average number of birds for the month which is the average of the number of hens on the first and last day of month.

(*ii*) *Hen day average*: It is obtained as the number of eggs laid during a given period (month) divided by the average number of birds during the period. The average number of birds during the month period is estimated as the average number of hen per month. This is obtained by dividing the actual number of hen days with number of days in that month.

(*iii*) *Hen – housed average*: This is estimated as the total number of eggs produced during a given period by the number of bird in the flock at the beginning of the period. This is usually obtained on yearly basis. Thus total number of eggs produced during the year is divided by the number of birds in the flock in the beginning of the year.

20.4.5 Genetic classification of characters

There is another classification of characters which is based on their mode of inheritance, particularly on the basis of the number gene loci controlling the character, type and causes of variation among individuals of a population. Such characters are grouped as qualitative and quantitative characters.

Qualitative and quantitative characters: Some characters are controlled by one or few pairs of genes with their major effects which are not affected by environment, in general, making possible to classify the genotypes accurately based on phenotypes and hence show discontinuous variation expressed in Mendelian ratios. Such characters are called as qualitative characters. There is another category of characters controlled by many gene pairs (poly genes) with their minor effects which are modified by the environment and show continuous variation. Such characters are called as quantitative characters or metric traits.

These two types of characters (qualitative and quantitative) differ from each other with respect to the way the characters are measured, number of gene loci affecting the character and the extent of the effect of gene on the character, modification of gene's effect by the environment, type of variation shown by the character, description of a population for the character, inheritance pattern and the method of study used for the character:

1. Qualitative Characters: These characters are also called as *attributes*. They have the following characteristics:

- The characters are simply measured / recorded with naked eyes or by chemical test or other test. The observation on an individual is called the phenotype. The phenotype is assigned a rank or value to each individual according to the phenotype of the individual. There is no quantitative measurement for these characters.

- These characters are controlled by one or few pairs of genes with major effect of each gene and hence these genes are called as major genes. Each pair of gene has an effect large enough to cause discontinuity even in the presence of segregation of genes at other loci. It is easy to estimate the effect of evolutionary forces, mating system and population size changing the genetic structure of population.

- These characters are not usually affected by environment, but only in very few cases. The classical examples of environmental modification are the flower colour of Chinese Primrose, colour pattern in rabbit, sex limited and sex influenced characters, occurrence of diabetes, etc.

- These characters show discontinuous variation. The major effect of gene and no effect of environment to modify the genotypic value of the character cause discontinuity in phenotypic values. The data collected on these characters is binomially distributed as the individuals of a population belong to different distinct classes. This is because the individuals are recognized as belonging to one or the other group *viz.* red and white coat colour animals, tall and dwarf plants etc. This makes it possible to classify the genotypes accurately based on phenotypes and all the individuals of a population are grouped into discrete classes. For example, coat colour of sheep is recorded as black and white.

- A population is described for a qualitative character by estimating the percentage or proportion (ratios or frequencies) of individuals of each genotypic / phenotypic class. This is done by summarizing the raw data collected on different individuals (black and white coat colour of sheep) after counting the numbers of animals according to the phenotypic class *viz.* black or white colour.

- Mendelian analysis is applied to study the inheritance pattern of these characters. This is because the Mendelian ratios are expressed by these characters. The study of inheritance pattern of these characters is called as Mendelian genetics. The chi-square test of association is applied between the observed and expected ratios based on Mendelian inheritance to test the significance of departure in observed and expected ratios.

2. Quantitative characters: These characters are also called as *metric traits*. They have the following characteristics-

- The characters are measured in metric units like gm., kg, cm, days. These are the quantitative measurements to measure the character in a quantitative way assigning a numerical value to the phenotype of the character. Thus, the observation recorded on an individual in metric unit is called as phenotypic value and the character is called as quantitative or metric trait.

- These characters are controlled by many gene pairs (*i.e.* genes present at many loci). Each gene has small effect and such genes are called as minor genes for their minor effect or also called as poly genes because the genes are many in numbers to influence a character. These are also called as polygenic characters.

- All these characters are affected by environmental factors which cause modification in genotypic value assigned by the genes to the character. For example, milk production is affected by the diet given to a cow.

- The characters show continuation variation in phenotypic values of different individuals. The continuation variation is caused by the segregation of genes at many loci affecting the character and environmental modification of genotypic value of the character, for example, milk production, lactation length, body weight, wool production etc. The distribution of individuals follows the normal distribution.

- A population is described for a quantitative character by estimating the population parameters *viz.* the mean, variance and covariance.

- The statistical methods are applied to study the inheritance pattern of these characters due to the continuous variation. Mostly, the analysis of variance is conducted to test the significance of differences attributed to different genetic and environmental factors. As a result a new branch of genetics has been developed to study the inheritance pattern and called as quantitative genetics or biometrical genetics.

20.5 FARM RECORDS

An animal farm is dynamic and complicated enterprise having the objective of increasing the productivity and profit. It is dynamic for the changes that occur every moment which include change in the number of animals in milk and dry, number of animals open and pregnant, age structure of animals, daily production and requirement of inputs (feed, fodder, medicine, labour etc). It is complicated as it requires sound planning for synchronization of all inputs and activities related to each other. In order to achieve the objective of increasing the productivity and profit, it is essentially required for an efficient manager to keep track on all changes, activities and requirements for which he has to keep records and get useful information for taking decisions about selection of genetically superior animals on the basis of their breeding values.

20.5.1 RECORD KEEPING

The animal farm data is being maintained for the past many decades on sheets and registers. More recently, the use of PC has come on the scene for maintaining the data. This enables faster analysis and quick results for monitoring the various farm activities etc. However, in spite of excellent potential of maintaining data on PC, there is still importance and existence of recording the data on registers. The data maintenance in either way has many reasons as the records are useful to the farm manager.

Importance of record keeping: The records are kept for the following purposes.

- This helps to compare the between herd performance within breed as well as to make breed comparison.
- This is helpful in evaluation, culling and selection of animals for breeding purposes which in turn bring the genetic improvement of future generations.
- To know the pedigree and history of each animal pertaining to the production, reproduction and health performance.
- On the basis of the production level, close management and feeding levels can be provided.
- This also helps in fixing the price of the animal.
- Animals with optimum level of performance for the breed can be registered in central herd book and thus the herd and breed registration programmes can be implemented effectively.
- This also helps in research and development planning.
- To check the proper growth of young stock by weighing the animals at proper interval and recording the body weight. This will help in culling the animals with poor growth and late maturity.
- To avoid duplication in allotment of numbers to the young animals.
- To calculate the input/output relationships by keeping record on expenditure on feed and fodder.
- To know the financial status of the farm by keeping the sale register, death register, stock books, cash books etc.
- To know the health status by keeping records of the daily treatment of animals.

20.5.2 Types of records

The important records kept at animal farm are listed below:

1. Pedigree and history sheets- This contains the information of each animal regarding its brand number, its breed/genetic group, its dam and sire number, grand dam and grand sire number, date of birth/purchase, production and reproduction information of each lactation *viz.* lactation number, date of service, service sire number, number of services for conception, date of calving, types and sex of calf born, birth weight of calf, calf number, milk yield in 305 days, total lactation milk yield, peak yield, date of drying etc.
2. Birth register- This contains the information about date of birth, sire and dam number, birth weight, calf number and sex, date of death/disposal etc.
3. Growth register- The body weight recorded at different ages from calf hood age.
4. Daily treatment register- this contains the animal number of sick animal, name of disease/disorder, treatment given.
5. Livestock register- This has the information about the number of animals present.
6. Daily milking register- This contains the information about the amount of milk produced by each cow in the morning and evening.

7. Milk feeding register- About the amount of milk fed to the calves in case of weaning and also to orphan calves.

8. Feed and fodder register- To work out the expenditure on feed and fodder.

9. Service register, pregnancy report.

10. Disposal register: Sale/mortality register including post-martum report.

11. Cash book.

Pedigree and history sheets: The pedigree is a document about the origin of an individual (X), having the names or numbers of its ancestors (both paternal and maternal side) in the past few generations.

The pedigree is prepared in two ways. The first is the bracket style pedigree and the second is the arrow style pedigree. In preparing pedigree of any style, an individual is designated by the letter X, his sire by S and the dam by the letter D. The ancestors in the preceding generations are also designated by any suitable letters. Each ancestor is desig-nated by only one letter irrespective of its frequency of appearance in the pedigree.

The pedigree contains all the ancestors which are shown by arrow diagram (path diagram) by drawing the arrow from sire (S) to X as well as from dam (D) to X with the arrows pointing toward the individual X. The ancestors in one generation back are located and the arrows are drawn from the ancestors to the S and D with arrows pointing to the S and D from the ancestors of the preceding generation. This same procedure of putting arrows is completed for all the back generations.

Maintenance of pedigree is useful to study the genetic properties of a population, to formulate the animal breeding plan for genetic improve-ment and to avoid inbreeding. In modern scientific age, the pedigree, in addition to containing the names of the animal's ancestors, also contain objective informa-tion on performance traits. Such information on performance is useful to estimate the breeding value based on pedigree record. This facilitates early selection of superior animals for genetic advancement.

Pedigree depth: The number of ancestral generations in the pedigree is called as pedigree depth. The pedigree depth is a useful concept. It is taken as a base to define the genes identical by descent, the genetic relationship between any two relatives and to define inbreeding in relation to base (reference) population. The depth of pedigree is considered up to 4-6 generations due to less than 1% genetic contribution of ancestral genes after 6th generation and hence the maximum limit of considering an ancestor/common ancestor in the pedigree of an individual is up to 6 generations. The common ancestor is said to be remote or distant ancestor when it appears in the pedigree of the individual beyond six generations past considering its genetic contribution less than one parent.

20.6 DATA CORRECTION

The phenotypic value for milk, meat, egg, wool and reproducing ability is the joint product of genotype of the animal and environment received by it during the course of development and expression of any character. The environmental factors can be grouped as under:

(i) Random environment which includes sampling error and error in data recording.

(ii) Environmental factors which may be year and season when production start, heat/cold stress, feeding level and standards, health status, management level.

(iii) Physiological status *viz.* service period, dry period, lactation length, age and body condition etc.

20.6.1. Purpose of data correction

The above environmental factors influence greatly the ability to produce and reproduce and therefore mask the true breeding value of the animal. It is thus proper and logical to remove the effect of any environmental sources of variation and after eliminating or minimizing the effect of environmental factors, the data should be subjected to genetic analysis so as to obtain the accurate estimate of the genetic parameters. This is called as the correction or adjustment of data for environmental factors. Therefore, the purpose of adjustment of data is to minimize the environmental variation. The reduction in environmental variance increases the heritability and thereby the accuracy and response to selection is increased. Therefore, the adjusting records for environmental factors make the selection more effective. Theses adjusted records are the sum of the genetic and random effects. Therefore, the test of the validity of correction is that error variance (s_e^2) should be less on adjusted data.

20.6.2. Methods of data correction

The following four methods which can be used to adjust the data are:

1. Least squares method: The most important and commonly used method to adjust/correct the data is the least squares constant method given by Gacula(1968).This method adjust the data for several factors at the same time by the additive process *viz.* herd year, season, age etc. The least squares constants of different levels of each effect are estimated by the method of least squares technique of ANOVA (Harvey, 1966). To remove the environmental effects from the data, the correction factor for each source of variation irrespective of their significance are used. The correction factors (L.S. constants) are added to the original record of each individual with sign changed and the resultant quantity is the adjusted record as:

$$y = y_i - CF$$

where, y = adjusted record; y_i = unadjusted or original record

CF = sum of all the correctors factors (L.S. constants)

2. Ratio method: It is also known as gross comparison method which uses the simple average of all animals of each level taken as a ratio of the base level. The resulting factor is used for adjusting the record of animal in that level. Mathematically, the corrected record (y_i) is:

$$y_i = X_i / R_i$$

where, X_i is the unadjusted record

R_i is the conversion ratio taken as $\overline{x}_h / \overline{x}_i$

$\overline{x}_i$ is the level of record to be adjusted

b is the basal record

$_b$ is the average of base record

$\overline{x}_i$ is the average of record in i^{th} level

Therefore, the ratio factors for adjusting the record to a common base are obtained by dividing the mean of a base record by the means of other lactation records to be adjusted. Taking the milk production of cows which calved in two seasons with different age at first calving considering the base record of winter calving at 4 years of age will take the following farm:

	Summer calving			Winter calving		
	3 yrs	4 yrs	5 yrs	3 yrs	4 yrs	5 yrs
	X_{S31}	X_{S41}	X_{S51}	X_{w31}	X_{w41}	X_{w51}
	X_{s3n}	X_{s4n}	X_{s5n}	X_{w3n}	X_{w4n}	X_{w5n}
Mean	$\overline{x}_{s3}$	$\overline{x}_{s4}$	$\overline{x}_{s5}$	$\overline{x}_{w3}$	$\overline{x}_{w4}$	$\overline{x}_{w5}$

The different ratio factors will be: $\dfrac{\overline{X}_{w4}}{\overline{X}_{s3}}$, $\dfrac{\overline{X}_{w4}}{\overline{X}_{s3}}$, $\dfrac{\overline{X}_{w4}}{\overline{X}_{s5}}$, $\dfrac{\overline{X}_{w4}}{\overline{X}_{s5}}$, $1,\dfrac{\overline{X}_{w4}}{\overline{X}_{s5}}$, for the respective 6 groups of cows

3. Difference method: it is used when the level of any effect is only two like sex effect. The mean value of either sex is adjusted to that of the other sex by adding the average of the difference between two sexes.

4. Regression method: it is used in two ways *viz.* simple regression and partial regression.

(a) The simple regression is fitted when there is only one factor for which the adjustment is made *viz.* milk yield is to be adjusted for 300 days lactation length as:

$$\hat{y} = \overline{y} \pm b \, (x - _ix)$$

where, $\hat{y}$ is the adjusted milk yield; $\overline{y}$ is the average milk yield

X_i is days in lactation ; $\overline{x}$ is the average lactation length

$$b = b \, y \, x = \frac{\Sigma xy}{\Sigma x^2}$$

(b) The partial regression coefficients are estimated and used when data is to be adjusted for two or more factors like, age, season, year etc. a partial regression coefficient describes the effect of one independent variable on the dependent variable keeping constant the effects of all other independent variables. The prediction equation is an under:

$$Y - y = by_{1.2}(x_1 - \overline{x}_2) + by_{2.1}(x_2 - \overline{x}_2)$$

where, y = milk yield; X_1 = age; X_2 = lactation length

20.7. STANDARDIZATION OF RECORDS

The records on some animals are excluded before statistical analysis of data to study the effects of various environmental factors like year/period, season and parity of lactation etc. as well as before subjecting the data to genetic analysis for estimating genetic parameters. Such records are on the following animals.

- The incomplete records due to death or culling of the cows/buffaloes during lactation.

- The records on animals which dried up before 100 days of lactation, depending on the nature of the traits under study *viz.* such records are excluded for 305 days/lactation milk yield, lifetime milk production, lactation length and dry period whereas included for birth weight, gestation period, service period.

- The milk records following abnormal calving. The effects of abnormal calving on various production and reproduction traits are studied separately.

20.8. DATA GENERATION

The information recorded/available in history cum pedigree sheets of the animal are used to generate a number of economic characters. The description of these traits has been given under section 20.4.4

Chapter 21

Livestock Breeds

Since beginning, man has tried to exploit many species of plants and animals having desirable characteristics for mankind. However, man has domesticated a mere handful of about 40 species of animals for his own use like milk, meat, eggs, skin, wool/fiber, dung/manure and draught power. The various breeds and distinct animal types have been developed through selection and breeding practices and quest for development of need based animal types in different agro-climatic conditions having acquired specific morphological and physical characteristics grouped as adaptation. The farm animals of India have the unique features of adaptation to adverse climatic factors, diseases and parasites of tropics, survival on inadequate quantities of feeds, fodder and water. With these, the genetic diversity of different animal species has resulted due to the process of evolution over thousands of years during wild and domestic stages and for the efforts made by man to meet the market demand in the present day context.

21.1 PLACE OF DOMESTIC ANIMALS IN ANIMAL KINGDOM

The animal kingdom is divided into 10 phyla. The domestic animals come in phylum *Chordate (animals with backbone, it* has 4 sub-phylum) and in its sub-phylum *Vertebrate* or craniata (with spinal column). The vertebrate has two super-class *viz.* Agnata (animals with no jaw and no lateral appendages) and Gnathostomata. The super class *Gnathostomata* has 6 classes (mammals, aves, amphibian, reptile, osteichthyes and chondrichthyes or pisces.

The domestic animals come under class *Mammalia*, sub-class *Eutheria*, order *Ungulata* (hoofed mammals), sub-order *Artiodactyles.* The sub-order artiodactyle has a number of *families viz.* Hippopotamidae, Suidae (non-ruminant) and the families of ruminants.

The families of ruminants include Camenelidae (camels), Cervidae (Deer), Giraffidae (giraffes), and the Bovidae (hollow horned). The family *Bovidae* has the sub-family *Bovinae* which comprises the genus-Ovis (sheep), the genus-Capra (goat), the genus Bovis or Bos (Cattle) and antelope.

The genus *Bos* has four sub-genera or groups which are the followings:

* *Taurine (Bos)* group which includes the zebu cattle (*Bos indicus,* humped cattle) and hump less cattle (*Bos taurus*);

* *Bubaline (bubalus)* group which includes the Buffaloes: The buffaloes are kept in two sub-groups *viz. Bubalina* (Asian buffaloes) and *Syncerina* (African buffaloes). The bubalina sub-group includes: B. *bubalus bubalis* (Indian water buffalo, the *arni*), B. *bubalus* depressicornis (*Anoa*) and B. *bubalus mindorensis* (*tamarao*). The Syncerina sub-group has Cape buffaloes (*B. bubalus caffer*) and Congo buffaloes;

* *Bisontine* (Bison) group which includes the B. *grannies* (yak), B.*bison* (American bison) and B. *bonasus* (European bison);

* *Bibovine* (Bibos) group which includes the B. *gaurus* (gaur), B. *frontalis* (mithun or gayal), and B. *sondaicus* (banteng);

The domestic cattle descended from the aurochs and divided into two sub-group *viz.* zebu cattle of Asia and Africa (humped cattle), and the exotic cattle or hump less cattle predominantly of Europe. The hump less cattle also occur in Africa and East-Asia. The six species of cattle (Aurochs, mithun, yak, banteng, gaur and kouprey) can interbred but the male crossbreds are almost infertile and hence the breeding can be made with crossbred females.

The place of other species of domestic animals (Sheep, goat, camels, horse and pigs) in animal kingdom has been given together with giving the details of various breeds of these species in this chapter.

Species and its further division into Breeds: The morphological and physiological characters of animals had been the basis of dividing the animal kingdom. The breeds within a species are not included in the *Linnean Taxonomic classification* and hence the binomial nomenclature of different breeds of a species is same.

The first division of animal kingdom is into *species* based on reproductive discontinuity (reproductive isolation) which means that the animals of two species do not interbred or the progeny of two species do not produce fertile progeny on their mating together. For example, horse and ass are two species and on mating produce viable but sterile progeny, except few cases. The inter species crosses have variable fertility, *e.g.,* the female progeny produced by mating cattle and yak are fertile but their males are sterile. However, fertile male progeny of cattle and yak can be produced by the back crossing to cattle. Therefore, the reproductive discontinuity as the dividing criteria between species in some cases is liquidated. A species is further divided into sub-species or breeds.

Breed : The *breed* is a group of animals of the same species having same origin (related by descent) and having common characteristics like general appearance, body colour, features, size, configuration etc. A breed is a genetic entity developed over a long period of time as a result of planned mating and selection. Therefore, a breed has certain well defined physical conformation different from other breeds in the vicinity and has distinct local names.

Different breeds have both qualitative and quantitative characters which differ for one breed to the other. These qualitative characters are morphological such as colour and type of hairs, shape and size of horns, presence or absence of some body parts like zebu cattle are humped whereas European breeds are humpless. The quantitative characters can be considered as differences in type and size of body, milk production and fat percent in milk etc. However, there is no clear cut dividing line between breeds for quantitative characters for the reason that these traits are polygenic.

There is further division of a breed into *strains* based on isolation from each other due to geographical conditions or due to different aims of breeding the animals. This is called as strain breeding. The strain is developed by a breeder for certain specific purpose *viz.* egg size, egg weight, growth rate in poultry.

The *line* is a group of individuals of a breed or strain that are more closely related to each other due to interbreeding. A line is called as *inbred line* when the F is reached at least 0.375 as a result of two generation of FS mating. The FS or HS group of animals in cattle is the sire-dam or sire family and all the individuals of a family are equally closely related to each other.

Breed formation in the past

The different breeds of domestic animals and birds have developed over thousands of years during the process of evolution in wild and domesticated stages through intensive system of production to meet the market demands. They must have gone through stress conditions for developing different genes and gene combinations in them regarding their adaptability to different geo-climatic conditions, resistance to disease and parasites, heat tolerance and utilization of locally available roughages.

The natural selection and genetic drift caused breed differences within a species. Secondly, man's attempt to influence the evolutionary processes of domestic animals by selecting the animals with desirable characteristics, migrating and crossing to produce desired animals over a period of time also cased breed differences. All these caused the evolution of new types of animals with distinct appearance within a species. The natural selection increases an individual's fitness over others and this develops a stable population in a given environment. Thus, the geographically isolated groups eventually form the breeds. The migration plays its role to develop new type of animals with different survival and reproduction. Highly specialized breeds and strains have been produced and developed in developed world by recent advances in animal breeding techniques of genetic manipulation (A.I. with frozen semen and E.T.T.). These recent techniques have made easy, fast and inexpensive to introduce new genetic resources all over the globe.

The local agro-climatic conditions and socio-cultural preferences have played significant role in evolving the Indian breeds of livestock. Indian people prefer milk with high fat content and this has caused the buffalo breeding for milk. The type of cattle breeds evolved in the country had been the utility based. Thus, draft, milk and dual purpose breeds have been evolved. The major factors in development of breeds of different livestock had been the farmer's preference, geological isolation, natural selection and the intended utilities.

Cattle breeds of India: Oliver (1938) was perhaps the first to survey the important breeds of cattle in India. India has excellent quality of draft cattle which significantly contributed as draft animal power. Besides it, there are some dual purpose breeds which also provide good quality bullocks besides milk. The cattle breeds of India have been divided into milch breeds which produce good quantity of milk (about 2000 kg per lactation) but not produce good quality bullocks, dual purpose breeds (nearly 1000 kg) which produce fairly medium amount of milk and produce bullocks of good draught powers, and the draft breeds providing only a little amount of milk (around 500 kg) but their bullocks are good for draft purposes.

The different breeds under different categories of utility, their home tract, body colour, adult body weight (males and females) and the milk yield (kg) in lactation have been given in the tabular from below.

No.	Name of Breeds	Breeding tract	Body colour	Body wt. (kg) M, F	LMY (kg)
			Milch Breeds:		
1	Gir	Gujarat	Red or with white spots	550, 400	2000
2	Rathi	Rajasthan	Brown	380, 325	1600
3	Red Sindhi	Organized farms	Dark Red	460, 320	1600
4	Sahiwal	Organized farma in Pb, Haryana, Bihar, MP, WB	Red	540, 400	2200
5	Tharparkar	Rajasthan	White or grey	540, 350	1800
			Dual purpose breeds:		
6.	Deoni	M.S., Karnataka	White	590, 350	900
7.	Haryana	Haryana, U.P., Rajasthan	White	500, 350	800
8.	Kankrej	Gujrat & Rajasthan	Silver or iron grey	550, 430	1000
9.	Krishna Valley	M.S., Karnataka	White or Greyish white	500, 340	750
10	Malvi	M.P.	– do–	430, 340	900
11	Mewati	U.P. & Rajasthan	- do-	400, 350	900
			Draft Breeds:		
12	Amritmahal	Karnataka	White or grey	500, 330	500
13	Bachur	Bihar	-do-	385, 320	400
14	Bargur	T.N.	Light grey	340, 295	350
15	Dangi	M.S.	White with red/black spots	365, 295	450
16	Gaolao	M.S., M.P.	White or grey	430, 330	500
17	Hallikar	Karnataka, A.P., T.N.	Grey	450, 300	550
18	Kangayam	T.N.	White or grey	500, 330	550
19	Kenkatha	U.P., M.P.	Grey	340, 295	600
20	Kherigarh	U.P.	White or grey	475, 320	400

contd...

contd...

No.	Name of Breeds	Breeding tract	Body colour	Body wt. (Kg) M, F	LMY (Kg)
21	Khillar	M.S., Karnataka	White	500, 345	450
22	Nagauri	Rajasthan	White or grey	400, 340	600
23	Nimari	M.S., M.P.	Brownish red	390, 320	400
24	Ongole	A.P.	White or grayish	550, 440	650
25	Punganur	A.P.	White, light grey	200, 130	500
26	Ponwar	U.P.	Brown or black with white patches	320, 295	450
27	Kandhari	M.S.	Red , Dark red	-	550
28	Siri	Sikkim, W.B.	Brown or black with white patches	450, 350	500
29.	Umblicheri	T.N.	Grey	-	500
30	Vachur	Kerala	light red or black	150, 95	550

The crossbred strains/breeds of cattle evolved at different places in India are as under:

(i) Cattle breeds evolved in India

1.	Taylor	1875	Shorthorn and Jersey with local cows around Patna
2.	Jersind	1953	Jersey x Red Sindhi at A.A.I. ,Allahabad
3.	Brownsindh	1955	BS x red Sindhi at A.A.I.,Allahabad
4.	Sunandini	1964	BS, HF, Jersey x non descript at Munar (Kerala)
5.	Frieswal	1987	3/8 – 5/8 HF x Sahiwal at Military Dairy Farms
6.	Karan Fries	1980	HF,BS,Jersey x Tharparkar cows at NDRI, Karnal
7.	Karan Swiss	1980	Brown Swiss x (Sahiwal & Red Sindhi) at NDRI, Karnal.
8.	Jerthar	1958	Jersey bull x Tharparkar cows at NDRI Bangalore
9.	Vrindavani	1980	HF,BS, Jersey x Haryana at IVRI, Izatnagar

(ii) Cattle breeds developed in other countries

10.	Australian Milking zebu:	20 to 40% zebu blood (Sahiwal, Red Sindhi) and rest from Jersey at CSIRO, Australia
11.	Australian Frisian Sahiwal	Sahiwal bulls x H.F. cows of Queensland Deptt.
12.	Jamaica Hope	80% Jersey, 15% Sahiwal and 5% Holstein blood, at Jamaica
13.	Mambi	¾ HF and ¼ zebu in Cuba
14.	Siboney	5/8 HF and 3/8 zebu in Cuba
15.	Pitanquieras	5/8 Red poll and 3/8 zebu in Brazil
16.	Santa Gertrudis	Brahman bulls x Shorthorn cows at Taxes, in 1851.

contd...

contd...

17.	Brangus	Brahman x Angus
18.	Beefmaster	Brahman x Shorthorn x Hereford
19.	Charbray	Brahman x Charolais
20.	Guernsey	Brown and white cattle of Brittany x Barindlecattle of Normand

Exotic Breeds of Cattle: The exotic breeds of cattle are of dairy type and beef type. Some of the important breeds are as under:

Dairy Breeds of Exotic Cattle

1. Holstein Friesian: Originated in two provinces of Netherland (North Holland and West Fries land). Animals were migrated to Netherland from Europe, and imported into USA by Dutch settler in 1621. They have large size body with black and white colour. Adult body weight 800 kg of males and 550 kg of females. Best milk producer of the world with world record of about 19000 kg in a lactation and an average production of 6000-7000 kg with low fat % (3-3.5%). AFC is of about 30 months and C.I. of 13-14 months.

2. Brown Swiss: Native of Switzerland. The animals were introduced into United States in 1869. The animals are fairly of large size body with light brown to grey colour. Adult body weight 700 kg of males and 500 kg of females. These are medium milk producer with average 5000 kg milk per lactation. AFC and CI equal to HF cattle.

3. Jersey: Originated on the island of Jersey (English Channel). The animals were imported into England in 1811 and into United States in 1850. Smaller size cattle with reddish grey to brown colour. Adult body weight about 600 kg males and 400 kg females. AFC and CI of Jersey cows are almost equal to HF cattle. Average herd average of 4500 kg milk per lactation with 4.5% fat.

4. Ayrshire: Developed in Ayrshire country of Scotland during 18th century by crossing many strains of cattle. They have medium size body with cherry red to brown colour. Adult males weight about 700 kg and females 500 kg. They have outward and upward horn. These were imported into United States in 1882. The AFC and CI are almost equal to HF cattle.

5. Guernsey: Home tract is Guernsey in the channel Island. Developed by Brown and white cattle of Brittany and large brindle cattle of Normandy. The animals have smaller body with yellowish brown colour weighing about 600 kg males and 400 kg females. Herd average 4000 kg milk per lactation

Beef Breeds of exotic cattle

1. Hereford: Developed in Hereford (English) during 1800 and imported into USA in 1817. The animals are red in colour with white face and compact body, weighing 600 kg males.

2. Short horn: Developed during 18th century in Durham country in England. The animals are solid red in colour or red with white markings or white or roan.

3. **Angus**: Originated in Aberdeen country of Scotland. The animals are completely black body with polled head and compact body. The animals of this breed were introduced in USA in 1873.

4. **Santa Gertrudis**: Developed in taxes (USA) by crossing short horn with Brahmin crossbred bull (any Indian breed is known as Brahmin in USA). This breed has 5/8 short horn and 3/8 Brahmin. Deep cherry-red colour body.

5. **Brangus**: Developed in USA by crossing Angus with Brahmin.

6. **Beef Master**: Developed in USA by crossing Hereford with Brahmin bulls and shorthorn bulls. The Brahmin inheritance is about 15%.

21.2 BUFFALO BREEDS

Buffaloes have docile nature and superior genetic potential. Buffalo is low input animal in terms of management and nutrition, disease resistance, able to thrive under stress and a blessing to India contributing milk, draught power and manure.

Buffalo has been originated in the Indian sub-continent from wild *arni (Bubalus arnee)* which are still found in forests of Assam. The buffaloes are placed in group *bubalus* of the genus *Bos.* The group bubalus has two sub-groups *viz.* Bubalina (Asian buffaloes) and Syncerina (African buffaloes).

Bubalina group: This group is of *Asian wild buffaloes* which comprise three distinct species *viz.* the arni, *Bubalus arnee and* its descendant, the water buffalo, *Bubalus bubalis;* the Anoa (*Bubalus depressicornis*) and the tamarao, *Bubalus mindorensis.*

The arni (Indian or Asiatic buffalo, *Bos bubalus bubalis*) is a very large animal of northern India and closely associated with water. The domestic buffaloes have been developed from the original wild ancestor, the arni. The domestic buffaloe have been given its specific name, *bubalis* after Linnaeus. The most distinguishing feature is the shape of its horns, separated by a wider space on the forehead as compared to African buffaloes.

The anoa (Celebes Dwarf buffalo, *Bos Bubalius depressicornis)* is the smallest among the bubaline group, confined to northern Celebes, Indonesia. Its hide is exceptionally thick. This animal is very ferocious in nature and attacks even without provocation. This is widely hunted and is in the danger of extinction.

The tamarao *(Bos Bubalus mindorensis)* is a dwarf buffalo and is the native of the island of Mindoro in the Philippines and these buffaloes are intermediate between arni and anoa of Celebes in many ways. These are nocturnal, moving late in the afternoon and returns back in the morning from forests. The animals remain on rest during day time. Tamarao buffaloes do not take a dip in water pool or in a water stream, like other members of the bubaline group and they dislike rain. These animals are known for fierceness, having the habit to attach on man fiercely. It is on the verge of extinction.

According to Macgregor, there are two main groups of domestic water buffaloes (*Bubalus bubalis) viz. riverine group* and *swamp group*, depending on their habitat and genomic structure.

The *river buffaloes* have 50 chromosomes and massive body, prefer to enter clear water, primarily used for milk production and also used for meat production and draft power. These are black animals, with long face, small girth and bigger limbs. The horns of river buffaloes grow downward and backward, then curve upward in a spiral. River buffaloes have been developed in many dairy breeds by selection in India and Pakistan.

Swamp buffaloes have 48 chromosomes, with marshy land habitats, primarily used for draught power (work animal) in paddy fields and also used for meat and milk, found in South-East Asian countries with some animals in NE states of India. The swamp buffaloes are grey at birth but becomes slate blue later and are heavy bodied with large belly, flat forehead, prominent eyes, short face, short tail, wide muzzle and comparatively long neck than river buffaloes. The horns of swamp buffaloes grow outward and curve in a semicircle, but always remain on the plane of forehead.

Syncerina group: This group is of *African buffaloes* which are of two types-Cape buffalo, *syncerus caffer caffer*, the black buffalo of Southern Africa, northern to Ethiopia and Somalia and second is the smaller red type buffalo (Congo buffalo) of western Uganda, Cango to north-Angola. The *Cape* variety of the African buffalo has 52 chromosomes. The Cape buffaloes are still wild. *Congo buffalo* or West African buffaloes have 54 chromosomes

Buffalo population: The total population of buffalo in India (FAO, 2008) was 111 million out of which the breedable buffaloes were 55 million. Asia has 97% (171 million) and India 57% of the world buffalo population of 177 million followed by Pakistan (14%), China (13%), 2% each in Philippines, Nepal, Vietnam, and Myanmar, and 1% each in Indonesia, Thiland and Laos with some population in Burma, Bangladesh, and Sri Lanka. About 95% buffaloes in the world are found in 10 countries of Asia.

Buffalo Breeds: About 70% of the total buffaloes in India are non-descript and rest 30% population is classified into 19 breeds according to FAO but into 15 breeds according to ICAR. The Indian buffaloes (*river group*) have been classified into 5 groups:

(i)   Murrah group-Northern type: Murrah and Nili-Ravi, and Kundi of Pakistan

(ii) Gujarat-Western type: Surti, Jafrabadi, Mehsana and Banni

(iii) Uttar Pradesh-Bhadawari and Tarai

(iv) Central India- Nagpuri, Pandharpuri, Manda, Jerangi, Kalahandi and Sambalpuri

(v)   South India-Toda, Godawari and South Kanara

S.No.	Name of Breeds	Breeding tract	Body colour	Body wt. (kg)M,F	LMY (kg)
			Well recognized breeds of Indian buffaloes:		
1.	Murrah	Haryana, Pb.	Black	570, 450	2000
2.	Nili-Ravi	Pb.	Black with white markings (forehead, muzzle, legs, tail)	600, 450	2000
3.	Mehsana (Murrah x Surti)	Gujrat	Black to grey with white markings (face, legs tails)	500, 450	1700
4.	Surti	Gujrat	Brown	500, 380	1300
5.	Jafarabadi	U.P., M.P.	Black with white patches (face, legs)	600, 450	1800
6.	Bhadawari	U.P., M.P.	Copper colour	475, 400	900
7.	Nagpuri	M.S., A.P., M.P.	Black with white patches	520, 380	1000
8.	Toda (Berari, Ellichpuri)	T.N.	Grey skin (face, legs tails)	380, 380	800
9.	Pandharpuri	M.S.	Black with white marking on forehand, legs and tail switch	-	1100
10.	Marathwari	M.S.	Black	-	900
11.	Banni	Gujarat	Black	-	2000
12.	Chilika	Odisha	Brownish black	-	500
13.	Kalahandi	Odisha	Black grey	360	800

Besides above breeds of buffaloes, the following are known locally buffalo population in India:

S.No.	Name	Breeding Tract	Utility
1.	Tarai	U.P.	Milk, Meat and Draught
2.	Gojri	H.P.	Milk
3.	Diara	Bihar	Milk
4.	South Kanara	Karnataka	Milk and Draught
5.	Dharwari	Karnataka	Milk
6.	Godawari	A.P.	Milk
7	Parlakhemundi	AP and Odisha	Meat and Draught
8.	Jerangi	AP and Odisha	Meat and Draught
9.	Sambalpuri	M.P. and Odisha	Meat and Draught
10.	Kuttanad	Kerala	Meat and Draught
11.	Manda	Odisha	Meat and Draught
12.	Kujang	Odisha	Milk and Draught
13.	Assamese	Assam	Milk and Meat
14.	Swamp	Assam	Milk and Draught
15.	Sikmese	Sikkim	Milk and Draught
16.	Mizorami	Mizoram	Milk and Draught
17.	Manipuri	Manipur	Milk and Draught

Some of the above poplulations (Parlakhemundi, Manipuri, Mizorami and Swamp) have 48 chromosomes.

21.3. MITHUN BREEDS

The Mithun or gayal (Bos or Bibos frontalis) is a bovine species and also called as mountain cattle, or 'ceremonial ox' of north eastern hill region. Mithun originated in Himalayan region about 4500 years ago. It resembles domestic cattle in most of the physical, production and reproduction traits, but it has no hump, its dewlap is small, tail is short and horns are short stumpy. It has an oily sweat gland which acts as insect repellent. There are different views about its origin: it is the domesticated gaur, or a hybrid of crossing of gaur bull and zebu cow, or descendent from wild Indian bovine. The mithun resemble with gaur in appearance, colour, body features and some habits and produce fertile crossbred progeny when crossed with gaur whereas it differs from zebu cattle in hemoglobin, transferrin and blood group genotypes. It is thus considered more to have originated from the gaur. It has 58 (2n) chromosomes.

Mithun cows produced 1.4 to 3.5 kg milk per day with 4 to 5% fat, 3.5 to 4% protein, 4-5% lactose and 0.8% ash.

The birth weight of mithun varies from 20 to 30 kg, adult body weight from 450-500kg, body height 1.2 to 1.4 m. Sexual maturity is attained at 2.5 to 3 years, G.P 272 days, S.P. 60-180 days, productive life 12 years and total life span is 20 years. In India, there are two different types of mithun which are-

1. Arunachalees : They are found in Arunachal Pradesh. These are small in size than Nagamees. They have been developed for milk and meat.

2. Nagamees : They are found in Nagaland and Manipur. They are heavier in size and weight (400-500kg). They are good for beef (meat) purpose having long and massive body.

Hybrids of Mithun: In order to increase the milk production, the mithun males have been mated with cows. The hybrids are given names according to the species involved and sex of ancestor.

The F_1 males produced by crossing cows with mithun bull are called as *Jatsa* which are good for draft purpose while the F_1 females are called *Jatsamin* which produce milk higher than Mithun cows.

The backcross is also made by crossing Jatsamin with Mithun male and the two sexes are called as *Nupsa* (male progeny) and *Nupasmin* (female progeny).

Siri cows were crossed with mithun bulls in Bhutan. The male and female F_1 are called *Jechha* and *Jessam*.

A farm of mithun has been setup by govt. of Arunachal Pradesh at Kambi in West Siang Distt., whereas the National Research Center on mithun of ICAR has been working in Phek Distt.of Nagaland.

21.4 YAK BREEDS

Yak (Bos grunniens) is locally known as *banchour* for wild and *chour-gau* for domestic variety. It was originated in China which still has 90% of the total world yak population of 14 million.

No breed of yak has been described so far. But Indian yaks have been described of four types, *viz.* Ladakh yaks, Himachal Yaks of Spiti valley, Arunachal yaks of Twang and Dirang region, and Sikkim yaks which are of two types depending upon rearing community *viz.* Bho and Aho yaks.

The yaks interbred with bison, banteng, gayal, zebu and European cattle. The chromosome number of yak is 60 (2n).

21.5. SHEEP BREEDS

The sheep belong to the genus Ovis, of sub-family bovinae, of the family Bovidae (hollow horned). The species of the genus ovis are aries (domestic sheep), Canadensis and vignei, etc.

The total population of sheep in India (50.8 million) ranked fourth in the world in 1992 accounting for about 4.57% of the world sheep population whereas in 1997 sheep population in India was 56.47 million and ranked 3rd in the world with 5.3% of the world and 14.0% of the Asia sheep population. The sheep population in India registered an annual growth rate of 2.2% during 1987 to 1992 and 0.74% during 1951 to 1992 with an annual increase of 0.288 million. During 2007 the sheep population was 71.5 million.

Sheep population and wool production in different regions of India.

Traits	North Temperate	North Western	Southern Peninsular	Eastern
Sheep population (m)	3.63 (7.2%)	21.45 (42.2%)	20.86 (41.1%)	4.85 (9.5%)
Wool production. (m kg)	4.8 (10.5%)	29.8 (65.4%)	9.1 (20.0%)	1.9 (4.1%)

The sheep rearing is done for various purposes *viz.* fine wool, carpet wool, mutton and pelt production. Accordingly there are different breeds of sheep serving different purposes. The following are different types of sheep breeds according to *type of wool fibers:*

(i) Fine wool- Fine quality fleece producing breeds.

(ii) Medium wool - The fleece is medium in fineness and length.

(iii) Long wool – The wool fibers are coarse and long approaching 30 cm.

(iv) Carpet wool- It is coarse, wiry and tough.

(v) Fur – Pelts are used for fur purposes.

(vi) Crossbred wool – produced from crossing of long wool and fine wool breeds.

Sheep Breeds of India: In India there are 42 breeds of sheep which have been classified according to the *geographical regions of their home tract (breeding tracts):*

S.No.	Name of Breeds	Breeding tract	Body wt. (kg) M, F	Utility
North Temperate Region Breeds:				
1.	Bhakarwal	J & K		Carpet wool
2.	Changthangi	Laddakh	38, 34	-do-
3.	Gurej	Kashmir		-do-
4.	Karnah	-do-		Apparel wool
5.	Poonch	-do-		Carpet wool
6.	Kashmir Merino	-do-		Apparel wool
7.	Gaddi	H.P.	25, 24	Carpet wool
8.	Rampur Bushari	-do-	29, 25	-do-
North-western arid and semi arid breeds				
9.	Chokla/ Shekhawati	Rajasthan	34, 24	-do-
10.	Nali	-do-	34, 24	-do-
11.	Bikaneri / Magra	-do-	27, 24	-do-
12.	Pugal	-do-	31, 27	-do-
13.	Jaisalmeri	-do-	28, 27	Meat & Carpet wool
14.	Malpura	-do-	40. 24	-do-
15.	Marwari	-do-	30, 26	-do-
16.	Sonadi	-do-	38, 21	-do-
17.	Patanwar/ Kuchi/ Kathiawari	Gujarat	33, 26	Carpet wool
18.	Hisardale (Merino x Magra)	Hisar	54, 34	Apparel wool
19.	Munjal (Nali x Lohi)	Pb., Haryana	60, 40	Meat & Carpet wool
20.	Muzaffarnagri	Western U.P.	50, 40	-do-
21.	Jalauni	U.P.	40, 30	-do-
Southern Peninsular region breeds:				
22.	Deccani	M.S.	34, 34	Meat, wool
23.	Bellari	Karnataka	35, 27	Meat, carpet wool
24.	Hasan	Karnataka	25, 23	Meat
25.	Kenguri	-do-	32, 27	-do-
26.	Mandya/Bannur	-do-	34, 24	-do-
27.	Nellore	A.P.	36, 30	-do-
28.	Ramnad white	T.N.	30, 22	-do-
29.	Trichi black	-do-	25, 18	-do-
30.	Vempur	-do-	34, 28	-do-
31.	Nilgiri	-do-	30, 25	Apparel
32.	Coimbatore	-do-	25, 20	Meat & Carpet wool
33.	Kilakarsar	-do-	30, 21	-do-
34.	Madras Red	-do-	36, 23	-do-
35.	Mechari	-do-	35, 22	-do-
Eastern Region Breeds:				
36.	Tibetan	Arunachal		Carpet wool
37.	Garole	W.B.		Meat
38.	Shahbadi	Bihar	37, 21	Meat
39.	Chota Nagpuri	-do-	20, 20	Meat & Carpet wool
40.	Ganjam	Orissa		-do-
41.	Balangir	-do-	24, 18	-do-
42.	Bonpola	Sikkim		-do-

Indian breeds of sheep can also be classified according to the utility for which they are reared:

Wool Breeds: Bikanei, Bhakarwal, Marwari, Karnah, Gurej, Kaskmir Merino, Jaisalmeri, Deccani, Hassen, Rampue Bushair,

Meat Breeds: Jalauni, Mandya, Nellore

Dual Purpose: Lohi, Kachi,

Crossbred strains of sheep evolved in India:

Sl. No.	Name	Exotic inheritance	Crosses made with place
1	Bharat merino	75%	Chokla and Nail ewes with Rambouillet and Merino rams at Avikanagar
2	Avivastra	50%	Chokla and Nali ewes with Rambouillet and Merino rams at Avikanagar
3	Avikalin	50%	Malpura ewes with Rambouillet ram at Avikanagar
4	Avimans	50%	Malpura and Sonadi ewes with Dorset and Suffolk at Avikanagar
5	Indian Karakul	75%	Marwari, Malpura and Sonadi with Karakul at CSWRI, Bikaner
6	Kashmir Merino	50-75%	Gaddi, Bhakarwal and Poonchi with Merino & Rambouillet rams
7	Nilgiri synthetic	62.5-75%	Nilgiri ewes with Merino and Rambouillet rams at TNVASU,Sandynallah (TN)
8	Patanwadi synthetic	50%	Patanwadi with Rambouillet & Merino rams at GAU, Dantiwara
9	Hissardale	75%	Bikaneri ewes with Australian Merino rams at GLF, Hisar.

Exotic sheep breeds developed by crossbreeding are as under:

1	Columbia	Rambouillet ewes x Lincoln rams
2	Panama	Lincoln ewes x Rambouillet rams
3	Targhee	Rambouillet x Long wool breed
4	Romnelet	Romney Marsh rams x Rambouillet ewes
5	Suffolk	Dark face South Down rams x Norfolk strain
6	Corriedale	Merino ewes x Lincoln rams

Exotic Breeds of sheep: The important exotic breeds of sheep according to the purpose have been classified in following 5 groups-

l. Fine wool breeds

1. Merino: Developed in Spain, imported into America in 1801 and used to develop America merino from which delaine merino were developed. Rams weigh about 60 kg and ewes about 45 kg. Rams shear 10 kg wool and ewes shear 7 kg wool in a year. Medium size breed with short head. Ewes are polled and rams have spiral horns. Imported to India from USSR.

2. Rambouillet: Originated at Rambouillet in France by importing merino from spain and hence descended from Spanish merino. Imported into America in 1840. Rams weigh about 100 kg and ewes 70 kg and produce about 5 kg long fleece annually. Large breed with wide head and well balanced horns curving backward and upward. Ewes polled, rams have large spiral horns or polled. Imported to India from USA.

ll. Medium wool breeds

3. Corriedale: Developed in New Zealand by crossing merino ewes and Lincoln rams and also Licester rams subsequently. Lt has 50% merino inheritance and rest Lincoln and Licester. Introduced in America in 1914. It is hardy breed producing wool and mutton. Rams weigh 80 kg and ewes 60 kg. Rams and ewes produce 7 kg and 5 kg wool per year. Dual purpose breed (meat and wool). Imported in India from Australia.

4. Hampshire: Originated in England by crossing two strains of sheep. Imported in USA in 1840. Large breed good for mutton as well . Rams weigh more than 100 kg and ewes about 80 kg producing 4 kg fleece annually.

5. Cheviot: Developed in Scotland by infusion of black face high land, Licester, South down and Merino. Rams weigh 80 kg and ewes 60 kg producing 4-5 kg wool per year.

lll. Long coarse wool breeds

6. Lincoln: Developed in Lincoln (England) in 1749 with licester inheritance. Largest breed in the world. Rams weigh about 150 kg and ewes about 100 kg, fleece is 25-38 cm long often twisted into spiral.

7. Licester: originated in the licester (England) in 1960 by Robert bakewell from local sheep by close inbreeding. Rams weigh 120 kg and ewes 80 kg fleece coarse, 15-25 cm long with annual yield of 5 kg.

8. Romney Marsh: Originated in the masshy district of kent (England) resistant to parasites, medium size breed weighing about 100 kg rams and 70 kg ewes.

IV. Fur breeds

9. Karakul : Native to Bokhara (central Asia), derive it name after the area karakul (black lake0. Originated by crossing two breeds of this area. Medium size breed, rams weighing 80 kg and ewes 60 kg, producing light white fleece of low grade of about 3 kg annually. Poor in mutton quality. Produces fur pelt from prematurely born lambs or killed soon after birth. Imported in India in 1975.

V. Mutton breeds

10. Dorset: Medium size, origin exactly not known but developed through selection, native of England (Dorset region). These are horned and polled strains, wool is short and 2.75 to 3.2 kg, rams weighs 80kg and ewes 60 kg imported in India in 1973 at CSWRI, Avikanagar.

11. Suffolk: Native of England (Suffolk region0. Developed from dark faced south down rams used in Norfolk strain. It is black faced free from wool. Both rams and ewes polled, though rams have scars, short, dense and fine fleece of 2.75 to 3.75 kg annually. Ewes are prolific, mature rams weigh 100 kg and ewes 80 kg. Imported in India in 1973 at Avikanagar (Rajasthan).

12. South down: Originated in south down hills of England and developed by selection and inbreeding of local sheep. It is hornless and mutton breed rams weigh 80 kg and ewes 60kg producing 3-4 kg wool per year.

21.6. GOAT BREEDS

The goat is placed under genus *Capra* of sub-family bovinae of family bovidae. There are 5 species of goat: *C.hircus* (Benzoar), *C.falconeri, C.caucasia, C.pyrenaica* and *C.ibex.* The true goats are *C.hircus.*

The goat population in India was estimated to be about 120.6 million in 1997 and 128 million in 1999-2000. India stands 2nd in goat population. The goat population had increased by 3.6% (1.66 million) annually during 1951 to 1992 which is higher than cattle, buffalo and sheep. This is in spite of fact that the current mean rate of slaughter of goats is about 40% and mean mortality rate of about 15%. Goat population was 140.5 million in 2007.

There is a history of migration of goats from western Asia to India from 7500-7000 BC onwards. The geographical isolations due to hilly rivers, high mountains, dense jungles and deserts etc., physical environment, ecological constraints, social preferences, diseases resistance and the natural selection had resulted into the origin of present day Indian breeds over thousands of generations.

There are nearly 20 well recognized breeds of goats in India constituting 20-25% of the total goat population and the remaining populations are considered as mixed and non-descript. Different breeds have been developed in India to produce milk, meat and fiber (pashmina, mohair) through natural selection and have been adapted to diversified agro- climatic conditions based on their utility. The good quality fibers are produced by the goat of temperate Himalayan region and they possess the finest quality *under coat* called *pashmina.* Primarily milch type breeds which are large in size are found in North and North-Western region of the country. The goat breeds found in Southern and peninsular region of India are of *dual purpose* for milk and meat.

Goat Breeds of India: The different goat breeds of different regions (Breeding tracts) of India have been listed as under:

S.No.	Name of Breed	Home tract	Utility
North Temperate Region Breeds:			
1.	Changthangi	Ladakh	Fibre (Pashmina)
2.	Chegu	Lahul	Pashmina & meat
3.	Gaddi	H.P.	Long fibre & meat
North semi-arid region breeds:			
4.	Jamunapari	U.P.	Dual purpose (milk & meat)
5.	Barbari	U.P.	Dual purpose (milk & meat)
6.	Beetal	Punjab	Dual purpose (milk & meat)
North-west arid region breeds:			
7.	Marwari	Rajasthan	Milk, meat & hair
8.	Jhakrana	Rajasthan	Milk & meat
9.	Sirohi	Rajasthan	Milk & meat
10.	Kutchi	Gujarat	Milk & meat
11.	Mehsana	Gujarat	Milk & meat
12.	Surti	Gujarat	Milk & meat
13.	Zalawadi	Gujarat	Milk, meat & hair
14.	Gohlwadi	Gujarat	Milk & meat
Southern Peninsular region breeds:			
15.	Osmanbadi	M.S.	Milk & meat
16.	Sangamneri	M.S.	Milk & meat
17.	Malabari	Kerala	Milk & meat
18.	Kanaiadu	T.N.	Meat
Eastern region breeds:			
19.	Ganjam	Odisha	Meat
20.	Bengal	W.B.	Meat

Exotic Goat Breeds: The followings are some of the important exotic breeds of goats-

1. Saanen: Originated in Saane and Simental valleys of Switzerland. White or light cream colour body, have pendulous udder, produces about 100 kg milk per lactation. Legs are short in comparison to body, ears erect pointing forward, bucks weigh about 80 kg and Does about 60 kg. Male horned but female hornless.

2. Alpine: Originated in France, large animals, long legged and hardy, small ears and Roman nose. Bucks weigh 80 kg and Does 60 kg.

3. Anglo Nubian: Development unknown. Largest and heaviest breed among all European breeds of goats. Produce 3 kg milk daily.

4. Toggenburg: Originated in Toggenburg valley of Switzerland, very hardy and high milk producing breed, light brown to dark chocolate body colour. Small white ears pointing forward, white strips on both sides of face. Bucks weigh about 70 kg and Does about 60 kg.

5. Boer: It is from Southern Africa and exported to many countries. Mean birth weight 4.0 kg, adults weigh about 30 kg, produce 1.4 kg milk daily.

6. Angora: Originated in Angora region in Asia Minor (Turkey). It is fiber (Mohair or pashmina) producing breed. Bucks weigh 60 kg and Does 40 kg. Average yield of *mohair* is 3.0 kg annually.

7. Orenberg: Originated from local goats of Kazakhistan and later on improved by selection for quality of pasmina. Bucks weigh 70 kg and Does 40 kg. Produce *pashmina fibers* of 16 microns diameter and 5.5 cm in length of about 300 gm per year.

21.7 CAMEL BREEDS

The camel is placed under the genus camelus of the family camenelidae of ruminants. There are two species of camel *viz.* *C.bactrianus* (Asian two humped camel) and *C. dramedrius* (Asian one humped camel).

The camel is known as ship of desert, capable of travelling long distances on sandy soil carrying man and material. It utilizes various adaptive mechanisms particularly the water retention and ability to thrive on water deprivation. It provides energy to agricultural operations.

The north-western African countries have about 73% (14.6 million) of the total world camel population (19.2 m), about 28% (5m) are present in Asia and rest 1.36% (0.225) population in rest of the world. In India, Rajasthan contributes about 70 % of the country's camel population with more number of camels in 10 districts (Barmer, Churu, Bikaner, Hanumangarh, Jodhpur, Jaisalmer, Jhunjhunu, Nagaur, Sikar and Ganganagar). India stands third possessing 1.5 million camel after Somalia (6.2 m) and Sudan (2.9 m).

In India, there are 4 well defined breeds of camel. These are Bikaneri, Jaiselmeri, Mewari and Kuchchhi. The other breeds are the crosses of different breeds *viz.* Marwari, Mewati, Shekhawati and Sindhi are such breeds of camel.

1. Bikaneri: This breed was evolved by selective crossbreeding of different breeds like Sindhi, Baluchi, Afgan and Thari. Its home tract is Bikaner and adjoining districts. It is used for draft, meat, milk, hair, hides and dung. The birth weight is 42 and 39 kg for two sexes, gestation length 382 days, AFC 1885 days, adult body weight 580- 650 kg, milk yield in 300 days as 1650 kg with average of 5.5 kg per day. Animal hair production 0.8 kg.

2. Jaisalmeri: Its home tract is Jaisalmer district. It is riding camel. It has been descended from Tharparkar camel of Sindh region. The birth weight is 37 kg and 34 kg, adult weight of 50-590 kg. Milk yield 2-4 kg/ day and hair production 0.758 kg.

3. Mewari camel: Originated from hill camel of Punjab. The hilly region of Udaipur is the home district.

4. Kuchhi camel: The habitat is Kuch region of Gujarat. They are heavy, well built. Birth weight 35 and 32 kg, adult weight 630 and 500 kg. Females are good milk producers producing 4-6 kg per day.

21.8 EQUINE GENETIC RESOURCES

The equines belong to the genus Equus under family Equidae (non-ruminants) of the sub-order Perissodactyles. The equines comprise horses (E. caballus), donkeys or Ass (*E. asinus*) and their inter-species hybrids- the mules and zebra, having made significant contribution to mankind.

They are used for draft/cart and as riding animal. The riding horses were being used to carry the urgent messages before mechanized transport in old days. They had important role to win battles in the warfare. The ponnies, donkeys and mules had significant contribution in agriculture as a means of transport, draft and pack purposes.

There had been a sharp decline in horse and donkeys population in India for the reason of mechanization, small land holdings and abolition of Jagirdaris. The horse breeding in India remained in the hands of Rathores of Marwar and Maharawals of Kathiawar whereas donkeys were kept by rural poor for their livelihood.

(1) Horse (*Equus caballus*) breeds: There are 6 horse breeds in India. These are Kathiawari, Marwari, Spiti, Manipuri, Zanskari and Bhutia. Besides these, the FAO has also recognized 3 more breeds of horse *viz*. Chummarti, Deccani and Sikang.

There had been the dilution of horse breeds for indiscriminate breeding. The crossbreeding between kathiawari and Marwari horse has considerably reduced the number of purebreds. There are only about 13% (3000) purebred Marwari horses out of total population of horse (23300) in its home tract. Likewise, the purebred population of Kathiawari horses in its home tract (Gujarat) is about 56% (8000) of the total horse population of 12900.

1. Marwari horse: It has Arab blood. Home tract is Marwar, and Mewar region of Rajasthan. Used for riding, sports and safari

2. Kathiawari horse: Evolved from wild horses of Katiawar and Arabian horses. Home tract is Rajkot, Junagarh, Bhavnagar, Amreli and Surendernagar in Gujarat. Used for riding, sports and safari.

3. Spiti horse: Spiti ponnies originated in Tibet and found in Spiti region of Lahaul, Kullu and Kinnaur districts of H.P. It is riding and pack animal

4. Zanskari horse: The home tract is Leh and Ladakh area. Used for draft, transport and riding.

5. Manipuri horse: These ponnies are found in Manipur and Assam, have been originated from Asiatic wild horse and Arabian horse. They are beautiful, hardy and known for fastness.

6. Bhutia horse: Resemble to Tibetian pony. Home tract is Sikkim and Darjeeling. Used for riding and pack animal.

(2) Ass or donkeys (*Equus asinus*) breeds

The donkey is low input requirement animal and available everywhere. The modem domestic asses have mainly originated from Nubian race of Africa. Three distinct types of Indian donkeys, Indian wild asses and kiang. The Indian wild asses are found in the Rann of kutchh of Gujrat known as Kulau. The kiang is found in Sikkim and Ladakh.

(3) Mules

The mules are the species hybrid producing by crossing of horse mare (female) with jack as a stallion (male ass). The mules are used as draft, pack and transport animals particularly in hills. Adult bodyweight of draft mules vary from 450 to 650 kg with 155 to 172 cm height. The size, speed, strength and sprit are of horse type whereas sure footedness, lack of excitability, endurance (patience) and ability to survive on poor feed came from ass side.

21.9. PIG BREEDS

The pig belongs to the genus *Sus* (non-ruminant) of family Suidae of the sub-order Artiodactyles. There are two species of pigs *viz. S. scrofa* (wild boar) and *S. domestica* (domestic swine).

The age at fist conception in pigs is 300 days, gestation length of 112 days, farrowing interval 227 days, litter size at birth and weaning 6-8 and 2 to 6, average litter weight at birth and weaning 4.5 to 4.8 and 21 to 42 kg. The average birth weight is 0.52 kg and average weaning weight (8weeks) ranged from 6 to 12 kg.

Indian breeds of pigs: No attempt has been made to catalogue the pig breeds in Iindia and hence there is no definite breed. However, the following types/breeds of domestic pigs are known:

States	Breed/Local Name	States	Breed/Local Name
Assam	Doom, Pigmy hog	Bengal	Ghungroo
Sikkim	Lepchamoun	Odisha	Golla, Burudi
Manipur	Manipuri Desi	Kerala and TN	Angamali
Meghalaya	Niang Megha		
Nagaland	Suho	U.P., Bihar, M.P. and Punjab	Non-descript (Desi)
Mizoram	Zovwak		
Tripura	Dome, Mali		
Andman	Andman Wild		
Nicobar	Nicobari pigs		

Exotic breeds of pigs: The important exotic breeds of pig are-

1. Yorkshire: Originated in Yorkshire (England), developed from Leicester pigs by Robert Bakewell, migrated into USA in 1893, good bacon breed, white in colour with small ears tilting to the front, broad face.

2. Berkshire: Developed in Berkshire (England) by crossing of English hog with sows of Chinese and Siamese and Siamese origin. Large and long body with arched narrow back, adult weight 500 kg, black coat with white feet.

3. Duroc: Originated in USA, developed by crossing pigs from Africa, Spain and portuga. Light golden to dark red in colour, body medium length with tall lege, adult weight 450 kg, for boars and 350 kg for sows.

4. Tamworth: Developed in Ireland, colour varies from golden to dark red, large, long and tall breed known for lean meat.

5. Landrace: developed in Ireland, colour, long narrow body.

6. Poland China: Originated in USA, developed by crossing Berkshire, Hampshire, tamword and some inheritance of Russian pigs, large, long body with long legs, black colour with white extremities, prolific but poor mothering ability.

Crossbreds:

Minnosota No.1	Tamworth x landrace
Minnosota No.2	Poland China x Yorkshire
Beltsville No.1	Poland China x Landrace
Beltsville No.2	Yorkshire x duroc x landrace x Hampshire
San pierre	Berkshire x chester white
Palouse	Landrace x chester white
Mary land No.1	Berkshire x landrace

21.10 POULTRY BREEDS

The term poultry implies a wide variety of birds of many species *viz.* chickens (fowls), pigeons, guinea fowl, pea fowl, turkeys, ducks, geese, swans, pheasants, quail, ostriches, emu and others. The alternate term to the poultry science is the *ormithology* which is the study of birds other than poultry. The distinguish feature of poultry is the presence of comb. Fowls have relatively high breathing and pulse rate as well as high body temperature (105°C to 109°C) than other domestic animals.

The fowl is said to be domesticated in Iran during 800 B.C. and spread by Persians to western Asia and Mediterranean. Later on, it was immigrated to America and Australia.

Classification of the class Aves:

Phylum: Chordata; **Sub-phylum:** Vertebra

Class: Aves (Feathered, warm blooded with 4 chamber heart)

Sub-class: Neonithes (without teeth)

Order: I. Galliform

Family: Maleagridae Phasianidae

Genus: Maleagris Coternix Phasians Gallus Pavo

Species: *M. galloopavo C. japonica P. colchicus* **G.domestica** *P.cristatus*

Common name	Turkey	Jap. Quail	Pheasant	Fowl P.	Fowl
Chrom. No.	80	78	82	78	80

Order II. Ansariform (Aquatic birds)

Family : Antidae

Genus :	Aser	Cairina	Anas
Species :	A. anser	C. maschata	A. platyrhynchos
Common name :	Goose	Muscovy	Duck
Chrom. No.	80	80	80

The wild fowl have four species *viz.* red jungle fowl (Galls gallus), Gray jungle fowl (*Gallus sonnerati*), Ceylon jungle fowl (*G. lafayetti*) and Java jungle fowl (*G.varius*). The domestic breeds of fowl have been originated from red jungle fowl (Gallus gallus) and it has 5 sub-species which are murghi, gallus, jaboullei, bankiva and spadaceus. The Indian red jungle fowl is *Gallus gallus murghi*.

The further grouping of fowl is into-

- **Classes** include a group of breeds, depending upon the region of their development *viz.* Asiatic, English, American, Polish, French, Mediterranean etc.
- **Breeds** within class, a group having some common characters (shape, body weight),
- **Varieties** within breed are different in colour pattern, comb type, feather pattern and colour, shape. Different breeds have a number of varieties *viz.*
  Plymouth Rock has barred, white, buff varieties;
  RIR has single comb and Rose comb varieties;
  Wyandott breed has silver, golden, white, buff, black varieties;
  Leghorn breed has black, white, buff, silver, red, rose comb varieties,
  Cornish breed has white, dark, buff varieties;
  Minorca breed has white, black, buu, rose comb varieties;
  Brahama breed has dark, light, buff varieties;
  Cochin breed has white, black, buff varieties
- **Strains** within variety are developed by a particular breeder.
  Classification based on utility (Purpose): The fowl are classified into different types based on their utility (economic value and fancy purpose) *viz.* egg types, meat types, dual purpose, game birds, ornamental (fancy) and bantam (dwarf)
- **Egg** Type breeds: Leghorn, Minorcas, Anconas, Campines
- **Meat** Type: Cornish, Aseel, Chittagong, Brahama, Cochins, Langshan
- **Dual purpose:** RIR, Plymouth Rock, Australorp, Orpington, Dorking, Wyandotte
- **Fancy** birds: Andalusian, Hamburg, Polish
- **Pleasure** birds: Brahama, Cochin
- **Game** birds: Aseel

Classification based on origin: The fowl breeds are classified based on their origin *viz.* Asiatic, American, English, Mediterranean and continental European. The important breeds of these classes are:

American Class: Rhode Island Red, New Hampshire, Plymouth Rock, Whyandotte

English Class: Australorp, Orpington, Sussex, Cornish

Asiatic Class: Brahama, Cochin

Mediterranean: Leghorn, Minorca, Ancona, Andalusian, Spanish

Continental: Capine, Lakenvelder

Native Vs. Exotic Breeds: The fowl are also classified as *native* breeds, *exotic* breeds and *crossbred* genotypes.

The *native birds* (chickens) show great variability in their morphological traits *viz.* body weight(dwarf, normal and heavy weight), plumage pigmentation (blackish, brownish), plumage distribution (nacked neck, frizzle, crest, philopody), comb shape (single, rose, pea, and walnut), shank and skin (green and blue pigmentation) and melanin deposition in skin, meat, internal organs and bones in some birds like Kadakanth. The major morphological maker genes creating these variations increase the adaptability to tropical environment.

Some of the examples of the effects of major genes are as under:

1. Frizzle gene (f) is incomplete dominant causes curling of feathers and reduced feathering with side effect of improving the convection ability.

2. Dwarf gene (dw) which is sex linked recessive reduces the body size by 10-30% and has its effect on metabolism (reduction), fitness (increase) and disease tolerance.

3. Slow feathering gene (K) is sex linked dominant causes delay of feathering with side effect on reduction in fat deposition and protein requirement, increases heat loss and delay the immune response mechanism.

4. Naked neck gene (Na) is incomplete dominant with direct on loss of neck feathers and secondary feathers and its side effect to increase convection ability and fitness but reducing the hatchability.

5. Pea comb gene (P) is dominant and causes the compact comb size with side effect to improve the convection ability.

Breeds of Desi Fowl (chickens)

S.No.	Name of Breed	Habitat	Important character
1.	Faverolla	Kashmir	Feathered comb
2.	Punjab brown	Pb & Hariana	Meat type, brown plumage
3.	Brown Desi	U.P.	Layer type, single comb
4.	Ankaleshwar	Gujrat	Poor prod., Single comb
5.	Bursa	Gujarat & M.S.	No particular character
6.	Frizzle	Hot & humid, NE region	Curved feather
7.	Necked neck	-do-	Necked neck
8.	Kadaknath	M.P.	Black pigmentation
9.	Aseel	A.P.	Game bird, biggest size
10.	Denki	A.P.	Males good fighters, long neck & legs
11.	Kalasthi	A.P.	Resemble to Denki but small size
12.	Ghangus	A.P. & Karnataka	Small size, small comb & wattles
13.	Tellichery	Kerala	Small bird with black skin
14.	Haringhata black	W.B.	Good layer, small size black bird
15.	Daothigir	Assam	Fairly heavy bird
16.	Miri	Assam	Small size black bird
17.	Chittong/Malay	NE region	Large heavy bird, fighting nature
18.	Nicobari	Nicobar islands	Short legs, small size, good layer
19.	Teni	All over country	Small, all purpose, tolerate high temp.
20.	Titri		Small, yellow beak & neck, poor layer

Crossbred genotypes evolved

Name	Breeds involved		Purpose	Place
	Male	Female		
Vanraja	Cornish	Synthetic popul.	Dual	ICAR, Hyderabad
Grama priya	Synthetic	White Leghorn	Dual	ICAR, Hyderabad
Giriraj			Dual	UAS, Bangalore
Krishna	Synthetic		Egg	JNKVV, Jabalpur
Yamuna	Kadaknath	New Hampshire		
Kalinga brown	Selective breeding of RIR		Egg	CPBF, Bhubeneshwar
Gramalakshmi	Australorp	White Leghorn	Dual	KAU, Kerala
Nandanam	RIR population			TNVASU
Cari gold	Selective breeding RIR		Egg	CARI, Izatnagar
Cari-Nirbhik	Aseel cross		190 eggs	CARI, Izatnagar
Cari-Shyam	Kadaknath cross		210 eggs	CARI, Izatnagar
Upcari	Frizzle cross		220 eggs	CARI, Izatnagar
Hitcari	Nacked neck cross		220 eggs	CARI, Izatnagar

Exotic breeds of poultry

1. **Leg horn:** Mediterranean breed, number one egg producer (250-230 eggs annually), developed in Spain, white in colour with yellow skin and single combed.

2. **Minorca:** largest and heaviest Mediterranean breed, also called as red faced black Spanish, produces large white eggs, long wattles and white ear lobes with black beak, shanks and tons.

3. **Australorp:** Developed in Australia, single combed with black plumage and white skin, good egg producer

4. **Sussex:** developed in England, single comb, white skin, good layers, small size bred.

5. **Orphington:** England breed, single comb, colour are black, white, blue or buff, moderate egg producers

6. **Cornish:** developed in England, white in colour, heavy breed, excellent meat breed

7. **Phode island red:** developed in Rhode island (USA), heavy breed, single combed, dark red colour, good strains for both egg and broilers

8. **Plymouth Rock:** American breed, heavy breed with strains for both eggs and broilers, single combed with yellowish skin.

9. **New Hampshire:** American breed, chestnut red plumage, single comb, good producer of large brown shelled eggs.

10. **Brahman:** Asiatic breed developed in India, pea comb, have there varieties *viz.* light, dark and buff.

OTHER POULTRY SPECIES

The other poultry species are Guinea flow (*Titri*). Ducks, Turkeys, Japanese quail (*Batter*), Pigeons, Ostrich and Emu. They constitute only a small segment of the total poultry population. These species except pigeons were introduced in the country (India).

1. Guinea fowl (*Numeda meleagris geleata*): Titari or Tatiri

It is native of France. Its population rank third after chickens and ducks. It is known by different local names *viz. Titri* in northern plains, *Chittra* in western parts and *China (cheena) murgi* in south and eastern parts. They are raised in Punjab, UP, Bihar, Rajasthan, MP, Maharashtra, Orissa, AP, TN and Karnataka.

The breeds of guinea fowl are based on the plunge colour variation. The main varieties recognized world over are: *Pearl, Lavender and White*

In India, Pearl guinea fowl are most common. They have dark- grey feathers with white spots having pearl like appearance. The lavender birds have light grey spotted feathers. The feathers of these two varieties are often used for ornamental purpose. The white variety has completely white plumage.

CARI Izatnagar (Bareilley) maintains all the 3 varieties while pearl is maintained at PAU Ludhiana. USA Bangalore, central poultry breeding farm, Bhubneshwar and AVM hatcheries, Coimbatore.

A new variety of guinea fowl has been developed at CARI, Izatnagar and named as Guncari. It produces about 100-120 eggs per year from March to October and hence seasonal layer. The egg weight is about 40gm and hatchability is about 70% the body weight at 12 week is about 1.0 kg.

The CARI Izatnagar has also developed 3 crossbred genotypes for meat *viz. Swetambari, Kadambari, Chitambari.*

The guinea fowl has a low input requirement with excellent foraging scavenging potential; they have flocking instinct hence offer no management problems. They are raised for egg and meat.

2. Ducks (*Anas platyrhynchos*)

They are popular in Assam, W.B., Orissa, AP, TN, Kerala and Tripura. They are reared for egg and meat. There is Central Ducks Breeding Farm at Hessarghata (Bangalore) and few farms of State Govt. and SAU's. The ICAR has undertaken a scheme for comparing pure and crossbred duck's egg and meat. Ducks constitute about 9% of the total poultry population. They are resistant to chicken diseases, exterminators of snails, harmful weeds and beetles. Asia is as a homeland of Mellard ducks but most of the modern breeds have been developed in Europe and America. The age at first egg is about 150 days.

The important Indian breeds of ducks are: *Nageshwari, Synthetic and Indian Runner.*

Chapter 22

Conservation of Germplasm

The different indigenous (Indian) breeds of different species of farm animals and poultry are getting diluted and facing degeneration. The *dilution of a breed* is in terms of purity of breed. There is a decline in the number of purebred animals conforming to the model attributes of the breed. Some of the breeds of farm animals of India are under risk of their extinction. The technical aspects of conservation of animal breeds were considered jointly in 1980 by FAO and UNDP and laid down definite procedures after analyzing the problems for establishing gene and data bank. The efforts in this direction in India were started with the establishment of National Bureau of Animal Genetic Resources, Karnal under the control of ICAR.

22.1 CONSERVATION Vs. PRESERVATION

The word conservation is closely related to the preservation. The preservation covers the continued maintenance of genetic variability and required when a breed reaches an endangered level of population or near to extinct. The preservation and multiplication may also be required in case of individuals having unique traits or exceptional genetic merit.

The conservation includes the preservation along with up-gradation (improvement) of the genetic potential and management of a breed for use in future. Thus, the conservation covers both continued maintenance of genetic variability, improvement and sustainable utilization by exploiting the genetic variability. The conservation can also be defined as the management of the biosphere of human use for benefits in present time together with maintaining its potential to meet the future needs.

22.2 REASONS OF THREAT TO GENETIC DIVERSITY

The following reasons can be assigned to affect the domestic animal diversity and dilution and declining trend of population of breeds-

(i) Breeding systems and breed dilution: There had been the unrestricted interbreeding among different breeds particularly in rural areas. The native animals/ breeds are crossed with improver breed for better economic returns. This has resulted into the mixture of various breeds. Bhadawari buffaloes have been upgraded with Murrah, Malpura and Chokla sheep breeds are getting Marwari inheritance through migratory flocks, intermixing of Jaisalmeri and Chokla sheep.

The efforts have been made to improve the productivity of indigenous breeds of livestock and poultry by introducing exotic germplasm of superior genetic merit through crossbreeding.

(ii) Geographical reorganization: The breeding tract and organized farms of some breeds of cattle (Sahiwal, Red Sindhi), buffalo (Kundi, Nili-Ravi), sheep and goats have gone to Pakistan.

(iii) Purpose based farming system: The different breeds are used to serve different purposes of the owner *viz.* milk and draft in case of dairy animals; meat, milk, wool/fiber in case of sheep and goats; eggs and meat for poultry, etc. The breeds that do not serve the purpose are neglected. They are either not kept or used for crossbreeding.

(iv) Economic viability: The economically useful breed is automatically conserved whereas the breeds with poor performance and declining economic returns to farmers are losing their existence *e.g.* Mewati, Kankatha, Kherigarh, Bachur cattle.

(v) Modern agricultural practices: There had been mechanization and a trend of growing cash crops due to increase of population pressure, reduction inland holding and common grazing area. This has caused the reduction in herd size, become uneconomic to keep a bull in the village. The waste land, grazing area and forests are declining. All these has resulted in reducing the requirement of draft power, genetic dilution and reduced performance of progeny.

22.3 NEEDS OF CONSERVATION

The native breeds need to be conserved for the following purposes-

(i) Genetic insurance: The native breeds have been developed over thousands of years as a result of evolutionary processes, have better adaptability to harsh climate, tropical diseases, heat tolerance with low management inputs in terms of health care, low quality feeds and fodder being capable to convert them more efficiently into animal products (milk, meat, wool, eggs). They are integral part of agriculture and have genes or gene combinations which are associated with adaptability and producing ability. Thus, loss of such germplasm means loss for specific adaptive traits which may be introduced in highly productive germplasm through new techniques of biotechnology. It is not known what might be needed in future and hence conservation of native breeds is required as a genetic security.

(ii) Scientific study: The conservation of native germplasm provides useful research material in genetics, biochemistry, physiology, morphology and anatomy, and to understand the process of evolution and domestication.

(iii) Economic potential in particular environment: The crossbred animals fail to exploit their genetic potential under Indian conditions of low input and harsh

climate, more sensitive to tropical diseases and their production level goes down beyond F_1's. Therefore, the conservation of native breeds is required to produce F_1's.

(iv) Environmental considerations: Every organism has its own role to play within an ecosystem and loss of germplasm has adverse effect on the ecosystem which is hazardous to the existence of mankind. The domesticated breeds are an integral part of our ecosystem.

(v) Cultural and ethical requirements: A breed if not economical, it should be preserved for cultural and public interest of historical importance, being part of natural heritage, culture and ecosystem. It is also required on ethical and moral grounds.

(vi) Energy source: India has a number of draft breeds of cattle. Non-renewable source of energy may be exhausted sooner or later. Under such condition, the animal draft power will be required.

22.4 PRINCIPLES OF CONSERVATION

The conservation should be based on the following principles-

(i) Population size: It should be of optimum size above the level of risk.

(ii) Type of animals: The animals for preservation/conservation must be of pure form, having special traits, and select the diverse stock.

(iii) Environmental conditions: The local adapted breeds should be maintained and conserved in the same location under the similar feeding, management and environmental conditions under which they had been kept traditionally.

(iv) Breeding methods: The genetic merit and diversity should be maintained using appropriate breeding methods.

22.5 METHODS OF CONSERVATION

The conservation may be done *in situ* as well as *ex situ* as follows-

(i) In situ conservation: This is the maintenance of live population of animals in their adaptive environment (native tract) where the animal population continues to evolve and develop for more sustainable use. There are two aspects of *in situ* conservation *viz.* active and passive. The active *in situ* conservation is equivalent to breed development through animal breeding programmes whereas the passive *in situ* conservation is the maintenance of live animal population within their environment.

The *in situ* method is best for conserving a breed by maintaining a large population size with proper breeding plan for genetic improvement so as the breed become economically viable. The breed so conserved can gradually adapt to the changing environment over time. However, it requires high cost to maintain large herd/flock. This method can better be used at organized farms of Govt. to ensure purity of breed and also can be used under field conditions maintained by large farmers and NGO's like BIAF, AFPRO, Goshalas and by forming breed societies. The conservation of breeds can be done by using MOET, nucleus herd both ONBS and CNBS and improving the management and environment.

The farmers maintain small herds/flocks and hence can not keep males of good quality and no data recording system is followed. These problems can be solved by forming breed societies with their Head from Govt. side/Animal breeders, covering the area under Herd Registration Scheme for data recording, giving incentives to farmers and provision of facilities of semen and A.I., health coverage, support price of animal products. The fodder resources can be increased to maintain grasslands, reseeding of grasslands and controlled grazing, planting high yielding species of plants for providing optimal lopping.

(ii) Ex situ conservation: The conservation in the native tract (*in situ*) is not practical for economically unviable population. The conservation can thus be done in an environment other than native, called as *ex situ conservation*. This can be done in two ways which are **in vivo** and **in vitro.**

(a) *In vivo* **method:** This is *ex situ* conservation of live animals in small numbers in a place away from home tract like organized farms, bull mother farms, zoo and breeding park. As the numbers of animals are small, the breeding population should be maintained so that inbreeding is minimal and the performance is improved over the years. A good system to conserve the precious germplasm as a self-sustaining unit is to keep the animals in live animal reservoir/breed safari/park. This may attract tourist industry and may serve as amusement parks if distinct breeds are kept.

(b) *In vitro* **method:** It is the storage of living cells for long period of time which includes deep freezing (cryopreservation) of sperms, oocytes, embryos and DNA.

22.6 AGENCIES FOR CONSERVATION PROGRAMME

There are a number of departments, institutions and other agencies involved in research/training/manpower development, animal health services, counseling and guidance to farmers. These agencies are-

1. Department of Animal Husbandry and Dairying, Govt. of India
2. I.C.A.R.
3. SAU's and Veterinary Colleges
4. State Animal Husbandry Deptt.
5. Non-Govt. Organizations (NGO) *viz.* gaushalas, dairy cooperatives, voluntary organizations (Bhartiya Agro-Industries Foundation, Urlikanchan, M.S.); Action for Food Production, Aligarh, U.P.; Raymond's Embryo Research Centre (M.P.); Janki Devi Bajaj trust (M.S.); NDDB; Deptt. of Biotechnology; Deptt. of Environment and private companies dealing with poultry sheep and goat keeping.

Chapter 23

Response to Selection

The *selection*, in animal breeding, is an outcome of the process of differential reproduction and survival of animals which may be natural or artificial or both. Artificial selection depends on the choice of the breeder to allow the animals to produce the next generation. Therefore, the selection is a process of giving opportunity to certain individuals in a population to produce next generation while others are denied. Artificial selection is man's activity and depends on the choice of the breeder which is objective specific *viz.* to bring genetic improvement (to produce animals of high genetic merit) in certain traits of interest like production of milk, meat, wool, egg, pashmina fiber, pelt production or for any other purpose like, draft, race, load carrying capacity, etc. The objective of selection is to bring genetic improvement in a character of interest and hence the character under selection is known as selection objective.

The genetic improvement through selection is possible by selecting the genetically superior animals present either in the herd (within herd improvement) or by introducing from outside the genetically superior animals (introduction or migration). The possibility within herd improvement through selection depends on the followings:

- Performance level of the herd,
- Variability in selection objective (character under selection), and
- Population size.

23.1 SELECTION DIFFERENTIAL

The most essential requirement of selection is the existence of the variability in the phenotypic values among the individuals of the population. The variability decides the availability of superior animals in the herd or flock. The superior animals are selected and the remaining ones are culled. Thus, selection divides the population in two groups *viz.* selected and culled (rejected). The selected group comprised the superior animals which have their mean performance (Ps) better than the mean of the whole population (P). The difference in the mean performance of the selected group over the population mean before selection, is called the *selection differential*, denoted by S. Therefore, the selection differential is the superiority of the selected parents (P_s) over the population mean (P). It is thus estimated as:

$$S = \overline{P}_s - \overline{P} \qquad \qquad ......23.1$$

The phenotypic superiority of the selected parents is due to the reason that the selected parents may have either better genes or they might have received better environment or both. Thus, the phenotypic superiority of selected parents (selection differential, S) corresponding to gene's effect and environmental effect has two main components *viz.*

S = Genetic part + Environmental part

= (Additive genetic + non-additive genetic) + environmental part

Intensity of selection: The selection differential (S) is taken in its standard measure called as the standardized selection differential. This is obtained by dividing the selection differential with the phenotypic standard deviation of the trait (σ_p). This *standardized selection differential* (S/σ_p) is called as the *intensity of selection*, denoted by *i*. Thus,

$$i = \frac{S}{\sigma_P} = \frac{\overline{P}_s - \overline{P}}{\sigma_p} \qquad \qquad ....23.2$$

The intensity of selection, *i*, is the selection differential in standard deviation units of the trait and it is the mean deviation of the selected animals in units of σ_P of the trait. Therefore, the intensity of selection (*i*) is the numbers of σ_P of the trait by which the mean of the selected group ($\overline{P}$s) is above the population mean ($\overline{P}$) before selection.

The standardized selection differential (intensity of selection) has its use and importance for the reason that the S can not be estimated unless the selection is actually done. It is also used to compare the different methods of selection and to compare the response to selection for different traits measured in different units.

The S in original units of measurement can be taken by multiplying i with standard deviation as: $S = i\ \sigma_P$ \qquad 23.3

The intensity of selection, *i*, is also taken from normal frequency distribution curve as:

$$i = \frac{Z}{p} \qquad \qquad ....23.4$$

where, p is the proportion of animals selected,

Z is the height of frequency distribution curve where the group of selected animals is represented.

Thus, Z is the height of normal curve at the point where the selected individuals with the lowest record fall. This point of separation is called the *point of truncation*. The Z is thus the height of ordinate at the point of truncation. There are tables available showing the ordinate and area of normal frequency distribution curve to obtain the value of Z. This is known as the selection differential under truncation which indicates that all the animals are selected with records above a certain level of the trait. Therefore, the selection differential in standard deviation units may be taken from mathematical properties of the normal distribution as:

$$i = \frac{S}{\sigma_p} = \frac{Z}{p}$$

The intensity of selection (*i*) is inversely proportional to the proportion selected (*i*). If the proportion selected (*p*) is given, the intensity of selection (*i*) can be estimated in terms of *p* the number of standard deviation of the trait (σ_p) by which the mean of selected individuals exceeds to the population mean.

The intensity of selection (*i*) can be known without actually measuring the superiority of parents. The approximate value of *i* can be obtained from *p* for any character based on their relationship after consulting a table of values.

Factors affecting selection differential (S): The following factors determine the intensity of selection (i) or selection differential (S) -

(i) Proportion selected (p): There is inverse relation between the two. The p depends on the breeding policy, population size and the capacity of the farm. The proportion selected (p) will be higher, if it is required to increase the herd size. This will decrease *i* and the S. On the contrary, *i* and S will be increased in a herd of constant size because fewer replacements are required. However, the population size put a restriction on *p*. In small herd size the chances of inbreeding are increased which have three fold effects *viz.* the inbreeding depression and random drift, reduction in genetic variability and thirdly reduction in *i* and S. On the contrary, in large size herd, the chances of inbreeding are less and *i* is increased. The increase in *p* can increase the herd size but it will reduce *i*.

The *p* is affected by number of replacement stock being available. This is the function of the *fertility, prenatal calf losses, sex ratio, post-natal calf mortality and culling,* etc. The S is also larger in litter bearing species. The constraint of low breeding efficiency and high calf mortality in genetic improvement can be over come by increase in herd size by increasing the capacity of the farm which will increase *i* and will reduce the inbreeding level. Secondly, the breeding efficiency can be increased and the calf losses can be reduced by providing better nutrition and better management practices. Further, the reproductive rate of superior animals can be increased by use of MOET technology. Thirdly, the dams of elite animals should be retained in the herd for longer time to get more calf crop of high genetic merit. This will increase the selection differential and hence the genetic gain.

Conclusively, the R increases with increase in S which in turn increases with decrease in *p*. However, more genetic variation is preserved in the population by higher p and results in greater cumulative R. The lower value of *p* does not result in maximum long term R. The chance to get lost the favourable alleles with small effect linked with unfavourable alleles is reduced with larger p compared to smaller p each generation.

(ii) Phenotypic Variability (σ_p): The phenotypic standard deviation of the selection objective is very important to influence the S and hence R. The S is small if the character is less variable and vice versa. The S increases with decrease in p and with increase in σ_p of the character. The σ_p depends on the breeding programme of the herd. The selective breeding increases the performance level of the herd and results fewer animals being available in the herd that exceed the herd average. This reduces the proportion of animals selected (p), σ_p and selection differential.

(iii) Sex of animal: The S is large in case of males. This is because fewer males are required for breeding and hence i will be higher in case of males.

(iv) No. of character under selection: The increase in number of traits under selection reduces the selection differential. This is because an animal may not be out standing in all the characters.

(v) Accurate measurement: Any error in recording a trait also influences the selection differential.

23.2 RESPONSE TO SELECTION

The selection, without creating new genes, changes the genetic structure of the population and hence brings the genetic changes by changing the frequency of genes and genotypes. The frequency of desired genes is increased in the progeny generation through selection of superior breeding stock at the expense of the frequency of undesirable or less favourable genes by culling the genetically inferior animals. The increase in frequency of desirable genes due to selection improves the performance of progeny generation over the parental generation. This is the *genetic effect of selection.*

However, the change in gene frequency can be estimated only for qualitative characters and it is not possible for quantitative traits for the well known reason of their polygenic nature. However, the effect of selection on quantitative characters is reflected in progeny performance and hence it is estimated by measuring the change in mean performance of progeny generation in comparison to the mean of parent generation. The change in performance of progeny generation in quantitative traits due to selection is known as *response to selection* or *genetic change* or *genetic gain* or *genetic improvement* or *genetic advancement.* This indicates the superiority of progeny generation due to selection over parent generation.

Predicted response to selection: The additive genetic part of phenotypic superiority of selected parents (selection differential) is more important and of more concern because it is the only additive genetic superiority which is transmitted to and shown by the progeny. The additive genetic superiority, measured as additive genetic variance (5^2_A) of the total phenotypic superiority (Ps – P), is thus responsive to selection. This additive genetic superiority or variability ((5^2_A) expressed as a ratio of total phenotypic superiority or variability (5^2_P) is known as the heritability, denoted by h^2. The heritability expresses the extent to which the phenotypic superiority of a trait shown by selected parents is transmitted to the progeny. Therefore, whole of the phenotypic superiority of selected parents (selection differential) is not transmitted but its portion equal to h^2 is transmitted and it is called as *response to selection*, denoted by R. This is actually the predicted response to selection (R) which equals to the expected inherited part of selection differential.

Different expression of predicted response to selection: The response to selection is predicted in different ways as under:

(i) R in terms of inherited part of selection differential (S) -

$$R = h^2 (\overline{P}_s - \overline{P})$$
$$= h^2 S \qquad\qquad ....23.5$$

(ii) R in terms of intensity of selection (standardized S) –

$$R = h^2 S$$
$$R/\sigma_P = (S/\sigma_P) h^2$$
$$= i\, h^2$$
$$R = i\, \sigma_P\, h^2 \qquad\qquad ....23.6$$
$$= i\, \sigma_P\, b_{AP} \qquad\qquad ....23.7$$

where, $i = S/\sigma_P$.

σ_P = phenotypic standard deviation of the trait

$b_{AP} = h^2$

(iii) R in terms of accuracy of selection -

$$R = i\, \sigma_P\, h^2$$
$$= i\, \sigma_P\, h\, h$$
$$= i\, \sigma_P\, (\sigma_A/\, \sigma_P)\, h$$
$$= i\, \sigma_A\, r_{AP} \qquad\qquad ....23.8$$

where, σ_A = Standard deviation of the breeding value of the trait,

r_{AP} = Correlation between B.V. and phenotypic value.

h = Accuracy of selection.

The response to selection (R) can be predicted from above formulae when the selection criterion is the individual's own record.

The above formula of predicting response to selection can be made generalized for other criterion of selection (selection based on performance records of relatives of the individual).

The genetic gain (ΔG) can be predicted based on any criteria of selection (I) by using the formula: $\Delta G = i\, \sigma_A\, r_{AI}$

where, r_{AI} is correlation between the selection criteria (I) and B.V. of
individual (A).

Factors affecting response to selection: The change in performance due to selection (response to selection) depends on the following factors-

(i) *Additive genetic variability in the trait* (σ_A). The additive genetic variability is the main stem and the raw material on which the selection acts. The magnitude of σ_A has direct relation with R. However, the genetic variability of the trait (σ_A) is determined by the population and the character as well by the breeding programme of the herd, whether there had been practiced the selective breeding in the herd for that trait. Thus, it is beyond the control of breeder. The nature and extent of genetic variability (additive genetic) present in the herd decides the choice of selection criteria for estimating the breeding value of a trait. The individual criterion of selection for within herd improvement is advocated for traits with high heritability and for those expressed in both the sexes at an early age (growth rate). The pedigree selection can be done for traits of high heritability. The selection based on performance of other relatives _viz._ progeny and collateral relatives (sibs) is effective when heritability of trait is low, the character is sex limited and for the characters which are not expressed in the life of animal (carcass traits, lifetime production, life span, etc.).

(ii) Intensity of selection (i): This depends on proportion of animals selected (p). The R will be more when p will be small. This is because when few animals will be selected (small p), the selection differential will be high and hence R will be high. The S is directly proportional to p. The genetic change occurs if some of the best animals are selected and hence p should be less which will increase i.

(iii) Accuracy of selection (r_{AP}): The B.V. for a trait is not observed directly but it is predicted and its prediction requires more sources of information for its accurate prediction. Thus, the estimate of true B.V. should be accurate so as its correlation with phenotypic value (r_{AP}), known as _accuracy of selection_, is more accurate. The r_{AP} depends on the magnitude of the heritability because $r_{AP} = h$. The R will be more if heritability is high and also the selection will be more effective (accurate).

The accuracy of selection should therefore be increased which is possible by increasing heritability (by reducing the environmental variance, use of multiple records, adjustment of data, accurate measurement of data and analyzing the data based on contemporary group means), use of combined selection to estimate B.V., making selection based on future performance (M.P.P.A.) of more number of records, and by keeping a population of large size.

(iv) Population size: The population size influences the R by way of influencing the inbreeding and genetic drift. The inbreeding is unavoidable in a population of small size. The inbreeding increases homozygosity and hence reduces the genetic variability and also the performance level due to inbreeding depression. Secondly, the genetic drift arises in small population due to sampling of genes. This may cause the loss of favourable alleles from the population and hence will reduce the R. Not only this, but it will affect the limit to selection which will reach more quickly.

(v) Generation interval: The generation interval is the time period between two generations with respect to the same stage of life cycle. It is defined as the average age of parents at the time of the birth of their offspring which also become parents.

The genetic response per unit of time (year) depends on the generation interval. This is because the genetic gain is observed in next generation, as the difference in mean performance of two generations. Thus, genetic gain will be higher in a population that breeds with younger animals than the population that breeds comparatively at later age. The genetic progress per unit of time (year) is important than genetic progress per generation.

The generation interval is *species specific* and depends on the *sex of animal* (higher for males because more number of their progeny are test recorded which are produced at later age of sire), *breeding age of animals* (age of the animal when reproduction starts) and the *selection schemes* (selection based on more number of offspring or more number of records per animal verses selection based on one offspring per animal with single record).

23.3 CORRELATED RESPONSE

The genetic correlation between two characters results the change in both the characters due to selection for either of the character. The amount and direction of associated change in second character (Y) as a result of change due to direct selection in first character (X) depends upon the size and sign of genetic correlation between the two characters. This associated change in the second character (Y) due to direct selection in first character (X) is called the ***correlated response to selection.*** This correlated response (C.R.) is for the second character (Y) for which no selection was made where as the change in character X for which selection was applied is called the ***direct response*** (D.R.) to selection. Thus, the C.R. indicates the change in correlated character (Y) when selection is applied for another trait X which is genetically correlated with Y.

The associated change in character Y will be equal to the regression of BV of Y on the BV of X (b $_{YX}$). This regression is

$$b_{A(YX)} = \frac{Cov_{A(YX)}}{\sigma^2_{A(X)}} = \frac{\sigma^2_{A(y)}}{\sigma^2_{A(X)}}$$

Therefore, the C.R. in character Y will be obtained as:

$$CR_{(Y)} = b_{A(YX)} DR_{(X)}$$

$$= r_{A(YX)} \frac{\sigma^2_{A(y)}}{\sigma^2_{A(X)}} [i_{(X)} h_{(X)} \sigma_{A(X)}]$$

$$= r_{A(YX)} \sigma_{P(Y)} h_{(Y)} h_{(X)} i_{(X)}$$

Thus, CR can be predicted if genetic correlation between two characters and their heritability estimates are known. The magnitude of CR also depends on the magnitude of genetic correlation. The causes of genetic correlation also influence the CR. The CR will be decreased if the genetic correlation is more due to linkage of genes. This is because the linkage between genes is diminished by crossing over and recombination. This will decrease the magnitude of genetic correlation.

Use of Correlated Response:

(i) Estimation of genetic correlation: The CR can be measured by conducting two way selection experiments. The selection is done for character X in one line and for character Y in other line. The DR and CR for each character is measured. The realized heritability estimates are obtained for both the characters from the response to selection. The genetic correlation can be estimated as the square root of the

product of the ratios of CR to DR for the two characters as-

$$\left(\frac{CR_Y}{DR_Y}\right)\left(\frac{CR_X}{DR_X}\right) = \left\{\frac{[ih_X\ r_A\ \sigma_{A(Y)}]}{[ih_y\sigma_{A(Y)}]}\right\}\left\{\frac{[ih_y\ r_A\ \sigma_{A(x)}]}{[ih_x\sigma_{A(x)}]}\right\}$$

$$= r^2_A$$

and therefore, $r_A = \sqrt{\left(\frac{CR_Y}{DR_Y}\right)\left(\frac{CR_X}{DR_X}\right)}$

(ii) Prediction of performance in another environment: The performance levels of a character in two environments are taken as two different traits. The genetic correlation between them decides the magnitude of CR in the trait under second environment where the population has to live based on selection of the trait under the first environment. When the CR is low the improvement should be carried out in the environment under which the population has to live whereas in case of high CR the improvement can be made in any environment.

(iii) Improvement through indirect selection: The CR is used to apply the indirect selection for making improvement in the traits for which direct selection is not possible for any reason.

Chapter 24

Bases of Selection

The breeding value (B.V.) of an animal is estimated based on different sources *viz.* animal's own performance or its relative's performance for selection objective (character under selection). The relatives of the animal under selection may be direct relatives (ancestors and progeny) or collateral relatives (full-sibs, half-sibs, aunts, niece, etc.). The various sources of information based on which the breeding value of an animal under selection is estimated are called by various names like *basis of selection or aids to selection or selection criteria.*

Various Bases (Criteria) of Selection: The animal itself is the smallest unit of selection and improvement. The animals own phenotypic value of a character is used to estimate its breeding value of that character. This criteria of selection is called the "individual selection" or *performance testing* because individual' own performance (phenotypic value) is taken as a measure of the breeding value (genetic merit) of the individual. This is also known as *mass selection* instead of individual selection if the individuals are kept together *en mass* for mating, *e.g. Drosophila* and *Tribolium,* in a bottle without keeping records of individual mating. But when the mating is controlled (records are maintained) as in case of large farm animals and mice etc., the term individual selection is used.

The estimation of B.V. of an individual based on ancestors performance (parents and grand parents) is known as pedigree selection and based on collateral relatives that are known as family members (sibs) is known as family selection, whereas based on its progeny performance is called as progeny selection or progeny testing.

The breeding value obtained from various basis of selection is called as the *estimated B.V. (E.B.V.) or probable B.V. (P.B.V.) or predicted B.V.* This estimated B.V. is really the genetic worth or breeding worth of an animal which is due to the average effect of genes possessed by the animal for a character and half of this genetic worth is transmitted to the progeny. Therefore, the estimation of this breeding value must be accurate which affects the efficiency of selection and hence the genetic improvement.

Why and when the relative's performance is required: The basis of estimating the B.V. of an individual from phenotypic value of its relatives lies in the fact that the relatives share the same genes possessed by the individual, to a certain extent depending on the degree of genetic relationship. The different selection criteria other than animal's own performance (relative's performance) are required under the following situations:

(i) When individual selection is not possible. The following are the traits for which individual selection is not possible:

- Sex-limited traits.
- Traits which can not be measured in living animals, like slaughter traits, longevity and lifetime production traits.

(ii) When selection is required at an early age but the trait under selection is expressed at later age of the animal. Under such conditions, the part records (part lactation milk yield to predict lactation milk yield), FLMY to predict milk yield in later lactations in dairy animals; part year egg production to predict the annual egg production in poultry; birth weight to gain in adult body weight may either be used or the information from relatives may be used to predict the B.V. of the trait.

(iii) When the heritability of the trait under selection is low, the individual selection is not effective. In this case, the information on relatives (family members like sibs) gives more reliable estimate of B.V. of the individual.

Male selection is more important to bring genetic improvement than female selection, though most of the economic characters in farm animals are expressed in females (sex-limited). The importance of male selection is because of the followings:

- Dams leave limited number of progeny whereas sires leave more number of progeny. The use of A.I. and frozen semen technology has further extended the extensive use of a sire in many herds.
- Only few sires are needed for breeding a herd for which selection is more intensive among males than females, making the selection differential larger.
- Sire selection is equally important in grading up and crossbreeding programmes, besides within herd improvement.

24.1 INDIVIDUAL SELECTION

The breeding value (B.V.) of an individual based on its phenotypic value can be estimated for one character (single trait selection) or for more than one character (multi trait selection).

24.1.1 Single trait selection

The phenotypic merit of the individuals under selection is determined and compared with the average merit of other individuals of the population kept under similar environmental conditions and that too at the same time. Therefore, the individuals are ranked relative to others under similar conditions.

Estimation of breeding value: The breeding value (B.V.) of an individual is estimated from the phenotypic value of that individual as a deviation from the population mean (selection differential) times the heritability of the trait. The value so obtained is known as the probable breeding value (PBV):

$$\text{PBV} = \overline{P} + h^2 (P_i - \overline{P})$$

where, $\overline{P}$ = population mean

P_i = phenotypic value of the individual

h^2 = heritability of the character

The PBV is the genetic superiority or inferiority of the individual in comparison to the population mean. The non- additive gene effects and the environmental factors bring the PBV nearer to the population mean rather than phenotypic value of the individual.

The accuracy of individual selection (r_{AP}) is the correlation of breeding value with phenotypic value which is equal to the square root of heritability (r_{AP} = h). Therefore, the accuracy of predicting the breeding value of an animal based on its own performance will be increased with the increase in h^2 of the trait and the progeny performance will come close to the phenotypic value of the selected parent rather than population mean.

Thus, efforts must be made to get accurate and higher estimate of heritability. This can be done by reducing the environmental variations. The accuracy of individual selection can further be increased by estimating the h^2 based on repeated records of the individual. The use of repeated records is helpful for traits of low h^2. The individual selection is applied when traits are expressed in the individual and when the h^2 of the trait is high.

Advantages of individual selection

1. The information on individuals to be selected is easily available.
2. It can be applied earlier to progeny testing. This is used when pedigree information are not available.
3. The generation interval is shorter by this method compared to progeny testing.
4. This gives a direct estimate of B.V. rather than on the basis of relative's performance and hence it is more accurate.
5. This allows a greater selection differential because the B.V. of all the individuals can be estimated while in case of P.T. only a few individuals may be progeny tested as parents.
6. This method minimizes the environmental effects because the individuals for selection are tested in the same environmental conditions.

Limitations of individual selection: The individual selection can not be applied under the following conditions:

1. When the traits are not expressed by the individual (sex limited traits).
2. When the traits are expressed in later life of the individual or after death of the animal.
3. When the traits have low heritability, the individual selection is not effective. This is because the accuracy of individual selection depends on r_{AP} which is equal to square root of heritability. The phenotypic value is a poor indication of B.V. if h^2 is low. Thus, low h^2 reduces the accuracy of selection.

In view of these limitations, the individual selection particularly for dairy animals, the male selection is not applied because the most important economic characters are sex-limited (expressed only in females) and also have low heritability estimates.

Aids to individual selection: There are certain situations when the information of relatives is not available, and also the h^2 of character is medium to low. This embarrasses the situation making the selection more difficult based on individual information. Moreover, it may also happen that early selection is required before the final trait is expressed by the animal.

When the h^2 of a trait is low, the repeated records of same animal can be used to improve the estimate of real breeding value or real producing ability. The use of part records and of genetic correlation among the traits will be helpful when earlier selection is to be practiced for the traits expressed later.

(1) Multiple records: Many traits are repeated several times in the life of the animal *e.g.* milk production, lactation length in dairy cattle and buffalo, egg production in poultry in poultry and fleece production in sheep. It is advantageous to predict the B.V. of an individual based on the mean of several records of that individual, if such records are available. The records on the same animal at different times show variation due to changes in the environment which prevail at the time the records are made, or sometimes due to error of measurement. The environment may include the internal conditions also *e.g.* endocrine functions of the animal, health of animal, as well as external conditions *e.g.* feeding, management and climate. As a result of environmental effects, the genetic potential of the animal is not fully expressed in a single record. The mean records will then give the better estimate of the real producing ability of the animal. The heritability based on n records is higher than based on single record. The heritability of n record with repeatability r is:

$$h^2{}_n = \frac{nh^2}{1+(n-1)r}$$

The relative efficiency of selection based on repeated records is always more efficient than single record. This can be obtained as the ratio of gain from selection on mean of n records over the gain from single record selection and it will be

$$= \sqrt{\frac{n}{1+(n-)r}}$$

(i) The above quantity increases with n for fixed value of r. Therefore, the more

the numbers of repeated records are used, the greater is the efficiency in terms of gain per generation.

(ii) Secondly, efficiency increases with decreases in the value of r, for any given value of n. Therefore, as r increases the use of more than one record in selection is not advantageous. This is because high value of r indicates that the temporary environmental influences were not important and permanent effects are more important. As such if r is high there is no use waiting even for the second record.

(iii) The use of many records in selection is useful for the traits which are more variable from time to time in the individual's life. The use of mean of records in selection result much of the gain with the use of second record and further if r is small the gain from using third or fourth record may be considerable. The main advantage of this method is that it is used for the traits influenced more by temporary environmental effects which are cancelled out in taking the average of more records.

Disadvantages of the method based on n records

(i) It results in increased cost because the animal has to be kept till more than one record is available.

(ii) It increases the generation interval and consequently will decrease the response (genetic gain) per unit of time in comparison to that based on single record.

(iii) The selection differential is reduced because the variation decreases in calculating the mean.

(iv) The individuals with several records constitute a highly selected group, within which further selection will be comparatively ineffective, if continuous selection is carried out every time. Thus, the final relative efficiency of the two methods must be based on the annual genetic gain which is obtained by dividing the genetic gain (ΔG) by the generation interval. Thus, the increased generation interval by the method of using several records will reduce the annual genetic gain. Therefore, using n records for selection is less efficient in terms of annual genetic gain.

(2) Use of part records: The part record selection uses the production performance for part period (*viz.* first 12 months growth, milk yield before lactation is completed). This method reduces the generation interval. The genetic gain (ΔG) by use of part record will be:

$$\Delta G = bAP\ (Xp - \overline{X})$$

where, b_{AP} is the regression of BV of total production (A) on part record

(P), $Xp - \overline{X}$ is the selection differential of part record.

Thus, for greater gain by use of part record, it is important that ^rAP be positive and greater for which the part production must be positively and highly correlated with total or full time production/record. If the genetic correlation is negative, net result of part record selection can be the reduction in total production.

24.1.2 Multi-trait selection (Methods of selection)

The economic value of an animal depends upon several characters. This is known as overall performance of total breeding value (net breeding merit) of animal. For example, a dairy cow will be more economical to maintain if she produces more milk with higher fat content for a longer period during lactation and remained dry for a shorter time between two successive calving. On the other hand, if a cow produces more milk daily of low fat content with shorter lactation length and goes dry earlier during the lactation, if will not be economical to maintain. Likewise, more profit from pigs depends on their fertility (litter size), mothering ability, growth rate, efficiency of food conversion, carcass quality, and also the resistance to various diseases. The economic production in poultry is affected by the annual yield of eggs, their weight, and age at egg production starts etc. The traits of direct economic importance in sheep are wool clip, fertility and lamb weight. Thus, the net economic value of an animal depends upon several traits. It is, therefore, essential to estimate the total breeding worth (net genetic merit) of an animal based on several characters. In practice, the breeding merit of an animal is often determined for several traits simultaneously and not for a single trait. It is required to improve the overall economic value based for several characters. This is known as multi trait selection.

Requirement of multi trait selection: The following information are required for improvement of several traits-

- (i) The *economic value* of the traits. This is measured as the amount by which each unit of variation in it actually raises or lowers an individual's value, known as the relative economic value of the trait.
- (ii) The *genetic significance* of the traits in terms of the heritability of the trait and genetic correlations among the traits.
- (iii) The method of selection chosen and the *number of traits* to be included in selection criteria also influence the efficiency of simultaneous selection of several traits.

Methods of multi trait selection: Simultaneous selection for many traits can be applied based on individuals' own performance by adopting any of the procedure of selection. These procedures are known as methods of selection and these are tandem selection or independent culling or selection index (Hazel and Lush, 1942).

1. Tandem selection: In this method of individual selection the individual traits are improved successively. The selection is practiced for one trait only at a time till the satisfactory improvement is achieved. After improvement in this trait the selection is started for improvement in the second trait till the goal is achieved to the desired level. Thus, trait A is improved first, then trait B, and so on.

- (1) This method is less efficient than other methods. The genetic progress per unit of time method is less and that too for great efforts. The average genetic improvement per generation in each of n traits (which are independent and equally important) would be only $\dfrac{1}{n}$ times which would be made if it was the sole object of improvement over the entire period of selection.

(2) Secondly, it requires more time for improvement in all the traits because while selection is being done for one trait, the other traits must wait.

(3) The efficiency of this method depends on the genetic correlation among the traits under selection. A desirable genetic association would lead improvement in other trait while selection is made for one trait. The low genetic correlation will decrease the efficiency of this method because there will be on correlated response in other trait. An undesirable genetic correlation between two traits, in which selection for an increase in desirable direction in one trait results in a decrease in the other trait, would neutralize the genetic progress made in any one trait.

2. Independent culling levels (ICL) method: This method of individual selection involves selection for two or more traits at a time. A minimum standard (level) for each trait is fixed and every animal to be selected must met this minimum standard fixed for each of the trait under selection. An animal fails to meet the minimum standard fixed for any one of the trait will be rejected, irrespective of the fact that how good the animal was in other trait. Thus, an animal is rejected in spite of its being exceptionally good in one trait but if it was little poor in another trait. The procedure of this method can be understood by the following table, taking an example in buffalo:

Traits	Standard set	Buffalo No.	
		1	2
AFC (moths)	42	40	44
Milk yield (kg)	1900	1950	2100
Fat (per cent)	7	7	6.5

Under the use of ICL method of selection buffalo No. 1 will be selected and no. 2 will be rejected because no.2 fails to meet the standard for AFC and fat percentage, though it was much superior in milk production, thus, the buffalo superior in milk production will be lost from the herd.

Advantage of ICL

(i) *Selection for more than one trait*: This method is superior to tandem selection because the selection is practiced for more than one trait.

(ii) *Culling at early age:* The practical and economic advantage of this method is that it allows to cull the animal earlier which are inferior in early expressed traits.

Disadvantages of ICL

(i) *No compensation for other traits:* The first disadvantage of this type of selection is that it does not permit superiority in some traits to compensate for deficiencies in other, and there is a possibility to cull genetically the superior animals.

(ii) *Culling level*: The method involve a tedious work to determine the optimum culling level.

(iii) *More emphasis to early expressed traits*: In case early selection, the early expressed traits get more emphasis rather than actually more important trait which are expressed later.

(iv) *Selection intensity*: It is reduced with increase in the number of traits. The effectiveness of this method depends on the level kept for each of the traits. In case of keeping low levels, very few animals are culled and it affects the genetic progress. On the other hand, if the standards are kept high, it results in gradual extinction of population because it will find difficult to have adequate replacements. This type of selection is used in selection of cattle for show purposes. The animal is selected with excellence in type, colour and conformation traits ignoring its performance for economic traits.

Selection intensity under I.C.L.: The ICL method decreases the selection intensity of the individual traits and hence it decreases the S (selection differential). The decrease is along the increases in number of traits. The intensity of selection for

n traits which are equally important and independent is equal to $\dfrac{1}{\sqrt{n}}$ compared to selection intensity for only one trait.

The genetic correlations among traits also influence the selection intensity. The negative genetic correlation among traits decreases more selection intensity in single trait because animal superior in one trait may be inferior in other trait. The effect of positive genetic correlation is less on the decreases of selections intensity because animal superior in one trait may also be superior in another trait. The decrease in selection intensity for including more number of traits suggests a restriction to a minimum number of traits. This is because animals superior in most of the economic traits are rate.

When the traits are equally important, the same culling level ($P_1 = P_2 = \sqrt{p}$) for each trait is most efficient for every value of p. On the other hand, when one trait is more important than the other, the culling proportions differ for the two traits in such a way that the culling proportion for less important trait is less to maximize H. For example, when first trait is twice as important as second trait, the culling level are 19 and 1 for p = 0.80; 47 and 6 for p = 0.50; and it is 76 and 18 for p = 0.20.

3. Total score or selection index: In this method of selection, the selection is practiced for several traits simultaneously based on an index. The selection index is an index of the net merit of an animal for many traits and obtained by adding the score for each trait into one figure. The scores are given to each animal according to the degree of superiority or inferiority in each trait. The animal with the highest score is then selected for breeding. The weightage is given to each trait depending on the economic and genetic significance of the traits. The amount of weightage given to each trait in relation to other traits determines the influence of each trait on the final index. Thus, this method encompass all the advantage and disadvantage of an animal by permitting extra merit in one trait to off set slight defects in another.

Construction of index selection: The discriminant function is used to construct a selection index based on the principle of fitting a multiple regression equation for predicting a dependent variable from two or more independent variables. Such a selection index should combine the different traits in a way that maximize the probability of progress in the aggregate economic value. The selection index will take the following from as:

$$I = \Sigma bi \; xi = b_1 X_1 + b_2 X_2 + \ldots\ldots\ldots + b_n X_n$$

where, X_i represent the phenotypic value for different traits

  bi are the multiple regression coefficients for different traits

The b_i and X_i values are put up in the index to obtain the index values for each animal and the animals with top highest index value are selected.

Genetic basis of selection index: The estimation of b_i values has the genetic basis, as the equations used to solve the b_i values contain the economic and genetic significance of all the characters under selection. The overall net genetic improvement (H) for several traits is the sum of the genetic gains made for several traits. Therefore, $H = \Sigma_{i=1}^{n} Gi$

The G_i is the additive genotype of the total economic value which is estimated as the sum of the breeding (genotypic) values of all traits $\overline{G}(i)$ weighted by their respective economic values (a_i). Thus, the aggregate genetic worth (H) of the individual as described by Hazel and Lush (1942) is:

$$\begin{aligned} H &= a_1 G_1 + a_2 G_2 + \ldots\ldots + a_n G_n \\ &= \Sigma a_i G_i = \text{Genetic worth of animal or, aggregate genotypic value.} \end{aligned}$$

where, a_i are the relative economic values. These measure the amount by which profit is expected to increase for each unit change in X_i a, is defined as the increment in profit occurring from improving the trait by one unit, independent of improvement in other traits. G_i are the expected values of X_i due to additive gene effect.

Estimation of b_i values: The b_i values of selection index are estimated by setting and solving the simultaneous equation whose number is equal to the number of traits under selection. The simultaneous equations use the economic value of each trait, genetic variances and co-variances as well as the phenotypic variances and co-variances of all the traits:

$$b_1 X_{11} + b_2 X_{12} + \ldots\ldots\ldots + a_n X_{1n} = a_1 G_{11} + a_2 G_{12} + \ldots\ldots + a_n G_{1n}$$
$$b_1 X_{n1} + b_2 X_{n2} + \ldots\ldots\ldots + a_n X_{nn} = a_1 G_{n1} + a_2 G_{n2} + \ldots\ldots + a_n G_{nn}$$

The equation in matrix from can be written as:

$$bP = aG$$

Therefore, $b = p^{-1}aG$

where, b = column vector of coefficients to be calculated,

  p^{-1} = inverse of phenotypic variance and covariance matrix,

  a = column vector for economic weights.

  G = genotypic variance and covariance matrix

The normal simultaneous equations for n traits under selection are set as follows:-

I. $b_1 V_{p1} + b_2 \; Cov \; P_{12} \quad + \; _{+} \ldots + b_{nc} ov P_{1n} = a_1 V_{A1} + a_2 \; Cov A_{12} + \ldots + a_{nc} ov A_{1n}$

II. $b_1 \; Cov P_{21} + b_2 \; V_{P2} \quad + \ldots + b_n \; Cov P_{2n} = a_1 \; Cov \; _{21} + a_2 \; V_{A2} + \ldots + a_n \; Cov A_{2n}$

N. $b_1 CovPn_1 + b_2 CovP_{n2} + ... + b_n V_{pn}$ $= a_1 CovA_{n1} + a_2 CovAn2 + ... + b_n VA_n$

The variances and co variances (both genetic and phenotypic) and economic values of all the traits under selection are required to be determined and put into the above equation to solve for bi values.

Difference between S.I. and ICL: The selection index method differs from ICL in that *(i)* the culling levels are flexible, *(ii)* each trait is weighted by a score and these scores for individual traits are summed to a total score (index value) for each animal, which is taken as selection criterion *(iii)* superiority in some traits can make up deficiency in others, unlike ICL which discards an animal failing to qualify in one trait regardless of its superiority in other trait.

In ICL method, the culling levels may be set too high for one trait and too low for other. In selection index method there may be mistakes in estimating genetic parameters and economic value. Moreover, the value of Z/P (intensity of selection) is not properly taken because the value of i (selection intensity) is taken from statistical tables based on normal curve which are not possible always in biological data that show little skewness. If the curve is skewed with long tail of distribution towards low merit, the progress is faster with mild selection and less rapid with intense selection than in a normal curve. On the contrary, if the curve is skewed towards high merit, the progress is less with high selection but more with intense selection. Lastly, the selection progress is also affected by animal health which affects p, and also by the breeders' ability to record the observations properly.

Practical problems in constructing selection index: *There are certain practical problems in construction of selection index and to recommend in general for all herds and even in the same herd over time.*

(i) The informations required are not available or accurate enough.

(ii) The relative economic importance of traits varies from time to time and in different locality and relative economic value of each trait is difficult to establish. It requires information in the long time price average and cost of production of the traits. Sometimes the economic values changes due to change in market demand.

(iii) The estimation of genetic and phenotypic parameters is not an easy job particularly when the number of traits included in the index exceeds three or more.

(iv) The selection index constructed cannot in general be recommended due to the following reasons -

    (a) The genetic and phenotypic parameters differ in different populations due to different genetic constitution of the herds and different management practices.

    (b) The genetic correlations have large sampling errors mainly due to small herd size.

    (c) The different management practices and environmental conditions at different places may cause the variability of the traits.

(v) The net merit (aggregate genetic worth H) of an animal has been assumed as an additive linear combination of characters each weighted by its relative

economic value. This assumption may not true always, and the economic value of a character either may not be linear or may not be independent of the other character. The example of milk yield and fat content may be taken. The net return for increate in milk production is not linear and fat content depends on milk yield of the animal. It is, therefore, advisable to use a model which covers both additive and non-additive functions.

Efficiency of S.I. over other methods of selection: Index is a more efficient method of selection than tandem and ICL methods of selection.

(*i*) It allows the selection of animals superior in some traits regardless of their inferiority in other traits and thus superior genotypes are saved for breeding.

(*ii*) All traits do not have equal heritability and so the same intensity of selection will not be expected to give proportionate improvement for each trait.

(*iii*) The selection index gives appropriate consideration to the correlation among traits. It also results in more genetic gain for the time and efforts made. The advantage of index method increases with the numbers of traits. Thus, it requires less time to bring about an overall improvement.

Hazel and Lush (1942) compared the efficiency of 3 methods in terms of genetic gain, assuming that the traits under selection are independent, have equal heritability, economic weight and variances.

The index selection based on equally important and uncorrelated traits is $\sqrt{n}$ times as efficient as the tandem method, regardless of intensity of selection. The progress in any one trait is only $\frac{1}{\sqrt{n}}$ times as efficient by the index selection as if selection was applied for that trait alone. The index is more efficient than tandem, the relative efficiency of the index increase with n (number of characters). The index is more efficient than ICL, its superiority rises with increasing n but falls with increasing i (intensity of selection).

The ICL method is more efficient than tandem selection for each trait at a time, the relative efficiency increases with the number of traits and intensity of culling. But ICL method is less efficient than index selection except in some cases of earlier selection.

The ICL method is always intermediate in efficiency between tandem and index selection method. However, the p value influences the efficiency of all the three methods. The lower value of p leads to more progress by all the 3 methods and the increase in progress is more rapid with ICL than other two methods. The relative efficiency of ICL method is closer to that of tandem method with mild selection but tends to be close to the total scores method when the selection is very intense.

24.2 PEDIGREE SELECTION

The selection criteria based on performance of ancestors (parents and grand-parents) is called as the *pedigree selection*. The pedigree selection adds very little to the accuracy of estimating the B.V. of an individual if the information on individual is available. The significance of pedigree is decreased when information are available either on the individual or its family members (sibs and progeny).

Practical difficulties to use pedigree selection:

- The ancestors' records are always not available.
- The records may be faulty due to stray mating.
- The pedigree records are destroyed with passage of time
- Most of the characters have low heritability

MERITS OF PEDIGREE SELECTION

- It is less costly as only compilation of pedigree is required.
- Allows selection at younger age and provides first hand information.
- It is helpful in multistage selection.
- It is useful for sex limited traits and those expressed in later life or after death of animal.
- It is helpful when two individuals have similar performance but one belongs to a better pedigree.

Demerits of pedigree selection

- There is a disadvantage of using pedigree selection that all animals of similar pedigree are culled out in spite of the fact that an individual may be of good merit and free from recessive allele causing the defect.
- Some pedigree gets undue favour irrespective of the true merit of the individual. Better environment is provided to the progeny of favoured pedigree.
- It introduces non random biases because pedigree records are for different environmental conditions.
- Pedigree selection provides no basis of selection among individuals which are descendants of the same ancestor.
- The accuracy is usually low.

Breeding value based on pedigree records: The B.V of an individual from its pedigree records is estimated by the selection differential of its relative ($\overline{P}$) from their contemporary ($\overline{P_c}$) mean times the regression of the genotype of the individual on the phenotypic mean of the relatives (b_{AP}) and added to the mean of the contemporary group as:

$$P.B.V = \overline{P_c} + b_{AP} (\overline{P} - \overline{P_c})$$

The b_{AP} is taken as rh^2 where r is the coefficient of relationship between the individual and its relative. In estimating the PBV of individual, the record (s) of one parent or both parents or parents and grandparents can be used. The accuracy of selection is increased based on information of ancestors (parents and grandparents) combined with individual's own record than based on individual's own record.

Accuracy of pedigree selection: The regression coefficients obtained indicated that the selection based on the performance of one parent is half as effective (accurate) as that based on individual's own performance. The values of regression coefficients are further reduced by half for each ancestor further back to the individual. The information on records of both the parents plus all of the four grandparents give lesser accuracy of selection (not exceeding 0.71) than based on individual's own record. However, the information on ancestors combined with individual's own performance give little higher accuracy of selection than based on individual's own record. The accuracy of selection based on single parent never exceeds to 0.50 and that based on either two parents or on parents and grandparents to 0.70.

It can be concluded that pedigree selection is only useful to select the individual before its own record is available. The pedigree selection is less accurate than individual selection or P.T. based on 4 or more progeny.

The accuracy of pedigree selection is expressed by the correlation of the genotype of the individual with the phenotypic mean of the ancestor used in selection. The ancestor records add little to the individual's own record for the traits of low heritability because the accuracy is increased with the increase of h^2 of the trait. Thus, for traits of low h^2 the information from ancestors without individual's own record results in low accuracy than individual selection.

The accuracy of pedigree selection is generally low because of the followings:
 (i) The different possibilities of combinations resulting from segregation of genes.
 (ii) It is not possible to predict as to which half of the dam's genes (better or inferior) are transmitted to the offspring because of the halving process and sampling nature of inheritance. Thus, the best lactation performance of the dam in sire evaluation is not a correct method.
 (iii) The different environmental conditions prevail at the time of taking records of ancestors and the individual.

24.3 FAMILY SELECTION

The selection criteria to estimate the B.V. of an individual may be the information (performance records) of its collateral relatives. The more closely related collaterals to the individual are likely to have more common genes possessed by the individual and can provide more accurate information about the individual. The more closely collaterals can be grouped as full sib families and half sib families. The family means of these collateral relatives form the basis to select the superior individual. .The procedure to estimate the B.V. of an individual on the basis of family mean is called the *family selection or sib selection* depending upon the inclusion or exclusion of individual's own record in estimating the family mean. The selection criteria is called as *family selection* when the individual's own record is also included to estimate the family mean but the selection criteria is called as the *sib selection* when the individual's own recorded is not included in estimating the family mean.

24.3.1 Sib selection

It is the selection of an individual based on its sib performance. The sibs may be full sibs or maternal half sibs or paternal half sibs. Thus, sib selection is of two type *viz.* full sib selection and half sib selection. The sib selection is practiced for the following traits for which the measurements on the individual are not available or recorded-

- Slaughter traits
- Sex limited traits
- Threshold traits like disease resistance
- Traits with low heritability in species with high reproductive rate so as many sibs are measured in short time.

The full sib (F.S.) selection is more accurate than half sib (H.S.) selection. However, the HS selection is favoured for the following reasons

- The half sibs are easily available in more numbers than F.S.
- The rate of inbreeding is more for F.S. selection than H.S. selection. The inbreeding counter balances the effect of selection.
- The F.S. correlation is more likely to be increased by c-effects (common environmental shared by F.S.). The intra class correlation (t) is rh^2 for H.S. and $rh^2 + c^2$ for F.S. where c^2 is the added contribution of maternal or common environmental effects. This reduces the accuracy of F.S. selection.

Estimation of breeding value: The breeding value (B.V.) of individual based on sibs performance can be predicted from multiplying the regression coefficient of the genotype of the individual on the average phenotypes of the sibs by the selection differential of the sibs as:

$$P.B.V. = \bar{P}_c + rh^2 \frac{n}{1+(n-1)t} (\bar{P}_s - \bar{P}_c)$$

where, $\bar{P}_c$ = average of contemporaries

$\bar{P}_s$ = average of sibs, n = number of sibs

r = coefficient of relationship between sibs and individual

= ½ for FS and ¼ for HS, if no inbreeding

t = Intra class correlation among sibs = rh^2

= ½ h^2 for F.S. and ¼ h^2 for H.S., if c-effects are not considered

h^2 = heritability of the trait

$$h^2 \frac{n}{1+(n-1)t} = h^2_s = \text{heritability of sibs}$$

The expected response to sib selection is: $R_s = i\sigma_p h^2 \sqrt{\dfrac{nr}{n\,[1+(n-1)t\ ]}}$

The accuracy of sib selection is taken as the correlation between the genotypes of the individual and the phenotypic average of sibs and estimated as:

$$\text{Accuracy} = r\sqrt{\dfrac{nh^2}{1+(n-1)t}}$$

24.3.2. Family selection

The selection criteria are known as family selection when based on the performance of the sibs plus the individual's own record. The family selection like the sib selection is of two types depending on the type of sib *viz.* full sib family selection and half sib family selection.

The **B.V. of individual based on family mean** is estimated as (Lasely, 1972)

$$\text{P.B.V.} = \overline{P}_s + h^2 \left[\frac{1-r}{1-r}(\text{PI} - \overline{P}_s) + \frac{1+(n-1)r}{1+(n-1)t} (\overline{P}_s - P_{Cs}) \right]$$

where, $\dfrac{1+(n-1)r}{1+(n-1)t} h^2 = h^2$ of family mean

$$h^2 \frac{1-r}{1-t} = h^2 \text{ of within family mean}$$

The expected response to family selection (R_f) is estimated as (Falconer, 1965)

$$R_F = i\sigma_p h^2 \frac{[1+(n-1)r]}{\sqrt{n[1+(n-1)t]}}$$

The accuracy of family selection (after Lasely, 1972) is estimated as:

$$\text{Accuracy} = \frac{h[1+(n-1)r]}{\sqrt{n[1+(n-1)t]}}$$

Common environment (c – effects): The environment effects which are different for different families but same for all members of the same family are known as common environmental effects denoted as c- effects (Lerner, 1950). The family members share common environment during pre- and post-natal stage. The c-effects thus create resemblance within family members over and above the resemblance due to having common genes and this contributes to the variance between families. This increases the intra class correlation (t) among family members. The c-effects are more for F.S. than for H.S. The F.S. are affected by maternal effects *viz.* liter mates in swine and full sibs have the same intra uterine environment *viz.* all the daughters of a sire being born almost at the same time and being reared together are likely to be subjected to similar environment conditions like climatic conditions, feeding regime and management practices etc.

When environmental similarities (c-effects) are present among family members, the intra class correlation among the phenotypic values of family member (t) is increased equal to the amount of c-effects as $(t=c^2)$, where c^2 is the portion of the total variation caused by difference in c-effects among families. This makes the denominator larger and hence the regression is deceased. Thus the c-effects decrease the accuracy of sib and family average

Factors affecting accuracy of sib and family selection: From the perusal of the formulae of the accuracy of sib selection and family selection, it is clear that the accuracy depends on the h^2 of the trait, coefficient of relationship between sibs and individual (r) , number of sibs (n) and the degree of correlation (t) between the phenotypes of sibs. The values of r and t are affected by inbreeding and common environments, respectively.

(i) The family selection and sib selection are superior for traits of low heritability. The relative efficiency of family and sib selection compared to individual selection decreases with increase in heritability. The accuracy of family and individual selection is equal at low h^2 values. The accuracy of sib selection compared to individual selection decreases along increase in h^2. When h^2 is greater than 0.5 the selection on any number of FS is never as accurate as individual selection and same is true for HS selection when h^2 is greater than 0.25.

(ii) The large size family brings the mean phenotypic value close to the phenotypic value. The effect of family size (n) on the relative efficiency of family and sib selection compared to individual selection is more when t (correlation between phenotypes of sibs) is low. The accuracy of sib selection just equals that of individual selection based on σ FS and 26 HS when h^2 is 0.1.

The accuracy of sib selection based on one FS is 0.5 h and based on one HS is 0.25 h. The accuracy of HS and FS selection as well as family selection is increased with increase in n but never exceeds more than 0.5 for HS ($\sqrt{0.25}$) and 0.71 for FS ($\sqrt{0.50}$) selection irrespective of n (number of sibs) and h^2 of the trait. The family and sibs selection based on large families results in greater response.

(iii) The type of family (HS and FS families) in terms of r between sibs also affects the accuracy of selection. The more close family members (FS and HS) provide more reliable information for selection compared to remote families. Thus, FS selection is more efficient than HS selection.

(iv) The degree of correlation between the phenotypes of sibs (t) affects the accuracy of selection. The family selection is better (even better than individual selection) when the phenotypic correlation (t) is low. When t is low, the h^2 will be low with little resemblance due to common environment. This makes the denominator smaller and consequently the regression is increased. Thus, with low values of t the accuracy of family selection is increased. It can be said that c – effects decrease the accuracy of family average. Family selection is preferred to individual selection for characters of low h^2 and when c- effects are not important. Individual selection is better than family selection when t is larger than 0.25 for the reason that family selection acts only on the variance between family means while the individual selection operators on whole of the additive genetic variance.

Advantages of family selection: *(i)* The family selection can improve the characters of low heritability in species with high reproductive rates (pig, poultry) so as to get many sibs in a short time.

(ii) The family selection does not allow the generation interval to increase.

(iii) Family selection is a support to individual selection because it is better to select an individual from a superior family.

Disadvantages or limitation of family selection: The family selection to be effective requires large family size and more number of families to avoid inbreeding as well as to increase the intensity of selection. In view of this it can be inferred that:

(*i*) The family selection is costly of space particularly when the breeding space and testing facilities are limited. The limited facilities reduce the intensity of selection.

(*ii*) The family selection as a unit of selection results in inbreeding and thus limits the genetic diversity. This is because only few families represent the next generation.

(*iii*) The F.S. family selection can only be applied in species with high reproductive rates to get large family size.

24.4 PROGENY SELECTION

The selection criteria for evaluating an individual based on his progeny performance is known as progeny selection or progeny testing (P.T.). The P.T. is the most important basis of selection. The selection criterion is the mean phenotypic value of the individual's progenies compared to the mean phenotypic value of the contemporaries. The P.T. is regarded as a form of family selection because progenies are the family members of each other.

Genetic principles of P.T.: The P.T. stems on the principle of like begets like. Each progeny inherits one-half of the genes from each parent due to the halving nature of inheritance. This is taken as the transmitting ability of the parent and hence the breeding value (B.V.) of the parent is twice the mean deviation of the progenies from the population mean. The progeny of an individual is thus the index of parents' genetic worth.

Second principle is the sampling nature of inheritance. Chance at segregation may result in any one or few progenies receiving a good or poor than average sample of genes from its parent. Different genes are transmitting by a parent to its different progenies due to segregation of genes and the probability to receive exactly the same set of genes by its different progenies is very low for polygenic traits. This segregation is called the Mendelian error. Therefore, the estimation of B.V. of an individual (parent) based on one or few progenies may be misleading. Moreover, the polygenic traits are influenced by environmental factors and the different progenies of the same parent do not get same environment. Thus, the environmental factors cause differences in the performance of different progenies. Therefore, the chance at segregation and environmental factors causes greater deviation from true B.V. of a parent. The effects of these two factors are balanced out by estimating the B.V. with the increase in progeny size.

Superiority of progeny selection over other criteria: The *individual selection* is not possible for traits of low h^2, sex limited traits and the traits expressed after death of the individual. Most of traits in farm animals have low h^2 and are sex limited. Thus these traits cannot be improved through individual selection and hence the B.V. is estimated based on the performance of relative (dam, sister and progeny).

The *pedigree selection* based on the performance of dam is not a good selection criterion for the reasons of halving process and sampling nature of inheritance. It is not possible to know as to which half (better or inferior) of dam's genes are transmitted to the progeny. Moreover, the pedigree selection is effective for traits of high h^2. In pedigree selection, only one sample is tested.

The *sib selection* has also low accuracy due to sampling nature of inheritance. Every individual receives different set of genes from its parent and the probability to set same get of genes by the individual and its sibs is very less.

Finally, the *progeny testing* gives the best and most reliable information about the genetic merit of parent (individual). This is because the P.T. is based on mean performance of many progenies. This overcomes the limitation of Mendelian error of gene segregation and hence provides the true estimate of B.V. of an individual.

Estimation of Breeding Value based on progeny performance: The B.V. of bulls based on its progeny performance is estimated by multiplying the regression of B.V. of parent on the phenotype performance of its progeny with the selection differential of the progeny of ith sire from the mean of the contemporaries (Pi -Pc). The regression (b) is taken as

$$b = \frac{r\,nh^2}{1 + (n - 1)\,t}$$

$$= \frac{0.5nh^2}{1 + (n + 1)\,t}$$

$$= \frac{2nh^2}{4 + (n - 1)\,h^2} \text{ since } t = 0.25h^2$$

where r = coefficient of relationship between individual and his progeny
   t = Intra class correlation among progeny which is ¼ for H.S.
   n = No. of progenies

Thus, the B.V. is estimated as: B.V. $= \overline{P_c}\,\dfrac{2nh^2}{4 + (n - 1)\,h^2}\left(\overline{P_i} - \overline{P_c}\right)$

Accuracy of P.T.: The accuracy of P.T. is expressed as the correlation of genotype of parent with the average phenotype of its progeny (half sibs) and hence it is estimated as r_{GP}- which is

$$r_{GP} = r_{AP} = 0.5\sqrt{\frac{2nh^2}{4 + (n - 1)\,t}}$$

Advantages of P.T.

- The P.T. is a better method for sex limited traits, the traits with low h^2, and slaughter traits in meat animals.

- The P.T. is useful to prove a sire whether he is free from any recessive gene. To identify the individual for harmful gene, it is mated to its sibs or progeny.
- The main feature and principal advantage of P.T. is the increased selection intensity. This is because the progeny are also used a parent and await selection at the time of testing their own parent. Thus P.T. is a modified from of family selection.
- Its accuracy increases with increase in progeny number.

Limitations of P.T.

- The time and cost required is the main limitation. It increases the generation interval.
- Due to larger generation interval the genetic gain per year is low.
- P.T. is effective on adequate number of progeny and hence requires high reproductive rate
- It requires to progeny test and compares many sires to find which one is superior. Thus selection is needed among sires under test. Thus, there is low selection intensity.

SIRE EVALUATION

The result of P.T. are expressed in the form of an index which is the index of the genetic worth of the sire and such an index is know a *sire index*. Based on sire index a numerical value is obtained which indicates the production ability of the sire. Since a number of sires are progeny tested, it requires ranking all the sires for their genetic worth so as to select the best sire.

Biasness in sire evaluation: The biases in obtaining the accurate estimate of sire's B.V. arise due to genetic and environmental differences. The genetic biasness is created by genetic differences among herds and due to the differences in genetic merit (production level) of the mates of the sire. The major environmental biases are introduced by the different environmental conditions (feeding, management, climatic factors etc.) of different herds where the year and season of calving create minor biasness. In order to eliminate or reduce these biases, it is better to adjust the data before estimating the B.V of sires. Considering all these points, a number of sire indices have been proposed by different workers.

Sire Indexing Methods: The different indices developed are for two purposes *viz.* indices which simply rank the sires and the indices which provide the estimates of breeding value of sires. The B.V. is estimated for indexing in a single herd a as well as for indexing in many herds.

The following notations have been used in the construction of various sire indices:

$\overline{D}$ = Average of all daughter of a sire under test

C_D = Average of contemporary daughters of the sire

$\overline{M}$ = Average of mate's of the sire or dams average

C_M = Average of contemporary mates

$\overline{H}$ = Herd average

n = Number of daughters

b = 0.5h², intra-sire regression of daughters on dam

(i) Simple daughter average index: Edwards (1932) proposed an index as:

I = $\overline{D}$

This index is the simplest measure in a single herd under similar environment. It does not take into account of the production level of mates of the bull (dams of he daughters) and hence subjected to bias.

(ii) Equi-parent or intermediate index: Hansson (1913) proposed this index which is alsoknown as Yapp's index (Yapp, 1925) also known as Mount Hope Index because it was first used at Mount Hope Farm in 1928. This index is

I = 2$\overline{D}$ − $\overline{M}$ = $\overline{D}$ + (D − $\overline{M}$)

This index makes adjustment of the variation in production level of the dam (mates). However, it over corrects the production level of mates allotted to different sires. This index is based on the assumption that the dams and daughters are raised under similar conditions so that the daughter dam difference reflects the sire effect otherwise it will reflect management practices in the herd.

(iii) Corrected daughter average index: Krishnan (1956) gave the following index-

I = $\overline{D}$ − b ($\overline{M}$ − $\overline{H}$)

= $\overline{D}$ − 0.5 h² ($\overline{M}$ − $\overline{H}$)

This index is superior to first two indices because it corrects for the production level of dams allotted to the sire over herd average. Thus, daughters average is corrected for the differences in the production level of their dams. The correction factor is b (M − H).

Searle (1964) compared the records of daughters with those of their herd mates in the same year and referred to as contemporary comparison or stable mate comparison. The sire's merit was estimated as-

$$I = \frac{\overline{H} + 0.5\,h^2\,n}{[1 + (n-1)\,0.25\,h^2]\,(\overline{D} - \overline{H})}$$

$$= \frac{\overline{H} + 2n\,h^2}{[4 + (n-1)\,h^2]\,(\overline{D} - \overline{H})}$$

(iv) Contemporary daughter average index: The contemporary comparison reduces the environmental variation de to herd, year and season. The records of the daughters of a sire are compared with the daughters of all other sires in the same herd calved in the same month or within two months.

The sire index proposed by Sunderasan *et al.* (1965) is-

$$I = \overline{H} + \frac{n}{(n+k)} (\overline{D} - \overline{C}_D)$$

Where, k = constant based on sire error variance

$$I = \overline{H} + \frac{n}{(n+12)} (\overline{D} - \overline{C}_D) - b(\overline{M} - \overline{C}_M)$$

The following sire index also takes into account the number of daughters and environmental conditions-

$$I = \overline{H} + 0.5\, h^2 \frac{n}{[1 + (n-1)\, t]} (\overline{D} - \overline{C}_D)$$

(v) *Corrected contemporary daughter average index*: The following sire index as an extension of the contemporary daughter average index has been proposed. This index besides adjusting the number of progeny and period variation also adjust for the differences in production level of dams allotted to different sires. The index is -

$$I = \overline{H} + 0.5\, h^2 \frac{n}{[1 + (n-1)\, 0.25\, h^2]} [(\overline{D} - \overline{C}_D) - b(\overline{M} - \overline{C}_M)]$$

(vi) *Least square constants*: The sire constants are obtained by least square technique which adjust the data for environmental effects including the non-orthogonality of data. The index obtained is

$$I = \frac{2nh^2}{4 + (n-1)h^2} (s_i)$$

Where, s_i is the sire constant for i^{th} sire

(vii) *Maximum likelihood method and REML*: The least squares method minimizes the error variance whereas the maximum likelihood (chance) method estimates the parameters by maximizing the logarithm of the likelihood function. The likelihood function is the likelihood of simultaneous occurrence of observations (variables). However, the ML estimates are biased because no account is taken of the degree of freedom in estimating the variance components. The ML method was improved by a method known as restricted maximum likelihood (REML) which takes care of the bias in estimates as well as avoids negative estimates of component of variance (Searle *et al.*, 1992). The variance components by REML are estimated based on residuals calculated after fitting by ordinary least squares from the fixed effects part of the model. This maximizes a marginal maximum likelihood. This is also called as the residual or marginal maximum likelihood

(viii) *Best Linear unbiased prediction (BLUP)*: The BLUP method of estimation was developed by Henderson (1949, 1973). This method is more powerful than the conventional selection index approach. This provides directly comparable estimates of the average breeding value of groups of animals born in different years. It takes into account the complications of non random mating, sires from more than one herd, environmental trends over time, herd differences for B.V. of dams and bias due to selection. Thus BLUP takes account of the fixed effects such as herd–year–season and it is applicable when mixed models are used whereas BLUE is applicable when all effects are fixed. The BLUP eliminates the non genetic biases in estimating B.V Not only this but it also removes the genetic biases taking into account the effects of non –random mating, genetic merit of dams and selection.

(ix) Other methods for multi-herds: Some methods for evaluation of sires used in more than one herd have also been proposed. These are Stable-Mate Daughter Average Index, Contemporary Daughter Average Index and Corrected Stable-Mate Daughter Average Index.

24.5 COMBINED SELECTION

The selection of an individual on the basis of two or more sources of information (selection criteria) is called as the **combined selection** or index selection. It is better to select on the basis of an index combining information from various relatives (dam, sibs and or progeny) with or without individual's own record. This is done by the technique of multiple regressions. These multiple regression coefficients are to be used as the weighting factors. All information available about each individual's breeding value combined into an index of merit is the optimal procedure for selection. Here two situations have been discussed.

1. Combining family mean and within family deviation: It is obvious that the phenotypic value of an individual (P_i) is composed of following two parts –

$$P_i = P_f + P_w$$

where, P_f is the deviation of its family mean from the population mean,

P_w is the deviation of the individual from the family mean called as the within family deviation.

The weight given to these two parts decides the procedure of selection. When equal weight is given to both the parts it is known as *individual selection*. When family mean is considered as a basis of selection giving no weight to the within family deviation (P_w), it is known as *family selection* whereas the selection on the basis of the within family deviation (P_w) alone giving no weight to the family mean (P_f), is known as *within family selection* and the best individual from each family is selected. Lastly, when the two components (P_f and P_w) are given different weights, the selection is known as **combined selection** or *index selection*.

The best estimate of an individual's breeding value on the basis of its phenotypic value is h^2P where h^2 is the regression of breeding value on phenotypic value. Now considering the two parts of the phenotypic value (P_f and Pw) which are uncorrelated and give independent information about the B.V., the expected breeding value of an individual can be estimated from multiple regression equation as:
$$E.B.V. = h^2{}_f P_f + h^2{}_w P_w$$

where, the two heritabilites ($h^2{}_f$ and $h^2{}_w$) are the weighting factors

(partial regression coefficients of family mean and within family deviation).

2. *Combining information from individual and all kind of relatives:* There are situations when information from the individual, its parents, FS, HS etc. are available and when the character on the individual is not measured (like sex limited characters). The individual can be selected to combine all information with an index. This involves the matrix method and takes the form of a multiple regression of B.V. on all the sources of information. The index so constructed is the best linear prediction of an individual's B.V. The index can be represented as:

$I = b_1P_1 + b_2P_2 + b_3P_3 + \dots\dots$

where, b's are the weighting factors (partial regression coefficients)

P's are the phenotypic values of different selection criteria.

Simultaneous equations equal to the number of sources of information are set. If the sources of information are three *viz.* the individual, the parents (dam) and paternal half sibs, then three simultaneous equations will be as follows.

$b_1 V_{P1} + b_2 \text{Cov } p_{12} + b_3 \text{Cov } p_{13} = V_{A1}$

$b_1 \text{Cov } p_{21} + b_2 Vp_2 + b_3 \text{Cov } p_{25} = \text{Cov }_{A21}$

$b_1 \text{Cov } p_{31} + b_2 \text{Cov } p_{32} + b_3 Vp_3 = \text{Cov }_{A31}$

Where 1 = individual, 2 = parents and 3 = PHS

The b's can be estimated after putting the values of variances and co variances.

24.6 INDIRECT SELECTION

The concept of *correlated response* (C.R.) is extended to indirect selection. The *indirect selection* means the direction selection applied to the character (X) other than the one (Y) that is to be improved. For example, the improvement is required in character Y but the direct selection is not possible due to some reason (low h^2 of trait, not possible or difficult to measure the character, costly to measure the trait and the traits expressed at later ages or after death of animal). In this situation (when character Y can not be selected directly), it is better to select for the character X having desired genetic correlation with character Y, so as there is C.R. in character Y. Thus, direct selection is done for character X when change is required in character Y rather than to go for direct selection for Y.

Basis of indirect selection: It is the high genetic correlation between the character under direct selection (X) and the character for which no direct selection can be done but requires improvement (Y) and hence the C.R. is the basis.

Applications of indirect selection: The indirect selection is advantageous or applicable under the following conditions-

- When C.R. is higher than direct response (D.R.) due to high genetic correlation and high heritability of the character under selection (h^2_X) than h^2y.
- When it is more difficult to measure character Y than character X, or when it is costly to measure the character Y (efficiency of feed conversion) than easily measured character X (growth rate). Like wise, the sex ration in offspring is a parental trait and can not be changed applying direct selection but the selection for blood pH may produce a correlated change in sex ratio.

- When desired character is sex limited but correlated character is measurable in both the sexes, a higher intensity of selection is possible by indirect selection.
- Indirect selection can be applied to reduce generation interval, *e.g.* selection based on part year production in poultry for egg production (X) will lead to a rapid genetic gain in annual egg production (Y).

Measurement of indirect selection: The character which is under direct selection (X) is called the secondary character because the primary character is one in which improvement is required (Y) through indirect selection. The relative efficiency of direct and indirect selection for character Y can be compared by calculating genetic gain and finding out the ratio of two gains (Q) as:

$$Q = \frac{CR_Y}{DR_Y} = \text{Gain by indirect selection / Gain by direct selection}$$

$$= \frac{r_A \, h_X \, h_Y i \, _X \, \sigma_{PY}}{h_Y i \, _Y \, \sigma_{AY}}$$

$$= r_A \sqrt{\left(\frac{h_X}{h_y}\right)}$$

Therefore, when Q is greater than one, the indirect selection is more effective. This is possible if the secondary character (X) has higher heritability than desired trait (Y) and also r_A is high.

Examples of indirect selection: Some of the examples of indirect selection are-

- Selection of weaner fleece for increased adult fleece
- Selection for birth weight or weight at early age for increased adult weight
- Selection based on par year record
- Selection based on FLMY for increased lifetime or subsequent production in later lactations.
- It is required to improve feed efficiency in pigs but the amount of feed consumed is difficult to measure and hence selection for gain will give the idea from good - r_A that as the gain is increased the feed conversion becomes better (less kg feed per kg gain)
- Selection for yearling body weight in sheep to increase fertility (lambs born) because the fertility has low heritability for which direct selection for fertility will be slow.

24.7 SELECTION FOR COMBINING ABILITY

The genotypic value of a quantitative trait has two component values *viz.* additive genetic value (breeding value) caused by additive effect of genes and non-additive genetic value known as gene combination value (GCV) caused by interaction effect of genes. These two component values (B.V. and GCV) produce the corresponding two components of genetic variation (additive genetic variance and non-additive genetic variance) in a random mating population. This total genetic variance present in a random mating population is distributed if the mating is non-random (inbreeding and out breeding). The distribution is made between lines and within line genetic variance on inbreeding whereas it is distributed between crosses and within cross genetic variance on crossing the lines.

Concept of GCA and SCA: When a line is crossed with several other lines, the total genetic variance between crosses can be partitioned into two observational components expressed as combining ability *viz.* general combining ability (GCA) of the lines and specific combining ability (SCA) of the cross.

The crossing of a number of lines among themselves in all possible combinations is called as the *diallel cross*. The analysis of a diallel cross provides the information on the nature and amount of genetic parameters as well as on the GCA of parents and the SCA of the crosses.

The mean performance of a line in all of its crosses with other lines (F_1's produced by crossing a line with several other lines) being expressed as deviation from overall mean of all the possible crosses is called the GCA. Therefore, each line has its GCA.

The cross of any two lines is expected to have its mean value equal to the sum of the GCA of both the lines. However, the mean value of the cross may deviate from this expectation due to the effect of non-additive gene action. This deviation is specific to any specific cross and hence called the SCA. Thus, *SCA is the performance of a particular cross of two lines expressed as deviation of this cross from the sum of GCA of the two lines.*

The GCA's are the main effects due to genic values which cause differences in GCA. The SCA is due to the interaction effect (dominance, over dominance and epistasis) which causes differences in SCA. This is the genetic basis of SCA. The deviation of true mean of a cross (X_{ij}) between any two lines (i^{th} and j^{th}) will contain the sum of their GCA's and SCA. Thus,

$$X_{ij} - \overline{X}$$
$$= GCA_i + GCA_j + SCA_{ij}$$

where, X = overall mean of all crosses (involving all lines in all possible combinations, general mean of all crosses in a diallel cross)

$$GCA_i = X_i - \overline{X}$$

$$GCA_j = X_j - \overline{X}$$

$$SCA_{ij} = X_{ij} - (\overline{X} + GCA_i + GCA_j)$$

The GCA's and SCA are uncorrelated. The between cross total genetic variance will thus be as: $\sigma^2_X = \sigma^2_{GCAi} + \sigma^2_{GCA\,j} + \sigma^2_{SCAij}$

$$= 2\sigma^2_{GCA} + \sigma^2_{SCA}$$

The variance of GCA is the additive genetic variance ($V_A + V_{AA}$) in the base population which is due to differences in additive gene effects whereas the variance of SCA (($V_D + V_{AD} + V_{DD}$) is due to the non-additive gene effects.

Further the variance of GCA increases linearly with F while the variance of SCA is expected to increase with higher powers of F and so the variance of SCA is more important as a cause of variation among crosses at higher degree of inbreeding (F). Therefore, it is advisable to use highly inbred lines for selection among crosses to be most effective. This means that when non-additive gene action has more influence on a trait, the most effective procedure of improvement is to produce highly inbred lines and then crossing them to find that which lines combine best.

The analogy between the causal components and observational components of the genetic variance making up the between cross variance is given below-

Causal components of variance   *Observational components of variance due to:*

Additive genetic variance
$$= F\ V_A + F^2\ V_{AA} + \ldots$$
$$GCA = \sigma^2_{GCA}$$
Non-additive genetic variance
$$= F^2\ V_D + F^3\ V^{AD} + F^4\ V_{DD} + \ldots$$
$$SCA = \sigma^2_{SCA}$$

1. Selection for GCA: Breeding plans for additive genetic differences:

The additive gene action (AGA) is responsible for differences of GCA. The AGA is indicated when the trait has high h^2 and show little or no heterosis or inbreeding depression. This means that average values for the parental, F_1 and F_2 generation are all the same with some individuals being superior over others. These best individual are best because they have more contributing or plus genes that act additively and not because of heterozygosity. It is therefore, essential to find the genetically superior individuals and select them for breeding. Thus, selection should be based on the individual's breeding value, because the correlation between the performance records of parental inbred lines and crossbred progeny is generally high for traits with large additive gene effects. To make selection effective for AGA, the environment variance should be minimized by keeping the environment standard as much as possible equal for all individuals. The superior individuals (progeny of superior parents) then must be mated with superior. Thus, the selection for the improvement of GCA is through individual selection with pedigrees, progeny tests or family selection without inbreeding as a supplement to make selection more accurate. Individual selection alone is more reliable when h^2 is higher. It is usually more important than any other basis of selection if h^2 is much higher than 0.25 and if the trait can be measured early enough on the individuals which are to be selected or rejected.

On the other hand, when h^2 is low or when the trait is sex limited or when the trait can not be measured early, and when the trait can not be measured on living animal, the other basis of selection are more important. The selection without inbreeding is effective because the variation in GCA is attributable to additive variance in the base population from which the lines or families are derived. The GCA of all the available lines can be measured and some selection can be applied so that a portion of lines can be discarded on the basis of their own performance before crosses are made. This is possible because there is some degree of correlation between a line's performance as an inbred and its GCA. The selected individuals are mated to individuals of their own population to produce the next generation.

The selection for GCA is repeated and thus it is knows as recurrent selection for GCA (Jenkins, 1940).

2. Selection for SCA: Breeding plans to exploit NAGA

The selection for SCA means to take advantage of hybrid vigour when NAGA is important. Selection on the basis of individuality is not effective when traits are largely affected by NAGA but the improvement in such traits depends upon heterozygosity through cross breeding resulting in expression of heterosis. Thus, the ultimate goal is to produce breeds or varieties each of which is homozygous which will produce the desired phenotypic merit when crossed with each other. This is important when epistatic effects and overdominace is important. The seed stock lines which are homozygous (produced by inbreeding) are selected and crossed with each other to get some special combination of genes which produce the desired phenotypic merit. The non-additive genetic variation is very important for viability and fertility in all species of farm animals.

(i) *When the SCA is caused by dominance*: The dominance gene action can be exploited by two ways: (a) Producing homozygous dominant individuals for the pairs of genes which affect the trait, and (b) Producing heterozygous individuals.

The production of homozygous dominant individuals has two difficulties *viz.* to distinguish homozygous dominants from heterozygotes, and that the quantitative traits are polygenic which reduces the probability of getting all dominant genes into one homozygous strain.

The second method of producing heterozygous individuals can be achieved by producing two complementary homozygous lines. Though the production of two complementary homozygous lines is problematic but can be achieved by directed process of selection during inbreeding. This will create maximum differences of gene frequency in two complementary lines. These lines are then crossed to obtain hybrids. However, for polygenic traits it is not possible to determine the genotype from phenotype *i.e.* which individuals are complementary homozygous. The identification of the complementary homozygous individuals (*i.e.* those which are homozygous in opposite ways) is done by the method of finding lines or breeds that combine and nick best when crossed. This method is used for production of hybrid seed corn as:

(1) The homozygosity of all pairs of genes is increased by inbreeding and forming several inbred lines.

(2) These lines are crossed and tested for their combining ability that which produce the best line cross progeny.

(3) The lines producing most superior progeny will be homozygous in opposite ways for many pairs of genes and will give greater heterozygosity in the progeny.

(4) The progeny are sold and not kept for further breeding, and these lines are kept pure to cross again and again to produce the line cross progeny.

(ii) *SCA caused by overdominance and epistasis*: The most practical procedure is the formation of different families, or lines and crossing them to find those which result in greatest hybrid vigour. Two breeding systems have been proposed when heterosis is largely dependent on over-dominance. The first is the recurrent selection proposed by Hull (1945) and the second is reciprocal recurrent selection proposed by Comstock *et al.* (1949). Both these methods involve progeny testing to take advantage of over-dominance.

(a) The recurrent selection proposed by Jenkins (1940) for improving GCA in plant breeding was modified by Hull (1945) to utilize SCA as well as GCA. This is thus called as Hull's modification of recurrent selection. This is the recurrent selection to inbred tester lines because it produces stock which combines best with inbred tester line or recurrent selection for SCA. This method has been used in poultry breeding. In this method a highly inbred line is selected as a tester (tester line) known to have good GCA. A large number of individuals (females) of a line under testing are crossed with this tester line and the resulting progenies are evaluated. The female individuals giving best results are selected based on performance of their test cross progeny and inter-mated. These selected females are then mated with males of their own line to produce next generation of parents to be tested with inbred tester line. The crossbred progeny are not used for breeding. The selected individuals from source population are mated again and again among themselves to produce next generation of complementary stock and also with the tester inbred line. The cycle is repeated to several generations. The complementary stock will be homozygous recessive and hence their test cross progeny will show greater heterosis. By this method the GCA of only one line and the SCA of cross is improved while the GCA of inbred tester is already determined. This method has also a practical difficulty to produce and maintain the highly tester line. Therefore, usually less inbred tester line is used which goes on changing genetically in subsequent generations and thus optimum gain is not achieved. Therefore, another method of selection is required.

(b) The reciprocal recurrent selection implies progeny testing of each of the two lines by crossing with each other. The crosses are made reciprocally and the parents are evaluated on crossbred progeny performance. The best parents of both lines are selected whereas the rest parents as well as all the crossbred progeny are culled. The crossbred progeny are used only to test the combining ability of parents. The selected parent are re-mated to members of their own line to produce the next generation of individual which are to be tested by mating them with individuals of the second line produced in the same way. They are evaluated based on their crossbred progeny and the best parents of both lines are selected to produce next generation of parents to be tested. Thus, RRS is a kind of progeny selection in evaluating the parents based on the performance of their progeny and producing the next generation based on selective breeding. This cycle is repeated over and over.

The start is made with the two breeds, lines or varieties differing in gene frequencies. Secondly, deliberate inbreeding is avoided as far as possible. This can be done by using all of the females as parents in their own lines and to intensify the selection among males. The disadvantage of this method is that the performance of the lines for the character under selection is reduced. This is because this method is used for the characters which respond to inbreeding depression and heterosis under directional dominance. Thus, the change of gene frequency by selection is towards the extremes and so the mean value of the lines declines. That is to say the selection is made to make both the lines homozygous for individual genes for loci which exhibit over dominance. However, with reciprocal recurrent selection, the decline in performance is not so deleterious as that of the effect of deliberate inbreeding because the inbreeding affects all loci and so mean value of all characters showing directional dominance declines whereas in reciprocal selection only the character under selection declines.

Comparing the recurrent selection and reciprocal selection, the recurrent selection is superior at the beginning. This is because of more differences in gene frequencies of inbred tester and reference population to be tasted compared to that between the two segregating population used in reciprocal selection. But the recurrent selection is inferior if the loci with additive or semi-dominant gene action prevail. This is because it is difficult to differentiate between homozygous dominant and heterozygous in test cross progeny if one allele is dominant over others. Thus the desirable dominant gene can not be accumulated in the segregating population.

The RRS is used for egg production in poultry and also encouraging results were obtained with maize.

Chapter 25

Mating Systems
I. Inbreeding

. The random mating does not change the genetic structure of population and hence it does not fulfill the objectives of genetic improvement of livestock. The mating systems based on the genetic relationship among mated pairs, are effective to change the genetic structure of population and are better tools for genetic improvement of farm animals.

25.1 GENETIC RELATIONSHIP

The concept of relatives and genetic relationship stems on the common origin of individuals of a population. The individuals of a population have been descended (originated) from their common ancestor and hence share common genes to some extent. The individuals sharing common genes directly or through their common ancestor(s) are said to be *relatives*.

The common genes (genes in common) shared by relatives are the duplicate copies of the ancestral genes and hence identical in origin, known as *genes identical by descent*. The common genes (identical genes) received by two or more individuals from their common ancestor(s) cause the genetic relationship between them. Therefore, the basis of genetic relationship between any two or more related individuals is that they have the common genes that are identical by descent.

The *ancestor* of an individual is one that appears in the pedigree of the individual (X) and hence inherited some of its genes to the total genetic make up of the individual. The ancestor is also called as *descent*. The individual (X) of the succeeding generation which received genes from the ancestor (descent) is called as *descendant*. The *common ancestor* of an individual is one that appears in both the sire's and dam's pedigree of the inbred individual (X). Therefore, the common ancestor contributes (inherits) its genes to the inbred individual (X) through its both the parents (dam and sire).

25.1.1 Types of relationship

The proportion of the identical genes shared by the relatives decides the *degree of relationship* between them. The relatives and relationship are of two types depending upon the way they receive the common genes.

1. *Direct relationship:* This is the relationship when an ancestor and its descendents are related to each other on direct genes donor-recipient basis. The ancestor transmits the genes while the descendent (progeny) receives. Thus, there exists a relationship of an individual with its ancestors (parents, grand parents) and with its progeny (son, daughter, grand-son, grand-daughter). This is called as the direct relationship because the genes are transmitted directly from ancestor to the descen-dents. The relatives with this relationship are called direct relatives and they belong to different generations.

2. *Collateral relationship*: The two individuals may not be related directly but they are related through their one or more common parents (ancestors) in the pedigree, due to genetic contribution of a common ancestor. The relationship between two animals due to genetic contribution of common parent (ancestor) in their pedigree is called the collateral relationship and the individuals having this relationship are called the collateral relatives. The collateral relatives may belong to the same generation (full sibs, half sibs, cousins) or to the different generations (uncle or aunt and nephew or niece).

25.1.2 Concept of gene's identity

Receiving and sharing of genes in common by relatives has the primary genetic consequence. This results the progeny of relatives (inbred individual) to inherit the same genes from each parent to some extent and hence leads to an increase in the probability of the inbred progeny to become homozygous. Therefore, the inbreeding increases the homozygosiy and this increase is at the expense of heterozygosity. The increase in homozygosity is due to increase in identical homozygotes that carry genes identical by descent and this leads to gene fixation.

The concept of *genes identical by descent* is basic to study the genetic relationship. The two alleles at a locus in an individual may be different (heterozygous state) or similar (homozygous condition) in structure (nucleotide sequence), function and origin. The similarity of allelic forms of gene at a locus is of two type *viz. identity by descent* and *identity by state and function.*

Two related individuals (having common ancestor) may receive replicate (duplicate) copies of the ancestral genes and transmit these replicates to their progeny, if they are mated. Thus, the inbred progeny produced from mating of such relatives may receive both genes at a locus that are replicated copies of one and the same gene of the ancestor and hence both alleles at a locus carried by inbred progeny have same origin. Two alleles that have originated by replication of gene in the ancestor are called the *genes identical by descent* or simply identical or *autozygous genes* (Cotterman, 1940). The individuals carrying the identical genes by descent at a locus are called the *identical homozygotes or autozygotes.*

The two alleles at a locus in homozygous state may not be identical by descent but they are of independent origin, though they are identical in structure and function. Such alleles in homozygous state at a locus are called the *genes identical in state* if they are identical in structure and function but not the replicates of the same allele of an ancestor within the depth of the pedigree. Such identity of alleles is called as *functional identity*. The homozygotes carrying alleles identical by state are called as the *independent homozygotes or allozygotes*. Cotterman (1940) called the alleles of allozygous individual as *allozygous* or *genes independent by descent* but identical in state (structure and function).

The genes identical by state (allozygous) are also identical by descent but through remote ancestor because they are also the replicates of the ancestral genes of the several past generations beyond normal pedigree depth which is considered to define the inbreeding. Thus, pedigree depth is taken as a base to define the genes identical by descent. The two alleles at a locus are identical by descent is a probability statement.

The concept of genes identical by descent is useful to measure the degree of relationship between mated pair and to measure the degree of inbreeding so as to avoid or advocate the inbreeding in future generation.

25.2 MEASUREMENT OF RELATIONSHIP

The relationship between the individuals of a population is measured in terms of the common genes they share. The relationship is expressed as the proportion of identical genes possessed by the relatives. This is the extra similarity in genes the relatives have over and above that present in the base population. Wright (1921) developed a measure for expressing the degree of relationship between two individuals that are related to each other directly or collaterally. He called this measure as the coefficient of relationship. This is defined as the *probable proportion of genes that are same (identical by descent) in two relatives due to their common ancestry, over and above that present in the base population*. The coefficient of relationship is a measure of the similarity of genotypes in two relatives (X and Y) and denoted by R_{XY}. The R_{XY} is taken as the correlation between breeding values of two individuals indicating that how much more alike are the breeding values in two related individuals compared to any other individuals in the same population. Thus, it is a measure and indication of the extra similarity in genes (percentage of genes held in common) the two individuals have due to their common ancestry, over and above that present in the base population. This is expressed either in percentage or in decimal fraction (proportion) ranging from 0 to 1.0.

1. R_{XY} of Direct relatives: The halving process of inheritance determines the genetic relationship between direct relatives. This means that each parent transmit only half of its total genes to each of the progeny. Therefore, the genetic relationship between a parent and its progeny is 50 percent. This means that the parent and offspring contain half of their total genes in common (genes identical by descent). This is the genetic relationship between a parent and its progeny. Due to the halving process of inheritance the similarity in genes (genetic relationship) is reduced to half every generation between an ancestor and its descendents. Thus, the genetic relationship between an individual and its direct relatives is as given here –

Generation	Relatives	Genetic relationship*		
1	Parent (Father/mother)	$= \frac{1}{2}$	$=$	50 per cent
2	Grand-parent	$= \frac{1}{2} \times \frac{1}{2}$	$=$	¼ or 25 per cent
3	Great grand-parent	$= \frac{1}{2} \times \frac{1}{2} \times \frac{1}{2} =$		1/6 or 12.5 per cent
4	Great-great-grand parent	$= (\frac{1}{2})^4$	$=$	1/ 16 or 6.25 per cent.

The genetic relationship indicates the proportion of ancestral genes to the total genes of progeny of subsequent generation. In this way, the genetic contribution of an ancestor is only 1.5 per cent of the total genes of the descendant of 6th generation, and the contribution is less than one percent (0.78) of ancestral gene to the total gene pool of the descendant of 7th generation.

Based on halving process of inheritance, ½ is taken as the base term and the number of generations between any two direct relatives is taken as the exponent term (*n*) to estimate the genetic relationship between them. Therefore, the genetic relationship between any two direct relatives (X and Y) is obtained as:

$$R_{XY} = (\tfrac{1}{2})^n$$

Thus, the extent of genetic contribution of the ancestor to the total genetic constitution of descendent depends on the number of generations between them.

2. R_{XY} for Collateral relatives: The collateral relatives have common ancestor in their pedigree through which they receive same genes (common genes or genes identical by descent) from their common ancestor. The common genes possessed by any two collateral relatives through their common ancestor cause the genetic relationship between them. Now it needs to explain that any two collateral relatives are related to what extent.

(i) Half sibs: Take an example of the genetic relationship between half sibs which is 25 per cent. The half sibs have one common parent (either mother or father). The common parent transmits only half of its gene to each of its progeny. For example, the individuals X and Y are two half sibs having the same father. The two half sibs (X and Y) may receive from their father the same genes or different genes. The probability of having received the same genes by two half sibs is 50 percent and the probability of receiving different genes is also 50 percent. The father is heterozygous for many gene loci and hence will produce two types of gametes carrying different genes. Due to sampling nature of inheritance, it is expected that one-half of the common genes received by X from its father would have also been received by Y from its father. Therefore, due to halving process and sampling nature of inheritance, the combined probability of receiving the same genes by two half sibs (X and Y) from their father is ½ x ½ = ¼ or 25 per cent. Thus, $R_{XY} = 0.25$.

(ii) Full sibs: Now, take that X and Y are full sibs, having both the parents (sire and dam) common. As a result of one common parent (father, sire) the similarity of genes between two half sibs was 25 percent. In this case of full sibs, another 25 percent of their genes (X and Y) transmitted by their common mother (dam) should be identical. Thus, 25 percent of genes in the body cells of full sibs are identical from sire side and likewise another 25 percent would be identical from mother side. Therefore, the total probability of having received the identical genes by two full sibs from common parents (sire and dam) is 25 + 25 = 50 percent. The formula

of estimating the coefficient of relationship becomes as: Σ ($\frac{1}{2}$) ^{n} after adding the contribution of two common parents to the total gene pool of the progeny, where " means to add (sum) the contribution of both parents and n indicates the number of arrows between two collateral relatives through common ancestor. This can be made clear by drawing arrow diagram of the pedigree of two full sibs (X and Y) as:

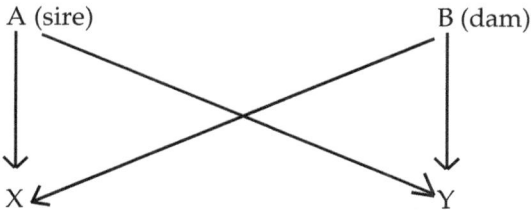

Pedigree no. 1 Full sibs

In this case, X and Y are two full sibs having both of their parents (sire, A; and dam, B) common. The two full sibs (X and Y) are connected by two arrows through their common sire (X$\leftarrow$ A$\rightarrow$Y) and likewise also by two arrows through their dam (X$\leftarrow$ B$\rightarrow$`Y). Thus, the contribution of common sire to transmit its common genes to the two full sibs is ($\frac{1}{2}$) ^{2} = $\frac{1}{4}$ and the contribution of common dam to transmit its common genes to the two full sibs is also ($\frac{1}{2}$) ^{2} = $\frac{1}{4}$. The contribution of both the common parents to transmit common gene will be added to obtain a combined estimate of the coefficient of relationship between two full sibs, making ($\frac{1}{2}$) ^{2} + ($\frac{1}{2}$) ^{2} = $\frac{1}{4}$ + $\frac{1}{4}$ = $\frac{1}{2}$. Thus, the relationship coefficient between two full sibs (R_{xy}) is $\Sigma(\frac{1}{2})^n$ = $\frac{1}{2}$. Here, n are the numbers of arrows connecting two collateral relatives through each common ancestor.

The estimation of coefficient of relationship between collateral relatives requires to prepare a table of all possible paths connecting two collateral relatives through each common ancestor including all individuals in between two relatives.

The coefficient of relationship between an individual and its collateral relatives are as given here:

Relatives	Coefficient of relationship (%)
(i) Full sibs (full brother or sister):	50
(ii) Half sibs (half brother or sister) :	25
(iii) Uncle / aunt, Niece/nephew (from FS family) :	25
(iv) -Do - (from HS family) :	12.5
(v) Progeny of FS from two families (Double first cousin, 4 common grandparents):	25
(vi) Progeny of FS from single family (2 common grandparents): (Single first cousin/first cousin/full cousin):	12.5
(vii) Half cousin (Progeny of half sibs, one common grandparent) :	6.25
(viii) Second cousin (progeny of full cousins) :	3.125
(ix) Third cousin (progeny of second cousins) :	0.78

The *coefficient of relationship* is the correlation between breeding values of two individuals. This is a numerical value which indicates that how much more alike are the breeding values of two related individuals compared to any other individual in the same population. Therefore, the *coefficient of relationship* is a measure of the percentage of common genes the two related individuals have due to their common ancestry.

Methods to Compute R_{XY}: The coefficient of relationship between two individuals (R_{XY}) is computed by the following methods -

1. Wright's Method of path coefficient: The coefficient of relationship is estimated from genic covariance and variance. The genic covariance measured from path coefficient is taken as:

$$\text{Genic Cov.}_{XY} = \Sigma(\tfrac{1}{2})^{n+n'} (1+ F_A)$$

where, $n+n'$ = number of generations connecting two individuals (X and Y) through common ancestor (A)

n = No. of generations from comnmon ancestor to X

n' = No. of generations from comnmon ancestor to Y

$\text{Genic } V_X = 1+ F_X$

$\text{Genic } V_Y = 1+ F_Y$

$$\text{Rxy} = \frac{\Sigma (\tfrac{1}{2})^{n+n'} (1+ F_A)}{\sqrt{(1+ F_X)(1+ F_Y)}} \text{—Collateral relatives}$$

$$\text{Rxy} = \frac{\Sigma (\tfrac{1}{2})^{n+n'} (1+ F_A)}{\sqrt{(1+ F_X)(1+ F_Y)}} \text{ Direct relatives}$$

$$= \frac{\Sigma (\tfrac{1}{2})^{n+n'}}{\sqrt{(1+ F_X)(1+ F_Y)}}$$

where, F_x = inbreeding coefficient of individual X,

F_Y = inbreeding coefficient of individual Y,

F_A = inbreeding coefficient of common ancestor

The above formula to compute the coefficient of relationship for direct relatives has been derived from the formula used for collateral relatives by multiplying both the numerator and denominator of the formula of collateral relatives with $\sqrt{(1 + F_A)}$. This is done because in case of direct relatives one of them (say X) is the ancestor (A) of the second individuals (say Y).

The above two methods of computing the coefficient of relationship between two individuals give the similar results.

Rules for tracing the paths: There are certain rules to trace the paths connecting two relatives through common ancestor. These rules are helpful to compute the contribution of each path when the relatives have more than one common ancestor and hence connected by more than one path through each common ancestor.

(i)   The path should connect the two related individuals either directly or through common ancestor. The path starts from one of the relative and ended to the second relative.

(ii) The path starting from one relative goes back to the common ancestor through various intermediate ancestors in the chain of segregation, if any and then the path takes turn to forward from common ancestor going to the second relative. The number of generations falling between one relative and common ancestor are denoted by n whereas those between common ancestor and second relative are denoted by n'. Therefore, the path first goes backward from one relative to the common ancestor and then comes forward (reverse) to the second relative.

(iii) Once the path started its reverse journey from the common ancestor, it will not go backward again. Thus, no reversal of direction in path is allowed except at the common ancestor.

(iv) No individual in the path is counted more than once. Thus, a path can not pass through the same individual twice.

Rules for considering the inbreeding coefficient of common ancestor as $(1 + F_A)$

There may be some pedigree in which the common ancestor is inbred. In this case, the term $(1 + F_A)$ is multiplied with the contribution of each path in which the inbred common ancestor falls as: $(\frac{1}{2})^{n+n'+1}(1+F_A)$. Thus, $(\frac{1}{2})^{n+n'+1}(1+F_A)$ is the contribution of each path.

(i)   The inbred ancestor in the pedigree should be the common ancestor. There are some pedigrees which had certain intermediate inbred ancestor (s) but they are not common ancestors. In this case, the inbreeding coefficient of intermediate inbred ancestor (s) is not considered. Therefore, unless the inbred individual in the pedigree is the common ancestor, its inbreeding coefficient is not multiplied with the contribution of the path.

(ii) The inbreeding coefficient of the common ancestor (F_A) gets multiplied with the contribution of each path and with all the paths in which the inbred common ancestor falls as the common ancestor and not as the intermediate inbred ancestor.

(iii) When the common ancestor is not inbred, $F_A = 0$ and hence $1+F_A = 1.0$. Therefore, it is then not multiplied.

(iv) The inbreeding coefficient of the parents (sire and dam) of the individual is not considered unless they have direct relationship and one become the common ancestor.

2. Coancestry method to compute R_{XY}: The coancestry of two individuals is the probability that two gametes taken at random, one from each individual, carry alleles that are identical by descent. This is the coefficient of relationship between two individuals. This has already been defined as the *proportion of genes that are same (identical by descent) in two relatives due to their common ancestry, over and above that present in the base population.* The relationship coefficient (R_{XY}) is taken as the correlation between breeding values of two individuals. This is the genetic correlation (R_{XY}) between two relatives and measured as:

$$R_{XY} = \frac{\text{Genic Cov (XY)}}{\sqrt{V_{G(x)} V_{G(y)}}}$$

The genic variance and covariance are computed from coancestries. This implies that coancestry is translated in terms of genic variance and covariance. The genic covariance between two individuals equals their coancestry.

Consider the following pedigree to compute the genetic correlation between two relatives (R_{XY}) as:

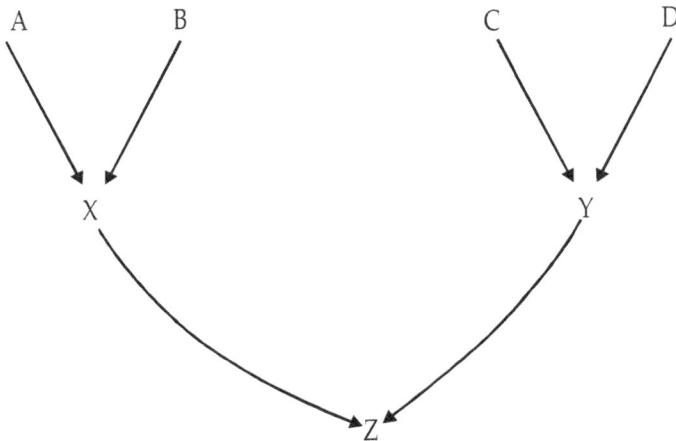

Pedigree no. 2

Genic Cov (XY) = Average covariance of either individual with
parents of other individual
= ½ (Sum of cov. of X with parents of Y)
= ½ (Cov.$_{XC}$ + Cov.$_{XD}$), if C & D are parents of Y
or ½ (sum of cov. of Y with parents of X)
= ½ (Cov.$_{YA}$ + Cov.$_{YB}$), if A & B are parents of X

Genic Var. (X) = Cov.$_{XX}$
= 1.0 + ½ (genic cov. of its parents)
= 1.0 + ½ (Cov.$_{AB}$) since A & B are parents of X
= 1.0 + 0 (if X is not inbred)
= 1 + F x (if X is inbred)

Genic Var. (Y) = 1 + F$_Y$

25.3 INBREEDING

The mating between relatives is known as inbreeding or genetic assortative mating. The inbreeding is a mating system when the mates are more closely related, up to 4-6 generations, than the average members of their population. The limit of relationship has been kept up to 6 generations, for the reason that the contribution of ancestral genes to the total genetic constitution of the descendent of 7th generation is less than one percent on the basis of halving nature of inheritance. Therefore, inbreeding is considered as the mating of animals related up to 4-6 generations. The progeny produced by mating of two relatives is called *inbred individual*. The extent to which an inbred individual carry the genes identical by descent is the degree of inbreeding which depends upon the degree of relationship between the parents of the inbred individual.

25.4 TYPES OF INBREEDING

The inbreeding is of two type *viz.* close breeding and line breeding.

25.4.1 Close breeding

This is the mating of more closely related individuals like parent-offspring mating (sire – daughter, son-dam), full sib mating. Sometimes, the half sib mating is also taken as close breeding. The close breeding has some *advantages* and hence it is done for some specific purposes which are:

- To develop highly inbred lines.
  - To discover undesirable and eliminate the recessive genes with the help of parent-offspring mating.
  - To get more uniform progeny

The close breeding has following *disadvantages*:

- Intensification of undesirable characteristics in the progeny
- Inbreeding depression, increase in reproductive failure and breeding problems
- Inbred are more susceptible to various diseases and adverse environment.

25.4.2 Line breeding

It is the mating of animals in which the relationship of an individual is kept as close as possible to particular ancestor in the pedigree, preferably male ancestor. This is the mating of distant relationship than those for close breeding. The descendants are mated to outstanding animals (sire) up to 3-4 generation. In this system, the mating may be between a fixed sire and his daughters and grand daughters. This develops the sire line as shown below-

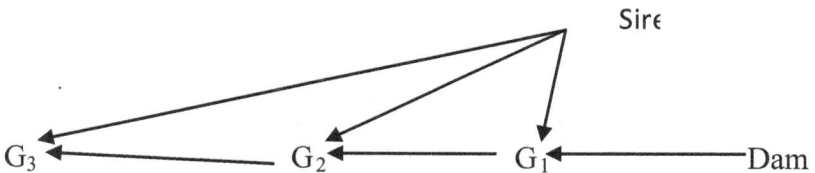

Fig. 25.1: Line breeding (mating of a sire to his offspring in successive generations).

The prime aim and advantage of line breeding is to maintain a high relationship to an outstanding ancestor for retaining its good proportion of genes. The relatives for their mating are selected because of a particular superior ancestor in their pedigree. The linebreeding thus incorporate into the linebred progeny the genes from the superior ancestor. The line breeding based on this logic can take various forms like parent-offspring mating in one generation or in subsequent generations continuously or after one-two generation as shown below-

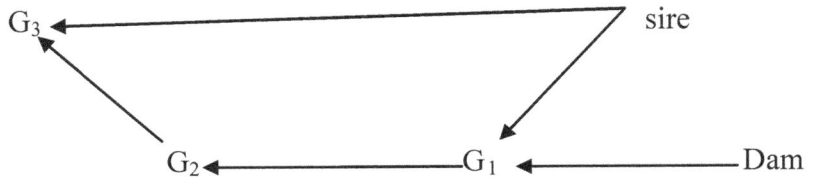

Fig. 25.2 Line breeding (mating of sire to grand daughter)

In view of various forms of line breeding, it is difficult to classify some mating either as close breeding or line breeding, *e.g.* sire-daughter mating is taken as close breeding but it concentrates effectively the hereditary material of the sire and hence logically should be called as line breeding.

Half sib mating as line breeding: The HS mating is practiced to avoid intense inbreeding and used as a system of line breeding when the ancestor is either dead or not available for breeding. In this situation half sib mating is the closest possible mating for line breeding.

Line breeding differ from ordinary inbreeding: The ordinary inbreeding and line-breeding differ with respect to the relationship of an individual (kept close in line-breeding) to a particular ancestor. Secondly, in ordinary inbreeding (other than line-breeding), the related parents are mated with no attempt to increase the relationship of the inbred progeny to any particular ancestor. Thirdly, the line breeding is practiced to retain or concentrate the genes of an outstanding animal.

When to use line-breeding: The line-breeding concentrates the inheritance of any one ancestor and hence it should be practiced in better than average herd, particularly when it is not possible to get good quality sires from outside. Secondly, there should be at least two sires in the herd in which the line breeding is practiced, to keep low the F_X. Thirdly, the line-breeding should be used in a purebred population of high genetic merit after identifying an outstanding individual based on progeny test. It should not be used in grade or commercial herds producing animals for the market. Line breeding is useful when the outstanding individual is dead or not available. Line-breeding must be directed through one or more of the sons of outstanding ancestor.

Why line-breeding is favoured: The line-breeding is favoured because it is less intense than close breeding and hence results in low ID as well as other dangers of close-breeding. Secondly, the line breeding perpetuates the special character of an outstanding animal or strain. This system is followed by selecting the best and culling the poor animals. In line-breeding, the sire daughter mating is not recommended but HS mating is done among the progeny of a particular sire. This is done to avoid intense inbreeding.

25.5 GENETIC EFFECTS OF INBREEDING

The effects of inbreeding on the genetic structure of population (genotype frequencies and gene frequencies) are as under:

1. Genotypic frequencies: The primary effect of inbreeding is that it increases the homozygosity. The frequency of homozygous genotypes is increased at the expense of the frequency of heterozygous genotypes. The inbreeding converts a proportion of heterozygotes into identical homozygotes.

Taking a single locus having two alleles (A and a) with their frequency as p and q, respectively, and inbreeding coefficient as F, the frequency of heterozygotes in an inbred population is equal to $2p_0q_0 (1 - F)$. Thus, the change in the frequency of heterozygotes due to inbreeding is $- 2 p_0q_0 F$. This is the loss of heterozygotes with inbreeding. The heterozygotes lost due to inbreeding ($2 p_0q_0F$) are converted into identical homozygotes. Thus, the rate of increase in homozygosity is $2pqF$ in each generation.

The conversion of heterozygotes into homozygotes ($- 2 p_0q_0F$) is equal for the two alleles (A and a) and hence the increase in AA homozygotes per generation is equal to pqF and the increase in *aa* homozygote is also equal to pqF. Therefore, the frequencies of 3 genotypes present under random mating ($p_0^2, 2 p_0q_0, q_0^2$) are changed after inbreeding and become as under:

Genotypes	Random mating (F=0)	With inbreeding coefficient F	
		Original + Change due to inbreeding	Co mplete inbreeding (F = 1)
AA	p_0^2	$p_0^2 + p_0q_0F$	p_0
Aa	$2 p_0q_0$	$2 p_0q_0 - 2 p_0q_0 F$	
aa	q_0^2	$q_0^2 + p_0q_0F$	q_0

The fraction of identical homozygotes equals to F and hence, F is the probability of identical homozygotes.

2. Gene frequencies: The inbreeding does not change the gene frequency, in spite of changing the genotype frequencies in favour of the homozygotes. This has been shown above. Therefore, the F is independent of gene frequencies.

Consequences of homozygosity: The increase in homozygosity due to inbreeding has the following consequences-

(i) The inbreeding changes the *genetic structure* of the inbred population by changing the genotypic frequencies, as shown above. With complete inbreeding (F = 0), the heterozygotes are completely eliminated from the population and the whole inbred population contains only the homozygotes. The frequencies of homozygotes for two alleles (dominant and recessive) become equal to the allelic frequencies of base population. The rate of decrease in heterozygosity depends upon the intensity of inbreeding. The rate of decrease in heterozygosity and consequent increase in homozygosity (F) per generation with full sib mating and also with parent-offspring mating is ¼ which means that ¼ of the non-identical genotypes are made identical. The net amount of increase in homozygosity is more in first generation and in early generations than in subsequent generations under any system of inbreeding. However, the rate of increase is same in all generations.

(ii) Frequency of recessive traits: Many recessive genes, being paired with

dominant genes remain hidden in out-bred population, are brought in homozygous condition by inbreeding. The inbreeding thus brings the recessive genes, present in rare frequency in a random mating population, to light by increasing the frequency of homozygous recessive individuals in the population. The recessive genes have unfavourable effect and hence the homozygous recessive individuals are undesirable.

(iii) The inbreeding leads to *gene fixation*, increases *prepotency* and increases the *phenotypic similarity* among relatives. The homozygosity is equally increased for dominant genes, thereby increasing the frequency of homozygous dominant animals which are more desirable than homozygous recessive animals. These inbred parents homozygous for dominant genes are more potent to transmit their characteristics to their progeny than non-inbred parents. This results the progeny to resemble their parents.

(iv) The inbreeding leads to *genetic differentiation between lines and genetic similarity within line*. The inbreeding has an effect on heritability to reduce it within lines because of decrease in additive genetic variance within line on inbreeding

25.6 PHENOTYPIC EFFECT OF INBREEDING

The inbreeding has adverse effect particularly on fitness traits like reproduction, vigour (vitality) and growth characters.

1. The inbreeding tends to depress the *growth rate* in farm animals. This results in small margin of profit when the live weight decides the profit in meat animals. The growth rate also affects the production and reproduction of animals.

2. The *reproductive efficiency* is reduced with inbreeding in farm animals. The reproductive efficiency is adversely affected in terms of delay in puberty and testicular development, reduction in gametogenesis (number of ova shed by females) and increase in embryonic losses, though with variable results.

3. The inbreeding into an increase in death rate and hence loss in *vigour*. The inbred animals are more susceptible to environmental conditions.

4. The inbreeding also causes the appearance of *genetic defects* which are mostly recessive in inheritance and remain hidden in outbred animals. The inbreeding brings them in homozygous state and hence abnormalities are expressed and identified.

5. The **phenotypic uniformity** among inbred animals is increased for the characters which are mono-factorial. The variations within inbred line are more due to environment than to genetic causes.

6. Change in Population Mean on Inbreeding: The population mean of productive traits is reduced due to inbreeding. This reduction in population mean is known as *inbreeding depression*. The population mean of a random mating population for single locus two alleles is: $Mo = a(p-q) + 2pqd$.

The M_O has two components attributed to the homozygotes which is $[a (p-q)]$ and to the heterozygotes $(2pqd)$. The proportion of two types of homozygotes in an inbred population is increased equally and hence do not contribute to the population mean whereas the proportion of heterozygotes in an inbred population is reduced to $2pq$ $(1-F)$ from $2pq$ in a random bred population, the population mean under inbreeding will be:

$M_I = a\ (p-q) + 2pqd\ (1-F)$

$\quad = a\ (p-q) + 2pqd - 2pqdF$

$\quad = Mo - 2pqdF$

This can be verified from weighted average of genotypic values, *i.e.* weighting the genotypic values with their frequencies as:

Genotypes	Freq. in inbred population	Genotypic values	Freq. x value
A_1A_1	$p^2 + pqF$	a	$a\ (p^2 + pqF)$
A_1A_2	$2pq\ (1-F)$	d	$d[2pq\ (1-F)]$
A_2A_2	$q^2 + pqF$	$-a$	$-a\ (q^2 + pqF)$

$\text{Total} = M_I = a\ (p^2 + pqF) - a\ (q^2 + pqF) + d[2pq\ (1-F)]$

$\qquad\qquad = ap^2 + apqF - aq^2 - apqF + 2pqd - 2pqdF$

$\qquad\qquad = a\ (p^2 - q^2) + 2pqd - 2pqdF$

$\qquad\qquad = a\ (p-q) + 2pqd - 2pqdF$

$\qquad\qquad = Mo - 2pqdF$

The quantity $(-2pqdF)$ as a result of the change of population mean on inbreeding is called the inbreeding depression. This indicates the followings-

- There is reduction in mean of an inbred population compared to the mean of a random mating population and it equals to the amount of $2pqdF$.
- The reduction in population mean is a function of gene frequencies, the degree of inbreeding and dominance among alleles.

7. Change of variance on inbreeding: The inbreeding besides reducing the mean value also changes the variance. The inbreeding influences both genetic as well as environmental variance.

(i) **Genetic variance:** The inbreeding differentiates the base population into distinct lines with extreme values of gene frequencies as 0 or 1.0, as a result of gene fixation. The different alleles become homozygous in different lines arising from base population. This results a decrease in the genetic variance within lines or families on inbreeding and the genetic differences between lines become more and the total additive variance (between and within, lines) increases. This means that the inbreeding leads to genetic differences between lines and genetic similarity within lines. The genetic variance within lines decreases because the gene frequencies at extreme values reduce the genetic component of variance. Redistribution of genetic variance is therefore one of the consequence of inbreeding leading to the increase of genetic component between the means of lines and a decrease of component within the lines.

(ii) **Environmental variance:** The genetic variance in inbred lines as well as in their hybrids (F_1) is negligible and hence the difference in phenotypic variance between them represents the environmental part of variance. The environmental variance differs between inbred and hybrids. The inbred are more susceptible to environmental factors than hybrids and hence the inbred individuals show more environmental variance than non-inbred. The environmental variation may be induced or variation may be in adaptability of individuals to different environments.

There are some characters like body temperature which are not influenced by environmental variations. This restriction of variation is known as homeostasis. There are some other characters like amount of sweat which vary with variation in ambient temperature. The relation of these two types of characters with fitness is different. The individuals for homeostatic characters are fittest by regulating their physiological functions in different environments while the individuals for second category of characters are fittest by varying according to environmental conditions. The homeostasis, which may be under the control of certain genes, may be reduced by inbreeding and hence it may leads to an increase in environmental variance of inbred. Secondly, the heterozygotes produce two enzymes which provide a biochemical diversity making the heterozygotes to be well buffered to different environments while the homozygotes produce only one enzyme.

25.7 GENETIC BASIS OF INBREEDING DEPRESSION

The inbreeding increases the homozygosity. The percentage of homozygous pairs of genes increases whereas the percentage of heterozygous gene pairs decreases under inbreeding. The inbreeding affects the phenotype depending on the kind of gene action involved in affecting a trait. The inbreeding causes a decrease in mean value (performance) of inbred population. The decrease in performance is known as *inbreeding depression* which is equal to 2pqdF. The genetic causes of inbreeding depression are as under-

(*i*) The change in mean due to inbreeding is greatest at intermediate gene frequencies (p = q = ½). This is because the inbreeding depression depends on the frequencies of heterozygotes which are maximum when p = q = ½. Therefore, if inbreeding is started in a completely heterozygous population (crossbreds produced by crossing two breeds or pure line), inbreeding depression will be maximum compared to that started in a population with high or low gene frequencies.

(*ii*) The inbreeding depression is linear to the degree of inbreeding, provided there is dominance.

(*iii*) The change in population mean under inbreeding depends on the type of gene action *viz.* additive or non-additive gene action (dominance, over-dominance and epistasis).

(*a*) In case of additive effect, there will be no change of mean on inbreeding. Therefore, no change in mean due to inbreeding is thus an indication of the additive gene action. *It is thus said that inbreeding is not always harmful.* The inbreeding is not harmful or detrimental for traits affected by additive gene action. Additive genes are made homozugous without regard to their effect. The additive genes are thus least affected by increase in homozygosity. Thus, homozygosity has no meaning for genes that act additively.

(*b*) In absence of dominance among alleles (d = 0), there will be no change in mean on inbreeding. Further, in case of polygenic traits, the dominance is not sufficient cause of inbreeding depression but it is the directional dominance. The inbreeding depression will occur only if the dominance effect of all or most of the loci are in the same direction (directional dominance). The dominance in both the directions in equal amount will not change the mean on inbreeding. The inbreeding will reduce the mean if favourable genes are dominant over their alleles which are not favourable. The reduction in mean of an inbred population is further due to the expression of recessive alleles, coming into homozygous condition, which remain hidden by dominant alleles in a non- inbred population. This results the average value of the population to reduce towards the value of recessive genotype. In case of complete homozygosity for dominant alleles that increase the value of the trait there will not be any decrease in the mean value of the character.

(*c*) The reduction in mean value is more in case of overdominance when heterozygotes exceed to that of the better homozygote. Since inbreeding increases the homozygosity at the expense of heterozygosity, a reduction in mean value of a trait affected by over-dominance of alleles is likely to occur if there has been no selection for more desirable individuals. The selection for desirable individuals means the selection in favour of heterozygotes and hence the increase in homozygosity will be at a lower rate in the inbred population. The inbreeding leads to a decline when dominance is positive (alleles with positive effects are dominant). This makes the heterozygotes to be superior than either homozygotes. Therefore, as long as d is positive, the inbreeding will produce a decline in mean value of the inbred population. Thus, the positive dominance (favourable alleles to be dominant – directional dominance) and over-dominance are the two possible causes of reduction in mean value on inbreeding.

(*d*) For polygenic traits, the epistasis without dominance does not produce a decrease in the mean. However, the dominance and overdominance with no epistasis decrease the mean on inbreeding proportional to F. The inbreeding effect may be quadratic in F if there is both dominance and epistasis. There is no inbreeding effect without dominance whether or not there is epistasis.

25.8 PHYSIOLOGICAL BASIS OF INBREEDING EFFECT

The inbreds are more susceptible to change in environment than outbreds. There is in general reduction in growth rate and fertility but an increase in death rates and susceptibility to diseases among inbreds. These effects seem to be the consequence of the combined effect of many pairs of recessive genes rather than at particular locus. The inbreeding changes the genotypic frequencies in favour of homozygotes and it leads to an increase in more pairs of recessive genes among inbred individuals, though the frequency of recessive gene does not change but they appear more often as homozygotes. The reason of detrimental effect of homozygous recessive genotypes may be the deficiency of essential enzymes that could have been produced by their dominant alleles or due to the production of abnormal proteins by recessive homozygotes. Not only this, but there could be some imbalance of some hormones among inbreds.

25.9 DISADVANTAGES OF INBREEDING (GENERAL)

- The frequency of undesirable recessive genes is increased
- Inbreeding results in inbreeding depression in growth, reproduction and vigour in terms of vitality

- Inbred animals are more susceptible to environmental changes

25.10 APPLICATIONS OF INBREEDING

The inbreeding is a useful tool in animal breeding. Though inbreeding results in inbreeding depression but there are certain cases when inbreeding is advantageous:

- Inbreeding is used to determine the genetic worth of an animal by mating a sire with his about 30 daughters. It will *test the sire for recessive alleles.*
- The inbreeding brings the undesirable recessive alleles in homozygous condition and it may thus help to uncover and eliminate recessive genes through selection.
- Inbreeding is used to *produce distinct families* within a breed. The selection between families for traits of low h^2 is more effective than individual selection.
- Inbreeding is used to *develop inbred lines* as a seed stock which can be crossed according to their combining ability. Thus, inbred lines can be used in crossbreeding programme. The highly inbred lines are also used in bioassay and other research work related to genetics, nutrition and physiology.
- Inbreeding should be used only for production of seed stock, to increase the purity of breeding animals.
- Inbreeding with selection had helped in the past to *develop several breeds* of livestock *viz.* Long horn cattle, Leicester sheep, Shire horse of Robert Bakewell and later on Merino, Rambouillet and South Down breeds of sheep were developed. Some strains of lab. animals (mice, guinea pigs, rabbit) had also been developed.
- Inbreeding also helps to know the *type of gene action* affecting a trait as the traits affected by additive gene action do not show inbreeding depression.
- Inbreeding also increases the homozygosity of favourable genes. The homozygotes with better performance can be selected for further breeding and these can pass on their superiority (own characteristics) to their progeny more successfully. This is called as the *prepotency* which is defined as the ability of an animal to stamp its own characteristics on the progeny. The prepotency depends on the homozygosity of dominant genes. Thus, the inbreds are more potent.

However, the inbreeding should be practiced with selection. Secondly, inbreeding should be practiced only for research work and for the production of seedstock, particularly the males. The inbred females should not be used in commercial herd because of their low performance than outbreds.

25.11 MEASUREMENT OF DEGREE OF INBREEDING

The primary consequence of inbreeding is the increase in homozygosity. The inbreeding increases the homozygosity due to the genes identical by descent (autozygosity). The genes identical by descent are the result of genetic similarly of the two gametes produced by two relatives mated to produce the inbred progeny.

The extent to which an inbred individual carry the genes identical by descent is the degree of inbreeding and measured by *coefficient of inbreeding*. The degree of inbreeding of an individual (degree of genes identical by descent) depends upon the degree of relationship between the parents of the inbred individual. The coefficient of inbreeding is a measure of the increase in homozygosity in an individual over the parents, due to inbreeding. The inbreeding coefficient also indicates the degree to which the parents of inbred individual are related.

Wright (1921) developed the procedure of path coefficient to measure the degree of inbreeding as the correlation between the genic values of the two uniting gametes and coined the term *"coefficient of inbreeding"*. The inbreeding coefficient of an individual (F) is a function of the common ancestry of its parents and expresses the extent to which the two parents of the inbred individual are related through their common ancestor. This is the degree of relationship by descent between two parents of the inbred individual. The common ancestry increases the probability of two alleles at a locus that they are identical by descent.

Another approach based on probability rules was given by Haldane and Moshinsky (1939), Cotterman (1940) and Malecot (1948). The inbreeding coefficient was defined by Malecot (1948) as *the probability that two genes at a locus in an inbred individual are identical by descent (autozygous)*. Thus, an individual with inbreding coefficient F has a probability equal to F that the two genes at a particular locus are identical by descent and a probability equal to 1-F is that they are not identical but independent in origin.

Methods to compute inbreeding coefficient (F_x)

The inbreeding coefficient (F) is a measure of increase in homozygosity in an individual over its parents, due to inbreeding and conversely a measure of decrease in heterozygosity (1 - F) compared to that of the parents or in reference of base population. The F is also an indication of the degree of relationship between the parents of inbred individual. Therefore, generally, the F of an individual (Z) is one-half of the coefficient of relationship between the parents of that individual (X and Y). There is thus similarity in calculation of these two coefficients. The inbreeding coefficient is estimated by calculating the relationship coefficient between the parents and dividing it by 2, provided neither of the parents nor the common ancestor are inbred. In this case, $F_Z = \Sigma(\frac{1}{2})^n / 2$. Thus, the numerator of the formula to calculate F is similar to that of the R_{XY}.

The following two methods to compute the inbreeding coefficient are in use-

1. Wrigh's method of path coefficient to compute F_X: Wright (1921) applied the approach of path coefficient analysis to compute the coefficient of relationship between two individuals (R_{XY}) and the inbreeding coeffi-cient of an individual (F_X):

$$Fx = \Sigma(\frac{1}{2})^{n+n'+1} (1+F_A)$$
$$= \frac{1}{2} \Sigma(\frac{1}{2})^{n+n'} (1+F_A)$$

where, $n + n' + 1 = N = $ Number of individuals in the path whose inheritance has been halved to produce the inbred individual.

n = No. of generations between sire of individual and common ancestor.

n′ = No. of generations between dam of individual and common ancestor

Σ = Summation over all the paths connecting sire and dam of the individual through the common ancestor and over all the common ancestors, if the common ancestors are more than one.

F_A = Inbreeding coefficient of the common ancestor for each path.

Rules for tracing the paths: The rules to trace the paths and the rules for considering the inbreeding coefficient (F) of common ancestor are the same as outlined earlier in estimating the coefficient of relationship. In estimating the F of inbred individual, the numbers of generations are counted connecting the sire and dam of the inbred individual through common ancestor.

Steps involved in estimating inbreeding coefficient:

(*i*) To find out the common ancestor and the common inbred ancestor

(*ii*) Tracing the different paths connecting the sire and dam of inbred individual through common ancestor

(*iii*) Contribution of each path is $(\frac{1}{2})^{n+n'+1}$ or $(\frac{1}{2})^{n+n'+1}(1+F_A)$ if the common ancestor is inbred

(*iv*) Summation of all the paths

2. Coancestry method to compute F: There is another way to consider the inbreeding coefficient of the progeny produced by mating of two relatives. The inbreeding coefficient of an individual depends on the amount of common ancestry (co-ancestry) of its parents, *i.e.* the common genes the parents have. Therefore, instead of inbreeding coefficient of the progeny, the degree of relationship by descent between its two parents (coancestry) is considered. This is called the *coancestry.* The *coancestry* of two individuals *is the probability that two gametes taken at random, one from each, carry alleles identical by descent.* Thus, the coancestry of any two individuals is identical with the F of their progeny.

The coefficient of relationship denoted as R_{XY} is the correlation between the genic values of two individuals and it is exactly twice the coefficient of consanguinity. Thus $R_{XY} = 2\varnothing_{XY}$. The inbreeding coefficient of the progeny (F_1) is identical to the coefficient of coancestry of its parents (f_{XY}).

Thus $F_Z = f_{XY} = \varnothing_{XY} = \frac{1}{2}R_{XY}$

. The coefficient of coancestry of two parents of an individual decides the inbreeding coefficient (F) of progeny and indicates the probability of identity of genes by descent in two individuals whereas the inbreeding coefficient indicates the probability of identity of genes at a locus in one individual.

The **coancestry is translated into genetic variance and covariacne** as:

(*i*) *Genic variance*: The genic variance of an individual is the genic covariance of the individual with itself. For example, the genic variance of an individual Z will be:

Var. z = Cov.zz

The genetic relationship of an individual with itself is 1.0. This is the genic variance of an individual which is 1.0 for non-inbred population but it is equal to I + Fz when the individual is inbred (Fz = inbreeding coefficient of the individual Z). Thus,

$$Var.\ z \quad = Cov.zz = 1.0 \qquad \text{when Z is not inbred}$$
$$= 1.0 + Fz \qquad \text{when Z is inbred.}$$
$$\text{Thus, } Fz = Var.z - 1.0 = f_{XY}$$
$$Var.z = 1.0 + ½ \text{ (Cov. of Z's parents)}$$
$$= 1.0 + ½ \text{ (Cov.}_{XY}) = 1.0 + f_{XY}$$
$$= 1.0 \text{ since Cov.}_{XY} = 0$$

(ii) Genic covariance: The genic covariance of two individuals (X and Y) is the average covariacne of individual with the parents of the other individual. Thus, genic covariance equals the coancestry. Therefore,

$$\text{Genic Cov }_{XY} = ½ \text{ (sum of Cov. of X with parents of Y)}$$
$$= ½ \text{ (Cov.}_{CX} + \text{Cov.}_{DX}),$$

where, C and D are parents of X.

$$\text{or} \qquad = ½ \text{ (sum of Cov. of Y with parents of X)}$$
$$= ½ \text{ (Cov}_{AY} + \text{Cov.}_{BY})$$

where, A and B are parents of Y.

$$Fz = fx_Y = ½ \text{ Cov.}_{XY} = ½ \text{ (genic cov. of the parents of Z).}$$

Chapter 26

Outbreeding

Based on genetic relationship, there is another mating system other than inbreeding, when the mated pairs are unlike in pedigree. This implies that the mated pairs have no common pedigree but they are genetically unrelated up to normal depth of pedigree and hence have the genetic effect opposite to inbreeding.

26.1. DEFINITION OF OUTBREEDING

The mating of unrelated individual is called as *outbreeding* or genetic disassortative mating. This is opposite or complementary to the inbreeding. It is defined in a comparative way that *outbreeding* is the mating of animals that are less closely related than the average of the population. But it is not easy to know the average relationship in a population. To overcome this, the term "unrelated" is used in place of "less closely related than the average of a population". Therefore, the term *outbreeding* is defined as the mating of unrelated animals. The unrelated animals are those which do not have a common ancestor for at least five generations in the pedigree.

26.2. FORMS OF OUTBREEDING

The unrelated individuals that are to be mated may belong to the same herd or different herds of the same breed (line crossing or out crossing), or may belong to different breeds (breedcrossing or crossbreeding) and even to different species (species crossing or species hybridization). Thus, out breeding can occur in several ways. There are mainly four different forms of out breeding which have been given as under.

26.2.1. OUTCROSSING

This is the mating system when the mating occurs among the unrelated individuals of the same breed. The progeny produced by out-crossing are called *outcross.* The mating of animals of the same breed or the mating of animals of two different lines of the same breed is also called as *linecrossing.* Thus outcrossing is also called as linecrossing.

The out-crossing system uses the selected best available sires on the females of a herd. This is the most common breeding system practiced by breeders and has been responsible for most of the changes in livestock breeds. The outcrossing system is effective to bring marked changes in livestock characteristics for the reason that genetic variability exists within breed. The out crossing within a herd by use of selected sires is also called as *selective breeding.*

The selection of sires and their use on the females of the same herd results in close herd and hence the replacements come from the same herd. An opinion is prevalent about the close herd that inbreeding increases rapidly in close herd. However, the increase in homozygosity within a herd or pure breed is reduced by use of selected sires and sometimes there is a selection of females too. The animals selected for breeding purposes are having less than average homozygosity because the performance of the selected animals is superior. This is because the inbreeding results in an increased homozygosity and tends to depress the performance. Therefore, the selected animals being superior in performance are expected to be less homozygous. Thus out-crossing with selection is capable for genetic change and improvement.

This system of selection and out-crossing is very effective for characters governed by additive effect of genes having high heritability. The out-crossing is practiced to exploit intra-herd genetic variability. This system is also effective that it does not allow the fixation of undesirable genes and hence brings improvement. However, the selection combined with out-crossing have an effect that the response to selection after 10-15 generation starts declining and in some cases after 20-30 generation of continuous selection the response is ceased. Thus a point of selection limit is reached. This point or level at which there is a failure to respond is often called as a *plateau* and the population is called *plateaued population.* Thus, plateau is a phenomenon when selection is not effective. The effectiveness of selection determines the usefulness of out-crossing. However, in large farm animals hardly any economically important trait has reached the point when selection is not effective system for genetic improvement particularly for traits which are largely under the control of genes with additive effects. The out-crossing combined with selection is effective to offset the adverse effect of inbreeding.

Top crossing: This mating system is a form of outcrossing and is like grading up system. It is the mating of females to the last male in the top side of the pedigree. It is also used to refer the mating of purebred (inbred) males with unrelated females. The concept of top crossing can better be understood with the help of an example. Suppose a new strain has been developed by crossbreeding (like Karan-Fries) and further improvement is required in this newly developed strain. It is then advisable to introduce pure exotic inheritance in the herd of new strain developed earlier. Therefore, the mating of females of new strain (Karan-Fries) with purebred sire of Holstein Friesian breed will be known as "top crossing"

26.2.2 CROSSBREEDING

The mating of animals from different established breeds is called breedcrossing or crossbreeding. The progeny produced is called *crossbred*. This system has major role in livestock improvement by developing new breeds of livestock. It is rightly said that purebreds of today were the crossbreds of yesterday.

Types of crossbreeding: The mating of individuals from different populations of strains, breeds or species is known as crossing. The crossing of different breeds (cross breeding) can be practiced in different ways depending upon the number of breeds and the manner of their crossing. The major forms of crossbreeding are the regular crossing and composite crossing.

1. **Regular or systematic crossing:** This is making of the same cross on regular basis to take the advantage of heterosis and complementarity. The crossing exploits non-additive gene effects through heterosis and the additive gene effects through complementarity when two or more characters complement each other. The amount of heterosis depends on the environment, genetic variability between two populations involved in crosses and the non-additive gene effects. The heterosis may be parental (maternal or paternal) referring to the performance of animals as parents and the individual heterosis referring to non-parental performance of the individual.

There are two basic methods of regular crossing. These are known as specific crossing and rotational crossing.

(i) **Specific crossing**: Depending on the number of breeds used in crossing and the manner they are crossed, the specific crossing is grouped as:

(a) **Two breed crosses**: Two pure breeds are crossed together. This may be restricted with crossing of only the purebreds (purebred crosses) or may be extended to crossing the crossbreds (*inter-se mating*) or with males of pure breed (*back crossing or criss crossing*). The *criss crossing* is similar to back crossing except that both the parental breeds (P_1 and P_1) are used alternately for mating of crossbred progeny in each generation.

(b) **Three breeds cross or triple crossing**: In this system of crossbreeding three breeds are used. The first cross animal (F_1) is mated to the animals of a third breed like (AB) female mated with males of third breed (C).

(c) **Four breed crosses or double two breed crosses:** This involves the crossing of crossbred females produced by crossing two breeds (A and B) with crossbred males produced from crossing another two breeds (C and D). Thus, mating is between (AB) x (CD). This is used to produce commercial progeny. This enables full exploitation of both *maternal and paternal heterosis* as well as *individual heterosis* for the reason that both parents are crossbreds.

(ii) **Rotational Crossing:** It the mating of three breeds in a rotational manner. The males of three breeds are used in regular sequence (rotation) in successive generations on crossbred females of the previous generation.

The major disadvantage of rotational crossing is that it does not allow any exploitation of complementarity. Thus, it is good to use when the population show little or no complementarity but show heterosis. However, it produces less benefit from heterosis. The advantage of rotational crossing is that all female replacements are obtained from crossing programme itself and do not therefore have to be specifically bred in a separate population as they do in case of specific crossing.

2. Composite crossing to produce synthetic breed: This is alternate to regular crossing. It is done by producing one or few crosses between two or more population to produce a single population having genes from each of the population. This single population of a mixture of various crossbred populations having various level of inheritance is called a *synthetic or composite population or breed*. This synthetic population is improved by selection within it.

In the past, this crossing system with selection has resulted in the development of new breeds *e.g.* Santa Gertrudis, Jamaica Hope, AMZ, Australian Friesian Sahiwal, Norwegian Red and white and Belmout Red breeds of cattle which were developed by selection within a synthetic. Similarly synthetic populations of cattle to evolve new breeds have been produced by crossing of Bos Taurus breeds with heat tolerance and tick resistance of Bos indicus breeds.

The theoretical expectation about heterosis is that $1-1/n$ of the heterosis present in F_1 is retained in F_2 and in subsequent generations (Dickerson, 1973) where n is the number of different populations contributing equally to the synthetic.

Advantages of Crossbreeding: It introduces new desirable genes in the herd. Secondly, it helps to evolve new breeds of livestock. A number of new breeds of different livestock species have already been developed. It also helps in the study of genetics of characters. Moreover, it exploits the hybrid vigour.

Disadvantages of crossbreeding: The breeding merit of crossbreds is reduced due to heterozygosity. It is costly as it requires maintenance of two or more breeds.

26.2.3. GRADING UP

This is one type of crossbreeding or breed crossing. In this mating system, the females of non-descript, scrub or native (deshi) type are mated with the sires of a pure breed. The crossbred females are back crossed to the pure bred sires to produce the progeny with 75% genes from the pure breed. The crossbred females in successive generations are mated with sires of pure breed. Thus, after six generations of crossing the graded females with the purebred sires produce the progeny which have 98.4% genes of the pure-bred. Thus, upgrading system of mating changes the genetic composition by transmitting good quality genes of an improved breed.

Grading up is *advantageous* to produce purebreds after a few generations. Secondly, it is less expensive. Thirdly, this system helps to prove the genetic potentialities of the sire.

This system (grading up) has *limitation*. The most important is that the genotype-environmental interaction puts a limit to use this system. This is because the purebred sire which gives good results in one environment may not give similar results in other environment. Mostly, the sires of a pure breed are brought from another environment.

26.2.4 SPECIES HYBRIDIZATION

The mating is between the animals of two species. This system is the widest (extreme) possible type of out-breeding. It is the least common form of outbreeding for the reason that animals of different species do not interbred. The following are the classical example of species hybridization:

Species hybrid obtained	Species crossed
1. Mule	Male ass (Jack) x females horse (Mare)
2. Hinny	Female ass (Jennet) x Male horse (Stallion)
3. Zebroid	Zebra-horse hybrid
4. Asbra	Ass x zebra (in Africa)
5. Pien niu	Cattle - yak cross in Tibet
6. Cattalo	American buffalo bull(B.bisson)x B.taurus
7. Jatsa(F_1 male)	Mithun x cow
8. Jatsamin (F_1 females)	Mithun x cow
9. Jechha (F_1 males)	Mithun x siri cow
10. Jessam (F_1 females)	Mitheun x siri cow

The species hybridization generally results in sterile progeny. The mules and Hinny are sterile. The crossing of Bison with B.taurus cow produces sterile males but the females are fertile. The cattalo was developed by backcrossing the females to bison and cattle.

26.3. EFFECTS OF OUTBREEDING

The out breeding is opposite to inbreeding and hence its effects on the genetic structure of population as well as on population performance are also opposite to inbreeding.

26.3.1. Genetic effects

The out breeding tends to increase heterozygosity and reduce homozygosity. This is because the mated pairs are genetically unrelated (having different alleles). The change in heterozygosity depends on the degree of relatedness of the mating animals. The crossbreeding results in a more rapid increase in heterozygosity than outcrossing or linecrossing. The important and peculiar characteristic of out breeding is that the maximum heterozygosity is attained in first generation outcrosses and goes on decreasing in subsequent generations by random mating of out-bred. The decrease in heterozygosity in subsequent generation is due to the segregation of genes which consequently increases the homozygosity. The two parents are homozygous (AA and aa) and their F_1 progenies are all heterozygous (Aa) while the F_2 progenies are produced in a ratio of 1 AA: 2 Aa: 1 aa. Further the crossbred of F_1 generation are likely to be uniform in traits related to physical fitness particularly when the parents are homozygous for different alleles of a particular pair. However, the out-bred cannot breed true because of heterozygosity and hence the selection is less effective among out-bred.

26.3.2. Phenotypic effect (Change of mean on crossing)

Heterosis: The out-bred are heterozygotes in which the effects of undesirable recessive genes are hidden by the effect of favourable genes (dominant). Therefore, the out-bred are more vigourus. The word heterosis was coined by Shull (1914) to describe the increased vigour of out-bred/crossbreds relative to their parents. One of the purposes of out breeding system of mating is to take the advantage of heterosis.

Definition of heterosis: The heterosis is defined as the amount by which the mean of F_1 generation, produced by crossing two breeds, exceeds to its better parent. Thus, the heterosis (H) is indicated as: $H = F_1 > P_1 > P_2$.

Positive and negative heterosis: The better parent is generally taken with greater mean and denoted by the letter P_1 where the poor parent is designated by the letter P_2. However, the better parent may also have low mean depending on the character under study like age at first calving, the cow with low AFC is a better cow and in this case P_2 having low AFC will be a better cow. When P_1 is a better parent (milk yield) the F_1 should have higher milk yield than its P_1 parent ($F_1 > P_1$). This is positive heterosis. On the contrary, when the P_2 is the better parent (AFC) the F_1 should have lower mean than its lower parent (P_2) and thus for heterosis to occur $F_1 > P_2$. This is negative heterosis. The positive heterosis is often called as hybrid vigour. In some cases, the crossbred progeny exceed even the better parent.

In animal breeding, the heterosis is taken as the superiority of F_1 progeny relative to the average of their parents (mid parent value). Thus, the heterosis is the amount by which the F_1 population mean exceeds to the mid-parent value. Therefore, the heterosis (H) is taken as the superiority of out-bred over the mean of their parents and hence:

$$H = F_1 > \frac{(P_1 + P_2)}{2}$$

The heterosis is the measure of the effect of out breeding.

Estimation of heterosis: The amount of heterosis is estimated by comparing the mean value of purebred and crossbred animals as follows:

Heterosis = mean of F_1 progeny – mean of parental breeds

This can also be expressed in terms of percentage as:

$$\% \text{ heterosis} = \frac{\text{mean of } F_1 \text{ progeny} - \text{mean of parent breeds}}{\text{mean of parent breed}} \times 100\%$$

Genetic basis of heterosis: The hybrid vigour is the opposite to the phenomenon of inbreeding depression. Thus, the heterosis is caused by heterozygosity created due to crossing of two purebred populations with different gene frequencies.

The magnitude of heterosis depends on the genetic diversity of the mated pairs and the type of gene action.

The genetic dissimilarity among mates decides the rate of heterosis. The more distantly unrelated mates have fewer genes in common and hence the outbred progeny is more heterozygous. The linebred progeny produced by two lines of the same breed is expected to be slightly more heterozygous (genetically divergent) than that of the average heterozygosity of the breed in general. The genetic diversity is more in F_1 progeny produced by mating of animals of two breeds. Further, the genetic diversity between a Zebu and European breed is more than the genetic diversity among different Zebu breeds of cattle or among different breeds of European cattle. Therefore, the crossbreds produced from European and Zebu breeds will be more heterozygous than the crossbred produced from mating between any two Zebu breeds or between any two European breeds. The magnitude of heterosis will be according to the genetic distance between mates.

Secondly, the non-additive gene action (dominance, over dominance and epistasis) causes the heterosis, like inbreeding depression. No heterosis is observed for traits governed by additive gene action. The traits showing heterosis are called often as heterotic traits.

When additive gene action affects the character, the mean of F_1 progeny is exactly the same as the mean of the parents if environmental variations are not taken into account. Thus, this type of gene action is not responsible for heterosis. The mean of F_1 progeny differ from the mean of the parents, if non-additive gene action is important. In this case the mean of the F_1 may even be higher than the better parent or lower than the inferior parent. When F_1 exceed the better parent it is called useful heterosis.

(i) **Dominance hypothesis:** The degree of heterosis depends upon the type of trait and type of mating. The early expressed traits, like survival and growth rate to weaning are more influenced. The traits which are more adversely affected by inbreeding also show greatest degree of heterosis. The mating of unrelated individuals (mating of inbred, two purebreds) shows heterosis in their offspring. There are two reasons for this.

The mated unrelated individuals are homozygous for different alleles. The animals of one purebred will be homozygous for some loci and the animals of other group (lines, purebred) will be homozygous for other loci. The crossing of lines or breeds (homozygous for different gene) produces the heterozygous offspring. The favourable dominant genes in the offspring will mask the unfavourable recessive. Thus the performance of the hybrid offspring will exceed to that of the average of the parents and sometimes exceed even to that of better parent. This is because each purebred will be homozygous for some loci (for favourable genes at some and for unfavourable genes at other). In case if one line or breed complements the other, the hybrid progeny will have more favourable genes than either parent. The effects of favourable genes are generally dominant to those of unfavourable genes. Thus, the performance of the hybrid progeny is superior to that of parental line. For example, for polygenic traits, one line or breed may be homozygous dominant for some pairs and homozygous recessive for another like AA, bb, CC and DD whereas the other breed may be respectively homozygous recessive and dominant at other loci aa, BB, cc, dd. On crossing them the resulting progeny (F_1) will carry dominant alleles at all loci and would be superior to both parents for that trait.

Secondly, the crosses between two lines or breeds having different gene frequencies (p and q) produce a lower frequency of recessive than the average of the two parents. This is because the frequency of the recessive in the cross will be qq' instead of ½ ($q^2 + q'^2$). For example if q = 0.2 and q' = 0.6, then qq' = 0.12 and ½ ($q^2 + q'^2$) = ½ (0.04 + 0.36) = 0.20. This is always true except for equal gene frequencies and so the performance of the hybrid is superior to the average of the parents.

(ii) **Over dominance theory:** The over dominance is the interaction between genes that are alleles and it results in the heterozygous individual being superior to the best homozygous parent. The crossbreeding result in superior animals if over dominance is important for the reason that the animal produced by crossbreeding has a maximum number of heterozygous loci. For example, for a gene locus, there will be three different genotypes such as A_1A_1, A_1A_2, and A_2A_2. If over dominance is present, the alleles A_1 and A_2 coming together (A_1A_2) produce a reaction which is not produced by them separately. The over dominance may be an important factor in heterosis for some character and in some species. Crow (1952) suggested that the dominance hypothesis can account for only 5% of the increase in performance in crosses of lines of corn. Thus, the dominance theory appears to be insufficient to explain heterosis in corn. However, heterosis observed in corn can be only due to over dominance at least at few loci affecting corn yield.

(iii) **Epistatic effects:** The epistasis is a phenomenon of interacting of genes which are not alleles. The epistasis is the effect of genes resulting from the new combination of genes from different loci. The different genes coming together in the hybrid interact with each other and produce greater effect than when they are alone in different parents. For example in plant, one dominant gene governs the length of internodes (long) and the other dominant gene governs the number of internodes. When these two dominant gene pairs are present in a single hybrid, they show their multiplication effect and result in offspring taller than expected from the average height of parental types having few but long internodes and having many but short internodes.

Physiological basis of heterosis: It has been concluded on the basis of some studies that crossbreds have a more efficient metabolic system. This may be due the reaction produced by the presence of genes in different combinations. The heterosis is the result of heterozygosity.

The heterotic effect seems to be the consequence of the combined effect of many pairs of genes rather than at particular locus. The outbreeding changes the genotypic frequencies in favour of heterozygotes and it leads to an increase in more pairs of heterozygous genes among outbred individuals and decrease in recessive homozygotes. The reason of beneficial effect of heterozygous genotypes may be through the production of essential enzymes or normal proteins that could have been produced by the dominant alleles.

26.4 COMPARISON OF HETEROSIS AND INBREEDING DEPRESSION

These are two genetic phenomena which are reverse to each other. The inbreeding depression is a consequence of inbreeding whereas the heterosis or hybrid vigour results from crossing of inbred lines or different races, breeds or varieties differing in gene frequencies.

The inbreeding tends to reduce the mean phenotypic value of characters closely connected to fitness in animals and lead to loss of general vigor and fertility. This loss or reduction in mean performance is known as inbreeding depression. The reduced fertility and vigour of inbred lines is restored on crossing the inbred lines. The progeny produced on crossing inbred or purebred lines (known as F_1) show an increase in performance (heterosis) of those traits that suffered a reduction from inbreeding.

The inbreeding depression is caused due to the increased homozygosity while the heterosis is the effect of increased heterozygosity.

The heterosis equals inbreeding depression in amount but with opposite sign. The inbreeding depression equals $-2pqdF$ and the amount of heterosis is dy^2 where d is the non-additive gene effect and y is the genetic diversity of mated parents (y $= p_1 - p_2 = q_2 - q_1 =$ difference in gene frequency). The y^2 should be equal to $2pqF$ to prove the heterosis equal to inbreeding depression. This can be proved from variance of gene frequency ($\sigma^2 q$) among lines as:

$$\sigma^2 q = \frac{pq}{2n} = pqF \text{ since, } \frac{1}{2n} = F.$$

Taking a population subdivided into many lines and that the pairs of lines are crossed at random, the mean squared difference of gene frequency between pairs of lines (y^2) will be equal to twice the variance of gene frequencies among the lines. Thus, $y^2 = 2 \sigma^2 q = 2pqF$. Therefore, the amount of heterosis (dy^2) equals to the inbreeding   depression ($-2pqdF$) with opposite sign. The observed inbreeding depression is linear to F.

The effect of inbreeding and outbreeding depends on dominance effect, gene frequencies and the degree of inbreeding ($-2pqdF$). If additive gene effects only influence a trait (d=0), no heterosis or inbreeding depression occur. Thus the loci without dominance and the traits which are highly heritable show no heterosis or inbreeding depression.

Further a change in the direction of more recessive alleles and the genes with intermediate gene frequencies (p = q = 0.5) have the greatest effect on the change of mean. However, it is not only the dominance effect but it is the directional dominance which influence heterosis. The absence of heterosis does not indicate the absence of dominance because the amount of heterosis depends on directional dominance. If some loci show dominance in one direction and others in other direction, their effects are cancelled out and as a result no heterosis is observed. The same is true for inbreeding depression.

The amount of heterosis increases with the degree of genetic differences between the two mated population and is limited by the barrier of inter-specific sterility. The amount of inbreeding depression, on the other hand, increases with the degree of genetic similarity between two populations mated.

Chapter 27

Breed Improvement and Development

The various animal husbandry programmes for livestock and poultry improvement have been launched by Govt. of India through different research and development organizations. As a result, the suitable *animal breeding techniques* of selection and multiplication of superior germ-plasm (sire evaluation, field progeny testing, ONBS, crossbreeding) have been developed besides development of infrastructure facilities for network of A.I., animal health coverage, improved management practices and milk marketing facilities (Cooperative dairies engaged in milk production, collection, processing and marketing) which have encouraged the farmers of the country.

As a result, India has come up on top position in milk production in the world. In 1950-51, the country's milk production was only 17.0 million tones which had increased to 7 folds of producing 121.8 MT of milk in 2010-11. After 1970, the annual growth rate of milk production of the country had been around 4 – 5 %. The improvement has also been made in wool production (27.0 million kg in 1951 to 45.6 million kg in 1995) and egg production (2881 million eggs in 1961 to 36600 million in 1998 and 65.5 billions in 2011). The annual per capita egg has increased from 7 in 1961 to 36 in 1998 and 66 in 2011.

The indigenous species are well adapted to local conditions. The zebu cattle, sheep and goat breeds of India have acquired, by natural selection, certain adaptation traits *viz.* high degree of heat tolerance, resistance to certain disease and parasites and ability to survive on low feed and fodder resources. However, based on population and production statistics, there is a wide gap between requirement and per capita availability of milk, meat, eggs and wool in the country. This is due to the reason that zebu cattle have low potential for milk production, the indigenous sheep have also low level of performance for fleece production, its quality and mutton production and also likewise is the poor production potential of indigenous goats, pigs and poultry. Thus, there is a scope for further increasing the production ability of our animal population.

The poor production potential for milk, meat, egg and wool of indigenous breeds of animals in India has been for the following reasons -Poor genetic potential and improper breeding policies

Poor environment which includes inadequate nutrition, lack of health coverage and poor husbandry/management practices.

27.1 BREED IMPROVEMENT

In view of poor genetic potential of indigenous livestock and poultry, it has been the challenge to the animal breeders to improve the production potential to a desired level without sacrificing the adaptational traits. The scientific breeding of animals based on formulation of *breeding plan* is required to change the genetic specifications of animals in accordance to human needs and hence to make the animals more useful to the man. The change in genetic constitution of population is thus the pre-requisite for the improvement and development of a new genotype (breed).

The *selection and mating systems* are two important components of breeding plan and the tools of animal breeder to work for bringing genetic improvement of a breed as well as to evolve a new breed to meet the human need.

The *population size* also plays its role mainly through selection. This is because the selection is more effective in a large population due to the reason of ample scope of selection in large population by culling the low producers, and secondly the large population is free from the effect of random genetic drift. Thus, there is low chance of being lost the best gene and the best genetic material and also there is no chance of inbreeding in large population.

Further, it is true that no mating system without selection is effective in changing the genetic structure of population. However, both selection and mating systems require the existence of genetic variability to bring genetic change. The combination of selection and mating system to change the genetic structure is called the breeding plan. The formulation of breeding plan involve the followings:

(i) To decide the type of production required (*selection objective) viz*. milk, meat, egg, wool etc. This depends on the choice of owner and the market demand.

(ii) The choice of breed as a foundation stock. This depends on the availability, utility and adaptability of the breed.

(iii) To make a decision whether genetic improvement is possible within herd / flock or introduction of superior genes from outside is required. The decision on this aspect (within herd improvement and introduction of superior genes from outside, migration) rests on the following factors:

(a) Performance level of the herd in comparison to other herd of the same breed or of other breed. This requires the accurate measurements and records, and the estimation of average values of economic trait (selection objective),

(b)   Genetic variability present in the herd / flock. This requires the estimation of heritability of the traits of importance, their relative economic values and genetic correlation among traits,

(c)   Comparative genetic value and adaptation of the introductions (G-E interaction),

(d)   Heterotic effect.

27.1.1. Selection

Now the important consideration is the *utilization of indigenous genetic resources* which implies the use of *breeding methods for optimum production* or to make genetic improvement to the desired level. This requires the formulation of proper *breeding plan* on scientific basis which involves selection and mating systems.

The genetic improvement through selection (response to selection, ΔG) per year depends on certain factors discussed already in chapter 23. It is required to give due consideration to these factors. In addition to these factors which affect response to selection, the following points should also be considered.

The selection criteria for males and females are little different. The culling of low producing dams, selection of replacement young females born to high yielding dams and breeding them (replacements) by proven sires have been recommended. Therefore, genetic improvement can be made by selection of outstanding dams of future sire and by selection of males that will produce future cows and bulls.

(*i*) **Selection of Females:** The selection of females should be done considering the following plan-

Basic requirements: The population under selection should have its *adaptability* to the environmental conditions under which selection is to be carried out. The *initial performance level* of the population under selection should also be considered. Moreover, the *vital characteristics* of foundation population (fertility, abortion, sex ratio, mortality rate, involuntary culling etc) and *genetic parameters* of important traits should be known. *Uniform environmental conditions* regarding feeding and management should be provided to the population under selection or the records should be adjusted for non-genetic factors

Culling of adult females: The adult females are culled for a number of reasons *viz.* reproductive problems (anoestrus condition, repeat breeding), utero-vaginal disorders (prolapsed of vagina/uterus, retention of placenta, metritis), udder problems (mastitis, teat block, blood in milk), other diseases and also for low production (voluntary culling).

Grouping of adult females: All the adult females at the farm are listed with their phenotypic values of economic traits and graded as elite (superior in most of the traits), superior (meeting standards equal to elite but inferior in one or more traits), good and poor. Adult females should be selected as per standard set for the traits and the number of young replacement(s) available to maintain herd strength.

B.V. estimation: The final selection of female is done on the basis of EPA or MPPA and expected B.V of trait under selection. In case of females of farm animals, the individual selection is advocated rather than based on performance of relatives. Further, the young females from inferior dams should be culled.

The *multi trait selection* is more effective instead of single trait selection. The index selection is more effective than other methods of multi trait selection.

The genetic improvement can be speed up by using *embryo transfer technology* (ETT) than normal way.

(ii) **Selection of males:** The individual selection of males in farm animals is not possible. Therefore, the males are selected on the basis of the performance of relatives.

Criteria of sire selection: Sire selection should be completed in 3 steps *viz.*

- Preliminary selection for breed characteristics and physical defects, body growth and health, maturity age, testicles, libido, semen production and quality as well as freezing ability of semen.
- Selection of young bulls based on pedigree. The expected predicted difference of young males based on dam's performance can be estimated as-

$$\text{E.P.D.} = \frac{0.5\,nh^2}{[1+(n-1)\,r]}\,(M-H)$$

where, n = number of records of dam of young male

Progeny testing of bulls for estimating the B.V. by any sire index method (See chapter 24 – Sire evaluation).

The *test mating* should be carried out for testing the bulls carriers of recessive genes so as such bulls are not used in PT programme. A sire can be proved free from recessive gene based on five normal progeny produced from mating of sire with homozygous recessive females.

The *effective use of PT* requires accurate progeny testing of bulls, intensive selection among PT bulls and extensive use of selected bulls. This requires large population and A.I. facilities.

It is better to run the associated or farmer's herd PT programme in view of small population in organized herds/flocks and the nucleus herd should be raised to provide the males of high genetic merit for breeding.

Mating Plan/System: A set of 10 bulls should be used in breeding the selected females, out of these 10 bulls, 2 should be tested and 8 should be under testing in each cycle of one or two calving interval, depending on herd size. The female should be mated with male as per the nature of genetic variability *viz.* additive or non-additive genetic. In case of high heritability, the breeding plan should be the individual selection with *selective breeding* (mating best with the best) and the information on pedigree, family and progeny tests should be taken as supplement to increase accuracy of selection or for sex linked traits or for traits expressed in later life or after death of animal. When the genetic variability of non-additive type is important, the selection alone is not important but selection for *combining ability* (inbreeding followed by selection or crossing the lines) should be practiced.

27.1.2. Introduction of superior germplasm

The introduction of new genes (migration) from outside is recommended when performance of a herd is lower than other herd(s) and the genetic variability within herd is very low.

The non-descript animals with low performance can be improved by upgrading. The genetic improvement of well known breeds can be done by selective breeding (within herd improvement as discussed above) or by introducing the genes of the same breed from another herd of high genetic merit (Out crossing) or of other breed (crossbreeding). The crossbreeding in the present day context is the migration / introduction of genes of genetically superior exotic breeds in indigenous breed. This exploits the breed differences and heterosis for the reason that there are large genetic differences between two types of breeds (Sub-species). Thus, the genetic superiority for milk production, fleece production, meat and egg production and their quality traits can be combined with good qualities (adaptive traits) of indigenous breeds through crossbreeding (complementarity in crossbreeding).

27.2 DEVELOPMENT OF SYNTHETIC POPULATION

It combines the inheritance of two or more breeds. The synthetic population can be produced by *inter-se* mating after having the desired level of exotic inheritance. The population is established by *inter se* mating of first crossbred generation (F_1), if equal proportion of two breeds is optimum. When higher proportion of exotic inheritance is desirable, the *inter se* mating is started after making backcrossing to exotic breed (forward cross) for one or two generations. But the *inter se* mating declines in performance in next generation because of the decrease in heterozygosity. The decline is more in a population produced by *inter se* mating of F_1 but the decline can be avoided by *inter se* mating of population with higher level of exotic inheritance. A synthetic population is stable in the first generation produced by *inter se* mating after which the genetic structure will remain constant in the absence of selection. The heterosis is exploited in individual, maternal and paternal traits in synthetic breed programme. Secondly, the population is self contained and needs no outside genetic material.

The amount of heterozygosity depends on number of breeds forming a synthetic population and the difference in gene frequency between breeds. The proportion of maximum heterozygosity retained in *inter se* population is:

$H = 1 - \Sigma p_i^2$ (Dickerson, 1973)

where, p_i is the proportional contribution from i^{th} breed

or, $H = 1 - \dfrac{1}{n} = \dfrac{n-1}{n}$ for n breeds each contributing equally $\dfrac{1}{n}$

$H = 2pq$ in case of two breed synthetic population

Where, p and q are the proportion of the two breeds.

The heterozygosity is maximum with equal proportion of breeds (p = q). However, the genetic distance among temperate breeds as well as among tropical breeds is less. Therefore, there is slight increase in heterozygosity by including another temperate breed in crossbreeding programme. This is the reason that little is gained by using more than one exotic breed and one native breed. The breeds to be crossed must be genetically different to exploit heterosis.

27.3 BREEDING FOR OPTIMUM PRODUCTION

The breed improvement programmes (selection and mating systems), in general, for utilization of indigenous species have been discussed above. However, it is not feasible to have a uniform breeding policy for different breeds of either species to bring genetic improvement in view of the followings:

  (i) Different utility of different breeds (milch, draft and dual purpose dairy animal; milch, meat, coarse & fine wool and pelt production sheep; milch, meat and mohair/pashmina production goats; egg, meat production poultry) and large number of breeds ,

 (ii) A larger proportion of population (70 – 80 %) being non-descript with their low performance potential compared to well recognized breeds,

(iii) Different agro-climatic regions,

(iv) Different socio-economic level of farmers (small, marginal and landless).

27.3.1 Breeding of Zebu Cattle

The following breeding strategies have been recommended for zebu cattle:

1. Grading of non-descript cattle with improver breeds: The bulls to be used for grading up should be the progeny of superior dams having lactation yield of more than 2000 kg for those belonging to milch breeds and around 1500 kg for dual purpose breeds. The different improver breeds should be used for different states *viz.*

- Sahiwal and Tharparkar bulls to be used on local cattle of Hariana, Pb, UP, MP and WB;
- Tharparkar bulls for Kerala Tripura, AP, Bihar and Rajasthan; Gir for Gujrat, Rajasthan and MP;
- Kankrej for local cattle of Gujrat; Ongole for MP and AP local cattle;
- Kangayam for Kerala local cattle;
- Hallikar for TN and AP local cattle;
- Hariana for local cattle in almost all the states except south India;
- Red Sindhi bull to be used on local cattle of HP, J & K, UP, Bihar, Orissa, Kernataka, Kerala, Assam and Arunachal Pradesh.

  The upgrading programme is expected to make an improvement in milk yield by more than 5 % annually.

2. Grading up or crossbreeding of non-descript cattle with exotic cattle: In genera, HF bulls should be used in irrigated plains (Delhi, Haryana, Punjab, UP, Gujrat and Karnataka and Jersey in hills and coastal areas (Assam, Arunachal, Bihar, Goa, HP, Haryana, Punjab, J & K, MP, MS, Orissa, Rajasthan, UP, WB, Manipur, Southern states. The level of exotic inheritance should be restricted from 50 – 75 %. Brown Swiss is being used in Kerala. A number of crossbred strains of cattle (*Frieswal, Karan-Fries,* etc)have been developed in India as well as in other countries of the world like Ayrshire, *Jamaica Hope, Guernsey, Santa Gertrudis, AMZ, Beefmaster, Brangus* (See chapter 21 on Cattle breeds).

3. Selective breeding within zebu breeds: The milch and dual purpose breeds need to be improved genetically by selective breeding through PT programme, preferably in their home tract in associated herds. The selective breeding is expected to yield about 1 % genetic gain per year in organized herds whereas the genetic improvement in farmers herd is expected to be higher around 10 %. The indigenous breeds of cattle need to be improved by selective breeding in their home tract. The ONBS with or without MOET is a better technique. The cattle breeds developed by selective breeding in India are: *Gangatiri breed* of U.P. developed from Harian breed, *Rathi breed* of Rajasthan developed from a mixture of Sahiwal, red Sindhi and Tharparkar, and *Krishna valley* of Karnataka, etc.

27.3.2 Breeding of Indian Buffaloes

The following breeding strategies are recommended for improvement of Indian buffaloes:

1. Grading up of non – descript buffaloes: The non-descript buffaloes should be upgraded with superior bulls of Murrah, Nili-Ravi, Mehsana and Surti breeds. This will increase the milk production by 2-3 times from about 500 kg milk of non-descript buffaloes to about 1200 kg of first generation upgrade buffaloes. The continuous upgrading of upgraded buffaloes in 4-5 generations will replace the non-descript low producing buffaloes by comparatively high producing upgraded buffaloes which may be very nearer in conformation to the well defined breeds. A breed of buffalo, *Godawari* of A.P. was developed by crossing non-descript buffaloes with Murrah bulls.

Murrah breed has been recommended for upgrading the non-descript buffaloes of the states of UP, Bihar, WB, Haryana, Rajasthan, MS, Karnataka, AP, and TN. The Surti is the breed of choice for Karnataka, Kerala, Gujrat and Rajasthan. The bulls of Murrah and Nili breeds have been recommended for Punjab.

2. Selective breeding for well defined breeds: The selective breeding under PT programme for buffaloes on large scale needs to strengthen the existing organized farms of well known breeds like Murrah, Nili Ravi, Mehsana, Surti, Bhadawari, Jaffarabadi and, Nagpuri. Some more farms should be developed for production of breeding bulls.

- Murrah breed is preferred as a breed of choice in the states of Haryana, Punjab, Western UP and Delhi.
- Surti breed is preferred in the states of Gujrat, Rajasthan and Karnataka.
- Mehsana and Jaffarabadi are the breeds of Gujrat whereas Nagpuri and Pandharpuri are the breeds of choice for MS.
- The Nili Ravi has good population in Punjab whereas Bhadawari are present in UP.

  It will also be better to run PT programme in associating different herds as well as farmer's herds to overcome the constraint of small population size.

The buffalo breed, *Nili-Ravi,* was developed through selective breeding of Murrah buffaloes in Punjab.

3. The crossbreeding of Indian breeds of buffaloes has resulted in development of new breeds *viz. Mehsana* which was developed by crossbreeding Murrah nd Surti breeds,

27.3.3 Breeding Sheep in India

The fleece weight and quality traits have medium to high heritability estimates and correlated with body weight. Thus, the animals can be selected on the basis of live weight. The selective breeding has made some improvement in average fiber diameter and medullation percentage in Deccani breed at Pune, Bikaneri and Lohi breeds at GLF, Hisar.

The non-descript population should be either upgraded with recognized breeds of the locality or may be crossed with fine wool exotic breeds (Rambouillet, Merino) and mutton breeds of exotic origin (Suffolk, Dorset).

Fine Wool: The crossbreeding breeding of native breeds of hilly region, Nali and Chokla of Rajasthan, Patanwadi of Gujrat and Nilgiri of TN with Rambouillet and Merino rams has been recommended for fine wool breeding keeping 50-75 % exotic inheritance. Some strains of sheep have been developed through crossbreeding programmes of native breeds of sheep with rams of exotic breeds *viz. Kashmir Merino* (cross of Gddi, Bhakarwal and Pooncchi ewes with Merino and Rambouillet rams), B*harat Merino* and *Avivastra* (Cross of Chokla and Nali ewes with Merino and Rambouillet rams), *Nilgiri synthetic and Patanwari synthetic* (cross of ewes of respective native breeds with Rambouillet rams) and *Hissardale* (Cross of Bikaneri ewes with Merino rams).

Carpet wool: Selective breeding for improving carpet wool in breeds like Marwari, Jaisalmeri, Magra and Pugal should be considered in development programme incorporating six monthly body weight, fleece weight and the North Indian carpet wool breeds (Nali, Magra). The inferior carpet wool breeds (Malpura, Sonadi, Jaisalmeri, Muzaffarnagri) may be improved for better carpet wool quality by crossing them with exotic fine wool or mutton breeds stabilizing the exotic inheritance around 50 % exotic inheritance. Similarly, the crossbreeding of other native breeds like Deccani, Bellary, Shahabadi and Chhotanagpuri with exotic breeds can improve the carpet wool quality. A sheep strain named as *Avikalin* has been developed by crossbreeding Malpura ewes with Rambouillet rams at CSWRI, Avikanagar for producing ideal carpet wool. *Munjal breed* of sheep was developed by crossbreeding of Nali and Lohi breeds in Punjab.

Mutton: The mutton production characters have medium to high heritability estimates and hence respond to selection but with low rate. Therefore, the *upgrading* of non-descript population and crossbreeding of inferior native breeds with superior native breeds or with exotic breeds can bring rapid improvement. The native breeds *viz.* Malpura, Sonadi, Muzaffarnagri, Mandya and Nellore should be crossed with Dorset and Suffolk. A crossbred strain of sheep for meet, *Avimans,* has been developed at CSWRI, Avikanagar by crossing Malpura and Sonadi ewes with rams of Dorset and Suffok breeds.

Pelt production: The crossbreeding of coarse carpet wool Indian breeds with Karakul has good potential for pelt production. The native breeds like Malpura and Sonadi can be used for crossbreeding. *Indian Karakul,* a crossbred strain has been developed at sub-station of CSWRI, Bikaner by crossing Marwari, Malpura and Sonadi ewes with rams of Karakul exotic breed. The *Karakul breed* of sheep was itself evolved by crossing two local breeds of the area Karakul (Black lake), Bokhara (Central Asia).

27.3.4 Breeding of Goats

The breeding of goat for improvement should be on the following lines:

- The non-descript females should be upgraded by mating with bucks of high genetic merit of recognized breeds for improvement of milk and meat.
- The selective breeding within recognized breed is recommended, involving farmer's flock by selecting best males from breeding tract of each breed of milch and meat type
- Crossbreeding for meat with Boer breed of South Africa has been undertaken in Maharastra in farmer's flock with encouraging results. This can be started in Rajasthan also because the Boer breed belongs to a relatively dry region. The crosses of local and Sirohi goats with Boer have shown good growth
- Crossbreeding is also recommended for pashmina fiber production in Changthangi breed in J&K state with exotic breeds from Russia (Orenburg which is best quality Pashmina breed).
- Conservation of indigenous breeds for milk (Jamnapari), meat (Black Bengal) and fiber (Chegu, Changthangi). The threatened breeds like Surti, Beetal, Jamnapari should be conserved.

It is also emphasized here that successful implementation of the long term programme also needs the coordination between different agencies *viz.* State A.H. Deptt.; NBAGR, Karnal; CIRG, Makdoom, Mathura (UP).

27.3.4 Breeding of Poultry

The *egg production traits* in poultry (egg number, size and weight) have moderate heritability estimates and hence respond to selection. Therefore, selection for these traits directly is recommended. However, an increase in egg number results in the production of smaller eggs of lighter weight.

The exploitation of non-additive genetic variation is also important through crossing different lines/breeds in all possible combinations. The recurrent selection and reciprocal recurrent selection should be used. On the basis of combining ability (GCA and SCA), the commercial hybrids are produced.

Direct selection for egg quality is advisable. The superior parents should be identified through performance testing and then by progeny testing. The best parents can be used in family selection and exploited by crossing.

The *meat production traits* have fairly high heritability estimates and hence respond to selection by performance testing. There should be selection against fat deposition inside the body cavity. The egg production in broiler is also important at least to multiply the highly selected birds. To meet this objective, a sire of meat line is crossed with a female of egg laying line to produce the market hybrid.

Some strains of poultry have been developed in India by selective breeding (*Cari Gold* and *Kalinga Brown evolved* from selective breeding of RIR) for egg production, and by crossbreeding for egg production (*Hitkari, Upkari, Krishna)* and for dual purpose (*Vanraja, Grama priya, Giriraj).*

27.4 LIVESTOCK IMPROVEMENT PROGRAMMES

The Govt. of India on the recommendation of the Royal Commission on Agriculture, established the livestock farms to produce and keep purebred indigenous animals for improving nucleus herds of different breeds and to produce bulls of superior genetic merit of known pedigree and performance. A number of livestock improvement programmes were also started by Govt. of India from time to time. A brief description of these programmes has been given here

27.4.1 CATTLE AND BUFFALO IMPROVEMENT PROGRMME

The breeding methods for improvement of dairy cattle and buffaloes are sire evaluation which could be in the organized herd or in association of field progeny testing programme, open nucleus breeding system (ONBS) and crossbreeding. The following programmes for improvement of cattle and buffalo in India were started from time to time for their implementation through Deptt. of Agriculture, I.C.A.R., Various State Govt., Military dairy farms, Dairy cooperatives and NDDB, Central Goshala Development Board, Central Council of Gosamvardhan.

1. Central Herd Registration Scheme (CHRS)

An international convention on registration of cattle was held in Oct. 1936 in Rome, with its main recommendation that there should be only one herd book for a single breed in each country so as to avoid the conflicting standards and methods of performance recording. The herd book contains a list of animals qualifying the breed characteristics and production standard and thus purebred animals of high merit are enlisted in the book. This helps the owners in sale and purchase of the breeding stock.

In India the herd books were started by 1949 for Sahiwal, Red Sindhi, Gir, Tharparkar, Ongole, Hariana, Kankrej and Kangayam breeds of cattle and Murrah breed of buffaloes. During third five year plan (1961-66) the breeding tracts of important breeds were covered to organize breeder's societies for taking up this work.

The animals to be registered should meet its breed characteristics like type, colour etc. in addition to producing minimum amount of milk in a lactation of 300days, depending on the breed of the animals *viz.* Sahiwal (1700 kg), Red Sindhi and Tharparkar (1400 kg), Gir and Hariana (1100 kg), Kankrej and Ongole (700 kg), Kangayam (500 kg) and Murrah (1400 kg). However, these criteria had been subjected to change over time depending on the utility of the breed and the improvement in production levels.

The CHRS was reviewed in 1998 and made the following recommendations:

- To register only elite animals which should be purchased by national semen grid or its state components for breed improvement. The criteria for declaring the elite animals were revised for Gir (3500kg & above), Hariana (2700kg & above), Kankrej (3000kg & above), Ongole (2500kg & above), Murrah (3800kg & above), Jffarabadi (4000kg & above), Mehsana (3300kg&above), and Surti (2900kg & above).
- To give incentives to the farmers for maintaining the elite calves and dams as : Rs. 1000 per year for the dam for 4 years, Rs. 2000 for male calf for first year
- To register the bulls used for natural service in the area.

2. Key Village Scheme

This scheme was launched in 1951 (first five year plan, 1951-56) on all India scale taking a key village unit covering an area having a population of 1000cows and buffaloes. The objectives of this KVS were to use superior germ plasm (bulls) together with increased production of feeds and fodder, to provide prophylactic measures against contagious diseases, management and marketing facilities for selling of produce, and to contact farmers for advising them. A key village block comprising 6 to 10 key village units was set-up with A.I. station to supply semen to the key village units. Presently there are 550 key village blocks.

3. Intensive Cattle Development Project

Another large scale programme, intensive cattle development programme (ICDP) was started in 1965 at 15 centres of 10 states for increasing milk production as an area developed approach and taking care of every aspect of cattle improvement *viz.* breeding, feeding, health care, extension activities and marketing facilities. The ICDP was located in the breeding tract of indigenous breeds of cattle and buffaloes. Each ICDP was linked with a dairy project for processing and marketing of milk. Each project covered up to one lakh breedable cows and buffaloes. At present about 130 ICDP's exist.

4. Progeny testing schemes

The P.T. programme was sponsored during IV *Plan (1969-74)* by Agr. Ministry with the aim of producing the superior bulls. The progeny testing schemes were run in single herd in the beginning, then in associated herds and under field conditions.

(*i*) **Single Herd P.T. Scheme:** In the third five year plan, a progeny testing scheme was started with the aim to identify and produce the bulls of high genetic merit on the basis of their progeny performance instead of their dam's yield, to achieve higher genetic gain through use of the tested/proven bulls with A.I. The P.T. scheme was started in the beginning at GLF Hisar for Hariana and Murrah breeds.

During Fourth Five Year Plan, the scheme was started at different centers for other breeds *viz.* Sahiwal, Red Sindhi, Gir, Hariana, Murrah and Surti breed.

(ii) **Associated herd P.T. Scheme:** The progeny testing scheme at none of these farms could yielded the desired results mainly for small herd size and hence a few sires could be tested. To overcome this problem of small herd size, it was thought to associate different herds and testing the sires simultaneously based on the data off all herds. This can allow testing more number of sires and that too in different herds. This also increases the accuracy of selection. However, this scheme posses the problem of G-E interaction which can be taken care of. The associated herd progeny testing programme has been started for 3 breeds namely Sahiwal, Hariana and Ongole. The lead institute for running this programme for Sahiwal breed is the NDRI Karnal (Haryana) by associating 3 herds of NDRI Karnal, SCBF Durg and SCBF Chakganjaria (Lucknow). This programme is being run through frozen semen technology. The P.T. of 6 bulls in each set is the target and 2 best bulls to be selected for nominated mating to bring improvement.

The associated herd P.T. programme was started by Project Directorate on Cattle, Meerut, for Hariana and Ongole breeds through associated herds in their breeding tract *viz.* The associated herds for Hariana breed are located at GLF Hisar, Gaushala at Bhiwani, jind and Kurukshetra whereas for Ongole breeds, the associated herds are located in Andra Pradesh at Lam (Gantur), Chintaladeri (Nellore), Ramtheertham (Parkasam) and Mahanandi (Kurnool).

The other important breeds of cattle are Tharparkar and Gir on which this programme was planned to be started by the PDC Meerut. A similar programme for improvement of Rathi, Deoni and Kankrej breeds was also under consideration under NATP at 3 centers *viz.* RAU Bikaner for Rathi, Marthwada Agricultural University Parbani for Deoni and Dudhsagar Research and Development Association Mehsana for Kankrej.

(iii) **Field P.T. Scheme**: The PDC Meerut has also taken up the field progeny testing of crossbred bulls during VIII five year plan at 3 station *viz.* BAIF Urlikanchan (MS), PAU Ludhiana (Pb) and Kerala Agri. University, Mannuthy. The Holstein Friesian crossbred bulls having 50 to 75% exotic inheritance with a minimum of 4500 kg dam's milk yield are to be tested in batches for a period of 15 months and each batch will comprise 30 crossbred bulls. Atleast 40 daughters will be reached per bull spread over different units.

5. All India Coordinated Research Project (AICRP)

The Indian Council of Agricultural Research (ICAR) had launched a coordinated research project in 1968 with the objective of evolving high yielding breeds of different species for different utility *viz.*, cattle and buffalo for milk production, sheep for wool and mutton production, goat for milk, meat and mohair production, pig for meat production and poultry for egg and meat.

(i) **Cattle:** The AICRP on cattle involved the crossing of indigenous cattle breeds (Hariana, Ongoel, Gir, Tharparkar, Sahiwal, Red Sindhi and Local) with superior exotic breeds (H.F., Jersey, Brown Swiss). The objective of AICRP on cattle was to know the optimum combination and level of exotic inheritance based on

assessing the production and reproduction performance of the crossbreds in different agro-climatic regions. A control frozen semen bank was also established at Hessarghata (Karnataka) for collection, processing, storage and distribution of frozen semen. On the basis of the result obtained it was found that crossbreds with $^3/_4$ exotic inheritance did not do better than half breds due to disease problem, availability of required input and management problem etc. The better performance had been of those having 50 to 62.50% level of exotic inheritance than all other crossbreds.

(ii) **Buffalo:** The AICRP on buffalo was started in 1970-71 by ICAR for two buffalo breeds *viz.* Murrah and Surti at two centers for each breed. The centers for Murrah were NDRI Karnal and PAU Ludhiana whereas for Surti the centers were university of Udaipur at Vallabhagar (Raj) and UAS Dharwar. The objective was the progeny testing to identify the bulls of high genetic merit and to distribute the young bulls, produced from mating of elite buffaloes with proven sires, to various farms in the country. Later on during V plan the field units were attached to each of the centers for testing of sires under transfer to technology.

The breeding plan for all the 4 centers was to maintain 220 breedable females and a separate elite herd of 70 buffaloes. It was decided to evaluate 8 bulls at each centre and finally selecting only 2 bulls out of these 8 bulls for mating with elite buffaloes. But in 1980 the technical programme was changed to test 12 sires in testing herd every 2 years in the two herds of Murrah as well as Surti breed and finally selecting 2 bulls for mating in elite herds of both centers as one unit for both the breeds. Coverage of 3000-4000 breedable females in field units attached to each centre was also recommended. The semen of each bull had to be used in testing herds and 2 field units. Further, in was recommended to freeze a minimum of 2000 does of semen of each bull and to distribute 700 does from each bull to the 2 herds and other associated herd.

In 1985, the technical programme was changed:

- To keep Surti breed at Dharwad and Vallabagar
- To keep Murrah at NDRI Karnal and to associate CIRB Hisar
- To change over to Nili Ravi buffaloes at PAU Ludhiana and associate herd of Nabha (Pb) centre of CIRB.

Under P.T. programme 68 bulls in 9 sets at CIRB Hisar, 53 bulls in 8 sets at PAU and 47 bulls in 8 sets at NDRI were evaluated till 1993. Three sets of Surti buffalo bulls at Vallabhnagar and 5 sets at Dharwad were progeny tested.

Network project on Buffaloes: During VIII plan period (1992-97), the efforts were made in collaboration with the various institutional and organized buffalo farms under the network project on buffaloes, w.e.f. 1993. The participating units taken for Murrah buffaloes were CIRB, Hissar (Haryana) with its substation Nabha (Pb); NDRI, Karnal; PAU, Ludhiana; HAU, Hissar; IVRI, Izatnagar and CBBF, Alamadi (TN) while those for Surti buffaloes were : RAU, Vallbnagar; GAU, Navsari and Anand; CCBF Dhamrod and UAS, Dharwad.

A total of 147 dams of Murrah and 56 of Surti breed from various units were selected as eite dams. All male calves born from elite mating were saved for breeding and kept at their respective units till 12 months of age. The bull rearing and semen processing units for Murrah were taken at CIRN, PAU and NDRI. In 1998, new centers added for Murrah buffaloes were GLF Hastinapur (UP), Banvasi and Mamnoor (AP).

During IX Plan (1992-2002), the network project included the improvement programme on:

- Murrah buffaloes at institute farms of PAU, HAU, NDRI, IVRI, CIRB and CCBF Alamadi (TN)
- Murrah buffaloes under field conditions around CIRB, PAU and NDRI.
- Institutional centres including field testing for different breeds were- Jaffarabadi buffaloes, Nili-Ravi buffaloes, Bhadawari buffaloes, Murrah buffaloes, Surti buffaloes, Godawari buffaloes, Pandhanpuri buffaloes, and Swamp buffaloes

6. Project Directorate on Cattle

The PDC, Meerut came into existence on 3rd November, 1987 with the following mandate:

- To evolve a new breed of cattle 'Frieswal' and other crossbred genotypes for high milk production.
- To improve indigenous breeds of cattle for milk and draft through progeny testing in collaboration of already existing organized herds and farmer's animals.

For indigenous breeds in their breeding tracts, so far Ongole and Hariana breeds have been taken up. The other breeds (Gir and Tharparkar) will be taken up in next phase.

7. Gaushalas

The gaushalas are mostly in existence in Northern states of the country based on religio-economic considerations. There is great scope of improving the livestock of Gaushalas if proper assistance and guidance is provided. These can be used as a source of *in situ* conservation of indigenous breeds and can also be associated in progeny testing programme. In fact, some of the goshalas are already participating in Associated Herd Progeny Testing Scheme. Moreover, Goshalas which maintain non-descript animals can be engaged in crossbreeding programme so as to enhance the milk production and genetic improvement. There are about 1020 organized Gaushalas in 21 states maintaining about 1, 30, 000 cattle and 1400 breeding bulls giving employment to about 11, 000 persons. Some of the Gaushalas have very good resources and facilities *viz.* a Gashala on an average, possessing about 60 ha of grazing area and 25 ha of cultivable land.

The Govt. of India in 1949 set up a central Gaushala Development Board for development of Gaushalas as centres for cattle breeding. Later on 1952, the Govt. of India established the Central Council of Gosamvardhana (CCG) to act as the coordinating and advisory body on cattle development. The CCG, after conducting a survey on the resources of Gaushalas, sponsored an *ad-hoc* scheme for their development and proper functioning. The technical assistance and guidance was extended to the Gaushala managements by appointing Gaushala Development Officers by some of the states. The assistance was given to 242 Gaushalas against the target for the development of 346 Gaushalas during second plan (1956-61). This scheme continued under Third plan to develop more Gaushalas. But the scheme was transferred to state sector from Fourth plan but allocation of funds received low priority. Due to paucity of funds, these gaushalas could made little impact on the improvement in milk production except very few Gaushalas located at Nasik, Urli Kanchan, Indore, Ahmednagar and Amritsar.

9. Milk Producer Cooperatives

Milk production in India is mostly in the hands of small farmers and landless people who keep small number of milch animals. Thus there had not been any cooperative among milk producers and the marketing of milk has been dominated by middlemen who used to get the major share of the benefit. Sardar Vallabh Bhai Patel, the iron man, realized this and he encouraged the small dairy farmers of Kaira District of Gujarat to form the milk cooperative in 1946. This also led the foundation of NDDB in 1963 and Operation Flood in 1969.

(i) **AMUL (Anand Milk Producer's Union Ltd.):** The first attempt, to form the diary cooperative by rural milk producers under the chairmanship of Sardar V.B. Patel was made to directly reach the urban market without middlemen. The milk producer's cooperative societies were formulated at village levels which were federated into a district union and this founded the AMUL with the objective to extend the help and guidance in milk production and marketing to the rural dairy household. The headquater of AMUL was located at Anand, a distict town of Kaira district (Gujarat). The district union (AMUL) represented all the village societies.

In 1948, the union had only 924 members of 13 village milk societies whereas in 1994-95 the union had 532670 members of 954 cooperatives with a collection of 229210 tons of milk. The Kaira union in 1964 set up balance cattle feed plant and started to provide the facilities of animal health care and breeding by running a semen production centre to make the A.I. facilities to the milk producers.

(ii) **Operation Flood:** The Anand pattern dairy cooperatives had been successful in organizing the milk production, procurement, processing and marketing. The Govt. of India realized the importance of dairy cooperatives for enhancing milk production and related aspects and established in 1966, the National Dairy Development Board (NDDB) at Anand in Gujarat as an autonomous body to extend the milk producers cooperatives in major milk sheds of the country. The NDDB in 1969 formulated the programme of operation Flood for the development of dairy industry in India.

The *objective of the operation Flood* was to supply the milk to urban consumers at stable and reasonable price. The operation Flood was implemented in three phases as:

Operation Flood - I (1970-80): It was launched in July 1970 following an agreement with the World Food Programme which provided skin milk powder (124000 tons) and butter oil (40000 tons) as an aid to finance. The emphasis was to set up the Anand pattern rural milk producer's cooperatives for procuring, processing and marketing of milk together with providing technical input services for milk production. About 13300 Dairy Cooperative Societies in 27 milk sheds with membership of 1.8 million farmers were organized.

Operation Flood – II (1980-85): This was for the sixth plan period (1980-85) and launched on October 1979 with an outlay of Rs.2730 million and designed to build on the foundation of Operation Flood –I with the following objectives.

- To build a viable self sustaining dairy industry involving about 10 million rural milk producer's families.
- To develop National Milch Herd of about 14 million crossbred cows and upgraded buffaloes.
- To develop link of rural milk sheds with major demand centers of urban population of about 150 million by establishing a National Milk Grid.
- During this OF-II, the number of village cooperative societies increased to 34500 enrolling 3.6 million farmer members in 136 milk sheds.

Operation Flood – III (1985-96): It was launched in 1985 with major emphasis to consolidate the achievements gained during the earlier phases by improving the productivity and efficiency of cooperative dairy sector. It also provided the animal health and breeding facilities for improving milk production. During OF-III, 72744 Dairy Cooperative Societies in 170 milk sheds of the country with 94 million members were organized.

The Operation Flood (OF) had contributed to sustained increase in milk production because of assured market of milk, remunerative price for raw milk to producers, technical input services including A.I., balanced cattle feed and emergency veterinary health services to the dairy farmers.

9. Nucleus Breeding Schemes

The low rate of genetic improvement of livestock have been due to the non-availabilty of sires of high genetic merit in required numbers, poor spread of A.I. due to lack of infrastructure, small size of farmer's herd, high cost of data recording and the selling of animals before time in field whereas in organized herds the main reasons ad been the less intense selection particularly for female side, small population size and putting more emphasis on non-productive traits, etc.

A new concept, considering the above situations, for changing the breed structure so as to increase the overall genetic merit of the breed, is the Nucleus Breeding Scheme. There are two nucleus breeding schemes *viz.* closed and open depending on the direction of gene flow. A nucleus herd is created which is used entirely for production of males for breeding in the population. A breed is structured in such a way that all the animals are not allowed to make contribution for genetic improvement but very few are given this opportunity. In India, a breed of livestock is kept at organized/institute farms known as organized herd and kept by farmers known as commercial herd/flock or village herd/flock which are very small to the extent of 1-2 animals per farmer but collectively the village herds constitute more population of animals of a breed. However, the genetic improvement can be made in organized herd for obvious reasons. An organized herd under NBS is divided in two group *viz*. Nucleus herd and test herd (multiplier). The nucleus herd is constituted of elite females of high genetic merit and is of a size of about 10-15 % top ranking females of the total herd. The aim is to maximize the genetic gain in nucleus herd to pass on to the test herd and village herds. The nucleus herd breeds its own male and female replacements and occasionally introduces a sire or dam from another nucleus herd. The test herd takes the males and sometimes the females from the nucleus herd to produce sufficient breeding stock to meet the demand of commercial herds. Thus the genetic gain achieved in nucleus herd is passed on from nucleus herd to multiplier and then to village herds.

(*i*) **Closed Nucleus Breeding Scheme (CNBS):** There is one way gene flow only with the direction from top to down herds *i.e*. Nucleus to Multiplier to commercial / village herds. Thus, the improvement is made in nucleus herd and passed on to multiplier and then to the commercial herds and the only source of cumulative genetic progress in village herd is that occurred in the nucleus herd. Therefore, the improvement in nucleus herd is essential otherwise no improvement will occur in other herds because the improvement in nucleus herd is transmitted to other herds.

The time taken in transfer of genetic progress from one herd (nucleus) to the next, known as improvement lag, can be reduced by adopting any of two practices *viz*. transferring of males and females of nucleus herd directly to the commercial herds and keeping the males and females in the lower herd for short time before replacing them with younger stock.

As there is no gene flow to the nucleus, it is so called as the Closed Nucleus Breeding Scheme which is mainly used in pigs and poultry to avoid the risk of introducing diseases in the nucleus flock.

(*ii*) **Open Nucleus Breeding Scheme (ONBS):** In this scheme, the gene flow is both ways *viz*. downward from nucleus to other lower herd (multiplier) and upward from lower herd to upper herd (nucleus) by introduction of superior animals from other herds. Therefore, the superior animals from commercial herd are introduced into nucleus herd. This reduces the rate of inbreeding in the nucleus herd and increases the genetic progress because the superior animals are also available with farmers. This scheme is mostly used in cattle, buffalo, and sheep.

The ONBS can run in an organized (pedigreed) herd at institutional (Govt.) farm or by forming breed societies. The ONBS is run by a group of breeders or breed societies and hence require a close cooperation between breeders. They cooperate in forming and running of ONBS for getting breeding bulls in turn from nucleus herd. This is called as *cooperative breeding scheme.*

The progeny generation of nucleus herd is reared, recorded and the males are evaluated on the basis of the performance of their sibs (paternal half sibs) and their own performance. The males with high genetic merit for trait under selection can be used in the base population (multiplier) for genetic improvement through natural service or A.I. The ONBS can be operated with and without E.T.T.

27.4.2 SHEEP IMPROVEMENT PROGRAMME

In 19th century, sheep development programme was started by the East India Company by importing exotic breeds (Cape Merino) of sheep for crossbreeding of indigenous sheep breeds around Poona.

In the beginning of 20th century, the crossbreeding of sheep with Romney Marsh was undertaken in plains and hilly areas. Crossbreeding with South Down and Cape Merino was undertaken in Punjab, Bengal, Madras and Mysore. Bikaneri ewes were crossed with Merino rams at GLF, Hisar and a fine wool breed with ¾ exotic inheritance was evolved by the name *'Hisardale'*.

After independence, in 1950, the research work on sheep and wool development in India was taken up on regional basis by ICAR. The emphasis was on selective breeding within native stock and the crossbreeding with exotic fine wool breeds. The regional centers were setup in the hilly areas of UP, Rajasthan and Deccan plateau to conduct crossbreeding trials during first five year plan.

During 2nd five year plan, the sheep breeding farms were established with the aim to produce stud rams along with wool extension centers in some states so as to improve the field stocks by providing breeding rams of superior merit.

During 3rd five year plan, the action taken were the strengthening of the existing sheep breeding farms, establishment of more number of sheep and wool extension centers, initiating a large scale sheep shearing and wool processing programmes in Rajasthan and other states of Northern and Western parts of the country.

Central Sheep and Wool Research Institute, Avikanagar (Malpura, Tonk Distt), Rajasthan was established under Central Govt. in 1962. The aims of establishing CSWRI were to undertake the fundamental and applied research in sheep production and wool utilization, and to provide post-graduate training in sheep and wool sciences. Later on, during 1965 the CSWRI was transferred under the administrative control of ICAR.

During 4th five year plan, there were 84 Sheep Breeding Farms, 52 sheep breeding centers and 602 sheep and wool extension centers in the country.

The GOI in 1970 constituted an *ad-hoc* committee on Sheep Breeding Policy. The committee suggested the importation of 20000 exotic sheep of fine wool breeds to produce breeding rams in order to meet the shortage of stud rams. Therefore, Indo-Australian Sheep Breeding Farm, Hisar was set up with 3000 ewes and 300 rams of Corriedale breed to meet the objective of producing and distribution of Corriedale stud rams to different states for crossbreeding work of improving wool and mutton production since Corriedale is a dual purpose breed. Lateron, The Corriedales were replaced by Rambouillet.

Two Sheep Breeding Farms in private sector in Maharastra and Gujrat were setup to produce purebred and crossbred rams.

Seven Central Sheep Breeding Farms in UP, MP, AP, Bihar, Karnataka and J&K were established for same purpose of producing pure and crossbred rams.

During 5th five year plan, some more breeding farms were established. Efforts were made for intensive sheep development through genetically upgrading the stock, providing health care, pasture improvement, machine shearing and wool processing.

AICRP on Sheep Breeding: The AICRP on sheep breeding were started in 1971 for fine wool and for mutton to achieve the objective by crossbreeding of native breeds with rams of exotic breeds. The centers for Fine Wool were located at CSWRI, Avikanagar (Raj.); GAU, Dantiwara; Sheep Breeding Farm, Tal and Sheep Breeding Research station, Sandynallah. The centers for mutton production were located at CSWRI, Avikanagar; APAU, Palampur (A.P.), CIRG, Makhdoom (U.P.); MPAU, Rahuri (M.S.) and TANVASU, Kuttupakkam (TN).

The above two coordinated research projects were merged into one project on Sheep Breeding in 1974 with the following objectives-

Fine wool component: 2.5 kg GFW/year with average fiber diameter of 22 microns and modulation percentage below 5.

Carper wool component: 2kg GFW/year with average fiber diameter of 30 microns and modulation percentage 25.

Mutton component: 30 kg live weight at 6 months of age with 50 % dressing percentage and feeding efficiency of 15-20 percent.

Net work project on sheep: In 1990, all the ongoing units of AICRP on Sheep Breeding were converted into Network Project on Sheep Improvement (NWPSI). The mandate of NWPSI was survey, genetic evaluation and conservation of sheep genetic resources of the country. The breeding policy for NWPSI is the selective breeding in indigenous sheep. The NWPSI is functioning at 5 farm based cooperating units and at 2 field based cooperating units for different breeds.

In addition to NWPSI, the State Agric. Universities and State sheep breeding farms are involved in genetic improvement of sheep but with the major constraint of small population size. Some NGO's like Nimbkar Agr. Research Institute, Phalton (MS) is also involved in sheep improvement programme. This institute is working on Garole sheep.

The CSWRI have also the breeding programmes for indigenous sheep in addition to the breeds (Chokla and Magra) under NWPSI. These are as:

• Genetic improvement of Magra sheep under field condition.

• Improvement of Malpura sheep for mutton production.

• Improvement of Awassi and Garole sheep for mutton, milk and prolificacy

In the state of Jammu and Kashmir, a new breed Kashmir Merino was evolved by crossbreeding of Gaddi and Bhakarwal breeds with rams of Delaine Merino, Soviet Merino and Rambouillet. This had 50-75% exotic inheritance.

27.4.3. GOAT IMPROVEMENT PROGRAMMES

All India Coordinated Research Project (AICRP) on *goats* was initiated in IV Five year plan (1971) with its basic theme to improve the production performance of Indian breeds of goats through crossbreeding with exotic breeds of temperate countries. Two separate projects were incepted *viz.*AICRP on goats for milk and AICRP on goats for fibre (pashmina and Mohair) production.

Three centres for *milk* were NDRI Karnal (Beetal goat); KAU, Mannauty, Trichur (Malabari gaots) for crossing these breeds with Alpine and Saanen and MPKVV Rahuri (Black Bengal) for crossing with Beetal and Jamnapari..

Three centres for *fibre* were: 2 for pashmina at IVRI Mukteshwer (Chegu) and Upesi Farm, Leh (Changthangi) and one for Mohair at MPKVV Rahuri (local Deccani) to cross with Angora goat.

The third component of *meat* was included in 1974 under AICRP on goats for meat. The centres for meat component were: CSWRI, Avikanagar (Sirohi x Beetal crossbreeding); Assam Agr. Univ., Khanapara, Gauhati (Assam local x Beetal crossbreeding); CIRG, Makhdoom (U.P.)

All the 3 AICRP on goats were merged into one project as AICRP on goats (milk, meat and fibre production) in 1974. The coordinating unit of AICRP on goats remained at CSWRI, Avikanagar upto Oct.1976 and lateron shifted to National Goat Research Centre (IVRI), Makhdoom (UP). The number of units increased to 9 in VI plan period (1980-85) which declined to 6 in VIII plan period.

Network Project on Goat: The AICRP on goats was transformed in 1990 into network project in VIII plan with its mandate to define, characterize and improve the productivity of native goat breeds incorporating farmers flock in the home tract through selective breeding based on progeny performance. All the ongoing 6 centres of AICRP on goats were converted into the centres of Network Project on goat improvement.

In 1993, the network centres at BAU, Ranchi; JNKVV Mhow: MPKVV, Rahuri and KAU, Trichur were delinked. The existing centres at Bikaner for Marwari goats in their home tract and Avikanagar for Sirohi goats in farm flock were allowed to continue.

During VII plan, three new centres were opened for Jamunapari and Barbari farm units at CIRG, Makhdoom, Jamunapari field unit (Chakarnagar ,Etawah, UP).

27.3.4. POULTRY IMPROVEMENT PROGRAMMES

The poultry improvement was first initiated by Christian Missionaries by establishing first poultry farm at Etah (UP) in 1912 and Katpadi (TN) during 1920's by importing exotic breeds from UK, America and Australia.

Selective breeding of indigenous fowl was started in 1940 at IVRI Izatnagar (UP). The crossbreeding was also initiated at the same time using exotic breeds.

A poultry development research scheme at Nagpur was undertaken by state Govt. in 1950's for evaluating the native chickens. The breeds evaluated were Aseel, Kadaknath, Black Bengal, Naked Neck and Haringhatta Black. The exotic breeds used for grading up/crossing were White Leghorn, Rhode Island Red, Black Australorp and New Hampshire.

During the last phase of first Five year plan, the major work of poultry research was started. Scientific poultry rearing was started (during1957, 2nd five year plan) by All India Poultry Development Project.

The Regional Poultry Farms were set up at Delhi, Simla, Bangalore, Bombay and Bhubaneshwar.

Govt. of India had started Poultry Breeding on a systematic basis at four Central Poultry Breeding Farms (Hissarghata, Bombay, Bhubaneshwar) to develop high egg producing hybrid strain and CPBF (Chandigarh) to develop fast growing broiler strains. The parent stock to hatcheries and hybrid chicks to farmers were being supplied by CPBF's.

Regional Feed Analytical Laboratories were set up to provide facilities to farmers and feed manufacturers at Chandigarh, Bombay, Bhubaneshwar and Hissarghata.

The Random Sample Test Units were set up at Bombay, Bhubaneshwar and Hissarghata, to conduct egg laying and broiler quality tests for providing useful information to poultry breeders on the performance of various participating stock.

The Poultry Extension cum Development Centers with a unit of 100 layers of improved breeds were set up at 269 places.

Central Training Institute for Poultry Production and Management was set up at Hissarghata to impart training to poultry farmers and Officers.

Central Duck Breeding Farm was also set up in Hissarghata for supply of day old dckings of high egg producing breed of Duck (Khaki Campbell) to varios states.

In late 1960's, the evaluation and improvement of indigenous fowl was started under PL-480 Project at two centers *viz.* PAU, Hisar and College of Agric. Udaipur.

In 1960's some private sector *viz.* Arbor Acre, Acre, Hi-Breed, Rani Shaver and Unichix entered into foreign collaboration of Canada, Czechoslovakia and USA for commercial production of genetically superior chicks. The poultry business had also attracted some othr poultry farmers vz. Venketeshwara hatcheries, Kegg farms, CSR Ross Breeders, Poona Pearls, etc. The other leading international brand names available in India are Hy-line, Hypeco, Anak, Hubbard, Tegel, Dekalb, Vencobb, H & N, etc. The commercial egg type stocks are result of three/four ways crosses of White Leghorn whereas the meat stock involved the crosses of Cornish, Plymouth Rock and synthetic broilers strains.

The ICAR had started/ established:

Improvement programme of native breeds like Aseel, Chittagong, Punjab Brown and *Desi*. The improvement programme also involved the upgrading of indigenous breeds with exotic breeds.

A number of projects in 1960's on improvement of poultry for egg production by family selection, breeding and hybridization; improvement of broilers by selection and crossbreeding; and the improvement of native chickens.

AICRP on Poultry Breeding for eggs and meat in 1971 at a number of research centers under SAU's and ICAR institutes. The programme for *egg production* was to test 13 layer strains at different centers with 3-4 strains at each centre, as purebreds and inter crosses together for their improvement through combined selection. Eight strains of *broilers* were tested, at different centers each with 3-4 strains, as purebreds as well as crosses and intra population selection based on 8 weeks weight.

Central Avian Research Institute, Izatnagar (UP) was established in 1974 to carry out research and training programmes on poultry.

Project Directorate on Poultry at Hyderabad (AP) was established for improvement of poultry.

Numerical Solved Problems

CHAPTER 23. RESPONSE TO SELECTION

Example 23.1: The base population had an average of 35.3 bristle numbers whereas the selected parents had 40.6. The numbers of bristle in offspring generation were 37.9. The h^2 from base population was 0.52. (a) How much progress is expected in next generation? (b) Find out the realized response. (c)

Solution:

Selection differential	=	40.6 – 35.3	= 5.3 bristle
(a) Expected progress (R)	=	h^2S	= 0.52 x 5.3 = 2.8
(b) Realized R	=	37.9 – 35.3	= 2.6

Example 23.2: The standard deviation (σp) of a trait was 3.35. The parents selected were 20 male and 20 females out of 100, giving the selection intensities as 1.40. Estimate the expected response taking $h^2 = 0.52$.

Solution: $R = i \, \sigma p \, h^2$

$$= 1.4 \times 3.35 \times 0.52 = 2.44.$$

Example 23.3: A herd of Sahiwal cows had the herd average of 2000 kg milk of first lactation and the average of the selected population was 2200 kg. The h^2 estimate of milk yield was 0.25 and generation interval was 5 years. Estimate the time required to double the milk production.

Solution: S = 2200 – 2000 = 200kg

$$R = h^2S = 0.25 \times 200 = 50kg$$

$$\frac{\Delta G}{year} = \frac{50}{5} = 10kg$$

No. of generation required to double the milk production = 2000/40 = 50 and hence 40 generation = 40 x 5 = 200 years.

Example 23.4: The average annual greasy fleece weight of a flock of sheep was 2.0 kg and the average increased to 2.5 kg, after culling the low producers. The EPA of 10 rams used in the flock was 3.0 kg and the average increased to 4.0 kg., after culling the low producers. Assum the h^2 of fleece production as 0.4 and generation interval from dam to daughter as 2 years and that of sire to daughter as 3 years. Estimate the expected genetic gain per generation and per year

Solution:

$$S = 2.5 - 2.0 = 0.5 \text{ kg (females)}$$

$$S = 4.0 - 3.0 = 1.0 \text{ kg (male)}$$

$$\Delta G \text{ (F)} = 0.4 \times .05 = .02 \text{ kg}$$

$$\Delta G \text{ (M)} = .04 \times 1.0 = .04 \text{ kg}$$

$$\Delta G \text{ per generation} = \frac{0.2 + 0.4}{2} = 0.3 \text{ kg}$$

$$\Delta G \text{ per year (F)} = \frac{0.2}{4} = 0.05 \text{ kg}$$

$$\Delta G \text{ per year (M)} = \frac{0.4}{3} = 0.13 \text{ kg}$$

$$\text{Average } \Delta G \text{ per year} = \frac{0.5 + 0.13}{2} = 0.09 \text{ kg}$$

Ans: ΔG per generation = 0.3 kg = 300 gm

ΔG per generation = 0.09 kg = 90 gm

Example 23.5: Find out the observed selection differential, standardized and weighted (effective) selection differential from the data given below with $\bar{P} = 1900$ units and $\sigma p = 16$ units.

Sire no.	1	2	3	4	5	6	Total
Trait	2000	1970	1920	1880	1850	1900	11520
No. of progeny (n_1)	10	15	12	14	11	8	70

Solution

(i)   Observed $S = \bar{P}_s - \bar{P} = \dfrac{11520}{6} - 1900 = 1920 - 1900 = 20.0$

(ii) Standardized $S = \dfrac{S}{\sigma p} = \dfrac{20}{16} = 1.25$

(iii) Weighted $S = \dfrac{\Sigma xini}{\Sigma ni} - \bar{Po} \quad \dfrac{134460}{70} - 1900 = 1920.8 - 1900 = 20.8$

Example 23.6: The following are the dates of birth in years for sires and dams of 10 calves born in the year 1975. Estimate the generation interval for sire and dam

Calf no.	1	2	3	4	5	6	7	8	9	10
Sire	1970	1968	1970	1968	1971	1970	1970	1972	1971	1970
Dam	1971	1967	1969	1970	1971	1968	1965	1970	1967	1967

Solution

G.I. = Year calf born − year sire or dam born

$$\text{G.I. = for sire} = \frac{(5+7+5+7+4+5+5+3+4+5)}{10} = 5.0 \text{ years}$$

$$\text{G.I. = for dams} = \frac{(4+8+6+5+4+7+10+5+8+8)}{10} = 7.0 \text{ years}$$

Example 23.7: The AFK of 500 goats is given here. Find out the G.I. for AFK.

Age at kidding (years)	2	3	4	5
No. of goats	140	70	100	90

Solution:

$$\text{Generation interval} = \frac{[(140 \times 2 + 70 \times 3 + 100 \times 4 + 90 \times 5)]}{500}$$

$$= \frac{1640}{500} = 3.28 \text{ years}$$

Example 23.7: The average milk yield of first lactation of a dairy herd was 1830 kg with a standard deviation of 550 kg., and heritability of the trait as 0.30. (a) Estimate the phenotypic variance and additive genetic variance. (b) Take the non additive genetic variance as 45375 kg ^{2}, find out the total genetic variance and the heritability in broadsense.

CHAPTER 24 (BASIS OF SELECTION)

The breeding value of an individual can be estimated based on single record in a herd, in many herds, as contemporary deviation and with average of multiple records of the individual.

I. Single record in a herd

Probable breeding value = $\overline{H} + h^2 (P_i - \overline{H})$

where, $\overline{H}$ = herd average, p_i = individual's record.

Example 24.1: A herd of Tharparkar cows had an average milk production of 6.0 kg with an h^2 estimate of 0.25. The average milk production of 4 cows is 9.0, 5.0, 8.5 and 6.0 kg. Estimate the B.V. of these cows.

Solution:

Cow no. 1	PBV = 6.0+0.25 (9-6)	= 6.75kg
Cow no. 2	PBV = 6.0+0.25 (5-6)	= 5.75kg
Cow no. 3	PBV = 6.0+0.25 (8.5-6)	= 6.62kg
Cow no. 4	PBV = 6.0+0.25 (6-6)	= 6.0kg

II. Single record in many herds

$$\text{P.B.V.} = \overline{H} + h^2 (P_{ij} - \overline{H}) + h^2{}_H (\overline{H}_i - \overline{H})$$

where, $\overline{H}$ = population mean (all herds)

$h^2 = h^2$ of the character

P_{ij} = j^{th} individual of ith herd

$\overline{H}_i$ = i^{th} herd mean

$h^2{}_H = h^2$ of the herd difference within population

Example 24.2 A population comprising two herds had average milk production of 1900 kg whereas the herd average of two herds was 2000 kg and 1800 kg. Consider 3 cows in each herd with their milk production as 1900, 2000, 2100 for one herd and 1700, 1800, 1900 for second herd, h^2 of milk production as 0.25 and $h^2{}_H$ as 0.12. Find out the PBV of all the cows and rank them.

Solution

$$\text{P.B.V.} = \overline{H} + h^2 (P_{ij} - \overline{H}) + h^2{}_H (\overline{H}_i - \overline{H})$$

Cow no. 1 of herd 1 = 1900+ 0.25(1900-2000) +0.12(2000-1900)

= 1900-25+12 = 1887 kg

Likewise the P.B.V. of all other cows can be estimated and their ranking may be done.

III. Single record with contemporary deviation

It is better to use the contemporary deviation to estimate the P.B.V. so as to remove the effects of year or season of calving. Thus

$$\text{P.B.V.} = \text{P.B.V.} = \overline{H} + h^2 (P_i - \overline{P_c})$$

Where, $\overline{P_c}$ is the contemporary average

The contemporary are those cows which calved within ± 2 months of the cow whose P.B.V. is to be estimated.

IV. Average of multiple records

When h^2 of a trait is low, it is better to select on the basis of family mean to increase accuracy but F.S. families are not available and thus the accuracy of individual selection can be increased by using multiple record on the same individual. This reduces the environmental effects and hence increases the accuracy of P.B.V.

$$\text{P.B.V.} = \overline{H} + \overline{H} + \frac{nh^2}{1 + (n-1)r}\left(P_i - \overline{H}\right)$$

where, n = no. of records, r = repeatability of the trait,

P_i = average of n records of ith individual,

$\overline{H}$ = herd average

Example 24.3: The average milk production of a flock of goats is 400 kg. The milk production of 4 outstanding goats in different lactations (L) is as under:

Goat no.	L_1	L_2	L_3	Average
1	460	480	-	470
2	390	400	450	413.3
3	480	-	-	480
4	400	380	-	390

Taking the h^2 as 0.25 and repeatability as 0.35, estimate the B.V. of all the 4 goats and rank them:

Solution

Goat no. 1: PBV = $400 + \dfrac{2 \times 0.25}{1 + 0.35}$ (470-400) = 425.92kg

Goat no. 2: PBV = $400 + \dfrac{3 \times 0.25}{1 + (2 \times 0.35)}$ (413.3-400) = 404.8kg

Goat no. 3: PBV = $400 + \dfrac{1 \times 0.25}{1 + 0.00}$ (480-400) = 420kg

Goat no. 4: PBV = $400 + \dfrac{2 \times 0.25}{1 + 0.35}$ (390-400) = 396.2kg

Therefore, the ranking of 4 goats for their P.B.V. is

Goat no.	1	2	3	4
Rank	1	3	2	4

Estimation of MPPA: Culling of low producers is essential to increase the productivity next year. This is done by estimating the 'most probable producing ability'. The estimation of MPPA is also essential under the situation where the h^2 estimate cannot be made for non availability of the information on pedigree of the individual. The MPPA is estimated as:

$$\text{MPPA} = \overline{H} + \frac{nh^2}{1 + (n-1)r}(P_i - \overline{H})$$

The contemporary deviation are used instead of herd average () in order to reduce the seasonal variation

Example 24.4: The average milk production in a lactation of Tharparkar herd is 1800 kg. The milk production of 4 cows in different lactations is given as under:

Cow no.	Lactation 1	Lactation 2	Lactation 3	Lactation 4	Average
1	1820	1860	-	-	1840
2	1810	1836	1860	1840	1816.5
3	1830	1850	1870	-	1850
4	1840	-	-	-	1840

Taking the repeatability of milk yield as 0.40, estimate the MPPA of all the cows and rank them.

Solution: Cow no.1: MPPA $= 1800 + \dfrac{2 \times 0.40}{1 + 0.4} (1840 - 1800) = 1822\text{kg}$

Cow no. 2: MPPA $= 1800 + \dfrac{4 \times 0.40}{1 + (3 \times 0.4)} (1816.5 - 1800) = 1813\text{kg}$

Cow no. 3: MPPA $= 1800 + \dfrac{4 \times 0.40}{1 + (2 \times 0.4)} (1850 - 1800) = 1813\text{kg}$

Cow no. 4: MPPA $= 1800 + \dfrac{4 \times 0.40}{1 + 0} (1840 - 1800) = 1816\text{kg}$

Ranking of the cows is as under

Cow No.	1	2	3	4
Ranking	1	4	3	2

Example 24.5: Construct an index to selection for fast growth and lean beef percentage from the information given below:

Traits	h^2	a	σ_P	r_g	r_p
Average daily gain	0.43	1.5	80.0	0.3	-0.1
Lean beef %	0.30	0.5	7.0		

Solution: The genetic and phenotypic variance and covariance are estimated as:

$\sigma^2_{p(1)} = (80)^2$	$= 6400 \qquad \sigma^2_{p(2)}$	$= (7.0)^2 = 49.0$
$\sigma^2_{A(1)} = h^2\sigma^2_{p(1)}$	$= 0.43 \times 6400$	$= 2752$
$\sigma^2_{A(2)} = h^2 \sigma^2_{p(2)}$	$= 0.30 \times 49$	$= 14.7$
$\text{Cov}_{P(12)} = r_p\, \sigma\, P_{(1)}\, \sigma\, P_{(2)}$	$= 0.1 \times 80 \times 7.0$	$= -56.0$
$\text{Cov}_{A(12)} = r_g\sigma_{A(1)}\sigma_{A(2)}$	$= 0.3 \times \sqrt{2752} \times \sqrt{14.7}$	
$= 0.3 \times 52.46 \times 3.83$	$= 60.27$	

(*i*) The b_i values will now be estimated as under

$$\begin{vmatrix} b_1 \\ b_2 \end{vmatrix} = \begin{vmatrix} V_{p(1)} & \text{Cov}_{p(12)} \\ \text{Cov}_{p(12)} & V_{p(2)} \end{vmatrix}^{-1} \begin{vmatrix} a_1\, V_{A(1)} + a_2\, \text{Cov}_{A(12)} \\ a_1\, \text{Cov}_{A(12)} + a_2\, V_{A(2)} \end{vmatrix}$$

Now putting the values, the equations will be:

$$\begin{vmatrix} b_1 \\ \\ b_2 \end{vmatrix} = \begin{vmatrix} 6400 & -56 \\ \\ -56 & 49 \end{vmatrix}^{-1} \begin{vmatrix} 1.5(2752) + 0.5\,(60.27) \\ \\ 1.5(60.27\ ) + 0.5(14.7) \end{vmatrix}$$

$b_1 = 0.63$ and $b_2 = 1.27$

(ii) The b_1 value can also be estimated by setting the two simultaneous equations and solving them as

$$b_1 V_{p(I)} + b_2 Cov_{p(12)} = a_1 V_{A(I)} + a_2 Cov_{A(12)}$$
$$b_1 Cov_{p(12)} + b_2 V_{p(2)} = a_1 Cov_{A(12)} + a_2 V_{A(2)}$$

Now putting the values, equations will be

$$6400\, b_1 - 56 b_2 = 1.5\ (2752) + 0.5(60.27)$$
$$-57.60 b_1 + 49 b_2 = 1.5\ (60.27) + 0.5(14.7)$$

Now multiply eq. 1 with 49 and eq. (2) with -56 and adding two equations will cancel the b_2. Thus b_1 value can be obtained. Then the value of b_2 can be estimated.

The selection index (I) will be:

$$I = 0.63 X_1 + 1.27 X_2$$

$$= 0.63\ (X_1 - \overline{X}_1) + 1.27(X_2 - \overline{X}_2)$$
$$I = X_1 + 2.01\ X_2 \text{ by dividing with } 0.63$$

$$= (X_1 - \overline{X}_1) + 2.015\ (X_2 - \overline{X}_2)$$

Example 24.6: Based on first lactation milk yield (kg) and age at first calving (months) using the following statistics, construct the selection index.

Traits	Economic value (a)	Genetic variance (V_A)	Phenotypic variance (V_P)	Cov. A	Cov.P
Milk yield(X)	1.00	84687.21	125423.41	-42.26	111.989
AFC (X_2)	-5.84	13.678	17.498		

The following two simultaneous equations to be set are:

$$b_1 V_{p\ (X1)} + b_2 Cov\ P_{(X1\ X2)} = a_1 V_{A\ (X1)} + a_2 Cov_{A\ (X1\ X2)} \quad ......(1)$$
$$b_1 CovP_{(X1\ X2)} + b_2 V_{p\ (X2)} = a_1 Cov\ A_{(X1\ X2)} + a_2 V_{A(X2)} \quad ......(2)$$

Putting the values given in the table we get the following two equations:

$$125423.41\ b_1 + 111.989\ b_2 = 1.00(84687.21) + (-5.84)\ (-42.26 \quad ....(1)$$
$$111.989\ b_1 + 17.498\ b_2 = 1.00(-42.26) + (-5.48)\ (13.678) \quad .....(2)$$

Multiplying equation (i) with 17.498 and equation (ii) with 111.989, we get

$$219465.1\ b_1 + 1959.58\ b_2 = -122.139 \times 17.498 \quad ....(3)$$
$$12541.54\ b_1 + 1959.58\ b_2 = -8493.06 \times 111.989 \quad ....(4)$$

Substracting equation (4) from equation (3) we get

$$2182112.6\ b_1 = 1499853.6\ b_1 = 0.687$$

Substituting the value of b_1 in equation (2) we get

$$111.989 \times 0.687 + 17.498 = -122.139$$
$$76.97 + 17.498\, b_2 = -122.139$$
$$b_2 = \frac{122.139 - 76.97}{17.498} = 2.581$$

$$\text{The index, } I_1 = b_1\, X_1 + b_1\, X_2$$
$$= 0.687\, X_1 - 11.378\, X_2$$
$$= X_1 - 16.56\, X_2$$

Progeny Testing

EXAMPLE 24.7: The average milk production of Sahiwal herd comprising 150 progenies of 6 sires is 2000 kg with h^2 of yield as 0.30. The average milk production and number of progenies per sire is given below. Rank all the sires based on their breeding value.

Sire no.	1	2	3	4	5	6
N	30	20	25	15	10	50
Ȳ	1600	2250	1840	2200	2300	2100

Solution: (*i*) Mean of contemporaries of all sires will be

(1) $\dfrac{300000 - 48000}{120}$ = 2100 kg   (2) $\dfrac{300000 - 45000}{130}$ = 1961.5kg

(3) $\dfrac{300000 - 46000}{125}$ = 2032 kg   (4) $\dfrac{300000 - 33000}{135}$ = 1977.7 kg

(5) $\dfrac{300000 - 23000}{140}$ = 1978.5 kg   (6) $\dfrac{300000 - 105000}{100}$ = 1950 kg

(*ii*) Regression of B.V. of sire on his progeny (b_{AP}) $= \dfrac{2nh^2}{4 + (n-1)\, h^2}$

Sire.	N	h^2	$\dfrac{2nh^2}{4+(n-1)h^2}$	
1	30	0.3	18/12.7	=1.41
2	20	0.3	12/9.7	=1.24
3	25	0.3	15/11.2	=1.34
4	15	0.3	9/8.2	=1.09
5	10	0.3	6/6.7	=0.89
6	50	0.3	30/18.7	=1.60

(*i*) Breeding value of sires = $\left(\overline{H} + \dfrac{2nh^2}{4 + (n-1)\, h^2}\, P_i - P_c \right)$

Sire 1. 2000 +1.41 (1600-2100) = 1295.0 kg

2. $2000 + 1.24 (2250-1961.5) = 2354.7$ kg

Like wise, the B.V. of other sires can be estimated.

Example 24.8: From the following data on milk yield construct the sire indices by different possible methods, taking h^2 of milk yield = 0.28, population mean = 1660.

Sire No.	Daughters records	Dam's Average	Contemporary Daughters	Contemporary Dams
1	1252, 1858, 2343, 1880, 1532,1743, 1274 1806,	1600	1930	1550
2	1780, 1510, 1781, 1561, 1744	1340	1710	1310
3	2455, 1498	1680	2200	1590
4	1581, 2051, 2100, 1583, 2155,2178	1500	2000	1490
5	1258,1616, 1424	1315	1510	1300

Solution:

(i) Simple daughter average index $(I_1) = \overline{D}$

Sire no.	1	2	3	4	5
Index(T_1)	1711	1675.2	1976.5	1941.3	1432.6
N_j	8	5	2	6	3

(ii) Equi-parent index $(I_2) = \overline{2D} - \overline{M}$

Sire No.	1	2	3	4	5
	1822	2010.4	2273.0	2382.6	1550.2

(iii) Corrected daughter average index $(I_3) = \overline{D} - 0.5h^2 (\overline{M} - \overline{P})$
   (Krishnan index)
   Sire No. 1: $1711.0 - 0.5(0.28) (1600 - 1660) = 1719.4$
   Sire No. 2: $1675.2 - 0.5(0.28) (1340 - 1660) = 1720.0$
   Sire No. 3: $1976.5 - 0.5(0.28) (1680 - 1660) = 1973.7$
   Sire No. 4: $1941.3 - 0.5(0.28) (1500 - 1660) = 1741.3$
   Sire No. 5: $1432.6 - 0.5(0.28) (1315 - 1660) = 1720.0$

(iv) $I_4 = \overline{H} + \dfrac{2nh^2}{4 + (n-1)\ h^2}(\overline{D} - \overline{H})$

   Sire No. 1: $1660 + \dfrac{2 \times 9 \times 0.28}{4 + (7)\,0.28} (1711 - 1660) = 1698.3$

Sire No. 2: 1660 + (1675.2 – 1660) = 1668.3

Sire No. 3: 1660 + (1976.5 – 1660) = 1742.8

Sire No. 4: 1660 + (1941.3 – 1660) = 1835.0

Sire No. 5: 1660 + (1432.6 – 1660) = 1576.0

(v) Dairy search Index $(I_5) = \overline{H} + \dfrac{n}{n+12}\left(\overline{D} - C_D\right) - b\left(\overline{M} - C_M\right)$

Sire No. 1: 1660 + (1711 – 1930) – 0.5x 0.28 (1600 – 1550)
$\qquad$ = 1600- 87.6 – 7.0 = 1565.4

Sire No. 2: 1660 + (1675.2 - 1710) – 0.14 (1340 – 1310)
$\qquad$ = 1660 - 10.2- 4.2 = 1645.6

Sire No. 3: 1660 + (1976.5 – 2200) - 0.14 (1680 – 1590)
$\qquad$ = 1660 – 31.9 – 12.6 = 1615.5

Sire No. 4: 1660+ 6/18(1941.3 – 2000) – 0.14(1500 – 1490)
$\qquad$ = 1660 – 19.5 – 1.4 = 1639.1

Sire No. 5: 1660 + 3/15 (1432.6 – 1510) – 0.14 (1315 – 1300)
$\qquad$ = 1660 – 15.5 – 2.1= 1642.4

CHAPTE 25: MEASUREMENT OF RELATIONSHIP AND INBREEDING

Exercise 25.1: Compute the coefficient of relationship between two individuals D and E from the following pedigree No. 3.

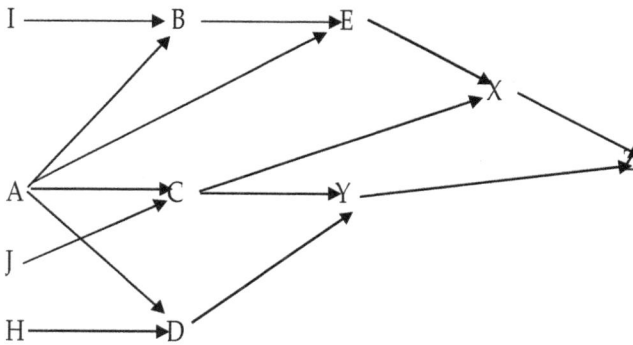

Pedigree no. 3.

Solution: In this pedigree, the common ancestor of the individuals D and E is A. The A is not inbred and hence $F_A = 0$. The individual D is also not inbred and hence $F_D = 0$ whereas the second individual (E) is inbred. It is thus required first to compute the inbreeding coefficient of E (F_E). The individual E is the product of sire-daughter mating.

$F_E = (½)^{1+1} = (½)^2 = ¼ = 0.25$ for AB path.

The relationship coefficient is estimated by putting the values of genic covariance and genic variance in the formula. The genic covariance between D and E is computed by tracing the paths, finding the contribution of each path and then summing over all the paths: $(\text{Cov}_{DE}) = \Sigma(½)^n (1 + F_A)$

Common ancestor and paths	n	Contribution of each path
DAE	2	$(\frac{1}{2})^2 = \frac{1}{4}$
DABE	3	$(\frac{1}{2})^3 = 1/8$

$$\text{Sum} = \frac{1}{4} + 1/8 = 3/8 = 0.375$$

$$R_{DE} = \sqrt{(\frac{1}{2})}^{\,n} (1 + F_A) / \sqrt{[(1 + F_D)(1 + F_E)]}$$
$$= 0.375 (1.0) / \sqrt{[(1.0)(1.25)]}$$
$$= 0.375 / \sqrt{1.25}$$
$$= 0.375 / 1.118 = 0.3354$$

Example 25.2: Find out the inbreeding coefficient of the individual Z from the pedigree No. 3 by Wright's method of path coefficient.

Solution: 1. To find out the *common ancestors and the common inbred ancestors*: In the above pedigree A and C, are two common ancestors neither of which is inbred. The X and Y are inbred but their inbreeding coefficient will not be considered because they are the sire and dam of Z, neither of the two is common ancestor nor direct relatives.

2. *Tracing of the paths between sire and dam of inbred individual through common ancestor*: The first path connecting X and Y is through the common ancestor C and this is YC X making n = 1 and n' = 1. Further X and Y are connected through A by 5 paths *viz.* YDACX, YDAEX, YDABEX, YCAEX and YCABEX. It is better to underline the common ancestor. Neither of the common ancestor (C or A) is inbred.

3. *Contribution of each path*: This is estimated as $(\frac{1}{2})^{n + n' + 1}$ or as $(\frac{1}{2})^{n + n' + 1} (1 + F_A)$ if the common ancestor is inbred . The contribution of path is then added to find out the F_Z as below:

Paths	Generations		Contribution of each path = $(\frac{1}{2})^{n + n' + 1}$
YCX	1	1	$(\frac{1}{2})^3 = 1/8$
YDAEX 2	2	$(\frac{1}{2})^5$	= 1/32
YDABEX 2	3	$(\frac{1}{2})^6$	= 1/64
YCAEX 2	2	$(\frac{1}{2})^5$	= 1/32
YCABEX 2	3	$(\frac{1}{2})^6$	= 1/64
YDACX 2	2	$(\frac{1}{2})^5$	= 1/32

$$F_Z = 1/8 + 1/32 + 1/64 + 1/32 + 164 + 1/32 = 0.25$$

Example 25.3: Find out the inbreeding coefficient of the individual X from the following pedigree.

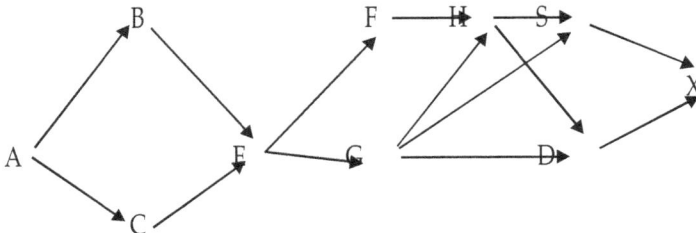

Pedigree no. 4

Solution: *(i)* Find common ancestor and common inbred ancestor: There are three common ancestors (E, G, H) in this pedigree, out of which only two (E and H) are inbred.

(ii) To compute F_X requires the F_E and F_H because E and H are two common inbred ancestors whereas the common ancestor G is not inbred and hence F_G is not required.

$$F_{E.} = (\tfrac{1}{2})^3 = 1/8 = 0.125 \qquad \text{for C AB path}$$
$$F_H{}' = (\tfrac{1}{2})^3 (I + F_E) \qquad \text{for G EF path}$$
$$= (\tfrac{1}{2})^3(1+1/8) = 1/8(9/8) = 9/64 = 0.1406$$

(iii) Tracing of paths and their contribution: The F_X will be computed as under:

Paths	Generation		Contribution of each path
	n	n'	
DHS	I	I	$(\tfrac{1}{2})^3 (1+0.1406) = 0.1426$
DGS	1	1	$(\tfrac{1}{2})^3 (1/8) = 0.125$
DGHS	1	2	$(\tfrac{1}{2})^4(1/16) = 0.0626$
DHGS	2	1	$(\tfrac{1}{2})^4 (1/16) = 0.0625$
DGEFHS	2	3	$(\tfrac{1}{2})^6 (1+ 0.125) = 0.0176$
DHFEGS	3	2	$(\tfrac{1}{2})^6 (1 + 0.125) = 0.0176$
			$F_X = 0.4278$

In this case, the F_H in the paths DGHS and DHGS has not been considered because the individual H is not the common ancestor in either of the path but H is the intermediate between parents (S and D) and the common ancestor (G). Secondly, the path DHFEGHS is not allowed because the individual H has come twice in this path.

Example 25.4: Prepare the variance-covariance chart (Table) from the given pedigree No. 3 and compute the inbreeding coefficient of the individual Z.

Solution:

1. Arrangement of all the individuals in the pedigree in N x N table. All the individuals in the pedigree are arranged in the N x N table according to their birth (generation) where N are the number of individuals. Among the individuals of base generation, only the common ancestors are taken in N x N table. The parents of all the individuals in the table are also written.

2. Computation of the variance and covariance: The inbreeding coefficients of the animals in the base generation are assumed to be zero. Secondly, the genic covariances of the animals of base generation are also assumed to be zero.

Variance-Covariance Table

Sire	-	A	A	A	A	C	C	X
dam	-	I	J	H	B	E	D	Y
Individuals	A	B	C	D	E	X	Y	Z
A		1.0 ½	½	1/3	¾	5/8	½	9/16
B			1.0	14	¼	¾	1/2	i/4	3/8
C				1.0	¼	3/8	11/16	5/8	21/32
D					1.0	3/8	5/16	5/8	15/32
E						5/4	13/16	3/8	19/32
X							19/16	½	27/32
Y	-						-	9/8	13/16
Z									5/4

(i) Genic variance of A and its genic cov. with other individuals:

Var. A = 1 + ½ (genic cov. of A's parents)

$\qquad$ = 1 + ½ (0) = 1.0 Since parents of A are unknown and hence unrelated

Cov_{AB} = ½ (coy of A with parents of B) -

$\qquad$ = ½ (Cov_{AA} + Cov_{AI}) = ½ (1.0 + 0) = ½

Since A & I are unrelated.

Cov_{AC} = ½ (Cov_{AA} + Cov_{AJ}) = ½ (1.0 + 0) = ½

Cov_{AD} = ½ (Cov_{AA} + Cov_{AH}) = ½ (1.0 + 0) = ½

Cov_{AE}=½ (Cov_{AA} + Cov_{AB}) = ½ (1.0 + ½) = ¾

Since A & B are parent offspring

Cov_{AX} = ½ (Cov_{AC} + $C0v_{AE}$) = ½ (½ + 3/4) = 5/8

Cov_{AY} = ½ (Cov_{AC} + Cov_{AD}) = ½ (½ + ½) = ½

Cov_{AZ} = ½ (Cov_{AX} + Cov_{AY}) = ½ (5/8 + ½) = 9/16

(ii) Genic variance of B and its genic cov. with others:

Var.B = 1.0 + ½ (Coy, of its parents A & I) = 1.0

Since A & I are unrelated

Cov_{BC} = ½ (Cov of B with both parents of C)

$\qquad$ = ½ (Cov_{BA} + Cov_{BJ}) = ½ (½ + 0) = ¼

Since B & J are unrelated

Cov_{BD} = ½ (Cov_{BA} + Cov_{BH}) = ½ (½ + 0) = 1/4

Cov_{BE} = ½ (Cov_{BA} + Cov_{BB}) = ½ (½ + 1.0) = ¾

Cov_{BX} = ½ (Cov_{BC} + Cov_{BE}) = ½ (¼ + ¾) = ½

Cov_{BY} = ½ (Cov_{BC} + Cov_{BD}) = ½ (¼ ÷ ¼) = 1/4

Cov_{BZ} = ½ (Cov_{BX} + Cov_{BY}) = ½ (½ + ¼) = 3/8

(iii) Genic variance of C and its genic coy, with others:

Var.C $= 1.0 + ½$ (Cov.of itsparents A & I) $= 1.0$

Since A & I are unrelated

$\text{Cov}_{CD} = ½ (\text{Cov}_{CA} + \text{Cov}_{CH}) = ½ (½ + 0) = ¼$

$\text{Cov}_{CE} = ½ (\text{Cov}_{CA} + \text{Cov}_{CB}) = ½ (½ + ¼) = 3/8$

$\text{Cov}_{CX} = ½ (\text{Cov}_{CC} \div \text{Cov}_{CE}) = ¼ (1.0 + 3/8) = 11/16.$

$\text{Cov}_{CY} = ½ (\text{Cov}_{CC} + \text{Cov}_{CD}) = ½ (1.0 + ¼) = 5/8$

$\text{Cov}_{CZ} = ½ (\text{Cov}_{CX} + \text{Cov}_{CY}) = ½ (11/16 + 5/8) = 21/32$

(iv) Genic variance of D arid its genic coy, with others :-

Var.D $= 1.0 + ½$ (Cov. of its parents A & H)

$\quad = 1.0 + ½(\text{Cov}_{AH}) = 1.0$

$\text{Cov}_{DE} = ½$ (Coy of D with parents of E)

$\quad = ½ (\text{Cov}_{DA} + \text{Cov}_{DB}) = ½ (½ + ¼) = 3/8$

$\text{Cov}_{DX} = ½ (\text{Cov}_{DC} + \text{Cov}_{DE}) = ½ (¼ + 3/8) = 5/16$

$\text{Cov}_{DY} = ½ (\text{Cov}_{DC} + \text{Cov}_{DD}) = ½ (¼ + 1.0) = 5/8$

$\text{Cov}_{DZ} = ½ (\text{Cov}_{DX} + \text{Cov}_{DY}) = ½ (5/16 + 5/8) = 15/32$

(v) Genie variance of E and its genie coy, with others:

Var.E $= 1.0 + ½$ (genic cov. of A B) $= 1.0 + ½ (½) = 5/4$

$\text{Cov}_{EX} = ½$ (Coy of E with parents of X)

$\quad = ½ (\text{Cov}_{EC} + \text{Cov}_{ES}) = ½ (3/8 + 5/4) = 13/16$

$\text{Cov}_{EY} = ½ (\text{Cov}_{EC} + \text{Cov}_{ED}) = ½ (3/8 + 3/8) = 3/8$

$\text{Cov}_{EZ} = ½ (\text{Cov}_{EX} + \text{Cov}_{EY}) = ½ (13/16 + 3/8) = 19/32$

(vi) Genie variance of X and its genie coy, with others:

Var.X $= 1 + ½$ (genie coy, of its parents C & E))

$\quad = 1 + ½ (\text{Cov}_{CE}) = 1.0 + ½ (3/8) = 19/16 - -$

$\text{Cov}_{XY} = ½ (\text{Cov}_{XC} + \text{Cov}_{XD}) = ½ (11/16 + 5/16) = i/2$

$\text{Cov}_{XZ} = ½ (\text{Cov}_{XX} + \text{Cov}_{XY} = ½ (19/16 + ½) = 27/32$

(vii) Variance of Y and its genie coy, with Z :-

Var.Y $= 1.0 + ½$ (genie coy. of Y's parents)

$\quad = 1.0 + ½ (\text{Cov}_{CD}) = 1.0 + ½ (¼) = 9/8$

$\text{Cov.}_{YZ} = ½ (\text{Co}_{XY} + \text{Cov}_{YY}) = ½ (½ + 9/8) = 13/16-$

(viii) Variance of Z :

Var.Z $= 1 + ½$ (genie coy, of the parents of Z)

$\quad = 1 + ½ (\text{Cov}_{XY}) = 1 + ½ (½) = 1 + ¼ = 5/4$

3. Computation of the F_Z

(i) $F_Z = ½$ (genic Cov. of the parents of Z)

$\quad = ½$ (genie $\text{Cov}_{XY}) = ½ (½) = ¼ = 0.25$

(ii) F_Z = Genic variance of Z - 1.0-

$\quad = 5/4 - 1.0 = 1.25 - 1.0 = 0.25$

$\quad = f\ xy$

EXERCISES

5.5 *(a)* Find out the inbreeding coefficient of individual X produced   from half sib mating with inbred sire (common ancestor).

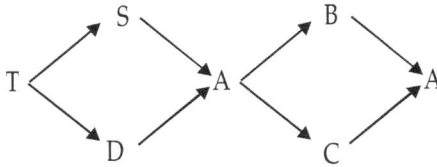

Ans. $F_X = 0.1406$

(b) Find out the inbreeding coefficient of individual Z produced from full sib mating with inbred sire.

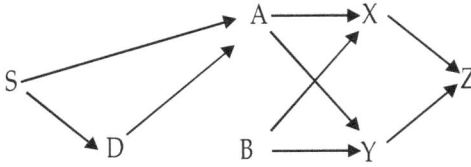

Ans. $F_X = 0.28125$

5.6. Work out the inbreeding coefficient of an individual X produced from sire daughter mating with inbred sire of HS mating.

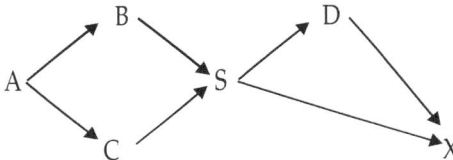

Ans. $F_X = 0.28125$

5.7. Compute the inbreeding coefficient of the individual X from the following pedigree.

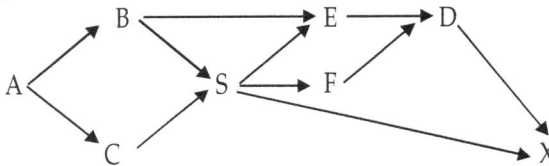

Ans. $F_X = 0.35193$

5.8. Work out the coefficient of relationship between two full sibs produced from sire daughter mating with inbred sire.

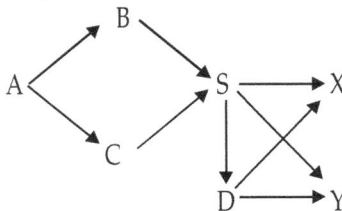

Ans. $R_{XY} = 0.6341$

5.9 Compute F_X from the following pedigree

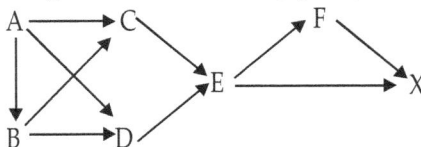

Ans. $R_{XY} = 0.6341$

Chapter 26: Out Breeding - Estimation of heterosis

Example 26.1 Calculate the heterosis in F_1 and F_2 progenies considering the milk yield in 300 days of Sahiwal and Holstein Friesian cows as 1600 and 5600 hg respectively and their F_1 crossbreds as 3900 kg.

Solution:

$$\text{Heterosis } F_1 = \overline{M}_{F1} - \overline{M}_P = 3900 \quad - \tfrac{1}{2}(5600 + 1600)$$
$$= 3900 - 3600$$
$$= 300 \text{ kg}$$

$$\% \text{ heterosis in } F_1 = \frac{300 \times 100}{3600} = 8.3$$

$$\text{Heterosis } (F_2) = \tfrac{1}{2} \text{ heterosis in } F_1 = \overline{M}_{F2} - \overline{M}_P$$
$$= \tfrac{1}{2} \times 300 = 150 \text{ kg}$$

Example 26.2. In a poultry farm, the average egg production per year in breed A was 150 and in breed B was 180. When two breeds were crossed reciprocally, the average egg production was increased to 200 eggs per year. Calculate the heterosis.

Solution: $\text{Heterosis } \overline{M}_{F1} - \overline{M}_P = 200 - \tfrac{1}{2}(150 + 180)$

$$= 200 - 165 = 35 \text{ eggs}$$

$$\% \text{ heterosis} = \frac{135 \times 100}{165} = 21.2$$